Advanced Signal Processing
A Concise Guide

About the Authors

Amir-Homayoon Najmi completed the Mathematical Tripos at Cambridge University and received the D.Phil. degree in theoretical physics from Oxford University. He taught at the University of Utah and conducted research in seismic imaging at the Shell Oil Bellaire Research Center in Houston before joining in 1990 the Applied Physics Laboratory at Johns Hopkins University, where he also holds a joint faculty appointment at the Johns Hopkins School of Medicine. In addition to publications in diverse areas of theoretical physics, geophysics, and biophysics, Dr. Najmi has published a graduate textbook on wavelets.

Todd K. Moon received his bachelor's degree summa cum laude in electrical engineering and mathematics from Brigham Young University, and he completed his Ph.D. in electrical engineering at the University of Utah. His research interests include statistical signal processing, digital communications, and error correction coding. Dr. Moon's publications include three graduate-level textbooks in signal processing and error correction coding.

Advanced Signal Processing
A Concise Guide

Amir-Homayoon Najmi, D.Phil.
The Johns Hopkins University Applied Physics Laboratory
The Johns Hopkins University School of Medicine, Neurology Department

Todd K. Moon, Ph.D.
Utah State University, Electrical and Computer Engineering Department

Mc Graw Hill

New York Chicago San Francisco Athens London
Madrid Mexico City Milan New Delhi
Singapore Sydney Toronto

Library of Congress Cataloging-in-Publication Data

Names: Najmi, Amir-Homayoon, author. | Moon, Todd K., author.
Title: Advanced signal processing : a concise guide / Amir-Homayoon Najmi, Todd Moon.
Description: New York : McGraw Hill, 2020. | Includes bibliographical references and index.
Identifiers: LCCN 2020018309 | ISBN 9781260458930 (hardcover) | ISBN 9781260458947 (ebook)
Subjects: LCSH: Signal processing—Textbooks.
Classification: LCC TK5102.9 .N36 2020 | DDC 621.382/2—dc23
LC record available at https://lccn.loc.gov/2020018309

McGraw Hill books are available at special quantity discounts to use as premiums and sales promotions or for use in corporate training programs. To contact a representative, please visit the Contact Us page at www.mhprofessional.com.

Advanced Signal Processing: A Concise Guide

1 2 3 4 5 6 7 8 9 LCR 25 24 23 22 21 20

ISBN 978-1-260-45893-0
MHID 1-260-45893-8

The sponsoring editor for this book was Lara Zoble, the editing supervisor was Stephen M. Smith, the production supervisor was Lynn M. Messina, and the acquisitions coordinator was Elizabeth M. Houde. The art director for the cover was Jeff Weeks.

AHN dedicates this book to

Aria Elizabeth, Audrey Grace, and Colin Maziar
Chris, Jonathan, Jeffrey, and Linda,

who have made it all worthwhile

TKM dedicates this book to

H. Kay Moon, who encouraged him to be curious

Contents

List of Figures

List of Tables

Acronyms

AAE Adversarial Auto Encoder

AIC Akaike Information Criterion

AR Auto-Regressive

ARMA Auto-Regressive Moving Average

BIC Bayesian Information Criterion

BNT Batch Normalizing Transform

CBF Conventional Beamformer

CNN Convolutional Neural Network

CPSD Cross Power Spectral Density

CRLB Cramer-Rao Lower Bound

CZT Chirp **Z** Transform

DCT Discrete Cosine Transform

DFT Discrete Fourier Transform

DMR Dominant Mode Rejection

DOA Direction of Arrival

DPSS Discrete Prolate Spheroidal Sequence

DTFT Discrete Time Fourier Transform

ELU Exponential Linear Unit

EMVDR Enhanced Minimum Variance Distortionless Response

ESPRIT Estimation of Signal Parameters by Rotational Invariance Techniques

FFT Fast Fourier Transform

FIA Fisher Information Approximation

FIR Finite Impulse Response

GAN Generative Adversarial Network

GAP Global Average Pooling

GCC Generalized Cross-Correlation

GRU Gated Recurrent Units

HAR Human Activity Recognition

ICA Independent Components Analysis

IID Independent Identically Distributed

IIR Infinite Impulse Response

LCMV Linearly Constrained Minimum Variance

LHP Left Half-Plane

LMMS Linear Minimum Mean Square

LMS Least Mean Square

LP Linear Prediction

LPC Linear Predictive Coding

LReLU Leaky Rectified Linear Unit

LSTM Long Short-Term Memory

LTI Linear Time-Invariant

MAP Maximum A Posteriori

MDL Minimum Description Length

ML Maximum Likelihood

MLE Maximum Likelihood Estimator

MMS Minimum Mean Square

MMSE Minimum Mean Square Error

MNIST Modified National Institute of Standards and Technology

MRA	Multi-Resolution Analysis
MRI	Magnetic Resonance Imaging
MSC	Magnitude Squared Coherence
MSE	Mean Square Error
MUSIC	MUltiple SIgnal Classification
MVD	Minimum Variance Distortionless
MVDR	Minimum Variance Distortionless Response
MVDSE	Minimum Variance Distortionless Spectral Estimate
MVU	Minimum Variance Unbiased
NAG	Nesterov Accelerated Gradient
NML	Normalized Maximum Likelihood
PCA	Principal Components Analysis
PEF	Prediction Error Filter
PHD	Pisarenko Harmonic Decomposition
PSD	Power Spectral Density
PSNR	Peak Signal to Noise Ratio
QMF	Quadrature Mirror Filter
ReLU	Rectified Linear Unit
RHP	Right Half-Plane
RIP	Restricted Isometry Property
RLS	Recursive Least Squares
RN	Recurrent Neuron
RNN	Recurrent Neural Network
ROC	Region of Convergence
SCC	Standard Cross Correlator
SCOT	Smoothed Coherence Transform
SETI	Search for Extra-Terrestrial Intelligence
SGD	Stochastic Gradient Descent
SINR	Signal to Interference + Noise Ratio
SLL	Side Lobe Level
SMI	Sample Matrix Inversion
SNR	Signal to Noise Ratio
SSS	Strict Sense Stationary
STFT	Short-Time Fourier Transform
SVD	Singular Value Decomposition
TLS	Total Least Squares
ULA	Uniform Linear Array
WGN	White Gaussian Noise
WSS	Wide Sense Stationary

Preface

It is difficult to exaggerate the impact of signal processing in the modern world. From space exploration to the discovery of oil and gas within the earth, and to medical and biomedical applications, signal processing of audio, speech, and video imagery has been at the forefront of a technological revolution begun by the advent of digital computers. Although signal processing has been taught mostly in electrical engineering (and nowadays electrical and computer engineering) departments, vast numbers of physicists and mathematicians have contributed to the growth of the subject and continue to use signal processing techniques in their work.

Signal processing theory and algorithms are firmly rooted in applied mathematics (Fourier, Laplace, Wiener, and Shannon were mathematicians first before becoming known as physicists and engineers!); indeed, it is impossible to study or contribute to advanced research in modern signal processing without a thorough understanding of the underlying mathematics. This text is designed to cover the fundamental mathematical ideas and methods so as to enable the students to study the latest developments in the subject and to use signal processing algorithms judiciously and effectively. The prerequisites for this text are undergraduate courses in multi-variable calculus, linear algebra, probability theory, and basic signal processing concepts including sampled signals and filters. We saw no necessity to include any computer code for two reasons. The implementation of algorithms and testing them on data is a valuable learning experience; secondly, the availability of sophisticated signal processing libraries makes the task of learning the mathematical theory underlying the algorithms even more important—understanding the mathematical theory allows for correct and effective use of the algorithms.

If a single equation and figure were to represent the subject of linear signal processing, without a doubt it would have to be the discrete time version of the convolution operation depicted in figure P.1, where $h[m]$ is the linear system impulse response, $x[n]$ is the input, and $y[n]$ is the output of the system. Much of

$$x[n] \longrightarrow \boxed{h[n]} \longrightarrow y[n] = \sum_{m=-\infty}^{+\infty} h[m] x[n-m]$$

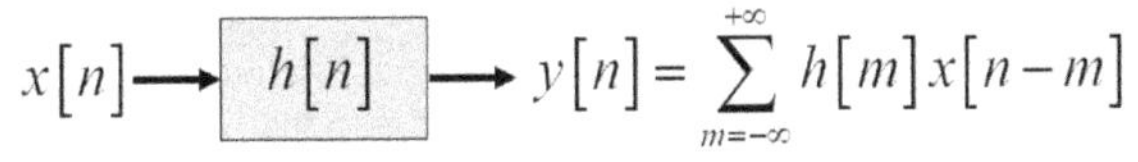

Figure P.1: A convolutional system in discrete time.

"advanced" signal processing is about the search for an "optimal filter" (which in all practical situations has a finite number of elements) satisfying an optimality criterion subject to some constraint(s). The optimality criterion is often expressed in terms of the minimization of some squared norm of an error sequence defined as the difference between the output $y[n]$ and a desired sequence $d[n]$. This formulation can be applied to a variety of problems depending on the relevant mathematical norm. For instance, without a statistical model we look for least squares solutions using the l_2 norm, although in some cases, e.g., compressive sensing, we may use the l_1 norm. When signals are stochastic in nature, we look for a minimum mean squared error (MMSE) filter, whose prime example is the Wiener filter that plays a central role in optimal linear filtering.

Other important concepts of advanced signal processing include "minimum phase" system functions and "spectral factorization". Figure P.2 shows a whitening filter constructed from the minimum phase spectral factor of a signal, and the white noise that is "informationally" equivalent to the signal. This is a natural way to describe ARMA (Autoregressive Moving Average) and, in particular, AR (Autoregressive) models for pole-zero system functions and figure P.1 can describe such parametric signal models in addition to the "forward linear prediction" model. The input to an AR model is a white noise sequence while in the

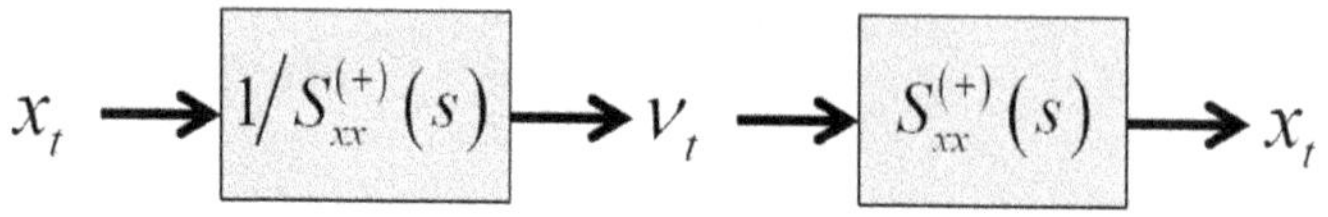

Figure P.2: Minimum phase spectral factor whitening filter.

forward linear predictor the input is a finite set of previous values of the output process, viz.,

$$\hat{y}[n] = \sum_{m=-\infty}^{n-k} a[m]\, y[n-m], \quad k \geq 1,$$

where k denotes the number of steps ahead for the prediction problem. The optimality criterion may be expressed as a (possibly constrained) minimization problem that involves a square Hermitian and positive definite "Toeplitz" matrix $\boldsymbol{R}$, namely, the "auto-correlation" or "auto-covariance" matrix. In the case of an autoregressive random process or a linear predictor of order P, the $P \times 1$ parameter vector $\boldsymbol{a}$ satisfies possibly the most important equation of statistical signal processing:

$$\begin{bmatrix} R_0 & R_{-1} & \cdots & R_{-(P-1)} \\ R_1 & \ddots & \ddots & \vdots \\ \vdots & \ddots & \ddots & R_{-1} \\ R_{P-1} & \cdots & R_1 & R_0 \end{bmatrix} \begin{bmatrix} a_1^{(P)} \\ a_2^{(P)} \\ \vdots \\ a_P^{(P)} \end{bmatrix} = - \begin{bmatrix} R_1 \\ R_2 \\ \vdots \\ R_P \end{bmatrix},$$

where R_l are elements of the process auto-correlation sequence. This equation is closely connected with the concept of minimum phase filters, and numerous applications, e.g., spectral estimation and adaptive beamforming, are based on analyzing the "eigenstructure" of the matrix of auto-correlation (or auto-covariance) values expressed by the decomposition

$$\boldsymbol{R} = \sum_{k=1}^{P} \sigma_k^2\, \boldsymbol{v}_k \boldsymbol{v}_k^+,$$

where σ_k^2 and $\boldsymbol{v}_k$ are the eigenvalues and eigenvectors of the auto-correlation (or auto-covariance) matrix, respectively. The estimation of the elements of $\boldsymbol{R}$ (and any other parameters in the model) is often performed by invoking the "maximum likelihood principle," which underlies much of optimal statistical signal processing. The Levinson-Durbin recursive solution to the above matrix equation is perhaps the most well known algorithm of advanced signal processing.

This text is mostly about optimal linear filters, the estimation of the auto-correlation (or auto-covariance) matrix and its eigenstructure, and applications to spectral analysis. We have also included some information theoretic concepts and their applications to signal processing, e.g., independent components analysis. We have also included a chapter on modern neural networks not only because they can be viewed as an extension of iterative techniques to solve an optimization problem for nonlinear activations applied to outputs of linear combiners or filters (in the case of convolutional neural networks), but because of an explosive growth in theoretical research and applications to many areas of traditional signal and image processing such as speech and image recognition.

We have made every effort to produce a self-contained text with accurate, consistent, and comprehensible mathematical notation that should be comfortable to electrical engineering, physics, applied mathematics, and computer science students. We have stated the most important results in the form of theorems with simple and sufficiently rigorous proofs. The topics covered in this text are central to the subject and provide a solid foundation for students who wish to continue their studies towards more advanced degrees, or to keep abreast of the latest theoretical developments with a view to applying them to their work. Any future corrections will be posted on `www.mhprofessional.com/najmi` and a number of homework problems will be available to instructors upon email request to `ahnajm@jhu.edu`.

A. H. Najmi, Columbia, MD
ahnajmi@jhu.edu

T. K. Moon, Logan, UT
Todd.Moon@usu.edu

Acknowledgments

AHN is grateful to the Whiting School of the Johns Hopkins University for providing him the opportunity to develop and teach an advanced signal processing course in the graduate EP program, and wishes to thank the Johns Hopkins University Applied Physics Laboratory (JHU-APL) for a Janney Fellowship to complete this textbook. Dr. Majid Rabbani (RIT, Rochester) provided valuable suggestions on section 5.6 and Mr. Brad Bazow (JHU-APL) provided the model in section 11.6. Mr. Benjamin Miller (JHU-APL) carefully read through most chapters and suggested improvements. Dr. Richard Pitre and Dr. Harry Cox read chapter 11 and provided some references.

Special thanks go to Mr. Stephen Smith of McGraw Hill who carefully edited the entire manuscript and provided guidance on the typesetting, which was performed by AHN in LaTeX.

Finally, chapter 12 has been written in collaboration with Mr. Patrick Emmanuel (JHU-APL); without his vast theoretical and practical knowledge of modern neural networks, the chapter would not have been completed. An early version of chapter 12 will appear in the 2020 Fall edition of the *Johns Hopkins University Applied Physics Laboratory Technical Digest*.

CHAPTER 1

Mathematical Structures of Signal Spaces

1.1 Introduction

A signal in this text refers to a time series, examples of which are the Dow Jones Industrial Average, a seismic event recorded on a seismometer, thermal noise on a magnetic induction sensor, and the sound of a whale in the ocean. Although most naturally occurring signals are in continuous time, they are digitized by an A/D converter into sequences that depend on a discrete time variable. We will discuss both continuous and discrete time signals since there are significant theoretical differences between the two. The single most important result that connects the discrete time sampled form of a signal to its progenitor continuous time version is the Shannon-Nyquist sampling theorem (see theorem 6 in section 1.9); the theorem states the conditions under which a bandlimited continuous time signal can be sampled with no loss in the fidelity of the original.

We denote the set of real numbers by $\mathbb{R}$, the set of integers by $\mathbb{Z}$, and the set of complex numbers by $\mathbb{C}$. We treat all signals with real or complex values as real or complex functions of either a continuous time variable $t \in \mathbb{R}$ denoted by $x(t)$ or x_t, or as a discrete time variable $n \in \mathbb{Z}$ denoted by $x[n]$ or x_n. Although many signals of interest are real quantities, some circumstances necessarily produce complex signals or pairs of real signals that can be treated as a single complex valued signal. For instance, the Fourier transform of a real signal is a complex signal. A second example is a signal-processing operation known as complex demodulation, whereby a real signal whose spectrum is limited to a small band of frequencies around a very high carrier frequency ω_c is shifted down in frequency, retaining the original shape of the spectrum but now centered around zero frequency (also known as DC). This process of shifting down the frequencies involves a multiplication of the original real signal with the two real functions $\cos(\omega_c t)$ and $\sin(\omega_c t)$ producing two real valued signals with a $90°$ phase difference, or equivalently, multiplication with the single complex exponential $\exp(i\omega_c t)$ producing a complex valued signal. A third example of complex signals is provided by magnetic resonance imaging (MRI), in which the circularly polarized changing magnetic field, the source of the MRI signal, results in two data streams with a $90°$ phase difference that are combined to form a single complex signal.

Whether continuous or discrete in time, signals can be divided into two general classes: finite energy non-random signals of known (or knowable origin) and finite power signals that are inherently random —the distinction has important consequences in signal analysis. As an example of a finite energy signal, consider the whale sound shown in figure 1.1. The sound begins some time after the recording starts and ends before the recording ends a finite amount of time later. This particular data appears to be a clean recording in the sense that very little, if any, "background" noise is present. It is a finite duration signal that follows a specific pattern produced by its emitter.

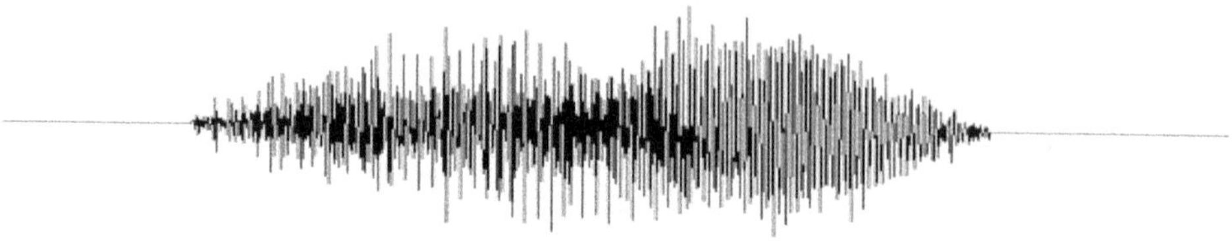

Figure 1.1: Bowhead whale sound.

A continuous time signal $x(t)$ or a discrete time signal $x[n]$ has finite energy if

$$\int_{-\infty}^{\infty} |x(t)|^2 dt < \infty \quad \text{or} \quad \sum_{n=-\infty}^{\infty} |x[n]|^2 < \infty . \tag{1.1}$$

Although the whale sound has finite duration, not all finite energy signals have finite time extent. For instance, the non-random real function (signal) $e^{-|t|}\cos(\omega_0 t)$ has infinite duration but finite energy. The space defined by the collection of all continuous time finite energy signals $x(t)$ (finite or infinite duration) is denoted by $L_2(\mathbb{R})$, while the space defined by all discrete time finite energy signals $x[n]$ (finite or infinite duration) is denoted by $l_2(\mathbb{Z})$.

Another example of a finite energy signal is the static dipole magnetic field generated by a fixed magnetic source in the absence of the Earth's geomagnetic, geologic, or any other possible additional sources of magnetic field, as shown in figure 1.2(a).

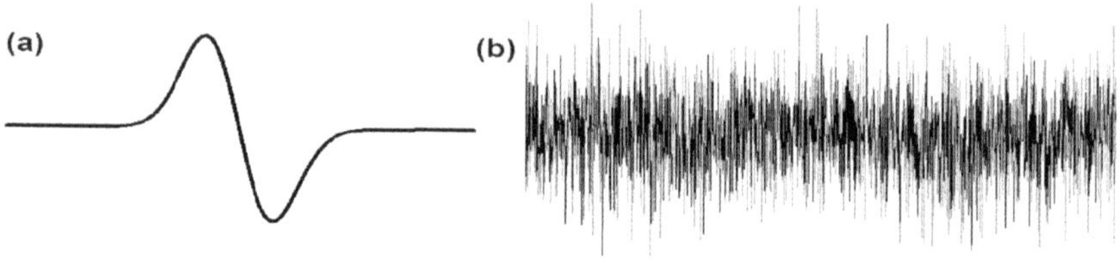

Figure 1.2: Magnetic field of a fixed dipole source (a) and random noise (b).

Now consider a "noise"-like infinite energy signal with no beginning or end, a finite portion of which is shown in figure 1.2(b). Although the natural processes producing this signal may be known, signal values at any time period are impossible to predict. For instance, consider thermal noise in a coil magnetometer; it depends on temperature (and other properties of the coil, such as its mass and diameter) and shows characteristics that are random. This randomness is a result of physical processes and not the same as artificial examples in which a number of deterministic signals may be combined together with some arranged probabilities, e.g., if the whale tosses a coin to decide whether to sing the first or the second half of its song. Infinite energy random signals must necessarily be described by their inherent statistical properties: the success of the description is directly related to how well those properties can be estimated. Random signals such as thermal noise have infinite energy simply because they have no beginning and no end; they are referred to as power signals, since their average energy in a time interval of length T has a finite limit as $T \to \infty$.

1.2 Vector Spaces, Norms, and Inner Products

The most basic mathematical structure of signal spaces is that of a *vector space* denoted by $\mathscr{V}$. Vector spaces of interest include continuous time (real or complex valued) functions $x(t)$, $t \in \mathbb{R}$, and discrete time (real or complex valued) sequences $x[n]$, $n \in \mathbb{Z}$, both of which are examples of vector spaces with infinite dimension. Real sequences with finite number of elements N comprise the vector space $\mathbb{R}^N$, while complex sequences of N elements define $\mathbb{C}^N$, both of which are examples of finite dimensional vector spaces.

A vector space $\mathscr{V}$ with elements denoted by $\boldsymbol{x}$ is defined with an operation of addition (more generally a commutative binary operation) and an operation of scalar multiplication with the elements of a mathematical structure known as a *field* $\mathbb{F}$. The only fields of importance to us are the real numbers $\mathbb{R}$ and the complex numbers $\mathbb{C}$ with the usual algebra; a complex vector space uses the field of complex numbers, whereas a real vector space uses the field of real numbers. A vector space [1] [2]:

- has an element called the zero, denoted $\mathbf{0}$, such that $\boldsymbol{x} + \mathbf{0} = \boldsymbol{x}$, $0\boldsymbol{x} = \mathbf{0}$,
- has a unique vector $-\boldsymbol{x}$ for every vector $\boldsymbol{x}$ such that $\boldsymbol{x} + (-\boldsymbol{x}) = \mathbf{0}$,
- is closed under the operation of scalar multiplication, i.e., if $\boldsymbol{x} \in \mathscr{V}$ then so is $\alpha\boldsymbol{x}$ for any $\alpha \in \mathbb{F}$, and $1\boldsymbol{x} = \boldsymbol{x}$,
- is closed under addition, i.e., if $\boldsymbol{x}, \boldsymbol{y} \in \mathscr{V}$, then any linear combination of the two vectors defined by $\alpha\boldsymbol{x} + \beta\boldsymbol{y}$, $\alpha, \beta \in \mathbb{F}$, is also a vector in $\mathscr{V}$.

The preceding properties also hold for vector spaces of functions whose elements are denoted by x or $x(t)$.

The *linear span* (or simply *span*) of a set of vectors $\boldsymbol{x}_k$, $1 \le k \le K$, is the space of all linear combinations of the set of vectors, i.e.,

$$\textbf{span}\,\{\boldsymbol{x}_k \,:\, 1 \le k \le K\} = \left\{ \sum_{k=1}^{K} c_k \boldsymbol{x}_k ;\; c_k \in \mathbb{F} \right\}, \tag{1.2}$$

which satisfies all of the requirements for a vector space (the linear span of a finite set of functions in a function space consists of all linear combinations of the form $c_1 x_1(t) + \ldots + c_K x_K(t)$). If the span of a set of vectors (or functions) is not the entire space from which those vectors (or functions) were chosen, then the span is a proper subspace of that vector space. When the span of a finite number of vectors is the entire space from which the vectors were chosen, then that space is a finite dimensional vector space, but if no linear span of any finite number of vectors is the entire space, then the space is an infinite dimensional vector space (function spaces are, in general, infinite dimensional).

A set of vectors $\{\boldsymbol{x}_k \,:\, 1 \le k \le K\}$ is said to be *linearly independent* if no vector in the set can be expressed as a linear combination of the other $K - 1$ vectors. Equivalently, the only possible way to represent $\mathbf{0}$ as a linear combination of a linearly independent set $\{\boldsymbol{x}_k \,:\, 1 \le k \le K\}$ is to use a 0 coefficient for every one of them, i.e., $c_1\boldsymbol{x}_1 + \ldots + c_K\boldsymbol{x}_K = \mathbf{0}$ if, and only if, $c_1 = \ldots = c_K = 0$. A similar definition applies to a finite collection of functions in a function space. For instance, the functions $1, t, t^2$ defined on a closed interval $[a, b] \in \mathbb{R}$ form an independent set, since the equation $c_1 + c_2 t + c_3 t^2 = 0$ for all $t \in [a, b]$ can only hold for $c_1 = c_2 = c_3 = 0$.

In a finite dimensional vector space, a non-unique linearly independent set of vectors exists such that all other vectors can be represented as a linear combination of the vectors in that set, which is then said to

[1] We use bold letters such as $\boldsymbol{x}$ to denote a sequence with either a finite or infinite number of elements belonging to finite or infinite dimensional vector spaces and non-bold letters such as x to denote functions $x(t)$ that belong to function spaces.

form a *basis*. The number of elements in any basis of a finite dimensional vector space is the same as in any other basis and is known as the *dimension* of the corresponding vector space.

We introduce the concept of *norm* to define convergence of sequences [3]. A *normed vector space* is equipped with a real valued and non-negative function, the norm, denoted by $\|.\|$, with the following properties:

- $\|\boldsymbol{x}\| \geq 0, \quad \|\boldsymbol{x}\| = 0 \Leftrightarrow \boldsymbol{x} = \mathbf{0}$
- $\|\alpha \boldsymbol{x}\| = |\alpha| \, \|\boldsymbol{x}\|$, where $\alpha \in \mathbb{F}$ and $\boldsymbol{x} \in \mathscr{V}$,
- $\|\boldsymbol{x} + \boldsymbol{y}\| \leq \|\boldsymbol{x}\| + \|\boldsymbol{y}\|$, known as the triangle inequality.

A sequence of functions $x_n(t)$, $n \in \mathbb{Z}$, is said to *converge in the mean* to the function $x(t)$ if $\|x_n(t) - x(t)\| \to 0$ for $n \to \infty$. The space $L_p(\mathbb{R})$, $p \geq 1$, is defined as the space of functions whose L_p norm,

$$\|x\|_p \equiv \left(\int_{-\infty}^{\infty} |x(t)|^p dt \right)^{1/p}, \quad p \geq 1, \tag{1.3}$$

is finite; $p = 2$ corresponds to the space $L_2(\mathbb{R})$. The space $l_p(\mathbb{Z})$ consists of sequences with infinite number of elements $\boldsymbol{x} = [\ldots, x_{-1}, x_0, x_1, \ldots]^T$ (super-script T denotes transposition) whose l_p norm,

$$\|\boldsymbol{x}\|_p \equiv \left(\sum_{n=-\infty}^{\infty} |x_n|^p \right)^{1/p}, \quad p \geq 1, \tag{1.4}$$

is finite (for sequences $\boldsymbol{x} \in l_2(\mathbb{Z})$ we use either $\|\boldsymbol{x}\|_2$ or $|\boldsymbol{x}|$ to denote the norm). The space $l_\infty(\mathbb{Z})$ is defined by the norm

$$\|\boldsymbol{x}\|_\infty \equiv \max |x_n| \tag{1.5}$$

The space of functions $L_\infty(\mathbb{R})$ is defined with respect to the L_∞ norm, namely,

$$\|x\|_\infty \equiv \inf \left\{ C > 0 : \; |x(t)| \leq C, \text{ for all } t \in \mathbb{R} \right\}, \tag{1.6}$$

where "inf" (the *infimum*) is the greatest lower bound. L_∞ functions (not used in our text, but with applications to economic theory) are known as essentially bounded functions, i.e., they are bounded except on a set of measure zero.

The limit $p \to 0$ of $\|\boldsymbol{x}\|_p$, denoted by $\|\boldsymbol{x}\|_0$, is not a norm but has applications to compressive sensing and sparse solutions of under-determined matrix equations; $\|\boldsymbol{x}\|_0$ counts the number of non-zero elements of $\boldsymbol{x}$. For example, if $\boldsymbol{x} \in \mathbb{R}^N$ and $\|\boldsymbol{x}\|_0 = s \ll N$, then $\boldsymbol{x}$ is said to be s-sparse [2], in which case it can be compressed by simply storing the s non-zero values together with their locations leading to a $2s/N$ compression ratio. The space of s-sparse vectors of dimension N is clearly not a subspace of $\mathbb{R}^N$ (the sum of two s-sparse vectors is not necessarily s-sparse!), and it is a difficult task to find the s locations of the non-zero values of an s-sparse vector. Compressive sensing (see section 1.14) is an attempt to solve this problem by designing a $k \times N$, $k \sim s$, structured random sensing matrix $\boldsymbol{A}$ (e.g., randomly selected rows of a discrete Fourier transform matrix) so that the measured $k \times 1$ vector $\boldsymbol{y} = \boldsymbol{A}\boldsymbol{x}$ is the compressed version of $\boldsymbol{x}$ and can be used to recover $\boldsymbol{x}$.

Although in a normed vector space we have the means for doing analysis and studying convergence, we are not guaranteed that converging sequences of vectors actually have a limit inside the space itself; when

[2] In reality, most signals $\boldsymbol{x}$ classified as s-sparse are those such that $\inf \|\boldsymbol{x} - \boldsymbol{s}\|_p \ll 1$ for all truly s-sparse vectors $\boldsymbol{s}$.

they do, we call this the *completeness property*, and the corresponding *complete normed vector space* is called a *Banach space* [3].

The *inner product* of two vectors or functions allows them to be projected along specific "directions." The inner product of any two complex functions in $L_2(\mathbb{R})$ is

$$\langle x, y \rangle = \int_{-\infty}^{\infty} x^*(t)\, W(t)\, y(t)\, dt, \tag{1.7}$$

where $W(t)$ is a positive weighting function; unless otherwise stated, it is 1 in this text. For complex valued vectors $\boldsymbol{x}, \boldsymbol{y}$, i.e., sequences with finite or infinite number of elements, the inner product is defined by

$$\langle \boldsymbol{x}, \boldsymbol{y} \rangle \equiv \boldsymbol{x}^+ \boldsymbol{y} = \sum_{n=-\infty}^{\infty} x_n^* y_n, \tag{1.8}$$

where super-script "+" denotes *Hermitian conjugation* (for vectors and matrices, this is equivalent to transposition and complex conjugation). These two examples illustrate the general requirements for an inner product:

- $\langle \boldsymbol{y}, \boldsymbol{x} \rangle = \langle \boldsymbol{x}, \boldsymbol{y} \rangle^*$,
- $\langle \boldsymbol{x}, c\boldsymbol{y} \rangle = c\, \langle \boldsymbol{x}, \boldsymbol{y} \rangle$,
- $\langle \boldsymbol{x} + \boldsymbol{y}, \boldsymbol{z} \rangle = \langle \boldsymbol{x}, \boldsymbol{z} \rangle + \langle \boldsymbol{y}, \boldsymbol{z} \rangle, \langle \boldsymbol{x}, \boldsymbol{x} \rangle \geq 0$,
- $\langle \boldsymbol{x}, \boldsymbol{x} \rangle = 0 \Leftrightarrow \boldsymbol{x} = \mathbf{0}$.

Note that the first two properties imply $\langle c\boldsymbol{x}, \boldsymbol{y} \rangle = c^* \langle \boldsymbol{x}, \boldsymbol{y} \rangle$ while the last two show that any inner product induces a norm $\sqrt{\langle \boldsymbol{x}, \boldsymbol{x} \rangle}$ known as the *induced norm*. Two vectors (or functions) are *orthogonal* if the inner product of the two vanishes. In addition, the induced norms of two vectors satisfy the *Cauchy-Schwarz inequality*

$$\left| \langle \boldsymbol{x}, \boldsymbol{y} \rangle \right| \leq \|\boldsymbol{x}\|\, \|\boldsymbol{y}\|\,. \tag{1.9}$$

A *Hilbert space* $\mathscr{H}$ is a *complete inner product space*; i.e., it is a Banach space whose norm is induced by an existing inner product, $\|\boldsymbol{x}\| = \sqrt{\langle \boldsymbol{x}, \boldsymbol{x} \rangle}$.

1.3 Orthonormal Vectors and the Gram-Schmidt Method

A set of vectors $\{\boldsymbol{x}_n\}$ is said to be *orthonormal* provided $\langle \boldsymbol{x}_m, \boldsymbol{x}_n \rangle = \delta_{mn}$, where δ_{mn} is the *Kronecker delta*: 1 if $m = n$ and 0 otherwise.

Given a set of vectors $\{\boldsymbol{x}_1, \ldots, \boldsymbol{x}_N\}$, the *Gram-Schmidt method* is used to obtain an orthonormal set $\{\boldsymbol{q}_1, \ldots, \boldsymbol{q}_M\}$ where $M \leq N$, such that both sets of vectors span the same vector space,

$$\textbf{span}\,\{\boldsymbol{x}_1, \ldots, \boldsymbol{x}_N\} = \textbf{span}\,\{\boldsymbol{q}_1, \ldots, \boldsymbol{q}_M\}, \quad M \leq N.$$

We begin the Gram-Schmidt algorithm by choosing $\boldsymbol{q}_1$ to be the *normalized* (unit norm) form of any one of $\boldsymbol{x}_k$, say $\boldsymbol{x}_1$, i.e., $\boldsymbol{q}_1 = \boldsymbol{x}_k / \|\boldsymbol{x}_1\|$. Next we choose $\boldsymbol{x}_2$, subtract its projection along $\boldsymbol{q}_1$, and normalize

[3] More technically, a Banach space is a normed vector space in which every convergent Cauchy sequence converges within the space; a Cauchy sequence $\boldsymbol{x}_n$ is defined by the requirement that $\|\boldsymbol{x}_n - \boldsymbol{x}_m\|$ can be made arbitrarily small for sufficiently large n, m. Although all convergent sequences are Cauchy, the converse is true only in a complete space.

the result, i.e., $\boldsymbol{q}'_2 = \boldsymbol{x}_2 - \langle \boldsymbol{x}_2, \boldsymbol{q}_1 \rangle \boldsymbol{q}_1$ and $\boldsymbol{q}_2 = \boldsymbol{q}'_2 / \|\boldsymbol{q}'_2\|$. This procedure of selecting the next $\boldsymbol{x}$ vector, subtracting the projections along the previous $\boldsymbol{q}$ vectors, and normalizing the result is continued until all $\boldsymbol{q}$ vectors are found. The Gram-Schmidt algorithm is summarized below.

- $\boldsymbol{q}_1 = \boldsymbol{x}_1 / \|\boldsymbol{x}_1\|$
- $\boldsymbol{q}'_n = \boldsymbol{x}_n - \sum_{l=1}^{n-1} \langle \boldsymbol{x}_n, \boldsymbol{q}_l \rangle \boldsymbol{q}_l$ and $\boldsymbol{q}_n = \boldsymbol{q}'_n / \|\boldsymbol{q}'_n\|$, for $n \geq 2$.

The standard Gram-Schmidt method is not numerically robust, as illustrated in the case of three vectors $[1, \epsilon, \epsilon]$, $[1, \epsilon, 0]$, and $[1, 0, \epsilon]$, with $\epsilon > 0$ small enough so that $1 + \epsilon^2 \approx 1$ to machine precision. In each stage of the standard method, we compute the projections of the present vector $\boldsymbol{x}_n$ with all the previous orthonormal vectors $\boldsymbol{q}_l$ and subtract all the projections. A modified and numerically stable method is to subtract the projection before making the next projection, i.e., the result of each subtraction is the input to the process of next projection and subtraction.

- $\boldsymbol{q}_1 = \boldsymbol{x}_1 / \|\boldsymbol{x}_1\|$ and then for $n \geq 2$:
- $\boldsymbol{q}'_n = \boldsymbol{x}_n$, $\boldsymbol{q}'_n \leftarrow \boldsymbol{q}'_n - \langle \boldsymbol{q}'_n, \boldsymbol{q}_l \rangle \boldsymbol{q}_l$, $1 \leq l \leq n-1$,
- $\boldsymbol{q}_n = \boldsymbol{q}'_n / \|\boldsymbol{q}'_n\|$.

1.4 Complete and Orthonormal Bases

The concept of an *orthonormal basis* in a finite dimensional vector space can be extended to an *infinite dimensional Hilbert space* $\mathscr{H}$ so that any element of a Hilbert space has a unique representation in terms of that basis. The decomposition of a function in terms of a basis set (analysis) and its inverse (reconstruction or synthesis) form the foundation of transform theory. For a complete and orthonormal basis denoted by $\phi_n, n \in \mathbb{Z}$, the analysis and synthesis operations are summarized in one formula,

$$x = \sum_n \langle \phi_n, x \rangle \, \phi_n, \tag{1.10}$$

with the inner product representing the analysis operation and the sum representing the reconstruction or synthesis operation. The right hand side is an infinite linear combination of the basis functions, and its convergence is guaranteed if $\|x\|$ is finite [4].

Theorem 1. *Given an orthonormal set of functions $\{\phi_n, n \in \mathbb{Z}\}$ in a Hilbert space $\mathscr{H}$, $\langle \phi_n, \phi_m \rangle = \delta_{nm}$, the infinite linear combination $\sum_{n=-\infty}^{\infty} c_n \phi_n$ converges if, and only if,*

$$\mathbf{c} = \left[\ldots, c_{-1}, c_0, c_1, \ldots\right]^T \in l_2(\mathbb{Z}). \tag{1.11}$$

If the condition (1.11) is met, then the series converges in the mean to a function $x \in \mathscr{H}$ (unique up to a set of measure 0) given by equation (1.10),

$$\left\| x - \sum_{n=-N}^{N} c_n \, \phi_n \right\| \to 0 \;\; as \;\; N \to +\infty. \tag{1.12}$$

Furthermore, Parseval's relation is

$$\|x\|^2 = \sum_{n=-\infty}^{\infty} |c_n|^2 . \tag{1.13}$$

Theorem 1 is the most widely used description of completeness of an orthonormal basis $\{\phi_n,\ n \in \mathbb{Z}\}$; the following summarizes all the equivalent definitions:

- For every x in $\mathscr{H}$, we have the expansion $x = \sum_{n=-\infty}^{\infty} \langle \phi_n, x \rangle \phi_n$.
- The only vector that is orthogonal to the set is the zero vector.
- The closure of the set is the entire space $\mathscr{H}$.
- For every x in $\mathscr{H}$, $\|x\|^2 = \sum_{n=-\infty}^{\infty} |\langle \phi_n, x \rangle|^2$, i.e., (1.13).
- For every x and y in $\mathscr{H}$, $\langle x, y \rangle = \sum_{n=-\infty}^{\infty} \langle x, \phi_n \rangle \langle \phi_n, y \rangle$.

An example of an orthogonal set of functions that is not complete is $\{\sin(nt),\ n \in \mathbb{Z}\}$, $t \in [0, 2\pi]$, since $\cos(kt)$, $k \in \mathbb{Z}$, is orthogonal to all members of the set.

If the functions $\phi_n(t)$, $n \in \mathbb{Z}$ and $t \in \mathbb{R}$, are a complete and orthonormal basis of $L_2(\mathbb{R})$, then the expansion coefficients for $x(t) \in L_2(\mathbb{R})$ (see equation (1.10)), namely,

$$c_n \equiv \langle \phi_n, x \rangle = \int_{-\infty}^{\infty} \phi_n^*(t) x(t)\, dt, \tag{1.14}$$

are the transform coefficients for $x(t)$. The coefficients c_n fully describe $x(t)$ in the sense that it can be reconstructed from the transform coefficients at almost every point t, i.e., all points except for a set of measure zero [4]. In other words, the transform relation is uniquely invertible in the form of equation (1.10). An important example of an orthonormal basis for $L_2(\mathbb{R})$ is the Haar basis, which together with the associated orthogonal wavelet transform is described in section 1.13.

A generalization of the orthonormal and complete sets is provided by frames that are over-complete; Parseval's relation for frames is an inequality involving frame elements ϕ_n, namely,

$$A\,\|x\|^2 \le \sum_n |\langle \phi_n, x \rangle|^2 \le B\,\|x\|^2, \tag{1.15}$$

where $A > 0$ and $B < \infty$ are the frame bounds. Redundant frame expansions provide more flexibility than orthonormal basis expansions in certain signal processing applications.

1.5 Linear Operators in Function Spaces

Given two Hilbert spaces $\mathscr{H}_1$ and $\mathscr{H}_2$, an operator (transformation) T between them is denoted by $T : \mathscr{H}_1 \to \mathscr{H}_2$. The operator T is linear if for all $x, y \in \mathscr{H}_1$ and $\alpha, \beta \in \mathbb{F}$ (where $\mathbb{F}$ is a field),

$$T(\alpha x + \beta y) = \alpha T x + \beta T y. \tag{1.16}$$

[4] The Gibbs phenomenon is such an example. Given a periodic function that is piecewise continuously differentiable with a countable number of discontinuities, then the function's Fourier series expansion converges to the value of the function everywhere except at the points of discontinuity.

To every linear operator there correspond two fundamental subspaces, namely, the range space $\mathscr{R}_T$ and the null space $\mathscr{N}_T$. The range space is the image of $\mathscr{H}_1$ under the transformation and this is a subspace of $\mathscr{H}_2$, while the null space is a subspace of $\mathscr{H}_1$ consisting of all functions that are transformed to the 0 function of $\mathscr{H}_2$:

- $\mathscr{R}_T = T(\mathscr{H}_1) = \{y \in \mathscr{H}_2 : \; y = Tx, \text{ for some } x \in \mathscr{H}_1\}$,
- $\mathscr{N}_T = \{x \in \mathscr{H}_1 : \; Tx = 0\}$.

The norm of an operator T, denoted by $\|T\|$, is defined in the following three equivalent ways using the L_p norm defined earlier (the operator norm is a subordinate norm induced by the L_p norm in function spaces):

- $\|T\| = \inf \{a \geq 0 : \; \|Tx\| \leq a \|x\| \text{ for all } x \in \mathscr{H}_1\}$,
- $\|T\| = \sup \{\|Tx\| \,/\, \|x\| \text{ for all } x\}$,
- $\|T\| = \sup \{\|Tx\| \text{ for all unit norm } x : \; \|x\| = 1\}$,

where "sup" (the *supremum*) is the least upper bound (see equation (1.6) for "inf"). An operator T is bounded if its norm is finite.

Each linear transformation T has an associated adjoint operator denoted by T^+, which is a linear transformation $T^+ : \mathscr{H}_2 \to \mathscr{H}_1$ and is defined by the equation $\langle y, Tx \rangle = \langle T^+ y, x \rangle$. When $T : \mathscr{H} \to \mathscr{H}$ and if $T^+ = T$, then the operator is *self-adjoint* or *Hermitian*.

A linear operator Q for which $\|Qx\| = \|x\|$ for all x is called a unitary operator. Thus, $\langle Qx, Qx \rangle = \langle x, Q^+ Qx \rangle = \langle x, x \rangle$, which leads to the unitarity relation $Q^+ Q = 1$.

The following relations hold for all bounded linear operators whose range and null spaces are closed:

- $\mathscr{R}_T$ and $\mathscr{N}_{T^+}$ are orthogonal complements in $\mathscr{H}_2$ (disjoint orthogonal subspaces whose union is the whole space), i.e., every $y \in \mathscr{H}_2$ has a unique decomposition $y = r_2 + n_2$, $r_2 \in \mathscr{R}_T$, $n_2 \in \mathscr{N}_{T^+}$, and $\langle r_2, n_2 \rangle = 0$.
- $\mathscr{R}_{T^+}$ and $\mathscr{N}_T$ are orthogonal complements in $\mathscr{H}_1$, i.e., every $x \in \mathscr{H}_1$ has a unique decomposition $x = r_1 + n_1$, $r_1 \in \mathscr{R}_{T^+}$, $n_1 \in \mathscr{N}_T$, and $\langle r_1, n_1 \rangle = 0$.

The preceding relations allow us to study the existence of solution(s) of linear equations involving bounded linear operators, as embodied in the Fredholm Alternative theorem [3, 5].

Theorem 2. *The equation $Tx = b$, $b \neq 0$, has a solution if, and only if, $b \in \mathscr{R}_T$, or equivalently, $\langle b, n \rangle = 0$ for all $n \in \mathscr{N}_{T^+}$.*

If a solution exists, then it is unique if, and only if, $\mathscr{N}_T = \{0\}$, i.e., the only solution to $Tx = 0$ is the trivial solution $x = 0$. If the null space $\mathscr{N}_T$ has other non-zero elements, then the solution is not unique: the solution whose norm is the smallest among all solutions (the *minimum norm solution*) is given by

$$x_{\text{mn}} = T^+(TT^+)^{-1}b. \tag{1.17}$$

If $b \notin \mathscr{R}_T$, then there is no solution, but a *least squares solution* x_{ls} exists that minimizes $\|Tx - b\|_2^2$ and is given by

$$x_{\text{ls}} = (T^+T)^{-1}T^+b. \tag{1.18}$$

1.6 Matrix Determinant, Eigenvectors, and Eigenvalues

A linear operator mapping two finite dimensional Hilbert spaces $\boldsymbol{A} : \mathbb{C}^N \to \mathbb{C}^M$ with the usual inner product and induced Euclidean norm can be represented as an $M \times N$ complex matrix $\boldsymbol{A}$ that maps the vector $\boldsymbol{x} = [x_1, \ldots, x_N] \in \mathbb{C}^N$ to $\boldsymbol{y} = [y_1, \ldots, y_M] \in \mathbb{C}^M$. If we denote the columns of $\boldsymbol{A}$ by $\boldsymbol{a}_j$, $1 \le j \le N$, then $\boldsymbol{Ax} = x_1\boldsymbol{a}_1 + \ldots + x_N\boldsymbol{a}_N$ is simply a linear combination of the columns of $\boldsymbol{A}$, and so $\mathscr{R}_{\boldsymbol{A}} = \mathbf{span}\{\boldsymbol{a}_1, \ldots, \boldsymbol{a}_N\}$, i.e., the range space $\mathscr{R}_{\boldsymbol{A}}$ is the column space of $\boldsymbol{A}$, and the dimension of the latter is the column rank (or just rank) of the matrix $r_{\boldsymbol{A}} = \dim\{\mathscr{R}_{\boldsymbol{A}}\}$. The null space is the set of all vectors $\boldsymbol{n} \in \mathbb{C}^N$ satisfying the linear equation $\boldsymbol{An} = \boldsymbol{0} \in \mathbb{C}^M$ and $\dim\{\mathscr{N}_{\boldsymbol{A}}\} = N - r_{\boldsymbol{A}}$.

The adjoint operator is now the Hermitian conjugate (complex transpose) matrix $\boldsymbol{A}^+$ whose elements are $[A^+]_{jk} = [A^*]_{kj}$, representing a linear transformation $\boldsymbol{A}^+ : \mathbb{C}^M \to \mathbb{C}^N$. It is easy to show that $\dim\{\mathscr{R}_{\boldsymbol{A}^+}\} = r_{\boldsymbol{A}}$ and $\dim\{\mathscr{N}_{\boldsymbol{A}^+}\} = M - r_{\boldsymbol{A}}$ [2].

An important quantity associated with a square matrix is its determinant. For a square $N \times N$ matrix $\boldsymbol{A}$, each element a_{ij} is associated with a minor M_{ij} that is the determinant of the $(N-1) \times (N-1)$ matrix obtained by deleting row i and column j in the original matrix $\boldsymbol{A}$. The determinant of $\boldsymbol{A}$, denoted by $|\boldsymbol{A}|$, is defined by the Leibnitz formula

$$|\boldsymbol{A}| = \sum_{\sigma} \epsilon_\sigma \, a_{1,\sigma(1)} \cdots a_{N,\sigma(N)}, \tag{1.19}$$

where the sum is over all permutations $\sigma(n)$ of the numbers $1, \ldots, N$, and ϵ_σ is the sign of each permutation σ: $+1$ for an even permutation and -1 for an odd permutation. The determinant can also be written in terms of the minors by the Laplace formula

$$|\boldsymbol{A}| = \sum_{i=1}^{N} (-1)^{i+j} a_{ij} M_{ij}, \quad \text{for any column } j, \tag{1.20}$$

where the minor M_{ij} is the determinant of the sub-matrix obtained by deleting the row i and column j of the original matrix $\boldsymbol{A}$. The adjugate matrix $\mathbf{adj}(\boldsymbol{A})$ is defined as the transpose of the matrix of co-factors $[\mathbf{adj}(\boldsymbol{A})]_{ij} = (-1)^{i+j} M_{ji}$ and is related to the inverse by

$$\boldsymbol{A}\,[\mathbf{adj}(\boldsymbol{A})] = |\boldsymbol{A}|\,\boldsymbol{I}\,. \tag{1.21}$$

Thus, the determinant in terms of the adjugate matrix is

$$|\boldsymbol{A}| = \sum_{i=1}^{N} a_{ij}\, \mathbf{adj}(\boldsymbol{A})_{ji}, \quad \text{for any } j. \tag{1.22}$$

Differentiating (1.22) with respect to a_{kj} we find $\partial|\boldsymbol{A}|/\partial a_{kj} = [\mathbf{adj}(\boldsymbol{A})]_{jk}$, which is often written in the form $\partial|\boldsymbol{A}|/\partial \boldsymbol{A} = [\mathbf{adj}(\boldsymbol{A})]^T$ and is known as *Jacobi's formula*. Using the inverse matrix relation (1.21) we find

$$\partial|\boldsymbol{A}|/\partial \boldsymbol{A} = |\boldsymbol{A}|\,\boldsymbol{A}^{-T} \;\leftrightarrow\; \partial|\ln \boldsymbol{A}|/\partial \boldsymbol{A} = \boldsymbol{A}^{-T} \equiv [\boldsymbol{A}^T]^{-1} = [\boldsymbol{A}^{-1}]^T. \tag{1.23}$$

An eigenvector $\boldsymbol{x}$ of the square matrix $\boldsymbol{A}$ is a solution of the eigenvalue equation $\boldsymbol{Ax} = \mu\boldsymbol{x}$, where μ is an eigenvalue (in general not unique). All N eigenvalues are found by solving the characteristic equation $|\boldsymbol{A} - \mu\boldsymbol{I}| = 0$, which is a polynomial equation of degree N in μ, derived by writing the eigenvalue equation in the form $(\boldsymbol{A} - \mu\boldsymbol{I})\boldsymbol{x} = \boldsymbol{0}$.

Theorem 3. *If $\boldsymbol{A}$ is Hermitian, its eigenvalues are real and eigenvectors associated with distinct eigenvalues are orthogonal. If, in addition, $\boldsymbol{A}$ is* positive definite, *i.e., $\boldsymbol{z}^{+}\boldsymbol{A}\boldsymbol{z} > 0$ for all $\boldsymbol{z} \neq \boldsymbol{0}$, then the eigenvalues are positive* [5].

1.7 Matrix Norms

Four important matrix norms used in linear algebra are the l_∞, l_1, l_2, and the Frobenius norms [6]; the last two norms are used in section 2.8 to address the problem of reduced rank approximations to matrices [7].

- The l_∞ norm may be shown to be the largest column sum of the magnitude of the matrix elements, i.e., $\|\boldsymbol{A}\|_\infty^2 = \left(\text{Max} \sum_{j=1}^{N} |a_{ij}|\right)^2$.

- The l_1 norm may be shown to be the largest row sum of the magnitude of the matrix elements, i.e., $\|\boldsymbol{A}\|_1^2 = \left(\text{Max} \sum_{i=1}^{N} |a_{ij}|\right)^2$.

- The l_2 norm is defined by maximizing $\|\boldsymbol{A}\boldsymbol{x}\|_2^2 \,/\, \|\boldsymbol{x}\|_2^2$, which is equivalent to maximizing the quantity $\|\boldsymbol{A}\boldsymbol{x}\|_2^2$ subject to the constraint $\|\boldsymbol{x}\|_2^2 = 1$. Defining the Hermitian and positive definite matrix $\boldsymbol{A}^{+}\boldsymbol{A}$ and its eigenvalues and eigenvectors through $\boldsymbol{A}^{+}\boldsymbol{A}\boldsymbol{x} = \sigma^2\boldsymbol{x}$, we find the square of the l_2 norm to be the largest eigenvalue σ_{max}^2; the latter is also known as the spectral radius and the square root of it is the largest singular value of the matrix $\boldsymbol{A}$. Thus $\|\boldsymbol{A}\|_2^2 = \sigma_{\text{max}}^2$ (computational algorithms output the singular values of a matrix in descending order so the largest singular value is usually denoted by σ_1, as in section 2.7).

- The Frobenius norm is $\|\boldsymbol{A}\|_F^2 \equiv \sum_{i,j=1}^{N} |a_{ij}|^2 = \sqrt{\textbf{Trace}(\boldsymbol{A}^{+}\boldsymbol{A})}$, where the trace of a square matrix is defined as the sum of its diagonal elements.

The l_∞, l_1 and l_2 norms are special cases of l_p, $p \geq 1$ and are subordinate (or induced) norms that can be defined by sup $\|\boldsymbol{A}\boldsymbol{x}\| \,/\, \|\boldsymbol{x}\|$, $\boldsymbol{x} \neq \boldsymbol{0}$; the Frobenius norm is not a subordinate norm. Theorem 4 follows from the definition of matrix norms [7].

Theorem 4. *Let $\boldsymbol{A}$ be an $M \times N$ complex matrix. Then*

- $\|\boldsymbol{A}\boldsymbol{x}\| \leq \|\boldsymbol{A}\| \, \|\boldsymbol{x}\|$ *for all matrix subordinate norms.*

- $\|\boldsymbol{A}\boldsymbol{x}\|_2 \leq \|\boldsymbol{A}\|_F \, \|\boldsymbol{x}\|_2$ *for the Frobenius norm.*

- $\|\boldsymbol{A}\boldsymbol{B}\| \leq \|\boldsymbol{A}\| \, \|\boldsymbol{B}\|$ *for all norms (subordinate and Frobenius).*

- $\|\boldsymbol{Q}\boldsymbol{A}\| = \|\boldsymbol{A}\boldsymbol{Q}\| = \|\boldsymbol{A}\|$ *for a unitary matrix $\boldsymbol{Q}$, and only for the l_2 and the Frobenius norms.*

- $\|\boldsymbol{A}\|_2 \leq \|\boldsymbol{A}\|_F \leq \sqrt{L} \, \|\boldsymbol{A}\|_2$, *where* $L = \min(N, M)$.

[5] A Hermitian $N \times N$ complex matrix $\boldsymbol{A}$ is *non-negative definite* (or *positive semi-definite*) if $\boldsymbol{z}^{+}\boldsymbol{A}\boldsymbol{z} \geq 0$, in which case there may be one or more zero eigenvalues. Similar definitions, with reversed inequalities, hold for negative definite and negative semi-definite matrices.

1.8 Solutions to $\boldsymbol{Ax} = \boldsymbol{b}$

Consider the matrix equation $\boldsymbol{Ax} = \boldsymbol{b}$, $\boldsymbol{x} \in \mathbb{C}^N$ and $\boldsymbol{b} \in \mathbb{C}^M$. When $M = N = r_A$, the matrix $\boldsymbol{A}$ is square and full rank and therefore non-singular. Thus, it is invertible, and the equation $\boldsymbol{Ax} = \boldsymbol{b} \neq \boldsymbol{0}$ has a unique solution, namely, $\boldsymbol{x} = \boldsymbol{A}^{-1}\boldsymbol{b}$. As noted in section 1.6, the equation $\boldsymbol{Ax} = \boldsymbol{0}$ has non-trivial solutions only if $\boldsymbol{A}$ has zero determinant, in which case there is an infinite number of solutions.

If $M \leq N$ and $r_A < M$, the null space $\mathscr{N}_A$ has dimension greater than zero (it contains non-zero vectors in addition to the mandatory $\boldsymbol{0}$), and so if a solution $\boldsymbol{x}_0$ to $\boldsymbol{Ax} = \boldsymbol{b} \neq \boldsymbol{0}$ exists, then it is not unique since $\boldsymbol{A}(\boldsymbol{x}_0 + \boldsymbol{n}) = \boldsymbol{b}$ for any non-zero $\boldsymbol{n} \in \mathscr{N}_A$. In this case, a solution can be found whose norm is the smallest of all solutions (see equation (1.17)); it is known as the *minimum norm solution*,

$$\boldsymbol{x}_{\text{mn}} = \boldsymbol{A}^+(\boldsymbol{A}\boldsymbol{A}^+)^{-1}\boldsymbol{b}, \tag{1.24}$$

and it satisfies $\boldsymbol{Ax}_{\text{mn}} = \boldsymbol{b} \neq \boldsymbol{0}$, with $\|\boldsymbol{x}_{\text{mn}}\| \leq \|\boldsymbol{x}\|$ for all $\boldsymbol{x}$ that are solutions of the same linear equation. The minimum norm solution is also the solution to the minimization problem,

$$\arg\min_{\boldsymbol{x}} \ \|\boldsymbol{x}\|_2^2, \quad \text{subject to} \quad \boldsymbol{Ax} = \boldsymbol{b}. \tag{1.25}$$

To solve (1.25) we introduce a *Lagrange multiplier* vector $\boldsymbol{\lambda}$ and minimize the quantity $\boldsymbol{x}^+\boldsymbol{x} + \boldsymbol{\lambda}^+(\boldsymbol{Ax} - \boldsymbol{b})$ without a constraint [6]. Differentiating with respect to $\boldsymbol{x}$ and setting the result to $\boldsymbol{0}$ gives $\boldsymbol{x}^+ = -\boldsymbol{\lambda}^+\boldsymbol{A}$, whose Hermitian conjugate is $\boldsymbol{x} = -\boldsymbol{A}^+\boldsymbol{\lambda}$. Substituting for $\boldsymbol{x}$ in the constraint $\boldsymbol{Ax} = \boldsymbol{b}$ gives $\boldsymbol{\lambda} = -(\boldsymbol{AA}^+)^{-1}\boldsymbol{b}$, which when substituted in the equation for $\boldsymbol{x}$ gives the minimum norm solution (1.24). To show that $\boldsymbol{x}_{\text{mn}}$ is the solution with the smallest norm, consider another solution $\boldsymbol{x}$ when $\boldsymbol{A}(\boldsymbol{x} - \boldsymbol{x}_{\text{mn}}) = \boldsymbol{0}$ and

$$(\boldsymbol{x} - \boldsymbol{x}_{\text{mn}})^+\boldsymbol{x}_{\text{mn}} = \left[\boldsymbol{A}(\boldsymbol{x} - \boldsymbol{x}_{\text{mn}})\right]^+(\boldsymbol{AA}^+)^{-1}\boldsymbol{b} = 0, \tag{1.26}$$

i.e., $\boldsymbol{x} - \boldsymbol{x}_{\text{mn}}$ is orthogonal to $\boldsymbol{x}_{\text{mn}}$. Then

$$\|\boldsymbol{x}\|^2 = \|\boldsymbol{x} + \boldsymbol{x}_{\text{mn}} - \boldsymbol{x}_{\text{mn}}\|^2 = \|\boldsymbol{x}_{\text{mn}}\|^2 + \|\boldsymbol{x} - \boldsymbol{x}_{\text{mn}}\|^2 \geq \|\boldsymbol{x}_{\text{mn}}\|^2. \tag{1.27}$$

The minimum norm solution (1.24) shows that $\boldsymbol{x}_{\text{mn}} \in \mathscr{R}_{\boldsymbol{A}^+}$, and as such, it must have the smallest norm of all solutions of the form $\boldsymbol{x} + \boldsymbol{n}$ with $\boldsymbol{n} \in \mathscr{N}_{\boldsymbol{A}}$ since $\mathscr{N}_{\boldsymbol{A}}$ is orthogonal to $\mathscr{R}_{\boldsymbol{A}^+}$.

When $r_A = N < M$, the equation $\boldsymbol{Ax} = \boldsymbol{b}$ has no solution if $\boldsymbol{b} \notin \mathscr{R}_A$. However, a unique vector that minimizes the l_2 norm $\|\boldsymbol{Ax} - \boldsymbol{b}\|_2^2$ is the least squares solution [7] and is given by (see equation (1.18))

$$\boldsymbol{x}_{\text{ls}} = (\boldsymbol{A}^+\boldsymbol{A})^{-1}\boldsymbol{A}^+\boldsymbol{b}. \tag{1.28}$$

To see the formula (1.28) consider the vector $\hat{\boldsymbol{b}} \in \mathscr{R}_A$ (i.e., $\hat{\boldsymbol{b}}$ is a linear combination of the columns of $\boldsymbol{A}$). The error vector $\boldsymbol{e} = \boldsymbol{b} - \hat{\boldsymbol{b}}$ has its smallest norm when the error is orthogonal to $\mathscr{R}_A$. We have already noted

[6] When finding the extrema of a function $f(\boldsymbol{x})$ subject to a constraint $g(\boldsymbol{x}) = c$, then the extrema exist on g provided that the gradients of both f and g at the extrema are collinear, i.e., $\nabla f(\boldsymbol{x}) = -\lambda \nabla g(\boldsymbol{x})$, for some constant λ known as a Lagrange multiplier. This can be seen most easily by considering an infinitesimal vector $\Delta\boldsymbol{x}$ at the extremum: this vector must lie in the surface $g(\boldsymbol{x}) = c$ and so it must be orthogonal to the gradient to the surface ∇g, but since $f(\boldsymbol{x} + \Delta\boldsymbol{x}) - f(\boldsymbol{x}) \approx \nabla f . \Delta\boldsymbol{x}$, the extremum condition ensures that the infinitesimal vector is also orthogonal to the gradient of f. Thus, the two gradients are collinear and we set the gradient of the combined function $f + \lambda g$ to zero. Setting the derivative of the combined function with respect to λ to zero yields the constraint. For multiple constraints we introduce a Lagrange multiplier vector with the same number of elements as the constraints and solve $\nabla(f + \lambda_1 g_1 + \ldots \lambda_L g_L) = \boldsymbol{0}$.

[7] The least squares method was first used by Gauss in 1801 to calculate the orbit of the asteroid Ceres, and he published it in 1810 after the discovery of the asteroid Pallas whose orbit he calculated by solving 11 equations for 6 unknowns. Legendre, however, beat Gauss to the publication by five years when he published a paper on methods to calculate orbits of comets.

that $\mathscr{N}_{A^+}$ is the orthogonal complement of $\mathscr{R}_A$ in $\mathbb{C}^M$, and so there is a unique decomposition $\boldsymbol{b} = \boldsymbol{r}_b + \boldsymbol{n}_b$ where $\boldsymbol{r}_b = \boldsymbol{A}\boldsymbol{b}$ and $\boldsymbol{n}_b \in \mathscr{N}_{A^+}$. Therefore, $\boldsymbol{n}_b = \boldsymbol{b} - \boldsymbol{A}\boldsymbol{b}$ and $\boldsymbol{A}^+(\boldsymbol{b} - \boldsymbol{A}\boldsymbol{b}) = \boldsymbol{0}$. Thus, we find the least squares solution if the matrix $\boldsymbol{A}^+\boldsymbol{A}$ is invertible. This means that $\hat{\boldsymbol{b}}$, i.e., the vector in the column space $\mathscr{R}_A$ that is closest to $\boldsymbol{b}$ (in terms of the l_2 norm of the difference vector), is given by

$$\hat{\boldsymbol{b}} = \boldsymbol{A}(\boldsymbol{A}^+\boldsymbol{A})^{-1}\boldsymbol{A}^+\boldsymbol{b}. \tag{1.29}$$

The two matrices $(\boldsymbol{A}^+\boldsymbol{A})^{-1}\boldsymbol{A}^+$ and $\boldsymbol{A}^+(\boldsymbol{A}\boldsymbol{A}^+)^{-1}$ that appear in the least squares and the minimum norm solutions are examples of a pseudo-inverse or generalized (Moore-Penrose) inverse [8] when the indicated inverses exist (which is the case when the matrix $\boldsymbol{A}$ has full column rank, i.e., when the dimension of $\mathscr{R}_A$ is N). The pseudo-inverse is denoted by $\boldsymbol{A}^\dagger$ and is defined by the following properties:

- $\boldsymbol{A}\boldsymbol{A}^\dagger\boldsymbol{A} = \boldsymbol{A}$.
- $\boldsymbol{A}^\dagger\boldsymbol{A}\boldsymbol{A}^\dagger = \boldsymbol{A}^\dagger$.
- $(\boldsymbol{A}^\dagger\boldsymbol{A})^+ = \boldsymbol{A}^\dagger\boldsymbol{A}$ and $(\boldsymbol{A}\boldsymbol{A}^\dagger)^+ = \boldsymbol{A}\boldsymbol{A}^\dagger$.

When $\boldsymbol{A}$ has full column rank, $\boldsymbol{A}_L^\dagger \equiv (\boldsymbol{A}^+\boldsymbol{A})^{-1}\boldsymbol{A}^+$ is a left inverse, since $\boldsymbol{A}_L^\dagger\boldsymbol{A} = \boldsymbol{I}$, and $\boldsymbol{A}_R^\dagger \equiv \boldsymbol{A}^+(\boldsymbol{A}\boldsymbol{A}^+)^{-1}$ is a right inverse, since $\boldsymbol{A}\boldsymbol{A}_R^\dagger = \boldsymbol{I}$, where $\boldsymbol{I}$ is the identity matrix. For a square non-singular matrix $\boldsymbol{A}$, both the left and right inverses are equal to the unique inverse $\boldsymbol{A}^{-1}$.

The condition number of a matrix defined by

$$\kappa_{\boldsymbol{A}} \equiv \|\boldsymbol{A}\| \, \|\boldsymbol{A}^\dagger\| \tag{1.30}$$

plays an important role in numerical computation of the least squares and minimum norm solutions. We will return to this topic in chapter 2.

1.9 Projections in a Hilbert Space

Given a Hilbert space $\mathscr{H}$ and a proper subspace $\mathcal{V}$, a *projection* onto $\mathcal{V}$ is a linear map $P : \mathscr{H} \to \mathcal{V}$ with the property $Px = v$ where $x \in \mathscr{H}$ and $v \in \mathcal{V}$. Clearly, the defining property is the equation $P^2 = P$. It is easy to show that the range space $\mathscr{R}_P$ and null space $\mathscr{N}_P$ of a projection are disjoint subspaces $\mathscr{R}_P \cap \mathscr{N}_P = \{0\}$ that are algebraic complements, i.e., $\mathscr{H} = \mathscr{R}_P + \mathscr{N}_P$ in the sense that every element of the Hilbert space x has a unique decomposition $x = v + n$ with $v \in \mathscr{R}_P$ and $n \in \mathscr{N}_P$.

A projection is orthogonal if its range and null spaces are orthogonal to each other, i.e., $\langle v, n \rangle = 0$ for all $v \in \mathscr{R}_P$ and $n \in \mathscr{N}_P$. An alternative and useful way to characterize a projection is the following result.

Theorem 5. *A projection operator P with range and null spaces $\mathscr{R}_P$ and $\mathscr{N}_P$ is orthogonal if, and only if, it is self-adjoint (Hermitian).*

To prove this, we write $x \equiv Px + (1 - P)x$ for all $x \in \mathscr{H}$. Clearly $Px \in \mathscr{R}_P$ and $(1 - P)x \in \mathscr{N}_P$; the latter follows from the fact that $P(1 - P)x = (P - P^2)x = 0$. If $P = P^+$, then

$$\langle Px, (1 - P)\,x \rangle = \langle x, P\,(1 - P)\,x \rangle = 0,$$

and so the range and null spaces of P are orthogonal. If the range and null spaces are orthogonal then the above equation implies that $P^+(1 - P) = 0$ and $(1 - P^+)P = 0$, and so $P = P^+$.

[8] Originally described by Moore in 1920 and rediscovered by Penrose in 1955.

Specializing to finite dimensional Hilbert spaces, a *projection matrix* $\boldsymbol{P} : \mathbb{C}^N \to \mathbb{C}^M$ is orthogonal if it is Hermitian. The least squares solution discussed in section 1.6 illustrates an example of an orthogonal projection: the vector $\hat{\boldsymbol{b}} = \boldsymbol{A}(\boldsymbol{A}^+\boldsymbol{A})^{-1}\boldsymbol{A}^+\boldsymbol{b}$ is the orthogonal projection of the vector $\boldsymbol{b}$ onto the column space of the matrix $\boldsymbol{A}$, and the projection matrix is

$$\boldsymbol{P}_A = \boldsymbol{A}(\boldsymbol{A}^+\boldsymbol{A})^{-1}\boldsymbol{A}^+. \tag{1.31}$$

An important example of an orthogonal projection is the space of bandlimited functions and the associated complete and orthonormal basis. A bandlimited function $x_\Omega(t) \in L_2(\mathbb{R})$ is one whose Fourier transform $X_\Omega(\omega)$ (see section 3.7) has support in $[-\Omega, \Omega]$ (i.e., the function is zero everywhere outside the interval), where Ω is a positive real number. The unitarity of the Fourier transform operation ensures that all bandlimited functions in $L_2(\mathbb{R})$ form a closed subspace $L_2^\Omega(\mathbb{R})$, which is a *reproducing kernel Hilbert space* [8]

$$x_\Omega(t) = \int_{-\infty}^{\infty} \frac{\sin\left[\Omega\left(t-t'\right)\right]}{\pi\left(t-t'\right)}\, x_\Omega(t')\, dt'. \tag{1.32}$$

Equation (1.32) defines the reproducing kernel as the function $D_\Omega(t-t') \equiv \sin[\Omega(t-t')]/\pi(t-t')$. As $\Omega \to \infty$, the reproducing kernel $D_\Omega(t-t')$ approaches the most famous reproducing kernel of all, namely, the Dirac delta distribution $\delta(t-t')$. To see the result (1.32), we note that the operator projecting a function in $L_2(\mathbb{R})$ onto the space $L_2^\Omega(\mathbb{R})$,

$$P_\Omega\left\{x\left(t\right)\right\} = \left(2\pi\right)^{-1} \int_{-\Omega}^{\Omega} X\left(\omega\right) e^{i\omega t} d\omega,$$

is an orthogonal projection (its range and null spaces are orthogonal). This operator, in the frequency domain, has the form

$$P_\Omega\left\{X\left(\omega\right)\right\} = X\left(\omega\right)\, \mathbf{1}_{[-\Omega,\Omega]} \equiv X_\Omega\left(\omega\right),$$

where the function $\mathbf{1}_{[-\Omega,\Omega]}$, defined to be 1 on the interval $[-\Omega, \Omega]$ and 0 elsewhere, has as its inverse Fourier transform the function $\sin(\Omega t)/\pi t$. The inverse Fourier transform of $X_\Omega(\omega)$ is the convolution of $x_\Omega(t)$ and $\sin(\Omega t)/\pi t$, which is precisely equation (1.32). A complete orthonormal basis for $L_2^\Omega(\mathbb{R})$ is provided by the functions

$$p_n(t) = \sqrt{2B}\,\mathrm{sinc}\left[2B\left(t - n/2B\right)\right], \tag{1.33}$$

where $\mathrm{sinc}(t) \equiv \sin\left(\pi t\right)/\pi t$, and we have defined the linear bandwidth variable $B \equiv \Omega/2\pi$. The Fourier transform of $p_n(t)$ is identically zero outside the closed interval $\left[-2\pi B, +2\pi B\right]$. Any bandlimited function $x_\Omega(t)$ can be expressed in terms of this basis,

$$x_\Omega\left(t\right) = \sum_{n=-\infty}^{\infty} c_n\, p_n\left(t\right), \tag{1.34}$$

with the expansion coefficients given by the inner product,

$$c_n = \langle p_n, x_\Omega\rangle = x_\Omega\left(n/2B\right). \tag{1.35}$$

Thus, we arrive at the sampling theorem [9] [9].

[9]Although the theorem is usually referred to as the Shannon sampling theorem, it was independently discovered by Nyquist, Whittaker, and Kotelnikov.

Theorem 6. *Let $x(t)$ be a continuous time signal whose Fourier transform $X(\omega)$ is zero outside the interval $\omega \in [-2\pi B, +2\pi B]$, i.e., the linear frequency content of $x(t)$ is limited to the band $[-B, +B]$. If ΔT is a positive constant such that*

$$0 < \Delta T \leq 1/2B, \tag{1.36}$$

then $x(t)$ can be uniquely reconstructed from its sampled values $x(n\Delta T)$, $n \in \mathbb{Z}$, by the interpolation formula

$$x(t) = 2B\,\Delta T \sum_{n=-\infty}^{\infty} x(n\Delta T) \frac{\sin[2\pi B(t - n\Delta T)]}{2\pi B(t - n\Delta T)}. \tag{1.37}$$

Condition (1.36) is usually written in terms of the sampling frequency $f_s \equiv 1/\Delta T$ in the form $f_s \geq 2B$ and stated as the requirement that the signal sampling frequency should equal or exceed twice the highest frequency of the signal.

Although theorem 6 is usually stated for a time function and temporal frequencies, it can be used in the spatial domain and for spatial frequencies. For example, consider an array of M omni-directional receivers spaced a uniform distance d apart on the positive x-axis with position vectors $\boldsymbol{r}_m = (md, 0, 0)$, $0 \leq m \leq M-1$, known as a uniform linear array (ULA) [10]. Let a source in the far field emit plane waves $A\,\exp(i\omega t - i\boldsymbol{k}\cdot\boldsymbol{r})$ with frequency $\omega > 0$ and amplitude A. As will be seen in section 11.1, conical symmetry of the line array enables us to consider the problem of plane waves incident on a ULA in a two dimensional geometry with the array axis (taken to be the x-axis) and the wave vector $\boldsymbol{k}$ in one plane. Without loss of generality, we may take the axis normal to the array as either the y or the z-axis. The plane waves arrive at the array elements with an angle (direction) ψ measured from the array (the positive x-axis) ranging in the interval $[0°, 180°]$ (propagating wave fronts have reflection symmetry with respect to the array axis), as depicted in figure 1.3. The dispersion relation is

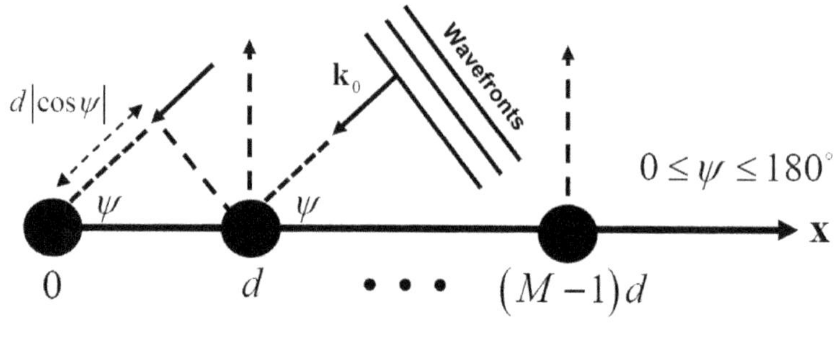

Figure 1.3: A uniform linear array (ULA) of receivers.

$$\omega = c\,|\boldsymbol{k}| = 2\pi c/\lambda, \tag{1.38}$$

where c is the wave phase velocity and λ is the wavelength. The arrival times at two adjacent receivers differ by $d\cos\psi/c$ corresponding to a time delay ($\psi > 90°$) or an advance ($\psi < 90°$) for a receiver relative to one on its left; for example, figure 1.3 shows the time difference is an advance for the receiver at position d relative to the receiver at the origin.

[10]We ignore some practical issues in the deployment of spatial arrays such as a seismic land array or a sonar array in the ocean whose positions can violate the linear array assumption and can exhibit large uncertainties.

The received signal $x_m(t)$ at the receiver position $(md, 0, 0)$ is $A\ \exp(i\omega t - i\boldsymbol{k}\cdot\boldsymbol{r}_m)$, which can be written in two equivalent forms

$$A\ \exp[i\omega(t - md\cos\psi/c)] = A\ \exp(i\omega t - 2i\pi md\cos\psi/\lambda). \quad (1.39)$$

For a fixed receiver position (fixed m), the left hand side of (1.39) shows a time difference $\tau_m = md\cos\psi/c$ with respect to the receiver at the origin. When sampled at a rate $f_s = 1/\Delta T$ with $t = n\Delta T$, $n \in \mathbb{Z}$, and $\omega = 2\pi f$, the received signal is of the form $A\ \exp(2i\pi kf/f_s)$ for some integer k (assuming that the time difference is an integer multiple of ΔT). According to the sampling theorem, the normalized linear frequency f/f_s must satisfy the condition

$$-1/2 < f/f_s < +1/2. \quad (1.40)$$

On the other hand, we may use the right hand side of (1.39) and consider the complex exponential at a fixed time t for all m, i.e., a "snapshot" of all the signals on the array receivers; ignoring the fixed time portion, we note that the spatially varying exponential $\exp(-2i\pi md\cos\psi/\lambda)$ appears as a spatial sampling of the plane waves with a normalized spatial frequency defined by

$$\tilde{f} \equiv d\cos\psi/\lambda\,. \quad (1.41)$$

This is the spatial equivalent of f/f_s with a spatial sampling distance d (the spatial equivalent of the sampling time $\Delta T = 1/f_s$). The sampling theorem requirement then translates to the following result for the normalized spatial frequency

$$-1/2 \leq d\cos\psi/\lambda \leq 1/2\,. \quad (1.42)$$

Thus, a half-wavelength array defined by $d = \lambda/2$ satisfies the sampling theorem requirement (1.42) and provides the maximum aperture $(M-1)\lambda/2$ for a given number M of elements. The wavelength λ is calculated from the array design frequency (also known as the array operating frequency) through $\lambda = c/f$; for instance, in a sonar array with $f = 1$ kHz and sound speed in water $c = 1500$ m/s, we find $\lambda = 1.5$ m and $d = 0.75$ m. A more practical reason for using half-wavelength arrays in the presence of spatially isotropic noise is the fact that noise samples on the sensors that are a half-wavelength apart are uncorrelated; this is an example of the concept of random fields in section 7.10. We will discuss optimal processing of uniformly spaced receiver arrays (beamforming and finding signal direction of arrival) in chapter 11.

1.10 The Prolate Spheroidal Functions

The space $L_2^\Omega(\mathbb{R})$ of bandlimited functions has no non-trivial function with compact support because a function cannot be simultaneously bandlimited and time-limited. The problem of finding the extent to which bandlimited functions are concentrated in the time domain, first put forward by Shannon, is to find functions $\psi(t)$ limited to the band $[-\Omega, +\Omega]$ that solve the constrained maximization problem

$$\underset{\psi(t)}{\arg\max}\ \frac{1}{\|\psi\|^2}\int_{-T}^{+T}|\psi(t)|^2\,dt \quad \text{subject to } \psi(t) \in L_2^\Omega(\mathbb{R}). \quad (1.43)$$

This problem was solved by Slepian [10], who showed that the solutions are eigenfunctions of a second-order singular Sturm-Liouville equation and form the *spectrum* of a symmetric integral operator with kernel

$D_\Omega(t-t')$ of equation (1.32). The eigenfunctions are known as *prolate spheroidal functions*, or Slepian functions; they are real bandlimited time domain functions that satisfy the following eigenvalue equation,

$$\int_{-T}^{+T} \frac{\sin[\Omega(t-t')]}{\pi(t-t')} \psi(t')\,dt' = \lambda\,\psi(t), \tag{1.44}$$

which upon a change of integration variable and scaling of the eigenfunctions produces the equivalent eigenvalue problem

$$\int_{-1}^{+1} \frac{\sin[TB(t-t')]}{\pi(t-t')} \psi(t')\,dt' = \lambda\,\psi(t). \tag{1.45}$$

Thus, the eigenvalues and eigenfunctions do not depend on T and Ω separately but rather on the time-bandwidth product TB. Defining a parameter $c \equiv T\Omega = 2\pi TB$, we denote the eigenfunctions and eigenvalues by $\psi^{(k)}(t;c)$ and $\lambda^{(k)}(c)$, $k = 1, 2, \ldots$, respectively. The eigenvalues are real and positive and the symmetric kernel $D_\Omega(t-t')$ satisfies all the conditions in Mercer's theorem [11] (another important application of this theorem is to the Karhunen-Loéve transformation—section 5.7).

Theorem 7. *Let $K(t,t')$ be a real valued, continuous, symmetric, and non-negative kernel defined on $[-T,+T]\times[-T,+T]$. Define the associated Hilbert-Schmidt integral operator*

$$\mathcal{A}f(t) = \int_{-T}^{+T} K(t,t')f(t')\,dt' \tag{1.46}$$

and let $\{\phi_k(t)\}$ be the orthonormal basis formed by the eigenvectors corresponding to the non-zero eigenvalues $\{\lambda_k\}$ of $\mathcal{A}$. Then

$$K(t,t') = \sum_k \lambda_n\,\phi_k(t)\phi_k(t'), \quad \forall t,t' \in [-T,+T], \tag{1.47}$$

where the series converges absolutely and uniformly in both variables.

Thus, $D_\Omega(t-t')$ has the absolutely and uniformly convergent series expansion in terms of its eigenfunctions and eigenvalues [12]:

$$D_\Omega(t-t') = \sum_{k=1}^{\infty} \lambda^{(k)}(c)\,\psi^{(k)}(t;c)\,\psi^{(k)}(t';c). \tag{1.48}$$

In addition, setting $t = t'$ in (1.48) and integrating over $[-\Omega,+\Omega]$ we find the absolutely convergent series

$$\sum_{k=1}^{\infty} \lambda^{(k)}(c) = \int_{-\Omega}^{+\Omega} D_\Omega(0)\,dt = 2T\Omega/\pi = 2c/\pi. \tag{1.49}$$

When arranged in descending order, $\lambda^{(1)} > \lambda^{(2)} > \ldots > 0$, the first $\lceil 2T\Omega/\pi \rceil$ eigenvalues are very close to 1, i.e., $\lambda^{(k)} < 1-\delta$, $0 < \delta \ll 1$ for $k < \lceil 2T\Omega/\pi \rceil$, while the rest, $k > \lceil 2T\Omega/\pi \rceil$, rapidly fall to 0.

The eigenfunctions (normalized on the entire real line) satisfy a remarkable set of orthogonality relations over the finite interval $[-T,+T]$ and $\mathbb{R}$:

$$\int_{-T}^{+T} \psi^{(l)}(t;c)\,\psi^{(k)}(t;c)\,dt = \lambda^{(k)}(c)\,\delta_{lk}, \quad \int_{-\infty}^{+\infty} \psi^{(l)}(t;c)\,\psi^{(k)}(t;c)\,dt = \delta_{lk}. \tag{1.50}$$

The dual problem of finding the extent to which time-limited functions are concentrated in the frequency domain is also solved by the eigenfunctions of equation (1.44) that have support in the interval $[-T,+T]$ with maximal concentration of energy in the frequency interval $[-\Omega,\Omega]$. They are denoted by $\psi^{(k)}(\omega;c)$ and satisfy the eigenvalue equation in the frequency domain:

$$\int_{-\Omega}^{+\Omega} \frac{\sin[T(\omega-\omega')]}{\pi(\omega-\omega')}\,\psi^{(k)}(\omega';c)\,d\omega' = \lambda^{(k)}(c)\,\psi^{(k)}(\omega;c), \quad k=1,2,\ldots\,, \tag{1.51}$$

which can be transformed into equation (1.45). The orthogonality relations are

$$\int_{-\Omega}^{+\Omega} \psi^{(l)}(\omega;c)\psi^{(k)}(\omega;c)d\omega = \lambda^{(k)}(c)\delta_{lk}, \quad \int_{-\infty}^{+\infty} \psi^{(l)}(\omega;c)\psi^{(k)}(\omega;c)d\omega = \delta_{lk}. \tag{1.52}$$

Discrete prolate spheroidal sequences (DPSS) with support in $[-N/2,+N/2-1]$ (assuming N to be even) and maximal energy concentrated in the frequency band $[-\Delta\omega,+\Delta\omega]$ are the inverse Fourier transforms of the prolate spheroidal functions defined for frequencies $\omega\in[-\pi,+\pi]$. Using normalized linear frequencies $\omega=2\pi f$ (assuming $f_s=1/\Delta T=1$ Hz), the prolate spheroidal eigenfunctions $\psi^{(k)}(f;c)$, $k=1,2,\ldots,N$, satisfy the eigenvalue equation

$$\int_{-\Delta f}^{+\Delta f} \frac{\sin[N\pi(f-f')]}{\sin[\pi(f-f')]}\,\psi^{(k)}(f';c)\,df' = \lambda^{(k)}(c)\,\psi^{(k)}(f;c), \tag{1.53}$$

with $c=\pi N\Delta f$, the orthogonality relations

$$\int_{-\Delta f}^{+\Delta f} \psi^{(l)}(f;c)\psi^{(k)}(f;c)df = \lambda^{(k)}(c)\delta_{lk}, \quad \int_{-1/2}^{+1/2} \psi^{(l)}(f;c)\psi^{(k)}(f;c)df = \delta_{lk}, \tag{1.54}$$

and the inverse Fourier transform relation

$$\psi_n^{(k)}(c) = \frac{1}{2\pi}\int_{-1/2}^{+1/2} \psi^{(k)}(f;c)\,e^{+2inf\pi}\,df, \quad 0\le n\le N-1. \tag{1.55}$$

Alternatively, DPSS are the eigenvectors of the matrix eigenvalue equation (see section 8.7)

$$\boldsymbol{D}\boldsymbol{\psi}^{(k)}(c) = \lambda^{(k)}(c)\,\boldsymbol{\psi}^{(k)}(c), \quad \left[\boldsymbol{D}\right]_{lm} = \frac{\sin\left[\pi\,\Delta f\,(l-m)\right]}{\sin\left[\pi(l-m)\right]}\;, \tag{1.56}$$

for $0 \leq l, m \leq N-1$, $k = 1, \ldots, N$, and $c = \pi N \Delta f$; note that $\boldsymbol{D}$ is symmetric and *Toeplitz* [11]. The left panel of figure 1.4 shows the first four discrete prolate spheroidal sequences with support in $[0, 99]$ with $c = 2.5\pi$ and eigenvalues 1, 0.9998, 0.9962, and 0.9522; the right hand side shows all 100 eigenvalues (on logarithmic x-axis) that rapidly approach zero.

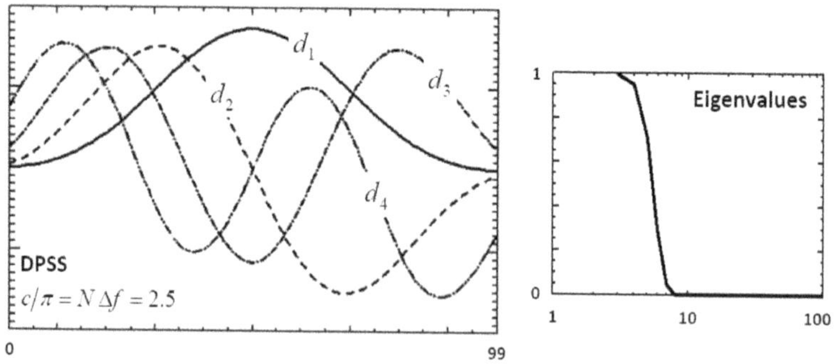

Figure 1.4: The first four discrete prolate spheroidal sequences with support in $[0, 99]$ and $c = 2.5\pi$ (left), and all 100 eigenvalues in descending order (right).

Prolate spheroidal functions have many applications in time series analysis, signal processing, and image processing; they also occur when solving the Helmholtz equation using the method of separation of variables in prolate spheroidal coordinates [13]. In signal processing applications, prolate spheroidal functions show that there are $2TB$ degrees of freedom associated with functions that are substantially time-limited to T seconds and bandlimited to B Hz. We will discuss the DPSS in section 8.7 when we introduce the multitaper method of spectral estimation.

1.11 The Approximation Problem and the Orthogonality Principle

The underlying ideas of optimal linear filters can be illustrated by the problem of approximating a vector in ordinary three dimensional Euclidean space $\mathbb{R}^3$. Referring to figure 1.5, suppose we wish to approximate a vector $\boldsymbol{b}$ with three non-zero components using two given vectors $\boldsymbol{a}_1$ and $\boldsymbol{a}_2$ that are not necessarily orthogonal but that both are in the subspace represented by the horizontal plane. Thus, we wish to approximate a vector with a non-zero z-component using a linear combination of the form $x_1\boldsymbol{a}_1 + x_2\boldsymbol{a}_2$. If we define a 3×2 matrix $\boldsymbol{A}$ whose columns are the approximating vectors $\boldsymbol{a}_1$ and $\boldsymbol{a}_2$, then the linear combination can be written as $\boldsymbol{Ax}$; the 2×1 vector $\boldsymbol{x} = [x_1, x_2]^T$ (superscript T denotes transposition) contains the unknown coefficients whose optimal values are found by minimizing the squared magnitude of the difference vector $|\boldsymbol{b} - \hat{\boldsymbol{b}}|^2$, where $\hat{\boldsymbol{b}} = \boldsymbol{Ax}$.

The solution to this least squares problem is found by the orthogonal projection of $\boldsymbol{b}$ onto the horizontal plane. To see this, let us define the error vector $\boldsymbol{e}$ as the difference $\boldsymbol{b} - \hat{\boldsymbol{b}}$. Now consider any non-orthogonal projection onto the horizontal plane, for instance $\boldsymbol{b}'$, resulting in a difference vector $\boldsymbol{e}'$. As seen in figure 1.5, the error vector $\boldsymbol{e}'$ is the hypotenuse of the (dashed) right-angle triangle on the left, and so its length always exceeds the length of the side represented by $\boldsymbol{e}$.

[11] A Toeplitz matrix has the property that elements along each diagonal are the same, i.e., the element at row i and column j depends on the difference $i - j$ and not on the individual row or column, e.g., the element at (i, j) is the same as the element at $(i-1, j-1)$ and at $(i+1, j+1)$, and so on.

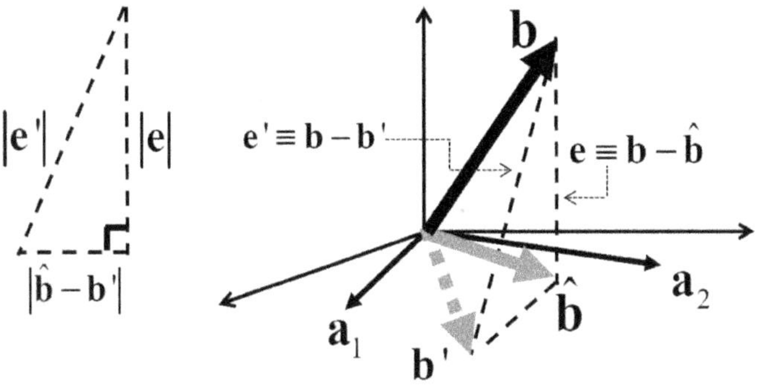

Figure 1.5: The best approximation to $\boldsymbol{b}$ using $\boldsymbol{a}_1$ and $\boldsymbol{a}_2$.

The geometrical solution of figure 1.5 can be expressed in terms of the least squares matrix formulation of section 1.8: $\boldsymbol{a}_1$ and $\boldsymbol{a}_2$ are the two columns of $\boldsymbol{A}$, and $\hat{\boldsymbol{b}}$ is the best approximation to $\boldsymbol{b}$ using only the columns of $\boldsymbol{A}$. The least squares solution $\boldsymbol{x}_{\text{ls}}$ is the vector of the coefficients $[x_1, x_2]^T$ in the linear combination $\hat{\boldsymbol{b}} = x_1\boldsymbol{a}_1 + x_2\boldsymbol{a}_2 = \boldsymbol{A}\boldsymbol{x}$.

An obvious corollary to the orthogonal projection construction is that the error vector for the least squares solution must be orthogonal to both $\boldsymbol{a}_1$ and $\boldsymbol{a}_2$; the generalization of this corollary to the approximation problem in a Hilbert space is known as the *orthogonality principle* and is described in theorem 8 stated for functions spaces.

Theorem 8. *Given a Hilbert space $\mathscr{H}$, a closed subspace $\mathcal{V}$, and a point $x \in \mathscr{H}$, there exists a unique point $v \in \mathcal{V}$ that is the orthogonal projection of x onto $\mathcal{V}$ and that minimizes the induced norm $\|x - v\|_2$.*

When specialized to the finite dimensional vector spaces $\mathbb{C}^N$ and $\mathbb{C}^M$, $N > M$, this approximation is known as the *full-rank least squares problem* whose solution was shown to be the result of an orthogonal projection in section 1.9. The orthogonality principle refers to the fact that the difference vector $\boldsymbol{b} - \hat{\boldsymbol{b}}$ has its shortest length when it is orthogonal to $\mathscr{R}_A$, as illustrated in figure 1.5: the columns of the matrix $\boldsymbol{A}$ are orthogonal to the error vector $\boldsymbol{e} = \boldsymbol{b} - \hat{\boldsymbol{b}}$ associated with the optimal solution $\boldsymbol{x}$ satisfying $\boldsymbol{A}\boldsymbol{x} = \hat{\boldsymbol{b}}$. The orthogonality principle written as $\left\langle \boldsymbol{a}_n, \boldsymbol{b} - \hat{\boldsymbol{b}} \right\rangle = 0$, $1 \le n \le N$, when substituted into $\hat{\boldsymbol{b}} = \boldsymbol{A}\boldsymbol{x}$ gives the following matrix equation, known as the *Normal equation* for the least squares solution:

$$\begin{bmatrix} \langle \boldsymbol{a}_1, \boldsymbol{a}_1 \rangle & \dots & \langle \boldsymbol{a}_1, \boldsymbol{a}_N \rangle \\ \vdots & \ddots & \vdots \\ \langle \boldsymbol{a}_N, \boldsymbol{a}_1 \rangle & \dots & \langle \boldsymbol{a}_N, \boldsymbol{a}_N \rangle \end{bmatrix} \begin{bmatrix} x_1 \\ \vdots \\ x_N \end{bmatrix} = \begin{bmatrix} \langle \boldsymbol{a}_1, \boldsymbol{b} \rangle \\ \vdots \\ \langle \boldsymbol{a}_N, \boldsymbol{b} \rangle \end{bmatrix}. \tag{1.57}$$

This is an equivalent way to write the least squares equation $\boldsymbol{A}^+\boldsymbol{A}\boldsymbol{x} = \boldsymbol{A}^+\boldsymbol{b}$, often written as $\boldsymbol{R}\boldsymbol{x} = \boldsymbol{p}$ defining the Hermitian and non-negative definite matrix $\boldsymbol{R}$ and the vector $\boldsymbol{p}$. The least squares solution is $\boldsymbol{x}_{\text{ls}} = \boldsymbol{R}^{-1}\boldsymbol{p}$ (provided the inverse exists) and the minimum error for the least squares solution is

$$\|\boldsymbol{e}_{\min}\|_2^2 = \langle \boldsymbol{b} - \boldsymbol{A}\boldsymbol{x}_{\text{ls}}, \boldsymbol{b} - \boldsymbol{A}\boldsymbol{x}_{\text{ls}} \rangle = \boldsymbol{b}^+\boldsymbol{b} - \boldsymbol{b}^+\boldsymbol{A}\left(\boldsymbol{A}^+\boldsymbol{A}\right)^{-1}\boldsymbol{A}^+\boldsymbol{b} = \|\boldsymbol{b}\|_2^2 - \|\hat{\boldsymbol{b}}\|_2^2. \tag{1.58}$$

Thus, the minimum error vector $\boldsymbol{e}_{\min}$, the vector $\boldsymbol{b}$, and the orthogonal projection $\hat{\boldsymbol{b}}$ satisfy Pythagoras's theorem in conformity with the orthogonality principle depicted in figure 1.5. Now using $\hat{\boldsymbol{b}} = \boldsymbol{A}\boldsymbol{x}_{\text{ls}}$ we have

$\|\hat{\boldsymbol{b}}\|_2^2 = \boldsymbol{x}_{ls}^+ \boldsymbol{R}\boldsymbol{x}_{ls}$, and so $\|\boldsymbol{e}_{\min}\|_2^2 = \boldsymbol{b}^+ \left(\boldsymbol{I} - \boldsymbol{A}\boldsymbol{R}^{-1}\boldsymbol{A}^+\right)\boldsymbol{b}$. We will discuss the statistical properties of the least squares solution in section 5.8.

If the subspace $\mathcal{V}$ of the Hilbert space $\mathcal{H}$ has an orthonormal basis $\{v_n\}$, then the orthogonal projection of any $x \notin \mathcal{V}$ onto $\mathcal{V}$ is given by

$$P_v : \mathcal{H} \to \mathcal{V} \;\Rightarrow\; P_v(x) \equiv v_x = \sum_k \langle v_k, x\rangle v_k. \tag{1.59}$$

This follows from the fact that the error $(1 - P_v)x$ is orthogonal to $\mathcal{V}$, i.e.,

$$\left\langle v_n, x - \sum_k \langle v_k, x\rangle v_k \right\rangle = \langle v_n, x\rangle - \sum_k \langle v_k, x\rangle\, \delta_{kn} = 0.$$

1.12 Orthogonal Projections and the Haar Scaling and Wavelet Functions

As described in theorem 8, an orthogonal projection is used in the problem of approximating a function to find an element of a given subspace that is closest to that function. An important concept in the theory and implementation of the discrete orthogonal wavelet transform is that of orthogonal projections onto *nested subspaces* that satisfy the *completeness* and *multi-resolution* properties. However, before formally introducing them, and to make the discussion in the previous sections more concrete, we consider particular sets of real functions in $L_2(\mathbb{R})$ and the spaces spanned by them. These functions are the *wavelet* and *scaling* functions, which have been widely researched and applied in recent decades [see [14] for a concise guide]. We will use the usual unweighted inner product for real and square integrable functions of $L_2(\mathbb{R})$:

$$\langle x, y\rangle = \int_{-\infty}^{+\infty} x(t)y(t)dt \tag{1.60}$$

Consider the function $\phi(t)$ shown in figure 1.6-a and two time shifted versions $\phi(t-1)$ and $\phi(t-n)$, $n \in \mathbb{Z}^+$. Since there is no overlap between these functions and each has unit energy, they form an orthonormal set. In fact, any two time shifted versions of $\phi(t)$ with unequal integer shifts (negative or positive) are orthogonal to each other. Thus,

$$\langle \phi(t-l), \phi(t-n)\rangle = \delta_{ln}, \tag{1.61}$$

and $\{\phi(t-n),\ n \in \mathbb{Z}\}$ is an orthonormal set whose span $\mathcal{V}_0$ is a subspace of $L_2(\mathbb{R})$ containing functions that are piecewise constant on unit integer intervals $[n, n+1]$, $n \in \mathbb{Z}$. Let $x(t) \in L_2(\mathbb{R})$ be an arbitrary

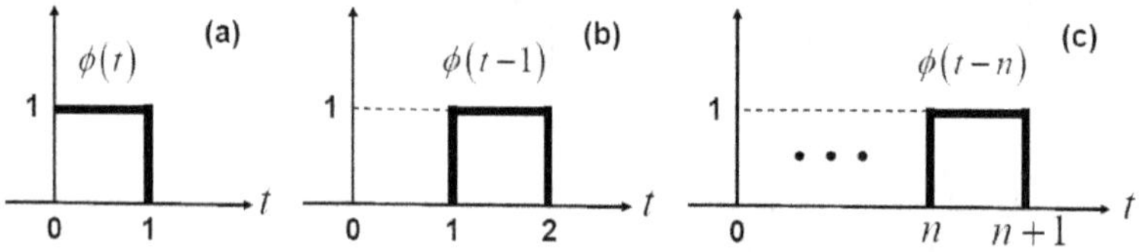

Figure 1.6: The function $\phi(t)$ and two time shifted versions.

function. Using equation (1.59), we find the orthogonal projection of $x(t)$ onto $\mathcal{V}_0$, denoted by x_0,

$$x_0(t) = \sum_n \langle \phi(t-n), x\rangle \phi(t-n) = \sum_n \left[\int_n^{n+1} x(t)dt\right] \phi(t-n). \tag{1.62}$$

$x_0(t)$ is the best approximation to the original function $x(t)$ using the subspace $\mathcal{V}_0$ and its basis functions $\phi_{0n}(t) \equiv \phi(t-n)$. The approximation can be refined by changing the scale using piecewise constant functions whose intervals of constancy are smaller than 1. We can double our resolution by using the functions $\phi_{-1,n}(t) \equiv \sqrt{2}\phi(2t-n)$ (the $\sqrt{2}$ factor is to ensure that these functions have unit norm). As illustrated in figure 1.7, $\phi_{-1,n}(t)$ form an orthonormal basis of a new subspace $\mathcal{V}_{-1}$, and we have the finer

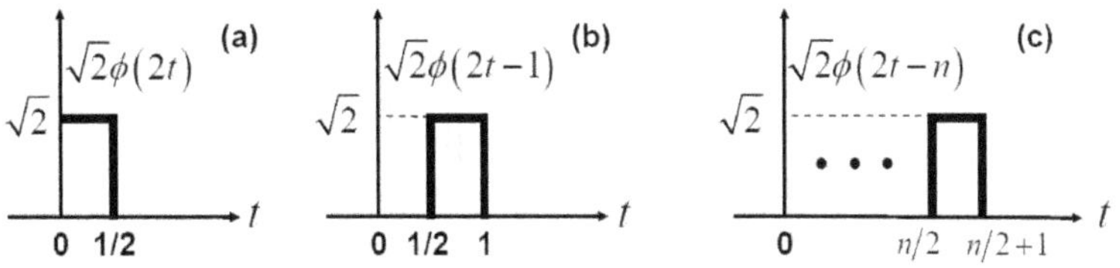

Figure 1.7: The function $\sqrt{2}\phi(2t)$ and two time shifted versions.

approximation

$$x_{-1}(t) = 2\sum_n \langle \phi(2t-n), x\rangle \phi(2t-n) = 2\sum_n \Big[\int_{n/2}^{(n+1)/2} x(t)dt\Big]\phi(2t-n). \tag{1.63}$$

The two functions $\phi(t)$ and $\phi(2t)$ satisfy an equation known as a *scaling equation*, and $\phi(t)$ is referred to as a *scaling function*:

$$\phi(t) = \phi(2t) + \phi(2t-1). \tag{1.64}$$

1.13 Multi-Resolution Analysis Subspaces and Discrete Orthogonal Wavelet Bases

The subspace $\mathcal{V}_{-1}$ includes all the functions in $\mathcal{V}_0$, and more. In addition, in going from a higher resolution approximation to a lower resolution approximation, we lose some detail, namely, $x_{-1}(t) - x_0(t)$. The detail is orthogonal to $\mathcal{V}_0$, and the space of all these details denoted by $\mathcal{W}_0$ is the orthogonal complement of $\mathcal{V}_0$ in $\mathcal{V}_{-1}$. Thus,

$$\mathcal{V}_0 \subset \mathcal{V}_{-1} = \mathcal{V}_0 \oplus \mathcal{W}_0, \;\; \mathcal{W}_0 \perp \mathcal{V}_0, \tag{1.65}$$

and the subspace $\mathcal{W}_0$ has an orthonormal basis that is formed by integer time translations of a *wavelet function* $\psi(t)$ illustrated in figure 1.8 (we will prove this result shortly).

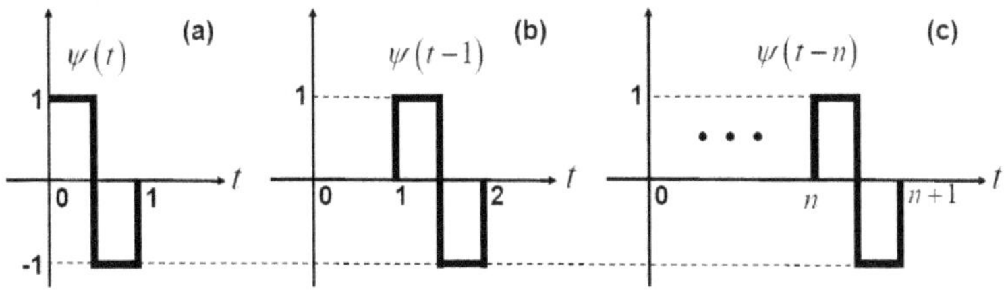

Figure 1.8: The function $\psi(t)$ and two time shifted versions of it.

Figure 1.9 shows a function $x(t)$, two approximations $x_0(t) \in \mathcal{V}_0$ and $x_{-1}(t) \in \mathcal{V}_{-1}$, and the difference $x_{-1} - x_0(t) \in \mathcal{W}_0$. Defining $\psi_{0n}(t) \equiv \psi(t-n)$ (including the wavelet function $\psi_{00}(t) \equiv \psi(t)$), it is easy

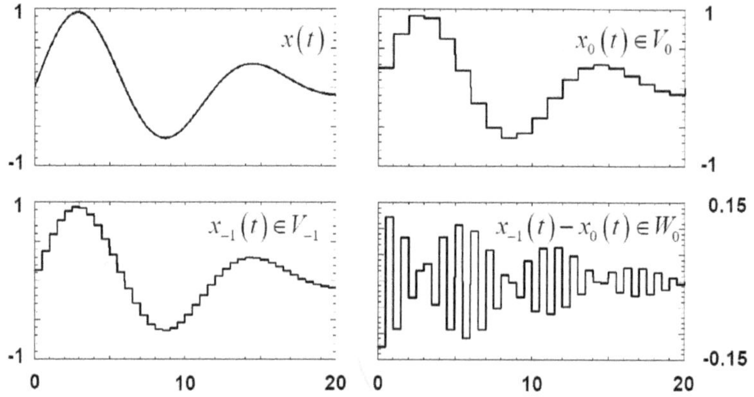

Figure 1.9: Function $x(t)$, scaling approximations, and the details.

to verify that all basis functions of $\mathcal{W}_0$ are orthogonal to all basis functions of $\mathcal{V}_0$, i.e., $\langle \psi_{0l}, \phi_{0n} \rangle = 0$. In addition, figures 1.6, 1.7, and 1.8 clearly demonstrate the *wavelet equation*, namely,

$$\psi(t) = \phi(2t) - \phi(2t-1). \tag{1.66}$$

The approximating subspaces $\mathcal{V}_0$, $\mathcal{V}_{-1}$, and the detail subspace $\mathcal{W}_0$ are parts of a much more general theory of multi-resolution nested approximating subspaces [15]. A set of closed subspaces $\mathcal{V}_m \subset L_2(\mathbb{R})$, $m \in \mathbb{Z}$, is said to satisfy the *nesting property* if

$$\mathcal{V}_m \subset \mathcal{V}_{m-1},\ m \in \mathbb{Z}. \tag{1.67}$$

A nested set of closed subspaces $\mathcal{V}_m$ satisfying the *completeness* properties (over-bar indicates closure)

$$\overline{\bigcup_{m\in\mathbb{Z}} \mathcal{V}_m} = L_2(\mathbb{R}),\quad \bigcap_{m\in\mathbb{Z}} \mathcal{V}_m = \{0\}, \tag{1.68}$$

is said to have the *successive approximation* property. For such a set of nested subspaces, we also have

$$L_2(\mathbb{R}) = \overline{\lim_{m\to-\infty} \mathcal{V}_m},\quad \{0\} = \lim_{m\to\infty} \mathcal{V}_m. \tag{1.69}$$

If $L_2(\mathbb{R})$ admits a nested set of subspaces with the successive approximation property, then any square-integrable function can be approximated by an orthogonal projection P_m onto the subspace $\mathcal{V}_m$. The process is iterative and is depicted in figure 1.10. An approximation at stage m produces two results: one is the coarser approximation at stage $m+1$, and the other is the detail that is lost between the approximations at stages m and $m+1$. The former is an orthogonal projection into $\mathcal{V}_{m+1}$, while the latter lies in $\mathcal{V}_m^{\perp}$. A nested set of successive approximation subspaces $\mathcal{V}_m$ is said to have the *multi-resolution* property, provided that

$$P_m(x) = x_m(t) \in \mathcal{V}_m \Leftrightarrow x_m(2t) \in \mathcal{V}_{m-1}. \tag{1.70}$$

The subspaces $\mathcal{V}_m$ are then said to form a set of *multi-resolution analysis (MRA)* subspaces [15]. The Haar subspaces $\mathcal{V}_m$ and $\mathcal{W}_m$ are an example of an MRA. To display the orthonormal bases for these subspaces let $\phi_{mn}(t)$ be the function that is zero for $t < 2^m n$ or $t > 2^m(n+1)$, and

$$\phi_{mn}(t) = 2^{-m/2},\quad 2^m n \le t \le 2^m(n+1),\quad m, n \in \mathbb{Z}. \tag{1.71}$$

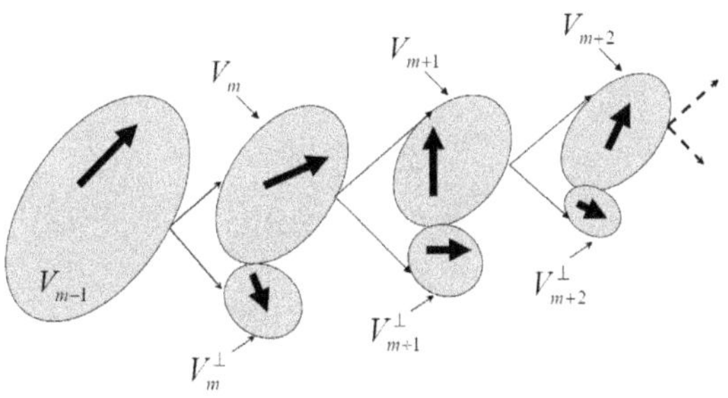

Figure 1.10: Approximation of a vector using nested subspaces.

For a fixed integer m these functions are orthonormal (those with different values of the integer shift n are nonzero on disjoint intervals, and hence orthogonal), i.e.,

$$\langle \phi_{ml}(t), \phi_{mn}(t) \rangle = \delta_{ln}, \tag{1.72}$$

and they span the subspace $\mathcal{V}_m$. Clearly, ϕ_{mn} are integer time-shifted and dilated (and normalized) versions of the Haar scaling function $\phi(t)$, i.e.,

$$\phi_{mn}(t) = 2^{-m/2}\phi(2^{-m}t - n), \quad \|\phi_{mn}\| = 1. \tag{1.73}$$

Given a function $x(t) \in L_2(\mathbb{R})$, the orthogonal projection of $x(t)$ onto $\mathcal{V}_m$, $m \in \mathbb{Z}$, is the approximation at resolution (or scale) m and is given by

$$x_m(t) = \sum_{n=-\infty}^{\infty} c_{mn}\phi_{mn}(t), \quad c_{mn} = \langle \phi_{mn}, x \rangle. \tag{1.74}$$

Using the scaling equation (1.64), we find the following recursion:

$$c_{mn} = 2^{-1/2}\left(c_{m-1,2n} + c_{m-1,2n+1}\right). \tag{1.75}$$

The scaling equation (1.64) is a special case of a general scaling equation that includes an infinite number of integer time shifts, namely,

$$\phi(t) = \sqrt{2} \sum_{n=-\infty}^{\infty} h_0[n]\,\phi(2t - n). \tag{1.76}$$

The Haar scaling equation (1.64) has only two terms with the corresponding coefficients given by

$$h_0[n] = 2^{-1/2}(\delta_{0n} + \delta_{1n}). \tag{1.77}$$

The approximation error (detail) at scale m is in the orthogonal complement of $\mathcal{V}_m$ in $\mathcal{V}_{m-1}$, denoted by $\mathcal{W}_m$. Thus, we have

$$\mathcal{V}_{m-1} = \mathcal{V}_m \oplus \mathcal{W}_m. \tag{1.78}$$

The space $\mathcal{W}_m$ is characterized by the error in the approximation at resolution (scale) m, i.e., the difference between the projections of the given functions onto the subspaces $\mathcal{V}_{m-1}$ and $\mathcal{V}_m$,

$$\boldsymbol{E}_m x \equiv \boldsymbol{P}_{m-1}x - \boldsymbol{P}_m x. \tag{1.79}$$

It can be shown [14] that for all $x \in L_2(\mathbb{R})$,

$$\boldsymbol{E}_m x = \sum_{n=-\infty}^{\infty} d_{mn} \psi_{mn}(t), \tag{1.80}$$

where the wavelet coefficients d_{mn} are

$$d_{mn} \equiv 2^{-1/2} \left\{ c_{m-1,2n} - c_{m-1,2n+1} \right\}, \tag{1.81}$$

and the wavelet functions ψ_{mn} are

$$\psi_{mn}(t) \equiv 2^{-1/2} \left\{ \phi_{m-1,2n}(t) - \phi_{m-1,2n+1}(t) \right\}. \tag{1.82}$$

Since $\boldsymbol{E}_m(x) \in \mathcal{W}_m$ (the orthogonal complement of $\mathcal{V}_m$ in $\mathcal{V}_{m-1}$) and $\mathcal{V}_{m-1} = \mathcal{V}_m \oplus \mathcal{W}_m$, we conclude that $\mathcal{W}_m$ is spanned by $\psi_{mn}(t)$. Now consider the wavelet function $\psi(t)$,

$$\psi(t) \equiv \psi_{00}(t) = 2^{-1/2} \left\{ \phi_{-1,0}(t) - \phi_{-1,1}(t) \right\}. \tag{1.83}$$

All the wavelet basis functions $\psi_{mn}(t)$ are found from the *mother wavelet* function $\psi(t)$ by time shifts $t \to t - n$ followed by a dilation and normalization $\psi(t-n) \to 2^{-m/2}\psi(t/2^m - n)$. The Haar wavelet equation (1.66) is a special case (series with two terms) of the more general wavelet equation

$$\psi(t) = \sqrt{2} \sum_{n=-\infty}^{\infty} h_1[n]\phi(2t-n). \tag{1.84}$$

For the Haar wavelet, the coefficients $h_1[n]$ have only two non-zero values given by

$$h_1[n] = 2^{-1/2}(\delta_{0n} - \delta_{1n}). \tag{1.85}$$

The definition of the Haar wavelets in terms of the Haar scaling function ensures that the wavelets form an orthonormal set at all scales and are orthogonal to the scaling functions at all scales $m \leq l$, i.e.,

$$\langle \psi_{mk}(t)\psi_{ln}(t) \rangle = \delta_{ml}\delta_{kn}, \;\; \langle \psi_{mk}(t)\phi_{ln}(t) \rangle = 0, \; m \leq l. \tag{1.86}$$

Equation (1.79), when rewritten in the form

$$\boldsymbol{P}_{m-1}x = \sum_{n=-\infty}^{\infty} c_{mn}\phi_{mn}(t) + \sum_{n=-\infty}^{\infty} d_{mn}\psi_{mn}(t), \tag{1.87}$$

has a simple interpretation. The approximation $\boldsymbol{P}_{m-1}x$ to a function $x(t) \in L_2(\mathbb{R})$ that is obtained by projecting it orthogonally onto the subspace $\mathcal{V}_{m-1}$ of a multi-resolution analysis space is the sum of two terms: one is the approximation to the function obtained by orthogonal projection onto the next coarser scale $\mathcal{V}_m$, represented by the term $\boldsymbol{P}_m x = \sum_{n=-\infty}^{\infty} c_{mn}\phi_{mn}(t)$, and the other is the difference between the two approximations represented by the term $\boldsymbol{E}_m x = \sum_{n=-\infty}^{\infty} d_{mn}\psi_{mn}(t)$. The latter is the detail lost in going from the finer scale $m-1$ to the coarser scale m.

Equation (1.79) and its equivalent form $\boldsymbol{P}_{m-1}x = \boldsymbol{P}_m x + \boldsymbol{E}_m x$, when recursively continued, give the result

$$\boldsymbol{P}_{m-1}x = \sum_{k=m}^{\infty} \boldsymbol{E}_k x = \sum_{k=m}^{\infty} \sum_{n=-\infty}^{\infty} d_{kn} \psi_{kn}. \tag{1.88}$$

Thus, each approximation subspace $\mathcal{V}_m$ is the direct sum of all the detail subspaces at lower resolutions (scales):

$$\mathcal{V}_{m-1} = \mathcal{W}_m \oplus \mathcal{W}_{m+1} \oplus \cdots = \bigoplus_{k=m}^{\infty} \mathcal{W}_k. \tag{1.89}$$

In addition, since $\boldsymbol{P}_m(x(t)) \to x(t)$ as $m \to -\infty$, we have the wavelet expansion equation, or the wavelet representation,

$$x(t) = \sum_{m=-\infty}^{\infty} \sum_{n=-\infty}^{\infty} d_{mn} \psi_{mn}(t), \tag{1.90}$$

where the wavelet transform coefficients d_{mn} are defined in equation (1.81). The wavelet representation is equivalent to the multi-resolution decomposition formula

$$L_2(\mathbb{R}) = \bigoplus_{m=-\infty}^{\infty} \mathcal{W}_m. \tag{1.91}$$

Using the scaling and the wavelet equations (1.76) and (1.84), we can write the following general recursive equations for the scaling and wavelet coefficients (cf. the Haar relations (1.75) and (1.81))

$$c_{mn} = \sum_{k=-\infty}^{\infty} h_0[k-2n]\, c_{m-1,k}, \quad d_{mn} = \sum_{k=-\infty}^{\infty} h_1[k-2n]\, c_{m-1,k}, \tag{1.92}$$

whose linear filter interpretation is the basis for the discrete orthogonal wavelet transform implementation. When the Haar coefficients (1.77) and (1.85) are used, we obtain the Haar wavelet; other sets of coefficients have $4, 6, 8, \ldots$ elements and define the Daub4, Daub6, Daub8, ... orthogonal wavelet systems to which we will return in section 3.13.

1.14 Compressive Sensing

Many signals and images of interest can be well approximated by compressed versions, and many schemes such as the DWT can perform efficient compressions. If a signal of interest is known to be sparse (in the original domain of acquisition or a transformed domain), then an important question is whether a method can be devised to acquire the compressed version of the data instead of acquiring the full uncompressed data and subsequently compressing it. Compressive sensing studies the conditions under which a compressed version of the signal can be acquired and used to reconstruct the original uncompressed signal by efficient l_1 minimization techniques [12] [17, 18]. As shown earlier (see the discussion following the definition of the l_p norm (1.4)), given an $N \times 1$ s-sparse vector $\boldsymbol{x} \in \mathbb{R}^N$, compressive sensing attempts to find a $k \times 1$ compressed version by designing a $k \times N$ structured random sensing matrix $\boldsymbol{A}$ with $k \sim s \ll N$ to obtain the compressed measured $k \times 1$ vector $\boldsymbol{y} = \boldsymbol{Ax}$. The reconstruction of the original s-sparse vector is then expressed as the optimization problem

$$\arg\min_{\boldsymbol{x}} \|x\|_0 \text{ subject to } \boldsymbol{Ax} = \boldsymbol{y}. \tag{1.93}$$

[12] Linear programming techniques apply to solutions of the real l_1 minimization problem, whereas the complex data case is the same as the second order cone program, which can be solved efficiently [16].

The solution should have the property that any two distinct s-sparse vectors $\boldsymbol{x}_1, \boldsymbol{x}_2 \in \mathbb{R}^N$ must have two distinct $k \times 1$ compressed vectors $\boldsymbol{y}_1 = \boldsymbol{A}\boldsymbol{x}_1 \neq \boldsymbol{y}_2 = \boldsymbol{A}\boldsymbol{x}_2$; this cannot be true, since $\boldsymbol{A}$ has rank $k \ll N$, and so it has a non-trivial null space. To solve the non-uniqueness problem, we might require that for $\boldsymbol{n} \in \mathscr{N}_{\boldsymbol{A}}$, $\boldsymbol{x} + \boldsymbol{n}$ is not s-sparse, but the best we can hope for is that $\mathscr{N}_{\boldsymbol{A}}$ has no $2s$-sparse vectors. This is formulated as the *restricted isometry property* (RIP) [19, 20] of the sensing matrix [13] $\boldsymbol{A}$: for each $s \in \mathbb{Z}^+$, we define the isometry constraints δ_s of a matrix $\boldsymbol{A}$ as the smallest numbers such that for all s-sparse vectors $\boldsymbol{x} \in \mathbb{R}^N$,

$$(1 - \delta_s) \|\boldsymbol{x}\|_2^2 \leq \|\boldsymbol{A}\boldsymbol{x}\|_2^2 \leq (1 + \delta_s) \|\boldsymbol{x}\|_2^2 . \tag{1.94}$$

The sensing matrix $\boldsymbol{A}$ is said to have the RIP if $0 < \delta_s < 1$. The RIP ensures that the norm of the measured vector stays very close to the norm of the original s-sparse vector and two distinct s-sparse vectors in $\mathbb{R}^N$ will result in two distinct measured vectors in $\mathbb{R}^k$.

Instead of solving the optimization problem (1.93) that is NP-hard when s is large, we relax the problem using the l_p norm for $0 < p \leq 1$. When $p < 1$, the objective function is non-convex and difficult to minimize, but for $p = 1$, it is convex and can be efficiently solved using linear programming techniques. Theorem 9 defines the solution to the optimization problem using the RIP [21].

Theorem 9. *Given $k \times 1$ vector $\boldsymbol{y} = \boldsymbol{A}\boldsymbol{x}$ where $\boldsymbol{x}$ is an s-sparse vector in $\mathbb{R}^N$, $s \ll N$, and $\boldsymbol{A}$ satisfies the RIP with $\delta_{2s} < 1/3$, the solution $\hat{\boldsymbol{x}}$ to the minimization problem*

$$\underset{\boldsymbol{x}'}{\arg\min} \ \|\boldsymbol{x}'\|_1 \quad \textit{subject to } \boldsymbol{A}\boldsymbol{x}' = \boldsymbol{y} = \boldsymbol{A}\boldsymbol{x} \tag{1.95}$$

is $\hat{\boldsymbol{x}} = \boldsymbol{x}$.

Theorem 9 implies that $\boldsymbol{x}$ can be fully recovered by finding the solution $\hat{\boldsymbol{x}}$ to the l_1 minimization problem (1.95) in the absence of any measurement noise. Theorem 10 shows the recovery conditions in the presence of noise [21].

Theorem 10. *Given $k \times 1$ measured vector $\boldsymbol{y} = \boldsymbol{A}\boldsymbol{x} + \boldsymbol{n}$, $\|\boldsymbol{n}\|_2 \leq \eta$, if $\hat{\boldsymbol{x}}$ is the solution to*

$$\underset{\boldsymbol{x}'}{\arg\min} \ \|\boldsymbol{x}'\|_1 \quad \textit{subject to} \quad \|\boldsymbol{y} - \boldsymbol{A}\boldsymbol{x}'\| \leq \eta, \tag{1.96}$$

then two positive constants c, d, depending only on δ_{2s} exist for which

$$\|\boldsymbol{x} - \hat{\boldsymbol{x}}\| \leq c\eta + d\,\sigma_s/\sqrt{s}, \quad \sigma_s = \inf \|\boldsymbol{x} - \boldsymbol{s}\|_2, \ \boldsymbol{s} \textit{ is s-sparse.} \tag{1.97}$$

Figure 1.11 shows the output of a convex l_1 optimization matching pursuit method applied to image data. The original image (a) was corrupted by randomly zeroing 50% of the pixels (b) with a *peak signal to noise ratio* (PSNR) of 12 dB [14]. Image (c) has a PSNR of 34 dB and shows nearly perfect reconstruction.

Structured random matrices are particularly useful for applications to signals and images; that is, we are often interested in reconstructing randomly sampled functions that have a sparse representation in terms of an orthonormal basis such as the discrete Fourier transform basis vectors (see section 3.9). Random partial discrete Fourier transform matrices are a special case, and theorem 11 describes how they can be used to construct a sensing matrix $\boldsymbol{A}$ that satisfies the RIP [17].

---

[13] The so called "null space property" is difficult to prove and has been replaced by the RIP.

[14] PSNR on a dB scale for an image is defined as $10 \log_{10}$ of the square of the image maximum divided by the variance of the image.

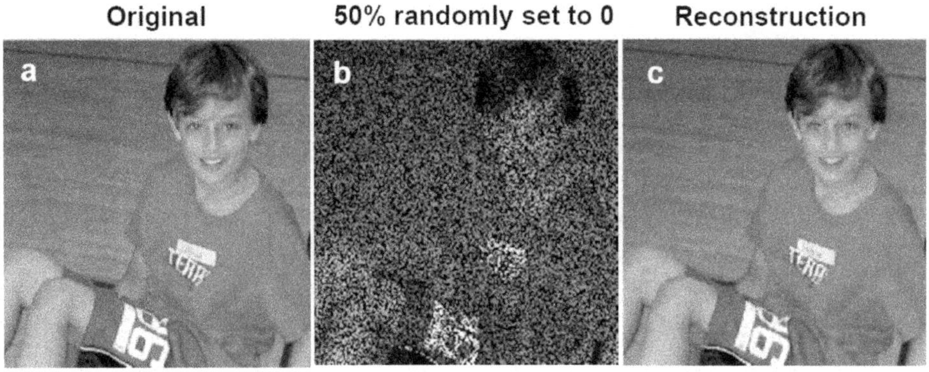

Figure 1.11: 512×512 image (a), randomly corrupted image (b), and reconstructed image (c).

Theorem 11. *Consider the $N \times N$ symmetric discrete Fourier transform matrix $\boldsymbol{F}$ (see section 3.9) whose (m, n) element is $\exp\big(-2i\pi mn/N\big)$. Let $\boldsymbol{A}$ be a $k \times N$ matrix whose k rows are selected by uniformly randomly selecting k rows of $\boldsymbol{F}$. If $k \leq C\ s(\ln N)^4$ for some constant C, then $\boldsymbol{A}$ satisfies the RIP with parameters s, δ_s, and $\delta_{2s} < 1/3$, with probability $1 - \exp(-k)$.*

Although probabilistic in nature, once a sensing matrix satisfies the RIP, it will work for all s-sparse vectors in $\mathbb{R}^N$. As an example consider 1024 samples of the signal $\sin(2\pi \times 0.07t) + \sin(2\pi \times 0.19t)$ sampled at 1 Hz, whose spectrum is sparse in the frequency domain (essentially two peaks at 0.07 and 0.19 Hz) as shown in figure 1.11(a). Let $\boldsymbol{F}$ denote the 1024×1024 symmetric DFT matrix whose elements are

$$\big[\boldsymbol{F}\big]_{kn} = \big[\boldsymbol{F}\big]_{nk} = e^{-\frac{2i\pi kn}{1024}},$$

and whose inverse is

$$\boldsymbol{F}^{-1} = \frac{1}{1024}\boldsymbol{F}^* = \frac{1}{1024}\boldsymbol{F}^+$$

where the last equality follows from the symmetry of $\boldsymbol{F}$ (see section 3.9).

The sensing matrix $\boldsymbol{A}$ is constructed by selecting k random rows of $\boldsymbol{F}^{-1}$ (since it is the spectrum of the signal that is sparse); we chose $k = 128$ for this data. The measured time series is the 128×1 vector found by multiplying the 128×1024 sensing matrix and the 1024×1 Fourier transform of the original signal. Once the solution $\hat{\boldsymbol{x}}$ to the l_1 minimization problem (1.93) is found (using $\boldsymbol{A}^T\boldsymbol{y}$ as the starting vector), the recovered signal in the time domain is the (real part) of $\boldsymbol{F}^{-1}\hat{\boldsymbol{x}}$. Figure 1.12(a) shows 5 minutes of the 17-minute signal with marked random samples and figure 1.12(b) shows the reconstruction of the same signal. Figures 1.12(c) and 1.12(d) show the spectra of the original and the reconstructed signals showing two peaks at 0.07 and 0.19 Hz.

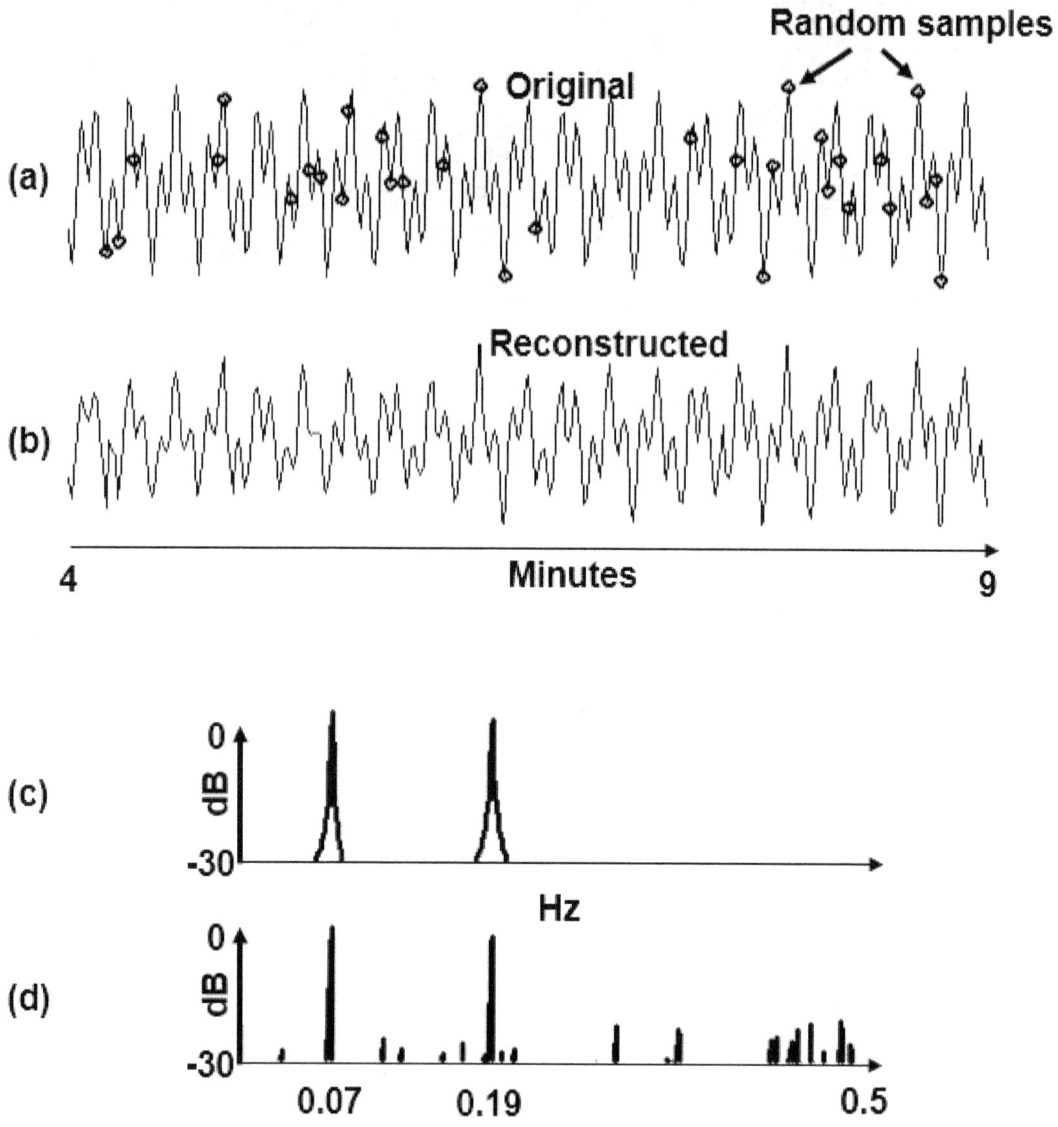

Figure 1.12: (a) and (c): Portion of signal and its spectrum. (b) and (d) Reconstructed signal and its spectrum using a 128×1024 sensing matrix (random samples used in the reconstruction are marked on the signal in (a)).

CHAPTER 2

Matrix Factorizations and the Least Squares Problem

2.1 Introduction

The formula $\boldsymbol{x}_{\text{ls}} = (\boldsymbol{A}^{+}\boldsymbol{A})^{-1}\boldsymbol{A}^{+}\boldsymbol{b}$ derived in section 1.8 does not offer the best method to compute the least squares solution of $\boldsymbol{A}\boldsymbol{x} = \boldsymbol{b}$ because of numerical instability arising from the possibility that the matrix $\boldsymbol{A}^{+}\boldsymbol{A}$ could be nearly singular.

In general, when solving the equation $\boldsymbol{A}\boldsymbol{x} = \boldsymbol{b}$, changes in the input matrix result in changes $\Delta\boldsymbol{x}$ in the solution $\boldsymbol{x}$ with

$$\|\Delta\boldsymbol{x}\| \,/\, \|\boldsymbol{x}\| \leq \kappa_{\boldsymbol{A}} \, \|\Delta\boldsymbol{A}\| \,/\, \|\boldsymbol{A}\| \,,$$

where $\kappa_{\boldsymbol{A}}$ is the condition number (1.30). The condition number of a matrix can be used to determine the numerical accuracy of solutions to the linear equation $\boldsymbol{A}\boldsymbol{x} = \boldsymbol{b}$; when a solution is calculated to d decimal places, then roughly the first $d - \log_{10}\kappa_{\boldsymbol{A}}$ decimals can be considered accurate. The situation is worse when solving least squares equations of the form $\boldsymbol{A}^{+}\boldsymbol{A}\boldsymbol{x} = \boldsymbol{b}$ because then the condition number of $\boldsymbol{A}^{+}\boldsymbol{A}$ is $\kappa^2_{\boldsymbol{A}}$, so the loss in accuracy is now $2\,\log_{10}\kappa_{\boldsymbol{A}}$.

QR factorization avoids matrix inversion and solves a full-rank least squares problem by triangularization. We will discuss two methods to achieve the QR form, namely, Householder reflections and Givens rotations; both methods are reasonably fast, but Givens rotations are about 50% slower than Householder reflections. QR factorization usually fails when $\kappa_{\boldsymbol{A}} \sim 1/\epsilon$, where ϵ represents machine precision. Another method to solve the least squares equation is Cholesky factorization, which fails when $\kappa_{\boldsymbol{A}} \sim 1/\sqrt{\epsilon}$. Finally, a systematic treatment for all ill-conditioned matrix equations is the singular value decomposition or SVD; it is a time consuming method, but it can handle all rank deficient problems and is indispensable in providing insight into many signal processing algorithms.

2.2 QR Factorization

Consider the Gram-Schmidt procedure by which we use a set of vectors $\boldsymbol{x}_1, \ldots, \boldsymbol{x}_N$ to construct an orthonormal basis $\boldsymbol{q}_1, \ldots, \boldsymbol{q}_{N'}$, $N' \leq N$, for the space spanned by the original set of vectors. We assume for now that the vectors $\boldsymbol{x}_n$ are linearly independent and that each has dimension $N \times 1$, i.e., $\boldsymbol{x}_n \in \mathbb{C}^N$, $1 \leq n \leq N$. In this case, the Gram-Schmidt procedure will produce the orthonormal set $\boldsymbol{q}_n$, $1 \leq n \leq N$, $\boldsymbol{q}^{+}_n \boldsymbol{q}_m = \delta_{nm}$.

To illustrate the concept of QR factorization [6, 22], we rewrite the Gram-Schmidt algorithm as a set of equations for the vectors $\boldsymbol{x}_n$ in terms of the orthonormal basis vectors $\boldsymbol{q}_n$ using the equation $\boldsymbol{q}'_n = \|\boldsymbol{q}'_n\|\, \boldsymbol{q}_n$

in section 1.3, i.e.,

$$\boldsymbol{q}_1 = \boldsymbol{x}_1/\|\boldsymbol{x}_1\|,\ \boldsymbol{x}_n = \sum_{l=1}^{n-1} \langle \boldsymbol{x}_n, \boldsymbol{q}_l \rangle \boldsymbol{q}_l + \|\boldsymbol{q}'_n\| \boldsymbol{q}_n,\ 2 \le n \le N. \tag{2.1}$$

For any matrix $\boldsymbol{X}$ and vector $\boldsymbol{c}$, the product $\boldsymbol{Xc}$ is a linear combination of the columns of the matrix $\boldsymbol{X}$; i.e., if $\boldsymbol{X} = [\boldsymbol{x}_1, \ldots, \boldsymbol{x}_N]$ and $\boldsymbol{c} = [c_1, \ldots, c_N]$, then $\boldsymbol{Xc} = c_1\boldsymbol{x}_1 + \ldots + c_N\boldsymbol{x}_N$. Thus, the linear combination on the right hand side of the above Gram-Schmidt algorithm for $\boldsymbol{x}_n$ can be written as a product of a matrix whose columns are the orthonormal vectors $\boldsymbol{q}_m$ and a vector whose elements are the various coefficients in the Gram-Schmidt algorithm:

$$\boldsymbol{x}_n = \left[\boldsymbol{q}_1, \ldots, \boldsymbol{q}_n\right]\left[\langle \boldsymbol{x}_n, q_1\rangle, \ldots, \langle \boldsymbol{x}_n, \boldsymbol{q}_{n-1}\rangle, \|\boldsymbol{q}'_n\|\right]^T. \tag{2.2}$$

Finally, using the identity $\boldsymbol{X}[\boldsymbol{y}_1, \ldots, \boldsymbol{y}_n] \equiv [\boldsymbol{X}\boldsymbol{y}_1, \ldots, \boldsymbol{X}\boldsymbol{y}_n]$ valid for a matrix $\boldsymbol{X}$ and column vectors $\boldsymbol{y}_k$, and using the stacked form $\boldsymbol{X} = [\boldsymbol{x}_1, \ldots, \boldsymbol{x}_N]$, we arrive at the following matrix equation

$$\boldsymbol{X} = \boldsymbol{Q}\begin{bmatrix} \|\boldsymbol{x}_1\| & \langle \boldsymbol{x}_2, \boldsymbol{q}_1\rangle & \langle \boldsymbol{x}_3, \boldsymbol{q}_1\rangle & \cdots & \langle \boldsymbol{x}_N, \boldsymbol{q}_1\rangle \\ 0 & \|\boldsymbol{q}'_2\| & \langle \boldsymbol{x}_3, \boldsymbol{q}_2\rangle & \cdots & \langle \boldsymbol{x}_N, \boldsymbol{q}_2\rangle \\ 0 & 0 & \|\boldsymbol{q}'_3\| & \cdots & \vdots \\ \vdots & \cdots & 0 & \ddots & \langle \boldsymbol{x}_N, \boldsymbol{q}_{N-1}\rangle \\ 0 & \cdots & 0 & \cdots & \|\boldsymbol{q}'_N\| \end{bmatrix} \equiv \boldsymbol{QR}, \tag{2.3}$$

where the columns of the square $N \times N$ matrix $\boldsymbol{Q}$ are the N orthonormal vectors $\boldsymbol{q}_n$, $1 \le n \le N$, and $\boldsymbol{R}$ is the upper triangular matrix of the indicated coefficients. Clearly $\boldsymbol{Q}$ is a unitary matrix, i.e., $\boldsymbol{Q}^{+}\boldsymbol{Q} = \boldsymbol{I}$ where $\boldsymbol{I}$ is the $N \times N$ identity matrix. The prescribed algorithm then computes one column of the matrix $\boldsymbol{Q}$ and one column of the upper triangular matrix $\boldsymbol{R}$ recursively starting with the first column.

The application of the Gram-Schmidt algorithm to the QR factorization of a full rank complex matrix $\boldsymbol{A}$ with dimension $M \times N$, $M > N$, produces a rectangular matrix $\boldsymbol{Q}$ that is $M \times N$ with only the unitarity condition $\boldsymbol{Q}^{+}\boldsymbol{Q} = \boldsymbol{I}$ (note that $\boldsymbol{Q}$ is not square and therefore has no inverse) and a square $N \times N$ upper triangular matrix $\boldsymbol{R}$. This is known as the "thin" QR factorization (since the number of rows of $\boldsymbol{Q}$ exceed the number of its columns), and the corresponding unitary $M \times N$ matrix is often denoted by $\boldsymbol{Q}_1$.

The usual form of QR factorization applied to the full rank complex matrix $\boldsymbol{A}$ with dimension $M \times N$, $M > N$, produces a square and unitary $M \times M$ matrix $\boldsymbol{Q}$ and will necessarily produce a rectangular upper triangular $M \times N$ matrix $\boldsymbol{R}$ whose last $M - N$ rows consist of zeros. The structures of the two factors in the QR decomposition of a full rank matrix of dimension $M \times N$ with $M > N$ are shown in figure 2.1.

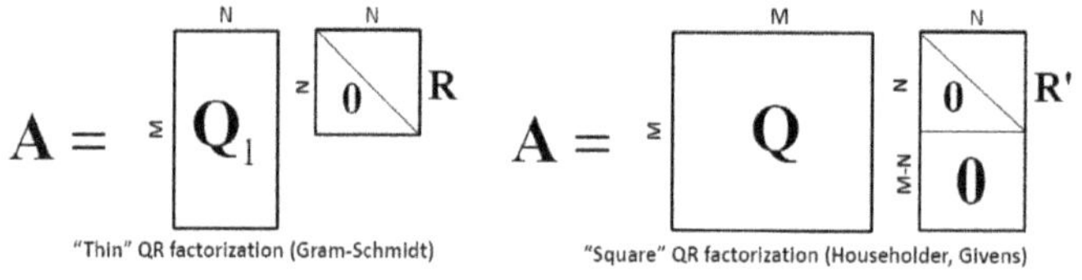

Figure 2.1: The "thin" and "square" QR factorizations.

The "square" QR form is found using Householder reflections [23] or Givens rotations [6, 22]: a Householder transformation produces a reflection through a line while the Givens transformation is a rotation; both sets of transformations leave the length of a vector unchanged and are used to shape a matrix into upper triangular form.

2.3 QR Factorization Using Givens Rotations

The main idea of QR factorization using Givens rotations [22] is illustrated in figure 2.2 where a 2×1 real vector $[a, b]^T$, $a, b \in \mathbb{R}$, is rotated (anti-clockwise) to the vector $[\alpha, 0]^T$ where $\alpha = \sqrt{a^2 + b^2}$. The 2×2 orthogonal rotation matrix is defined by the angle θ where $\tan\theta = b/a$.

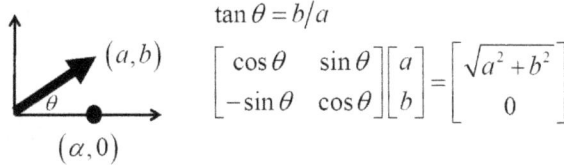

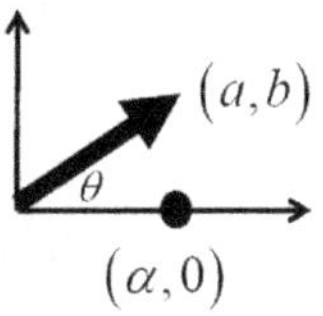

$$\tan\theta = b/a$$

$$\begin{bmatrix} \cos\theta & \sin\theta \\ -\sin\theta & \cos\theta \end{bmatrix} \begin{bmatrix} a \\ b \end{bmatrix} = \begin{bmatrix} \sqrt{a^2+b^2} \\ 0 \end{bmatrix}$$

Figure 2.2: Rotating $[a, b]^T$ to $[\alpha, 0]^T$ with no change in length.

More generally and for a complex 2×1 vector, we consider a 2×2 unitary complex rotation matrix to map the complex vector $[a, b]^T$, $a, b \in \mathbb{C}$, to the complex vector $[\alpha, 0]^T$, $\alpha \in \mathbb{C}$.

$$\boldsymbol{G} = \begin{bmatrix} c & s \\ -s^* & c^* \end{bmatrix}. \tag{2.4}$$

For $\boldsymbol{G}$ to be unitary, we must have $|c|^2 + |s|^2 = 1$ and $c = c^*$. Thus, we choose $c = \cos\theta$ and $s = e^{i\phi}\sin\theta$. It is then easy to see that for a complex 2×1 vector $\boldsymbol{x} = [|a|\, e^{i\beta_a}, |b|\, e^{i\beta_b}]^T$ if we choose $\tan\theta = |b| \,/\, |a|$ and $\phi = \beta_a - \beta_b$, we have $\boldsymbol{G}\boldsymbol{x} = [\sqrt{|a|^2 + |b|^2} e^{i\beta_a}, 0]^T$, and clearly $\|\boldsymbol{G}\boldsymbol{x}\| = \|\boldsymbol{x}\|$. For instance, for a real vector $\boldsymbol{x} = [\cos\theta, \sin\theta]^T$, $0 \leq \theta \leq \pi/2$, we have $\boldsymbol{G}\boldsymbol{x} = [1, 0]^T$.

Thus, for an $M \times 1$ vector $\boldsymbol{x}$ and integers p and q with $1 \leq p, q \leq M$, we can construct a Givens $M \times M$ rotation matrix $\boldsymbol{G}^{(p,q)}$ that will zero the element x_p, change the element x_q to $\sqrt{|x_p|^2 + |x_q|^2}$, and leave all other elements unchanged. This matrix has the following elements: 1's along the diagonal except for two locations $[\boldsymbol{G}^{(p,q)}]_{pp} = [\boldsymbol{G}^{(p,q)}]_{qq} = c$, and 0's everywhere else except for two locations $[\boldsymbol{G}^{(p,q)}]_{qp} = s$ and $[\boldsymbol{G}^{(p,q)}]_{pq} = -s^*$. For instance, consider the vector $\boldsymbol{x} = [1, 2, 3]^T$. The Givens rotation matrix to zero the last element ($p = 3$) and to change the first element ($q = 1$), leaving the second element unchanged, is

$$\boldsymbol{G}^{(3,1)} = \begin{bmatrix} 1/\sqrt{10} & 0 & 3/\sqrt{10} \\ 0 & 1 & 0 \\ -3/\sqrt{10} & 0 & 1/\sqrt{10} \end{bmatrix},$$

for which we have $\boldsymbol{G}^{(3,1)}\boldsymbol{x} = [\sqrt{10}, 2, 0]^T$. In computing the final form of the Givens rotation matrix, we used the equations $c = 1/\sqrt{1^2 + 3^2}$ and $s = 3/\sqrt{1^2 + 3^2}$. If the two elements of a vector affected by a Givens rotation matrix are denoted by a and b, then we use $c = |a| \,/ \sqrt{|a|^2 + |b|^2}$ and $s = e^{i\phi}|b|/\sqrt{|a|^2+|b|^2}$. To avoid possible overflow when computing the sum of squared magnitudes, we use the following procedure:

- if $|b| \geq |a|$, then $\gamma = |a| \,/\, |b|$, $s = e^{i\phi}/\sqrt{1 + \gamma^2}$, and $c = \gamma\,|s|$.
- if $|a| \geq |b|$, then $\gamma = |b| \,/\, |a|$, $c = 1/\sqrt{1 + \gamma^2}$, and $s = e^{i\phi}\gamma c$.

For an $M \times N$ complex matrix $\boldsymbol{A}$, we use Givens rotation matrices to shape it into an upper triangular matrix starting with the last element of the first column, i.e., we find the Givens rotation matrix $\boldsymbol{G}^{(M,M-1)}$ to zero $[\boldsymbol{A}]_{M,1}$ and to change $[\boldsymbol{A}]_{M-1,1}$. Then, we continue by constructing $\boldsymbol{G}^{(M-1,M-2)}$ for the first column of the product matrix $\boldsymbol{G}^{(M,M-1)}\boldsymbol{A}$, and so on, until all elements of the first column of the matrix

$\boldsymbol{G}^{(2,1)}\boldsymbol{G}^{(3,2)}\ldots\boldsymbol{G}^{(M-1,M-2)}\boldsymbol{A}$, except for the first element, have been transformed to zero. Next, we perform the same algorithm on the second column until all elements except for the first two have been transformed to zero, and so on. Figure 2.3 shows an example 5×3 matrix, the c and s coefficients for all Givens rotations, and the final $\boldsymbol{Q}$ and $\boldsymbol{R}$ factors.

$$
\mathrm{A} = \begin{matrix} 0.492667 & -0.937337 & 0.142382 \\ 0.547757 & -0.337883 & -3.61494 \\ 0.0767997 & 2.05533 & -1.19836 \\ 0.552353 & 2.46093 & 6.23741 \\ 0.382426 & 2.95410 & 4.07071 \end{matrix}
\qquad
C, S = \begin{matrix} 0.822173 & 0.569238 \\ 0.113576 & 0.993529 \\ 0.629448 & 0.777043 \\ 0.492667 & 0.870218 \\ 0.844550 & -0.535477 \\ 0.817658 & -0.575704 \\ 0.552353 & 0.833610 \\ 0.987460 & 0.157871 \\ 0.245310 & 0.969445 \end{matrix}
$$

$$
\mathrm{Q} = \begin{matrix} 0.492667 & -0.480668 & 0.177953 & -0.703257 & 0. \\ 0.547757 & -0.358349 & -0.577743 & 0.482466 & 0.0706229 \\ 0.0767997 & 0.475432 & -0.634321 & -0.431660 & -0.423525 \\ 0.552353 & 0.339055 & -0.480846 & 0.276885 & -0.521604 \\ 0.382426 & 0.547312 & 0.0311446 & -0.0982917 & 0.737271 \end{matrix}
\qquad
\mathrm{R} = \begin{matrix} 1 & 2 & 3 \\ 0 & 4 & 5 \\ 0 & 0 & 6 \\ 0 & 0 & 0 \\ 0 & 0 & 0 \end{matrix}
$$

Figure 2.3: Example of QR factorization using Givens rotations.

2.4 QR Using Householder Reflections

Another method for the "square" QR factorization of a complex $M \times N$ full rank matrix $\boldsymbol{A}$ with $M > N$ is to use Householder reflection matrices [23, 22]. In analogy to the Givens rotation matrices, a series of unitary $M \times M$ matrices $\boldsymbol{H}_k$ is constructed to multiply $\boldsymbol{A}$ on the left and to shape the matrix into upper triangular form $\boldsymbol{R}$, by introducing zeros in column k below the diagonal but retaining the zeros in the previous columns.

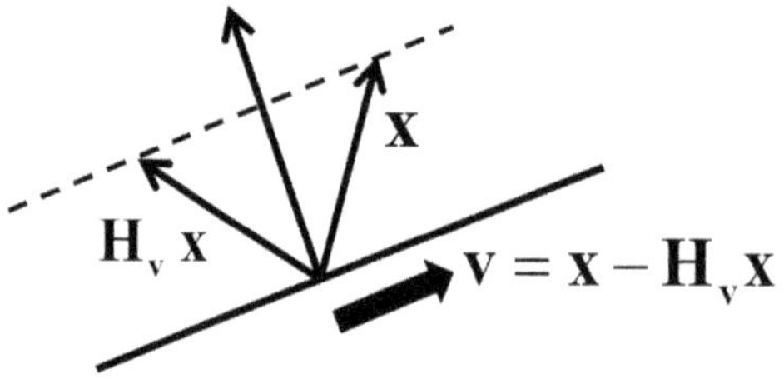

Figure 2.4: The geometric interpretation of the Householder matrix $\boldsymbol{H}_{\boldsymbol{v}}$.

The Householder reflector matrix $\boldsymbol{H}_{\boldsymbol{v}}$ reflects a given vector $\boldsymbol{x}$ through the orthogonal projection of that vector to $\boldsymbol{v}$ as shown in figure 2.4. Since the orthogonal projection of $\boldsymbol{x}$ onto $\boldsymbol{v}$ is given by $(\boldsymbol{v}^{+}\boldsymbol{x})\boldsymbol{v}/\,\|\boldsymbol{v}\|_2^2$ we find

$$\boldsymbol{H}_{\boldsymbol{v}}\boldsymbol{x} = \boldsymbol{x} - 2(\boldsymbol{v}^{+}\boldsymbol{x})\boldsymbol{v}/\,\|\boldsymbol{v}\|_2^2 = (\boldsymbol{I} - 2\,\boldsymbol{v}\boldsymbol{v}^{+}/\|\boldsymbol{v}\|_2^2)\boldsymbol{x}, \tag{2.5}$$

and so we obtain

$$\boldsymbol{H}_{\boldsymbol{v}} = \boldsymbol{I} - 2\,\boldsymbol{v}\boldsymbol{v}^{+}/\|\boldsymbol{v}\|_2^2 = \boldsymbol{I} - 2\,\hat{\boldsymbol{v}}\hat{\boldsymbol{v}}^{+}, \tag{2.6}$$

where $\hat{\boldsymbol{v}} = \boldsymbol{v}/\|\boldsymbol{v}\|_2$ is the corresponding normalized vector. Note that $\boldsymbol{H}_{\boldsymbol{v}}$ is clearly Hermitian and unitary: $\boldsymbol{H}_{\boldsymbol{v}}^{+} = \boldsymbol{H}_{\boldsymbol{v}}$, and $\boldsymbol{H}_{\boldsymbol{v}}^{+}\boldsymbol{H}_{\boldsymbol{v}} = \boldsymbol{I}$. This matrix has the important property that for a non-zero vector $\boldsymbol{x} = [x_1, \ldots, x_L]^T$, we can select an appropriate vector $\boldsymbol{v}$ of dimension L that depends on $\boldsymbol{x}$, and whose associated Householder matrix transformation of $\boldsymbol{x}$ is a vector, all of whose elements except for the first are equal to zero, namely,

$$\boldsymbol{H}_{\boldsymbol{v}}\boldsymbol{x} = a\boldsymbol{e}_1, \quad \boldsymbol{e}_1 \equiv [1, 0, \ldots, 0]^T. \tag{2.7}$$

We now show how to find the required vector $\boldsymbol{v}$ for a given vector $\boldsymbol{x}$. Since $\boldsymbol{H}_{\boldsymbol{v}}$ is unitary, we must have $\|\boldsymbol{H}_{\boldsymbol{v}}\boldsymbol{x}\|_2 = \|\boldsymbol{x}\|_2$, and so the equation $\boldsymbol{H}_{\boldsymbol{v}}\boldsymbol{x} = a\boldsymbol{e}_1$ implies that $a = \pm\|\boldsymbol{x}\|_2$. Now $\boldsymbol{H}_{\boldsymbol{v}}\boldsymbol{x} = a\boldsymbol{e}_1$ is equivalent to $\boldsymbol{v} = (\boldsymbol{x} - a\boldsymbol{e}_1)\,\|v\|_2^2\,/2v^+\boldsymbol{x}$, and so $\boldsymbol{v} \propto (\boldsymbol{x} - a\boldsymbol{e}_1)$ and its norm is proportional to $\|\boldsymbol{x}-a\boldsymbol{e}_1\|_2 = \sqrt{\|\boldsymbol{x}\|_2^2 + a^2 - 2a\,\boldsymbol{Re}\,[x_1]} = \sqrt{2a(a - \boldsymbol{Re}[x_1])}$. Thus, when normalizing $\boldsymbol{v}$ to construct the Householder reflection matrix, all of its elements except for the first must be divided by $\sqrt{2a(a - \boldsymbol{Re}\,[x_1])}$, which could result in a divergence if $a \approx \boldsymbol{Re}\,[x_1]$; the divergence is avoided by choosing $a = -\text{sign}(\boldsymbol{Re}\,[x_1])\|x\|_2$, which leads to

$$\boldsymbol{v} = \boldsymbol{x} + \text{sign}(\boldsymbol{Re}[x_1])\,\|\boldsymbol{x}\|_2\,[1, 0, \ldots, 0]^T \;\Rightarrow\; \boldsymbol{H}_{\boldsymbol{v}}\boldsymbol{x} = a\boldsymbol{e}_1. \tag{2.8}$$

A complex $M \times N$ matrix $\boldsymbol{A}$ whose first column is denoted by $\boldsymbol{a}_1$, with $M > N$, can be put in the "square" QR form by the application of N Householder reflection matrices in the following way:

- Let $\boldsymbol{x} = \boldsymbol{a}_1$ and calculate the first Householder reflection vector $\boldsymbol{v}_1 = \boldsymbol{x} + \text{sign}(\boldsymbol{Re}[x_1])\,\|\boldsymbol{x}\|_2\,\boldsymbol{e}_1$, where $\boldsymbol{e}_1 = [1, 0, \ldots, 0]^T$ has dimension M.

- Normalize $\boldsymbol{v}_1$ to obtain $\hat{\boldsymbol{v}}_1 = \boldsymbol{v}_1 / \|\boldsymbol{v}_1\|_2$.

- Calculate the first $M \times M$ Householder reflection matrix $\boldsymbol{H}_1 = \boldsymbol{I} - 2\,\hat{\boldsymbol{v}}_1\hat{\boldsymbol{v}}_1^+$.

- Denoting the product $\boldsymbol{H}_1\boldsymbol{A}$ by $\boldsymbol{A}_1$, we observe that its first column will have only one non-zero element in the first row, namely, $[\boldsymbol{A}_1]_{11}$, which is the first row of the final upper triangular matrix $\boldsymbol{R}$. The first element of the second column of $\boldsymbol{A}_1$, namely, $[\boldsymbol{A}_1]_{12}$, will be the first element of the second column of the final upper triangular matrix $\boldsymbol{R}$, namely, $[\boldsymbol{R}]_{12}$, while the second normalized reflection vector $\hat{\boldsymbol{v}}_2$ is constructed using elements $2, \ldots, M$ of the same second column of $\boldsymbol{A}_1$.

- The vector $\hat{\boldsymbol{v}}_2$ whose dimension is $M - 1$ is used to form a vector with dimension M, namely, $[0, \hat{\boldsymbol{v}}_2]$ to construct the second reflection matrix $\boldsymbol{H}_2$ of size $M \times M$. The matrix $\boldsymbol{H}_2$ has a block structure so that the product $\boldsymbol{A}_2 = \boldsymbol{H}_2\boldsymbol{A}_1$ will have exactly the same first column and first row as $\boldsymbol{A}_1$. Thus, the second column and the first two elements of the third column of $\boldsymbol{A}_2$ constitute the elements of the same position of the final upper triangular $\boldsymbol{R}$.

- Continue in the same way until the final matrix $\boldsymbol{R}$ is constructed: this should happen after N steps when $M > N$ (when $M = N$ only $N-1$ steps are needed). Thus, we will have $\boldsymbol{H}_{\boldsymbol{v}_N} \ldots \boldsymbol{H}_{\boldsymbol{v}_1}\boldsymbol{A} = \boldsymbol{R}$.

- The Hermitian $M \times M$ matrix $\boldsymbol{Q}$ is found by the product $\boldsymbol{Q} = \boldsymbol{H}_{\boldsymbol{v}_1} \ldots \boldsymbol{H}_{\boldsymbol{v}_N}$ and we have the QR factorization $\boldsymbol{A} = \boldsymbol{Q}\boldsymbol{R}$.

Figure 2.5 shows an example 5×3 matrix, three Householder vectors (not normalized), and the final matrices $\boldsymbol{Q}$ and $\boldsymbol{R}$. Figure 2.6 shows the three Householder reflection matrices for this example.

Thus, the method of QR factorization of a given complex matrix $\boldsymbol{A}$ of dimension $M \times N$ with full column rank and $M > N$ produces the factors $\boldsymbol{A} = \boldsymbol{Q}\boldsymbol{R}$, where $\boldsymbol{Q}$ is $M \times M$ and unitary while $\boldsymbol{R}$ is an $M \times N$ and upper triangular matrix whose last $M - N$ rows are zero vectors. QR factorization is often used when solving full rank least squares problems: the least squares solution $\boldsymbol{x}_{\text{ls}}$ minimizes the l_2 norm $\|\boldsymbol{A}\boldsymbol{x} - \boldsymbol{b}\|_2^2$ and will be discussed in section 2.5.

$$\mathbf{A} = \begin{array}{ccc} 1.2 & 0.95 & 1.0 \\ 2.0 & -0.75 & 3.5 \\ -3.0 & 1.5 & -2.0 \\ 1.1 & 4.0 & 1.1 \\ 2.0 & -2.25 & 2.1 \end{array} \qquad \begin{array}{l} \mathbf{v}_1 = 0.797091 \;\; 0.283016 \;\; -0.424524 \;\; 0.155659 \;\; 0.283016 \\ \mathbf{v}_2 = 0.000000 \;\; -0.755733 \;\; 0.192336 \;\; 0.550122 \;\; -0.298731 \\ \mathbf{v}_3 = 0.000000 \;\; 0.000000 \;\; 0.924990 \;\; 0.355375 \;\; -0.134541 \end{array}$$

$$\mathbf{Q} = \begin{array}{ccccc} -0.270707 & 0.258314 & 0.107286 & -0.914029 & -0.114158 \\ -0.451179 & -0.050546 & -0.827700 & 0.062633 & -0.323840 \\ 0.676768 & 0.153134 & -0.548007 & -0.269169 & 0.381801 \\ -0.248148 & 0.881935 & -0.006247 & 0.287088 & 0.279560 \\ -0.451179 & -0.359805 & -0.055245 & -0.075868 & 0.811277 \end{array} \qquad \mathbf{R} = \begin{array}{ccc} -4.43283 & 1.11892 & -4.42381 \\ 0 & 4.85031 & -0.01033 \\ 0 & 0 & -1.81654 \\ 0 & 0 & 0 \\ 0 & 0 & 0 \end{array}$$

Figure 2.5: Example of QR factorization using Householder reflections.

$$\mathbf{H}_1 = \begin{array}{ccccc} -0.270707 & -0.451179 & 0.676768 & -0.248148 & -0.451179 \\ -0.451179 & 0.839804 & 0.240294 & -0.0881078 & -0.160196 \\ 0.676768 & 0.240294 & 0.639559 & 0.132162 & 0.240294 \\ -0.248148 & -0.0881078 & 0.132162 & 0.951541 & -0.0881078 \\ -0.451179 & -0.160196 & 0.240294 & -0.0881078 & 0.839804 \end{array} \qquad \mathbf{H}_2 = \begin{array}{ccccc} 1 & 0 & 0 & 0 & 0 \\ 0 & -0.142264 & 0.290710 & 0.831491 & -0.451522 \\ 0 & 0.290710 & 0.926013 & -0.211617 & 0.114914 \\ 0 & 0.831491 & -0.211617 & 0.394731 & 0.328677 \\ 0 & -0.451522 & 0.114914 & 0.328677 & 0.821519 \end{array}$$

$$\mathbf{H}_3 = \begin{array}{ccccc} 1 & 0 & 0 & 0 & 0 \\ 0 & 1 & 0 & 0 & 0 \\ 0 & 0 & -0.711214 & -0.657438 & 0.248899 \\ 0 & 0 & -0.657438 & 0.747417 & 0.0956253 \\ 0 & 0 & 0.248899 & 0.0956253 & 0.963797 \end{array}$$

Figure 2.6: Three Householder matrices for the example of figure 2.5.

2.5 QR Factorization and Full Rank Least Squares

Here, we show how QR factorization can be used to solve for the least squares solution of the equation $\boldsymbol{A}\boldsymbol{x} = \boldsymbol{b}$ with $\boldsymbol{A}$ an $M \times N$ complex matrix of full column rank N, $M \geq N$, and $\boldsymbol{b} \notin \mathscr{R}_{\boldsymbol{A}}$. We use the QR factors

$$\boldsymbol{A} = \boldsymbol{Q}\boldsymbol{R} = \boldsymbol{Q}\begin{bmatrix} \boldsymbol{R}' \\ \boldsymbol{0} \end{bmatrix}, \tag{2.9}$$

where $\boldsymbol{Q}$ is $M \times M$ and unitary, $\boldsymbol{R}'$ is $N \times N$, and $\boldsymbol{0}$ denotes a rectangular $(M-N) \times N$ zero matrix. Assuming the l_2 norm and using the unitarity of $\boldsymbol{Q}$ including the defining property of a unitary transformation $\|\boldsymbol{Q}\boldsymbol{x}\| = \|\boldsymbol{x}\|$,

$$\|\boldsymbol{A}\boldsymbol{x} - \boldsymbol{b}\|_2^2 = \|\boldsymbol{Q}\boldsymbol{R}\boldsymbol{x} - \boldsymbol{b}\|_2^2 = \left\|\boldsymbol{Q}\left(\boldsymbol{R}\boldsymbol{x} - \boldsymbol{Q}^{+}\boldsymbol{b}\right)\right\|_2^2 = \left\|\boldsymbol{R}\boldsymbol{x} - \boldsymbol{Q}^{+}\boldsymbol{b}\right\|_2^2. \tag{2.10}$$

The right hand side is the sum of two squared norms corresponding to the first N and the last $M - N$ components:

$$\|\boldsymbol{A}\boldsymbol{x} - \boldsymbol{b}\|_2^2 = \left\|\boldsymbol{R}'\boldsymbol{x} - \boldsymbol{c}'\right\|_2^2 + \|\boldsymbol{c}\|_2^2, \tag{2.11}$$

where $\boldsymbol{Q}^{+}\boldsymbol{b} \equiv [c'_1, \ldots, c'_N, c_1, \ldots, c_{M-N}]^T$, defining the vectors $\boldsymbol{c}'$ and $\boldsymbol{c}$. The least squares solution $\boldsymbol{x}_{\text{ls}}$ minimizing $\|\boldsymbol{R}'\boldsymbol{x} - \boldsymbol{c}'\|_2^2$ is given by the equation $\boldsymbol{R}'\boldsymbol{x} = \boldsymbol{c}'$ which is an upper triangular set of equations whose solution is found by back-substitution for $[\boldsymbol{R}']_{nn} \neq 0$, $1 \leq n \leq N$,

$$x_N = \frac{[\boldsymbol{c}']_N}{[\boldsymbol{R}']_{NN}}, \quad x_n = \frac{b_n - \sum\limits_{l=n+1}^{N} [\boldsymbol{R}']_{nl}\, x_l}{[\boldsymbol{R}']_{nn}}, \quad n = N-1, \ldots, 1. \tag{2.12}$$

2.6 Cholesky Factorization and Full Rank Least Squares

Another method to solve for the least squares solution is to use the defining equation $\boldsymbol{A}^+\boldsymbol{A}\boldsymbol{x} = \boldsymbol{A}^+\boldsymbol{b}$. Then we write the positive definite matrix $\boldsymbol{A}^+\boldsymbol{A}$ as the product of a lower triangular matrix $\boldsymbol{L}$ (the *Cholesky* factor) and its Hermitian conjugate (which is necessarily upper triangular) [22], i.e., we find a lower triangular matrix $\boldsymbol{L}$ with $[\boldsymbol{L}]_{ij} = 0$ when $j > i$ so that $\boldsymbol{A}^+\boldsymbol{A} = \boldsymbol{L}\boldsymbol{L}^+$. The equation $\boldsymbol{A}^+\boldsymbol{A}\boldsymbol{x} = \boldsymbol{A}^+\boldsymbol{b}$ is then solved by first solving the lower triangular system of equations $\boldsymbol{L}\boldsymbol{y} = \boldsymbol{A}^+\boldsymbol{b}$ for $\boldsymbol{y}$ using forward-substitution

$$y_1 = \frac{[\boldsymbol{A}^+\boldsymbol{b}]_1}{[\boldsymbol{L}]_{11}}, \quad y_n = \frac{[\boldsymbol{A}^+\boldsymbol{b}]_n - \sum_{l=1}^{n-1} [\boldsymbol{L}]_{nl}\, y_l}{[\boldsymbol{L}]_{nn}}, \quad n = 2, \ldots, N, \tag{2.13}$$

and then solving the upper triangular system of equations $\boldsymbol{L}^+\boldsymbol{x} = \boldsymbol{y}$ for $\boldsymbol{x}$ using back-substitution

$$x_N = \frac{y_N}{[\boldsymbol{L}]^*_{NN}}, \quad x_n = \frac{y_n - \sum_{l=n+1}^{N} [\boldsymbol{L}]^*_{ln}\, y_l}{[\boldsymbol{L}]^*_{nn}}, \quad n = N-1, \ldots, 1. \tag{2.14}$$

Cholesky factorization can be thought of as finding the square root of a positive definite $N \times N$ Hermitian matrix $\boldsymbol{C}$; the simplest method to find the Cholesky factors is to write $\boldsymbol{C} = [\boldsymbol{c}_1, \ldots, \boldsymbol{c}_N]$ and $\boldsymbol{L} = [\boldsymbol{l}_1, \ldots, \boldsymbol{l}_N]$, after which the Cholesky factorization equation is equivalent to the following N equations for the columns of $\boldsymbol{L}$:

$$\boldsymbol{c}_1 = [\boldsymbol{L}]^*_{11}\boldsymbol{l}_1,\ \boldsymbol{c}_2 = [\boldsymbol{L}]^*_{21}\boldsymbol{l}_1 + [\boldsymbol{L}]^*_{22}\boldsymbol{l}_2, \ldots, \boldsymbol{c}_N = [\boldsymbol{L}]^*_{N1}\boldsymbol{l}_1 + \cdots + [\boldsymbol{L}]^*_{NN}\boldsymbol{l}_N. \tag{2.15}$$

First, $[\boldsymbol{L}]_{11} = \sqrt{[\boldsymbol{C}]_{11}}$ and $\boldsymbol{l}_1 = [\boldsymbol{L}]^{-1}_{11}\boldsymbol{c}_1$. Next, for $2 \le n \le N$ we have

$$[\boldsymbol{L}]_{nn} = \left([\boldsymbol{C}]_{nn} - \sum_{m=1}^{n-1} |[\boldsymbol{L}]_{nm}|^2\right)^{1/2}, \quad \boldsymbol{l}_n = [\boldsymbol{L}]^{-1}_{nn}\left(\boldsymbol{c}_n - \sum_{m=1}^{n-1} \boldsymbol{l}_m [\boldsymbol{L}^*]_{nm}\right). \tag{2.16}$$

A recursive method to find the Cholesky factors is based on the identity

$$\boldsymbol{C} \equiv \begin{bmatrix} a & \boldsymbol{u}^+ \\ \boldsymbol{u} & \boldsymbol{T} \end{bmatrix} = \begin{bmatrix} \sqrt{a} & \boldsymbol{0} \\ \boldsymbol{u}/\sqrt{a} & \boldsymbol{I}_{N-1} \end{bmatrix} \begin{bmatrix} 1 & \boldsymbol{0} \\ \boldsymbol{0} & \boldsymbol{C}' \end{bmatrix} \begin{bmatrix} \sqrt{a} & \boldsymbol{u}^+/\sqrt{a} \\ \boldsymbol{0} & \boldsymbol{I}_{N-1} \end{bmatrix}, \tag{2.17}$$

where $a = [\boldsymbol{C}]_{11}$, $\boldsymbol{u}$ is an $(N-1) \times 1$ vector composed of the second through the last elements of the first column of $\boldsymbol{C}$, i.e., $[\boldsymbol{C}]_{21}, \ldots, [\boldsymbol{C}_{N1}]$, $\boldsymbol{T}$ is the $(N-1)\times(N-1)$ sub-matrix of $\boldsymbol{C}$ starting with element $[\boldsymbol{C}]_{22}$ and ending with element $[\boldsymbol{C}]_{NN}$, and $\boldsymbol{C}' \equiv \boldsymbol{T} - \boldsymbol{u}\boldsymbol{u}^+/a$. If the Cholesky factors for the $(N-1) \times (N-1)$ Hermitian and positive definite matrix $\boldsymbol{C}'$ are known, i.e., $\boldsymbol{C}' = \boldsymbol{L}'\boldsymbol{L}'^+$, then equation (2.17) can be written as

$$\boldsymbol{C} \equiv \begin{bmatrix} \sqrt{a} & \boldsymbol{0} \\ \boldsymbol{u}/\sqrt{a} & \boldsymbol{L}' \end{bmatrix} \begin{bmatrix} \sqrt{a} & \boldsymbol{u}^+/\sqrt{a} \\ \boldsymbol{0} & \boldsymbol{L}'^+ \end{bmatrix} \equiv \boldsymbol{L}\boldsymbol{L}^+. \tag{2.18}$$

Thus, we can construct the Cholesky factors of the original $N \times N$ matrix $\boldsymbol{C}$ in terms of the Cholesky factors of the $(N-1) \times (N-1)$ matrix $\boldsymbol{C}'$.

An example of Cholesky factorization (using the recursive algorithm) for a real, symmetric, and positive definite 5×5 matrix is illustrated in figure 2.7, which shows the original matrix $\boldsymbol{C}$, the final Cholesky factor $\boldsymbol{L}$, and all four recursively calculated lower triangular matrices before the final result.

2.18611

2.39125	0
1.71280	2.18611

1.73575	0	0
0.812478	2.39125	0
1.03406	1.71280	2.18611

3.73210	0	0	0
1.53112	1.73575	0.000000	0
4.21057	0.812478	2.39125	0
5.35891	1.03406	1.71280	2.18611

C =

3.5	4.0	2.0	5.5	7.0
4.0	18.5	8.0	22.0	28.0
2.0	8.0	6.5	11.0	14.0
5.5	22.0	11.0	32.75	38.5
7.0	28.0	14.0	38.5	51.5

L =

1.87083	0	0	0	0
2.13809	3.73210	0	0	0
1.06904	1.53112	1.73575	0	0
2.93987	4.21057	0.812478	2.39125	0
3.74166	5.35891	1.03406	1.71280	2.1861

Figure 2.7: Cholesky factorization example.

Cholesky factorization of a Hermitian and positive definite matrix $\boldsymbol{C} = \boldsymbol{L}\boldsymbol{L}^+$ is sometimes written in an equivalent form $\boldsymbol{C} = \boldsymbol{L}_1\boldsymbol{D}\boldsymbol{L}_1^+$ where the lower triangular matrix $\boldsymbol{L}_1$ has 1s along its diagonal, and $\boldsymbol{D}$ is a diagonal matrix with real and positive numbers along the diagonal. The two forms are related to each other by $\boldsymbol{L} = \boldsymbol{L}_1\boldsymbol{D}^{1/2}$. Upper triangular Cholesky factors are not as prevalent in the literature but are defined analogously to lower triangular factors discussed here. The upper triangular forms are $\boldsymbol{C} = \boldsymbol{U}\boldsymbol{U}^+ = \boldsymbol{U}_1\boldsymbol{D}\boldsymbol{U}_1^+$, where $\boldsymbol{U}$ is an upper triangular matrix while $\boldsymbol{U}_1$ is upper triangular with 1s along its diagonal, and $\boldsymbol{D}$ is a diagonal matrix whose diagonal entries are the squares of the diagonal elements of $\boldsymbol{U}$, and $\boldsymbol{U} = \boldsymbol{U}_1\boldsymbol{D}^{1/2}$.

Cholesky factorization has many applications in efficient numerical solutions, e.g., adaptive filter implementations that often involve Hermitian and positive definite covariance or correlation matrices (see section (section 10.7), and in Monte Carlo simulations of systems with multiple correlation variables. Cholesky factorization is also used to generate stationary random signals with a specified covariance or correlation matrix (we will discuss random vectors, covariance and correlation matrices in chapters 5 and 6). The "white noise" random vector $\boldsymbol{\nu}$ with zero mean and unit variance has the identity matrix $\boldsymbol{I}$ as its correlation matrix. The Cholesky factor $\boldsymbol{L}$ of a specified correlation matrix $\boldsymbol{R}$ can be used to construct a new zero-mean random vector $\boldsymbol{y} = \boldsymbol{L}\boldsymbol{\nu}$ whose correlation matrix is $\boldsymbol{R}$.

2.7 Singular Value Decomposition (SVD)

When calculating the minimum norm or the least squares solutions to the equation $\boldsymbol{A}\boldsymbol{x} = \boldsymbol{b}$ using the generalized inverses $\boldsymbol{A}_R^\dagger$ or $\boldsymbol{A}_L^\dagger$, we must be aware that the Hermitian matrices $\boldsymbol{A}\boldsymbol{A}^+$ or $\boldsymbol{A}^+\boldsymbol{A}$ may be nearly singular in machine precision, in which case we refer to $\boldsymbol{A}$ as ill-conditioned. The condition number κ_A of a rectangular $M \times N$ matrix $\boldsymbol{A}$ can be defined as the ratio $\sigma_{\max}/\sigma_{\min}$, where σ_k^2 are the eigenvalues of the square matrix $\boldsymbol{A}^+\boldsymbol{A}$; this definition corresponds to using the l_2 norm of a matrix in the definition (1.30). When calculating the minimum norm or least squares solution, the relevant condition number is $\kappa_{\boldsymbol{A}}^2$. Singular value decomposition (SVD) of a matrix is perhaps the most important tool in numerical matrix algebra to deal with any problem involving ill-conditioned matrices [6, 22]. In addition, SVD is useful in image compression [7].

Theorem 12. *Every rectangular $M \times N$ complex matrix $\boldsymbol{A}$ can be factored as*

$$\boldsymbol{A} = \boldsymbol{U}\boldsymbol{\Sigma}\boldsymbol{V}^+, \tag{2.19}$$

where $\boldsymbol{U}$ is $M \times M$, $\boldsymbol{V}$ is $N \times N$, and both are complex unitary matrices, and $\boldsymbol{\Sigma}$ is a real and diagonal $M \times N$ matrix whose diagonal elements are (usually sorted) non-negative singular values of $\boldsymbol{A}$, namely, $\{\sigma_1 \geq \sigma_2 \geq \ldots \sigma_s \geq 0\}$ where $s = \min(M, N)$, as illustrated in figure 2.8. The columns of $\boldsymbol{U}$ are the

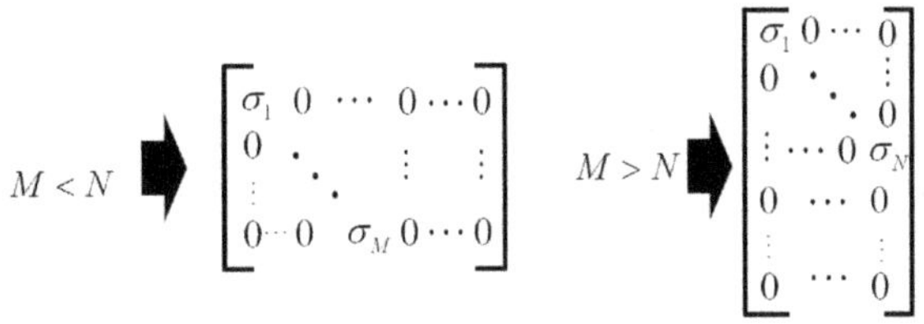

Figure 2.8: The diagonal matrix $\boldsymbol{\Sigma}$ in the SVD $\boldsymbol{U\Sigma V^+}$.

eigenvectors of $\boldsymbol{AA}^+$ *while the columns of* $\boldsymbol{V}$ *are the eigenvectors of* $\boldsymbol{A}^+\boldsymbol{A}$

$$\boldsymbol{AA}^+\boldsymbol{u}_m = \sigma_m^2 \boldsymbol{u}_m,\ 1 \le m \le M,\ \ \boldsymbol{A}^+\boldsymbol{A}\boldsymbol{v}_n = \sigma_n^2 \boldsymbol{v}_n,\ 1 \le n \le N. \tag{2.20}$$

Note that both $\boldsymbol{AA}^+$ and $\boldsymbol{A}^+\boldsymbol{A}$ are Hermitian and non-negative definite matrices and, therefore, have real and non-negative eigenvalues [1] as indicated by the notation σ^2. We can write the first set of M eigenvalue equations as a single matrix equation

$$\boldsymbol{AA}^+\left[\boldsymbol{u}_1,\ldots,\boldsymbol{u}_M\right] = \left[\sigma_1^2\boldsymbol{u}_1,\ldots,\sigma_M^2\boldsymbol{u}_M\right] = \boldsymbol{U}\begin{bmatrix} \sigma_1^2 & 0 & \cdots & 0 \\ 0 & \sigma_2^2 & \ddots & \vdots \\ \vdots & \ddots & \ddots & 0 \\ 0 & \cdots & 0 & \sigma_M^2 \end{bmatrix}. \tag{2.21}$$

A similar procedure can be applied to the second set of N eigenvalue equations to obtain

$$\boldsymbol{A}^+\boldsymbol{A}\left[\boldsymbol{v}_1,\ldots,\boldsymbol{v}_N\right] = \begin{bmatrix} \sigma_1^2 & 0 & \cdots & 0 \\ 0 & \sigma_2^2 & \ddots & \vdots \\ \vdots & \ddots & \ddots & 0 \\ 0 & \cdots & 0 & \sigma_N^2 \end{bmatrix}\boldsymbol{V}. \tag{2.22}$$

Both equations (2.21) and (2.22) can be written as

$$\boldsymbol{AA}^+\boldsymbol{U} = \boldsymbol{U\Sigma\Sigma}^T,\ \text{ and }\ \boldsymbol{A}^+\boldsymbol{AV} = \boldsymbol{V\Sigma}^T\boldsymbol{\Sigma}, \tag{2.23}$$

where $\boldsymbol{\Sigma}^T\boldsymbol{\Sigma}$ is an $N \times N$ diagonal matrix with diagonal entries $\{\sigma_1^2,\ldots,\sigma_N^2\}$, while $\boldsymbol{\Sigma\Sigma}^T$ is an $M \times M$ diagonal matrix with diagonal entries $\{\sigma_1^2,\ldots,\sigma_M^2\}$. The number of non-zero singular values is $L = \min(M,N)$, and the SVD can be written in terms of the individual singular values and eigenvectors,

$$\boldsymbol{A} = \sum_{l=1}^{L} \sigma_l \boldsymbol{u}_l \boldsymbol{v}_l^{\,+},\ \ \boldsymbol{U} = \left[\boldsymbol{u}_1,\ldots,\boldsymbol{u}_M\right],\ \boldsymbol{V} = \left[\boldsymbol{v}_1,\ldots,\boldsymbol{v}_N\right]. \tag{2.24}$$

The Frobenius and the l_2 norms of an $M \times N$ matrix can be written in terms of the singular values

$$\|\boldsymbol{A}\|_F^2 = \sum_{l=1}^{L}\sigma_l^2,\ \ L = \min(M,N),\ \text{ and }\ \|\boldsymbol{A}\|_2^2 = \sigma_1^2. \tag{2.25}$$

[1] If $\boldsymbol{H}$ is a square Hermitian matrix, then the eigenvalues λ corresponding to eigenvectors $\boldsymbol{x}_\lambda$ are real. Assume that the eigenvectors have unit length; then $\lambda = \langle \boldsymbol{x}_\lambda, \boldsymbol{H}x_\lambda\rangle$ and $\lambda^* = \langle \boldsymbol{Hx}_\lambda, \boldsymbol{x}_\lambda\rangle$. But since $\boldsymbol{H}$ is Hermitian, we have $\langle \boldsymbol{x}_\lambda, \boldsymbol{Hx}_\lambda\rangle = \langle \boldsymbol{Hx}_\lambda, \boldsymbol{x}_\lambda\rangle$, and so λ is real. Now consider the matrix $\boldsymbol{A}^+\boldsymbol{A}$, which is Hermitian. Then its eigenvalues (assumed normalized) are given by $\lambda = \langle \boldsymbol{x}_\lambda, \boldsymbol{A}^+\boldsymbol{Ax}_\lambda\rangle = \|\boldsymbol{Ax}_\lambda\|^2$, which is a non-negative quantity. Similarly, the eigenvalues of $\boldsymbol{AA}^+$ are non-negative.

2.8 SVD and Reduced Rank Approximation

If $\boldsymbol{A}$ is a square non-singular matrix, in which case $L = M = N$, its inverse is given by $\boldsymbol{A}^{-1} = \boldsymbol{V}\Sigma^{-1}\boldsymbol{U}^{+}$ (using the unitarity of $\boldsymbol{U}$ and $\boldsymbol{V}$), and so we use the following formula for any rectangular matrix:

$$\boldsymbol{A}^{-1} = \sum_{l=1}^{L} \sigma_l^{-1}\, \boldsymbol{v}_l \boldsymbol{u}_l^{+}. \tag{2.26}$$

This formula is the basis for the treatment of ill-conditioned matrices by the method of rank reduction. We observe that the true rank of a matrix $\boldsymbol{A}$ is the number of its non-zero singular values. From a practical computational point of view, however, we define the numerical rank of a matrix as the number of its singular values that significantly exceed 0 as represented by machine precision. For instance, consider the following 2×2 matrix and its SVD factors:

$$\boldsymbol{A} = \frac{1}{2\sqrt{13}} \begin{bmatrix} 1+5\varepsilon & 5-\varepsilon \\ 1-5\varepsilon & 5+\varepsilon \end{bmatrix} = \begin{bmatrix} 1/\sqrt{2} & -1/\sqrt{2} \\ 1/\sqrt{2} & 1/\sqrt{2} \end{bmatrix} \begin{bmatrix} 1 & 0 \\ 0 & \varepsilon \end{bmatrix} \begin{bmatrix} 1/\sqrt{26} & 5/\sqrt{26} \\ -5/\sqrt{26} & 1/\sqrt{26} \end{bmatrix}$$

The condition number is $\kappa_A = \varepsilon^{-1}$, and the inverse $\boldsymbol{A}^{-1}$ is

$$\boldsymbol{A}^{-1} = \frac{1}{2\sqrt{13}} \begin{bmatrix} 1+5/\varepsilon & 1-5/\varepsilon \\ 5-1/\varepsilon & 5+1/\varepsilon \end{bmatrix} = \begin{bmatrix} 1/\sqrt{26} & -5/\sqrt{26} \\ 5/\sqrt{26} & 1/\sqrt{26} \end{bmatrix} \begin{bmatrix} 1 & 0 \\ 0 & 1/\varepsilon \end{bmatrix} \begin{bmatrix} 1/\sqrt{2} & 1/\sqrt{2} \\ -1/\sqrt{2} & 1/\sqrt{2} \end{bmatrix}.$$

The vector $\boldsymbol{v}_l$ associated with an extremely small singular value is known as a *sensitive direction*. In the preceding example, there is only one sensitive direction, namely, the second column of $\boldsymbol{V}$, as can be seen below:

$$\boldsymbol{V}\Sigma^{-1} = \begin{bmatrix} 1/\sqrt{26} & -5/\sqrt{26} \\ 5/\sqrt{26} & 1/\sqrt{26} \end{bmatrix} \begin{bmatrix} 1 & 0 \\ 0 & 1/\varepsilon \end{bmatrix}.$$

A sensitive direction $\boldsymbol{v}_l$ can be neglected only if $\boldsymbol{u}_l^{+}\boldsymbol{b} = 0$, in the absence of which a general procedure to avoid small singular values is to simply reduce the numeric rank of the original nearly singular matrix by removing the small singular values. For instance, keeping the largest $K < L$ singular values, we terminate the sum in (2.26) at K. Thus, we have the generalized inverse or the SVD pseudo-inverse

$$\boldsymbol{A}^{\dagger} = \boldsymbol{V}\Sigma^{\dagger}\boldsymbol{U}^{+} = \sum_{k=1}^{K} \sigma_k^{-1}\, \boldsymbol{v}_k \boldsymbol{u}_k^{+}, \tag{2.27}$$

where $\Sigma^{\dagger}$ is a diagonal matrix with only K non-zero diagonal entries $1/\sigma_k$, $1 \le k \le K$. The reduced rank matrix is given by a similar formula

$$\tilde{\boldsymbol{A}} = \boldsymbol{U}\tilde{\boldsymbol{\Sigma}}\boldsymbol{V}^{+} = \sum_{k=1}^{K} \sigma_k\, \boldsymbol{u}_k \boldsymbol{v}_k^{\;+}, \tag{2.28}$$

where $\tilde{\boldsymbol{\Sigma}}$ is a diagonal matrix with only K non-zero diagonal entries σ_k, $1 \le k \le K$; this is known as the Eckart-Young theorem [24, 25]. In the above example, we find the following rank reduced approximation to the nearly singular 2×2 matrix $\boldsymbol{A}$,

$$\tilde{\boldsymbol{A}} = \frac{1}{2\sqrt{13}} \begin{bmatrix} 1 & 5 \\ 1 & 5 \end{bmatrix},$$

which clearly has rank 1; i.e., from a practical point of view the two columns of the original 2×2 matrix are linearly dependent. Theorem 13 justifies the rank reduction procedure expressed in (2.28).

Theorem 13. *If the $M \times N$ complex matrix $\boldsymbol{A}$ has rank L and SVD*

$$\boldsymbol{A} = \sum_{l=1}^{L} \sigma_l \boldsymbol{u}_l \boldsymbol{v}_l^{+}, \tag{2.29}$$

then for any $K < L$ the minimum l_2 distance of $\boldsymbol{A}$ to the set of all $M \times N$ matrices of rank K is given by σ_{K+1}, i.e., for all complex $M \times N$ matrices $\boldsymbol{B}_K$ of rank K the quantity $\|\boldsymbol{A} - \boldsymbol{B}_K\|_2$ is minimized when $\boldsymbol{B}_K = \boldsymbol{A}_K$ where $\boldsymbol{A}_K$ is the rank K approximation to $\boldsymbol{A}$:

$$\boldsymbol{A}_K = \sum_{k=1}^{K} \sigma_k \boldsymbol{u}_k \boldsymbol{v}_k^+. \tag{2.30}$$

To show this, we first prove $\|(\boldsymbol{A} - \boldsymbol{A}_K)\|_2^2 = \sigma_{K+1}^2$. Let $\boldsymbol{w}$ be the first singular vector of the matrix $\boldsymbol{A} - \boldsymbol{A}_K$, which can be written as a linear combination of the singular vectors $\boldsymbol{v}_l$ in the form $\boldsymbol{w} = \sum_{l=1}^{L} c_l \boldsymbol{v}_l$. Thus,

$$\|(\boldsymbol{A} - \boldsymbol{A}_{\boldsymbol{K}})\boldsymbol{w}\|_2^2 = \left\| \sum_{l=K+1}^{L} \sigma_l \boldsymbol{u}_l \boldsymbol{v}_l^+ \sum_{k=1}^{L} c_k \boldsymbol{v}_k \right\|_2^2 = \sum_{l=K+1}^{L} |c_l|^2 \sigma_l^2,$$

where we have used the orthonormality of the singular vectors $\boldsymbol{v}_l$. The minimum of this quantity subject to the constraint $\|\boldsymbol{w}\|_2^2 = \sum_{l=1}^{L} |c_l|^2 = 1$ occurs for $c_l = \delta_{l,K+1}$, and this leads to $\|\boldsymbol{A} - \boldsymbol{A}_K\|_2^2 = \sigma_{K+1}^2$. Next, the theorem is clearly true if the rank of $\boldsymbol{A}$ is $\leq K$. So let us assume that its rank exceeds K and that a matrix $\boldsymbol{B}_K$ of rank $\leq K$ exists that is closer to $\boldsymbol{A}$ (in the l_2 norm) than $\boldsymbol{A}_K$, i.e., $\|(\boldsymbol{A} - \boldsymbol{B}_K)\|_2^2 < \|(\boldsymbol{A} - \boldsymbol{A}_K)\|_2^2 = \sigma_{K+1}^2$. The dimension of $\mathscr{N}_{\boldsymbol{B}_K}$ is at least $N - K$, and so there exists a non-zero vector $\boldsymbol{w}$ that we take to have unit norm and that belongs both to **span** $\{\boldsymbol{v}_1, \ldots, \boldsymbol{v}_{K+1}\}$ and $\mathscr{N}_{\boldsymbol{B}_K}$. Then,

$$\|(\boldsymbol{A} - \boldsymbol{B}_K)\|_2^2 \geq \|(\boldsymbol{A} - \boldsymbol{B}_K)\, w\|_2^2 = \|\boldsymbol{A}\boldsymbol{w}\|_2^2,$$

since $\boldsymbol{w} \in \mathscr{N}_{\boldsymbol{B}_K}$ and $\boldsymbol{B}_K \boldsymbol{w} = \boldsymbol{0}$. Thus,

$$\|\boldsymbol{A}\boldsymbol{w}\|_2^2 = \left\| \sum_{l=1}^{L} \sigma_l \boldsymbol{u}_l \boldsymbol{v}_l^+ \boldsymbol{w} \right\|_2^2 = \sum_{l=1}^{K+1} \sigma_l^2 |\boldsymbol{v}_l^+ \boldsymbol{w}|^2 \geq \sigma_{K+1}^2 \sum_{l=1}^{K+1} |\boldsymbol{v}_l^+ \boldsymbol{w}|^2 = \sigma_{K+1}^2,$$

which, of course, contradicts the assumption that $\|(\boldsymbol{A} - \boldsymbol{B}_K)\|_2^2 < \sigma_{K+1}^2$, and so the theorem is proven. Theorem 13 also holds for the Frobenius norm.

2.9 SVD and Matrix Subspaces

The structure of the diagonal matrix $\boldsymbol{\Sigma}$ in the SVD of an $M \times N$ matrix $\boldsymbol{A}$ was shown in figure 2.8. As noted earlier the maximum number of possible non-zero singular values is given by $L = \min(M, N)$. Here we assume that the actual number of non-zero singular values is $K < L$ and partition the SVD matrix factors

$$\boldsymbol{A} = [\boldsymbol{U}_K, \boldsymbol{U}_0] \begin{bmatrix} \boldsymbol{\Sigma}_K & \boldsymbol{0} \\ \boldsymbol{0} & \boldsymbol{\Sigma}_0 \end{bmatrix} \begin{bmatrix} \boldsymbol{V}_K^+ \\ \boldsymbol{V}_0^+ \end{bmatrix}, \tag{2.31}$$

where $\boldsymbol{\Sigma}_K$ is a $K \times K$ diagonal matrix whose diagonal entries are the K non-zero singular values $\{\sigma_1, \ldots, \sigma_K\}$, and $\boldsymbol{\Sigma}_0$ is an $(M - K) \times (N - K)$ null matrix, as depicted in figure 2.9. The SVD is

$$\boldsymbol{A} = \boldsymbol{U}_K \boldsymbol{\Sigma}_K \boldsymbol{V}_K^+ = \sum_{k=1}^{K} \sigma_k\, \boldsymbol{u}_k \boldsymbol{v}_k^+. \tag{2.32}$$

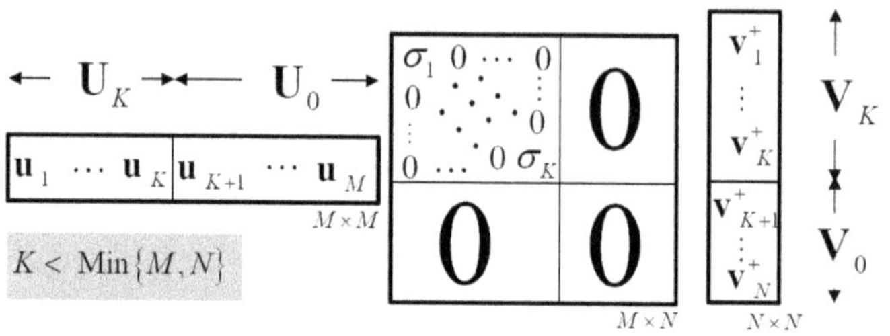

Figure 2.9: SVD of an $M \times N$ matrix with $K < \min(M, N)$ non-zero singular values.

The range space of $\boldsymbol{A}$ is the set of all vectors of the form $\boldsymbol{Ab}$. Using (2.32), we have

$$\boldsymbol{Ab} = \sum_{k=1}^{K} \sigma_l \left(\boldsymbol{v}_k^{+}\boldsymbol{b}\right) \boldsymbol{u}_k, \tag{2.33}$$

which is a linear combination of the K columns of $\boldsymbol{U}_K$. Thus,

$$\mathscr{R}_{\boldsymbol{A}} = \textbf{span}\left\{\boldsymbol{u}_1, \ldots, \boldsymbol{u}_K\right\} = \mathscr{R}_{\boldsymbol{U}_K}. \tag{2.34}$$

For the null space $\mathscr{N}_{\boldsymbol{A}}$, we observe that $\boldsymbol{Aw} = \boldsymbol{0}$ is equivalent to $\boldsymbol{U\Sigma V}^{+}\, \boldsymbol{w} = \boldsymbol{0}$, which on multiplying on the left by $\boldsymbol{U}^{+}$ and using the unitarity of $\boldsymbol{U}$ is equivalent to $\boldsymbol{\Sigma V}^{+}\, \boldsymbol{w} = \boldsymbol{0}$. Denoting the first K elements of $\boldsymbol{w}$ by $\boldsymbol{w}_K$, we arrive at $\boldsymbol{\Sigma}_K \boldsymbol{V}_K^{+} \boldsymbol{w}_K = \boldsymbol{0}$. The last equation is always satisfied by choosing $\boldsymbol{w}_K = \boldsymbol{0}$. So if $\boldsymbol{w} = \left[\boldsymbol{0}_K, w_{K+1}, \ldots, w_N\right]$, then $\boldsymbol{Aw} = \boldsymbol{0}$ if, and only if, $\boldsymbol{\Sigma}_0 \boldsymbol{V}_0^{+} \boldsymbol{w}_0 = \boldsymbol{0}$, where $\boldsymbol{w}_0 \equiv \left[\boldsymbol{0}_K, w_{K+1}, \ldots, w_N\right]$, and this equation is always valid, since $\boldsymbol{\Sigma}_0$ is a null matrix. $\boldsymbol{V}_0^{+} \boldsymbol{w}_0$ can be considered to be a linear combination of the columns of $\boldsymbol{V}_0$, and so we find

$$\mathscr{N}_{\boldsymbol{A}} = \textbf{span}\left\{\boldsymbol{v}_{K+1}, \ldots, \boldsymbol{v}_N\right\} = \mathscr{R}_{\boldsymbol{V}_0}. \tag{2.35}$$

Using similar arguments we find

$$\mathscr{R}_{\boldsymbol{A}+} = \textbf{span}\left\{\boldsymbol{v}_1, \ldots, \boldsymbol{v}_K\right\} = \mathscr{R}_{\boldsymbol{V}_K}, \quad \mathscr{N}_{\boldsymbol{A}+} = \textbf{span}\left\{\boldsymbol{u}_{K+1}, \ldots, \boldsymbol{u}_M\right\} = \mathscr{R}_{\boldsymbol{U}_0}.$$

Thus, the range and null spaces of the matrices $\boldsymbol{A}$ and $\boldsymbol{A}^{+}$ are given by the column spaces of various partitions of the $\boldsymbol{U}$ and $\boldsymbol{V}$ matrices in the SVD factorization of the original matrix $\boldsymbol{A}$.

Consider the least squares minimization of $\|\boldsymbol{Ax} - \boldsymbol{b}\|_2^2$. Using the SVD factors, the quantity to minimize is $\|\boldsymbol{U\Sigma V}^{+}\boldsymbol{x} - \boldsymbol{b}\|_2^2$, which on account of the unitarity of the matrix $\boldsymbol{U}$ is equivalent to minimizing $\|\boldsymbol{\Sigma V}^{+}\boldsymbol{x} - \boldsymbol{U}^{+}\boldsymbol{b}\|_2^2$. Defining $\boldsymbol{y} \equiv \boldsymbol{V}^{+}\boldsymbol{x}$ and $\boldsymbol{c} \equiv \boldsymbol{U}^{+}\boldsymbol{b}$, we are led to minimize $\|\boldsymbol{\Sigma y} - \boldsymbol{c}\|_2^2$: the minimization is now trivial since $\boldsymbol{\Sigma}$ is diagonal, and the solution is $\boldsymbol{y}_{\min} = \boldsymbol{\Sigma}^{\dagger}\boldsymbol{c}$. Thus, $\boldsymbol{x}_{\min} = \boldsymbol{V\Sigma}^{\dagger}\boldsymbol{U}^{+}\boldsymbol{b}$. This is, of course, the same solution as that given by the SVD pseudo-inverse $\boldsymbol{A}^{\dagger} = \boldsymbol{V\Sigma}^{\dagger}\boldsymbol{U}^{+}$ and discussed in section 2.8.

In light of the relation between the SVD eigenvectors and matrix subspaces associated with $\boldsymbol{A}$ and $\boldsymbol{A}^{+}$, we can summarize the SVD pseudo-inverse solution to the least squares problem, namely, $\boldsymbol{x}_{\min} = \boldsymbol{V\Sigma}^{\dagger}\boldsymbol{U}^{+}\boldsymbol{b}$, as follows (assuming a complex $M \times N$ matrix $\boldsymbol{A}$ with K non-zero singular values and $K < \min(M, N)$):

- The vector $\boldsymbol{U}^{+}\boldsymbol{b}$ is a linear combination of the columns of $\boldsymbol{U}$. Since the first K columns of $\boldsymbol{U}$ span $\mathscr{R}_{\boldsymbol{A}}$ while the last $M - K$ columns span $\mathscr{N}_{\boldsymbol{A}+}$, we have $\boldsymbol{U}^{+}\boldsymbol{b} = \boldsymbol{b}_K + \boldsymbol{b}_0$ where $\boldsymbol{b}_K \in \mathscr{R}_{\boldsymbol{A}}$ and $\boldsymbol{b}_0 \in \mathscr{N}_{\boldsymbol{A}+}$.

- Multiplication on the left by $\boldsymbol{\Sigma}^\dagger$ scales $\boldsymbol{b}_K$ and sets $\boldsymbol{b}_0$ to zero.
- Final multiplication on the left by $\boldsymbol{V}$ projects the result onto $\mathscr{R}_{\boldsymbol{A}^+}$ (the column space of $\boldsymbol{A}^+$ or the row space of $\boldsymbol{A}$).

2.10 SVD: Full Rank Least Squares and Minimum Norm Solutions

When solving $\boldsymbol{Ax} = \boldsymbol{b}$ for a rectangular matrix $\boldsymbol{A}$ with $\boldsymbol{b} \notin \mathscr{R}_{\boldsymbol{A}}$, the optimal solution is found through the minimization of $\|Ax - b\|_2^2$. We have identified two cases of interest: the full rank (over-determined) least squares solution given by $(\boldsymbol{A}^+\boldsymbol{A})^{-1}\boldsymbol{A}^+\boldsymbol{b}$, and the under-determined minimum norm solution given by $\boldsymbol{A}^+(\boldsymbol{AA}^+)^{-1}\boldsymbol{b}$. The SVD pseudo-inverse gives both solutions under appropriate conditions, as illustrated by the following simple examples.

Consider the over-determined set of equations

$$\boldsymbol{Ax} = \boldsymbol{b}: \quad \begin{bmatrix} \sigma_1 & 0 \\ 0 & \sigma_2 \\ 0 & 0 \end{bmatrix} \begin{bmatrix} x_1 \\ x_2 \end{bmatrix} = \begin{bmatrix} b_1 \\ b_2 \\ b_3 \end{bmatrix}.$$

$\boldsymbol{A}$ is already in its SVD form ($\boldsymbol{U}$ is the 3×3 identity matrix and $\boldsymbol{V}$ is the 2×2 identity matrix). Clearly, there is no solution unless $b_3 = 0$ (since the range space $\mathscr{R}_{\boldsymbol{A}}$ is spanned by two vectors $[1,0,0]^T$ and $[0,1,0]^T$). The null space $\mathscr{N}_{\boldsymbol{A}}$ consists only of the zero vector $[0,0,0]^T$. Setting $b_3 = 0$, we find the least squares solution $\boldsymbol{x}_{\text{ls}} = [b_1/\sigma_1, b_2/\sigma_2]^T$, which is precisely the solution provided by the SVD pseudo-inverse

$$\boldsymbol{x}_{\text{ls}} = \frac{1}{\sigma_1}\begin{bmatrix} 1 \\ 0 \end{bmatrix}\begin{bmatrix} 1 & 0 & 0 \end{bmatrix}\begin{bmatrix} b_1 \\ b_2 \\ b_3 \end{bmatrix} + \frac{1}{\sigma_2}\begin{bmatrix} 0 \\ 1 \end{bmatrix}\begin{bmatrix} 0 & 1 & 0 \end{bmatrix}\begin{bmatrix} b_1 \\ b_2 \\ b_3 \end{bmatrix}.$$

Next, consider the under-determined set of equations

$$\boldsymbol{Ax} = \boldsymbol{b}: \quad \begin{bmatrix} \sigma_1 & 0 & 0 \\ 0 & \sigma_2 & 0 \end{bmatrix} \begin{bmatrix} x_1 \\ x_2 \\ x_3 \end{bmatrix} = \begin{bmatrix} b_1 \\ b_2 \end{bmatrix}.$$

$\boldsymbol{A}$ is again in its SVD form ($\boldsymbol{U}$ is the 2×2 identity matrix and $\boldsymbol{V}$ is the 3×3 identity matrix). The null space $\mathscr{N}_{\boldsymbol{A}}$ is now non-trivial and is spanned by the vector $[0,0,1]^T$. The general solution is given by $[b_1/\sigma_1, b_2/\sigma_2, 0]^T + x_3[0,0,1]^T$ for arbitrary $x_3 \in \mathbb{R}$. The minimum norm solution is given by $[b_1/\sigma_1, b_2/\sigma_2, 0]^T$, which is precisely the solution given by the SVD pseudo-inverse

$$\boldsymbol{x}_{mn} = \frac{1}{\sigma_1}\begin{bmatrix} 1 \\ 0 \\ 0 \end{bmatrix}\begin{bmatrix} 1 & 0 \end{bmatrix}\begin{bmatrix} b_1 \\ b_2 \end{bmatrix} + \frac{1}{\sigma_2}\begin{bmatrix} 0 \\ 1 \\ 0 \end{bmatrix}\begin{bmatrix} 0 & 1 \end{bmatrix}\begin{bmatrix} b_1 \\ b_2 \end{bmatrix}.$$

Thus, the SVD produces the full rank least squares solution of an over-determined set of equations and the minimum norm solution of an under-determined set.

2.11 Total Least Squares

Although the name *total least squares* is relatively recent, the method has a long history in statistical sciences, where it is known as *orthogonal regression*, a special case of which was discussed in 1877 [26]! Applications of the total least squares problem include deconvolution and linear prediction.

The ordinary least squares solution minimizes the L_2 norm $\|\boldsymbol{Ax} - \boldsymbol{b}\|_2^2$ by projecting $\boldsymbol{b}$ onto the range space of the matrix $\boldsymbol{A}$, assuming that only $\boldsymbol{b}$ is subject to observation noise. In many problems, however, the elements of the matrix $\boldsymbol{A}$ can include observed data values that are necessarily subject to observation noise, too. For instance, when fitting a line $y_n = ax_n + b_n$ to a set of measured pairs (x_n, y_n), we have the least squares problem

$$\begin{bmatrix} x_0 & 1 \\ \vdots & \vdots \\ x_{N-1} & 1 \end{bmatrix} \begin{bmatrix} a \\ b \end{bmatrix} \approx \begin{bmatrix} y_0 \\ \vdots \\ y_{N-1} \end{bmatrix}.$$

Ordinary least squares is equivalent to solving the perturbed equation $\boldsymbol{Ax} = \boldsymbol{b} + \delta\boldsymbol{b}$ by solving for a minimum norm perturbation vector $\|\delta\boldsymbol{b}\|$ subject to the constraint $\boldsymbol{b} + \delta\boldsymbol{b} \in \mathscr{R}_{\boldsymbol{A}}$. This is described as "bending" the right hand side vector $\boldsymbol{b}$ with the aid of the perturbation vector that has the smallest possible norm, such that the "bent" vector now lies in the range space of $\boldsymbol{A}$. When the matrix $\boldsymbol{A}$ is also subject to observation noise, however, we try to bend both the matrix and the vector $\boldsymbol{b}$ such that the perturbed vector lies in the range space of the perturbed matrix while minimizing the norm of the combined perturbations, namely, the perturbation matrix augmented by an extra column which is the perturbation vector. Thus, the total least squares solution is found by solving the perturbed equation

$$(\boldsymbol{A} + \delta\boldsymbol{A})\boldsymbol{x} = \boldsymbol{b} + \delta\boldsymbol{b}, \tag{2.36}$$

with $\boldsymbol{b} + \delta\boldsymbol{b} \in \mathscr{R}_{\boldsymbol{A}+\delta\boldsymbol{A}}$ and minimizing the Frobenius norm (2.25) of the augmented matrix

$$\|[\delta\boldsymbol{A} \mid \delta\boldsymbol{b}]\|_F^2 = \sqrt{\textbf{Trace}\left([\delta\boldsymbol{A} \mid \delta\boldsymbol{b}]^+ \, [\delta\boldsymbol{A} \mid \delta\boldsymbol{b}]\right)}.$$

The minimization assumes that the noise variances in both $\delta\boldsymbol{A}$ and $\delta\boldsymbol{b}$ are of the same order; if they are not, then we consider minimizing the Frobenius norm of the scaled augmented matrix $[\delta\boldsymbol{A} \mid \gamma\delta\boldsymbol{b}]$. Equation (2.36) can be written in an equivalent augmented matrix form (Rouché-Capelli theorem):

$$\left[\boldsymbol{A} + \delta\boldsymbol{A} \,\middle|\, \boldsymbol{b} + \delta\boldsymbol{b}\right][\boldsymbol{x}^T \mid -1]^T = \boldsymbol{0}. \tag{2.37}$$

Thus, $[\boldsymbol{x}^T \mid -1]^T \in \mathscr{N}_{[\boldsymbol{A}+\delta\boldsymbol{A} \mid \boldsymbol{b}+\delta\boldsymbol{b}]}$. Let us assume that $\boldsymbol{A}$, which is $N \times M$, $N > M$, has full column rank M. Then the augmented matrix $[\boldsymbol{A} \mid \boldsymbol{b}]$ of size $N \times (M+1)$, has full column rank $M+1$ so long as $\boldsymbol{b} \notin \mathscr{N}_{\boldsymbol{A}}$. In order to have a unique solution $\boldsymbol{x}_{\text{TLS}}$, the augmented matrix $[\boldsymbol{A}+\delta\boldsymbol{A} \mid \boldsymbol{b}+\delta\boldsymbol{b}]$ must have rank M, yet it has $M+1$ columns; therefore, it must have a rank deficiency of 1 and the problem can be restated as finding the "smallest" augmented perturbation matrix $[\delta\boldsymbol{A} \mid \delta\boldsymbol{b}]$ that changes the augmented matrix $[\boldsymbol{A} \mid \boldsymbol{b}]$ of rank $M+1$ to a rank M augmented perturbed matrix

$$\left[\boldsymbol{A} + \delta\boldsymbol{A} \,\middle|\, \boldsymbol{b} + \delta\boldsymbol{b}\right] = \left[\boldsymbol{A} \,\middle|\, \boldsymbol{b}\right] + \left[\delta\boldsymbol{A} \,\middle|\, \delta\boldsymbol{b}\right]. \tag{2.38}$$

To find the best rank M approximation, we use the SVD to write

$$\left[\boldsymbol{A} \,\middle|\, \boldsymbol{b}\right] = \sum_{m=1}^{M+1} \sigma_m \, \boldsymbol{u}_m \boldsymbol{v}_m^+, \tag{2.39}$$

where we assume the smallest singular value σ_{M+1} to be unique (singular values are assumed to be in descending order). Using equation (2.28) of section 2.8, the best rank M approximation to the augmented perturbed matrix is

$$\left[\boldsymbol{A} + \delta\boldsymbol{A} \,\middle|\, \boldsymbol{b} + \delta\boldsymbol{b}\right] = \sum_{m=1}^{M} \sigma_m \boldsymbol{u}_m \boldsymbol{v}_m^+. \tag{2.40}$$

Thus, the correct perturbation is

$$\left[\delta \boldsymbol{A} \mid \delta \boldsymbol{b}\right] = -\sigma_{M+1}\, \boldsymbol{u}_{M+1}\, \boldsymbol{v}^{+}_{M+1}. \tag{2.41}$$

Since $\boldsymbol{v}_{M+1} \in \mathcal{N}_{[\boldsymbol{A}+\delta\boldsymbol{A} \mid \boldsymbol{b}+\delta\boldsymbol{b}]}$, and this null space has dimension 1, the solution vector of (2.41) must be a multiple of $\boldsymbol{v}_{M+1}$,

$$[\boldsymbol{x}^T \mid -1]^T = \alpha\, \boldsymbol{v}_{M+1}. \tag{2.42}$$

If the last component of $\boldsymbol{v}_{M+1}$ is zero, then there is no total least squares solution; if the last component is non-zero, then the total least squares solution is

$$\boldsymbol{x}_{\text{TLS}} = -\, \boldsymbol{v}_{M+1}[0 : M-1]/\boldsymbol{v}_{M+1}[M]. \tag{2.43}$$

If the smallest singular value in (2.39) is not unique, then suppose that the last L eigenvectors correspond to the smallest singular value, i.e., the null space $\mathcal{N}_{[\boldsymbol{A}+\delta\boldsymbol{A} \mid \boldsymbol{b}+\delta\boldsymbol{b}]}$ is now spanned by the L vectors $\boldsymbol{v}_{M+2-L}, \boldsymbol{v}_{M+3-L}, \ldots, \boldsymbol{v}_M, \boldsymbol{v}_{M+1}$, any of whose linear combinations is a possible solution. To find the solution with minimum norm, let us define a matrix $\boldsymbol{W} \equiv \left[\boldsymbol{v}_{M+2-L}, \boldsymbol{v}_{M+3-L}, \ldots, \boldsymbol{v}_M, \boldsymbol{v}_{M+1}\right]$ and consider a Householder reflection matrix $\boldsymbol{H}$ such that

$$\boldsymbol{W}\boldsymbol{H} = \begin{bmatrix} \alpha & \boldsymbol{x}' \\ \boldsymbol{0} & \beta \end{bmatrix}. \tag{2.44}$$

Defining the vector $\boldsymbol{z} \equiv [\boldsymbol{x}'^T, \beta]^T$, the Householder transformation maximizes the last component of $\boldsymbol{W}\boldsymbol{x}'$ subject to the constraint $|\boldsymbol{z}| = 1$. Thus, the minimum norm total least squares solution is [27]

$$\boldsymbol{x}_{\text{mnTLS}} = -\boldsymbol{x}'/\beta. \tag{2.45}$$

Figure 2.10 illustrates the difference between ordinary least squares and total least squares procedures: the former finds a line such that the sum of the squares of the vertical distances from the data points to that line is minimized, whereas the latter finds a line such that the sum of the squares of the normal distances between the points and the line (distances found orthogonal to the line) is minimized. In general, total least squares finds the closest hyperplane to the set of M points $[\boldsymbol{a}^T_m, b_m]^T \in \mathbb{C}^{M+1}$. Figure 2.11 shows the

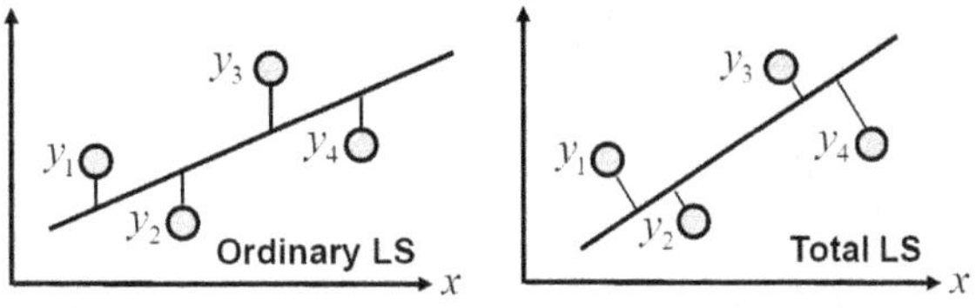

Figure 2.10: Distance measures in ordinary least squares and total least squares.

ordinary least squares and total least squares linear fits to life expectancy data as a function of the number of cigarettes smoked per day; this data has uncertainties in both dimensions, and so total least squares should provide a better model.

The conditioning of total least squares problems is worse than that of ordinary least squares [28]; i.e., total least squares is a "deregularization" of ordinary least squares, and the difference between the two methods grows with $1/(\sigma_M - \sigma_{M+1})$.

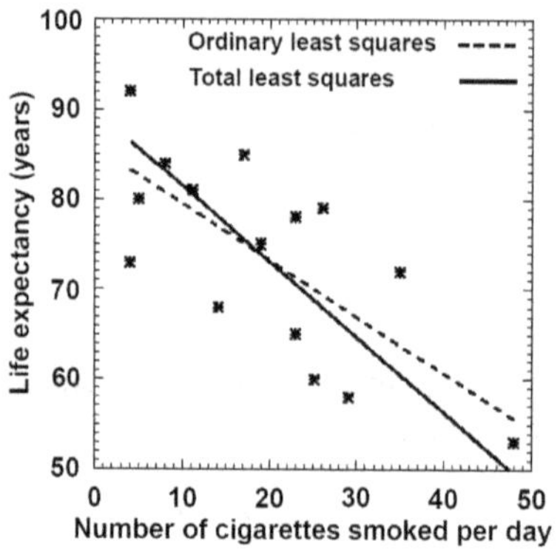

Figure 2.11: Life expectancy as a function of daily use of cigarettes.

2.12 SVD and the Orthogonal Procrustes Problem

Another application of SVD is to the orthogonal Procrustes problem [29]: given two $N \times M$ complex matrices $\boldsymbol{X}$ and $\boldsymbol{Y}$, find a unitary matrix $\boldsymbol{Q}$ to minimize the Frobenius norm $\|\boldsymbol{X}-\boldsymbol{QY}\|_F$ (section 1.6). The square of this norm is

$$\|\boldsymbol{X}-\boldsymbol{QY}\|_F^2 = \textbf{Trace}(\boldsymbol{XX}^+) + \textbf{Trace}(\boldsymbol{QYY}^+\boldsymbol{Q}^+) - 2\,\textbf{Trace}(\boldsymbol{XY}^+\boldsymbol{Q}^+),$$

and so the minimization of the left hand side is equivalent to the maximization of the last Trace operation. Using the SVD of the square $N \times N$ matrix $\boldsymbol{XY}^+$, namely, $\boldsymbol{U\Sigma V}^+$ (all matrices here are $N \times N$), we have

$$\textbf{Trace}(\boldsymbol{XY}^+\boldsymbol{Q}^+) = \textbf{Trace}(\boldsymbol{U\Sigma V}^+\boldsymbol{Q}^+) = \textbf{Trace}(\boldsymbol{\Sigma V}^+\boldsymbol{Q}^+\boldsymbol{U}), \tag{2.46}$$

where in the last equation we used the cyclic property of the Trace operation. Defining $\boldsymbol{Z} \equiv \boldsymbol{V}^+\boldsymbol{Q}^+\boldsymbol{U}$, it is clear that $\boldsymbol{Z}^+\boldsymbol{Z} = \boldsymbol{I}$, i.e., $\boldsymbol{Z}$ is unitary, and so we must have

$$\sum_{n=1}^{N} |\boldsymbol{Z}_{kn}|^2 = 1,$$

which implies that $|\boldsymbol{Z}_{kn}| \le 1$ for all k, n. Equation (2.46) then gives

$$\textbf{Trace}(\boldsymbol{XY}^+\boldsymbol{Q}^+) = \textbf{Trace}(\boldsymbol{\Sigma Z}) = \sum_{n=1}^{N} \boldsymbol{Z}_{nk}\sigma_n \le \sum_{n=1}^{N} \sigma_n, \tag{2.47}$$

with equality only when $\boldsymbol{Z} = \boldsymbol{I}$ or when $\boldsymbol{Q} = \boldsymbol{UV}^+$; this is the required unitary matrix for the orthogonal Procrustes problem.

CHAPTER 3

Linear Time-Invariant Systems and Transforms

3.1 Introduction

The mathematical characterization of linear time-invariant (LTI) systems rests on the concept of *convolution*, which in continuous and discrete time is given by

$$y(t) \equiv \int_{-\infty}^{\infty} h(\tau)\, x(t-\tau)\, d\tau, \quad y[n] = \sum_{m=-\infty}^{\infty} h[m]\, x[n-m]. \tag{3.1}$$

Referring to figure 3.1, we say that the output $y(t)$ of a continuous time linear time-invariant system is found by the convolution of the input function $x(t)$ with the system function $h(t)$ and similarly for discrete time sequences. A linear time-invariant system is defined by its system function $h(t)$ (system impulse response); the convolution is also known as a filtering operation on the input function. We will use the term "system function" to denote a transform (Laplace or Fourier transform) of the impulse response function. The convolution operation between two continuous time functions is denoted by $x * h(t) \equiv h * x(t)$, while that between two discrete time sequences is denoted by $x * h[n] \equiv h * x[n]$.

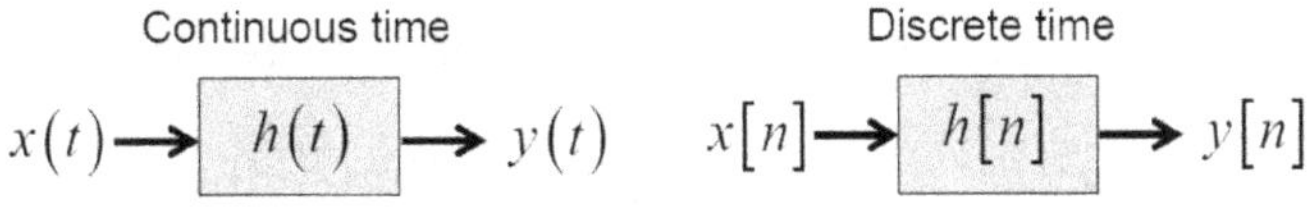

Figure 3.1: Continuous time and discrete time linear time-invariant systems.

A requirement for a linear time-invariant system is *stability*, i.e., a given bounded input to the system should produce a bounded output (known as BIBO stability): stability holds if, and only if, the system function is absolutely integrable (i.e., $h(t) \in L_1(\mathbb{R})$) in continuous time or absolutely summable (i.e., $h_n \in l_1(\mathbb{Z})$) in discrete time. A test for stability of a pole-zero system, known as the Schur-Cohn test, is described by theorem 47 in section 9.7 in the context of minimum phase property of prediction error filters.

An often desirable property of a linear time-invariant system is *causality*, i.e., the output function (sequence) at a given instant depends only on the present and past values of the input function (sequence): causality holds if, and only if, the system function vanishes for negative times, i.e., $h(t) = 0$ when $t < 0$ or $h[n] = 0$ when $n < 0$. An important concept is that of system invertibility: a system impulse response inverse $\tilde{h}(t)$ in continuous time or $\tilde{h}[n]$ in discrete time, should it exist, satisfies one of the following relations:

$$\int_{-\infty}^{\infty} \tilde{h}(t-t')\, h(t'-\tau)\; dt' = \delta(t-\tau), \quad \sum_{n=-\infty}^{\infty} \tilde{h}_{k-n} h_{n-l} = \delta_{kl}. \tag{3.2}$$

The problem of finding the inverse system function is quite difficult and often requires constraints to obtain a unique solution. A very useful constraint (and natural when we discuss systems in transform domains) is provided by the concept of a *minimum phase* system function: a linear time-invariant system impulse response h is minimum phase (or is said to have minimum group delay) if its inverse $\tilde{h}$ exists and both h and $\tilde{h}$ are stable and causal. Analogously, a *maximum phase* system function h is one whose inverse exists and both h and $\tilde{h}$ are stable and anti-causal. We will discuss the properties of minimum phase signals using transforms.

Analysis of linear continuous time-invariant systems is often performed using the Laplace transform because of the fact that the function e^{st}, $s \in \mathbb{C}$, is an eigenfunction of such systems, i.e., the input e^{st} to such a system produces an output $H(s)e^{st}$ where $H(s)$, known as the system response, is the Laplace transform of the system function $h(t)$. In discrete time systems, the analogous eigenfunction is z^n, $n \in \mathbb{Z}$ and $z \in \mathbb{C}$, and the corresponding transform is known as the **Z** transform. The Laplace transform evaluated on the imaginary axis $s = i\omega$, $-\infty < \omega < +\infty$, is the Fourier transform, while the **Z** transform evaluated on the unit circle $z = e^{i\omega}$, $-\pi < \omega < \pi$, is the discrete Fourier transform.

3.2 The Laplace Transform

The Laplace transform of $x(t)$, denoted by $X(s) \equiv \mathcal{L}\{x(t)\}$, is defined by the following integral (when it exists):

$$X(s) = \int_{-\infty}^{\infty} x(t)\, e^{-st}\, dt, \quad s \in \mathbb{C}. \tag{3.3}$$

The set of s values for which $X(s)$ converges absolutely cannot be the entire real s line and is, in general, of the form of a strip $a < \textbf{Re}\,(s) < b$ for $a, b \in \mathbb{R}$. We assume that the strip always includes the imaginary axis so that the Fourier transform of $x(t)$ is guaranteed to exist. Using this definition, it is easy to show that the Laplace transform of a convolution integral is the product of the individual Laplace transforms

$$y(t) = h * x(t) \quad \Leftrightarrow \quad Y(s) = H(s)X(s). \tag{3.4}$$

Equation (3.4) is known as the convolution theorem. In addition, it is easy to see that if $x(t) = e^{st}$ in figure 3.1, then the output is

$$\int_{-\infty}^{\infty} e^{s(t-\tau)} h(\tau)\; d\tau = e^{st} H(s), \tag{3.5}$$

and so e^{st}, $s \in \mathbb{C}$, is an eigenfunction of the linear time-invariant system depicted in figure 3.1 (left).

The inverse Laplace transform $x(t) = \mathcal{L}^{-1}\{X(s)\}$ when the region of convergence (ROC) of $X(s)$ is $a < \textbf{Re}\,(s) < b$ is given by the Bromwich integral

$$x(t) = \frac{1}{2i\pi} \int_{c-i\infty}^{c+i\infty} X(s)\, e^{st}\, ds, \tag{3.6}$$

with $a < c < b$. $X(s)$ is an analytic function of the complex variable s [1] and, in general, possesses poles p_n in the complex plane. If the ROC of $X(s)$ is to the right of all the poles of $X(s)$, i.e., $\boldsymbol{Re}(s) > \max\{\boldsymbol{Re}(p_n)\}$, then the inverse transform $x(t)$ is a causal function, i.e., $x(t) = 0$ for $t < 0$. If, on the other hand, the ROC of $X(s)$ is to the left of all the poles of $X(s)$, i.e., $\boldsymbol{Re}(s) < \min\{\boldsymbol{Re}(p_n)\}$, then the inverse transform $x(t)$ is an anti-causal function, i.e., $x(t) = 0$ for $t > 0$. These two results follow from the fact that the semi-circular part of the Bromwich contour can be closed to the right for $t < 0$, and since the region contained by the contour integral has no poles within it, the integral must vanish, which together with the fact that the integral on the semi-circular contour must also vanish as the radius of the circle $\to \infty$ leads to the vanishing of the inverse transform for $t < 0$.

For example, figure 3.2 shows a case when $X(s)$ has poles in the left half-plane (LHP) to the left of the ROC. To evaluate the Bromwich integral for $t < 0$, the semi-circle is closed to the right giving the result that $x(t)$ is a causal function: $x(t) = 0$ for $t < 0$. When $t > 0$, the semi-circle is closed to the left, thus enclosing the poles ensuring a non-zero function for $t > 0$. Similarly, when the ROC is to the left of all the poles, i.e., poles are in the right half-plane (RHP), and for $t > 0$ the contour is closed to the left and the Bromwich integral is zero ensuring a stable and anti-causal function of time. Conversely, if $x(t)$ is causal,

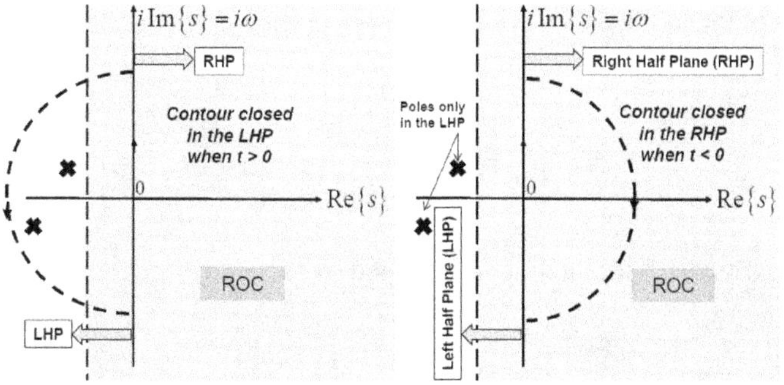

Figure 3.2: The complex s-plane: poles in the LHP are indicated by crosses.

then its Laplace transform has no poles in the RHP, while an anti-causal function's Laplace transform has no poles in the LHP. A function $x(t)$ that is neither causal nor anti-causal has poles in the LHP and the RHP, in which case the ROC is a strip in between both sets of poles excluding them all.

A useful and tractable system function $h(t)$ is one whose Laplace transform $H(s)$ is a ratio of polynomials, i.e., it has zeros and poles in the complex s-plane

$$H(s) = \prod_{m=1}^{M} (s - z_m) \Big/ \prod_{n=1}^{N} (s - p_n), \quad M < N, \tag{3.7}$$

where z_m denotes the zeros and p_n denotes the poles. Since $M < N$, we use partial fractions to write

$$H(s) = \sum_{n=1}^{N} A_n / (s - p_n). \tag{3.8}$$

[1]A complex function of a complex variable is analytic in a region if its first derivative exists at every point of that region. A necessary and sufficient condition for a complex function $X(s) = U(s) + iV(s)$ of a complex variable $s = w + i\omega$ to be analytic is $\partial U/\partial w = \partial V/\partial \omega$ and $\partial U/\partial \omega = -\partial V/\partial w$ (these are the Cauchy-Riemann equations), where the partial derivatives exist and are continuous in an open set containing the region of differentiability [30].

Consider each term separately and denote its associated inverse transform by $h^{(n)}(t)$: there are two possible stable inverse transforms depending on the sign of the real part of p_n:

$$\boldsymbol{Re}(p_n) < 0 \Rightarrow h^{(n)}(t) = \theta(t) A_n e^{p_n t}, \tag{3.9a}$$

$$\boldsymbol{Re}(p_n) > 0 \Rightarrow h^{(n)}(t) = \theta(-t) A_n e^{p_n t}, \tag{3.9b}$$

where $\theta(t)$ is the Heaviside step function: 0 for $t < 0$ and 1 for $t \geq 0$. The ROC for each term is $\boldsymbol{Re}(s) > \boldsymbol{Re}(p_n)$ when $\boldsymbol{Re}(p_n) < 0$ and $\boldsymbol{Re}(s) < \boldsymbol{Re}(p_n)$ when $\boldsymbol{Re}(p_n) > 0$. The ROC for $H(s)$ is, therefore, the infinite vertical strip that is between the LHP and RHP poles and that includes the imaginary axis. Thus, a causal pole-zero system function is stable when all of its poles are in the LHP, i.e., every pole has a negative real part, while an anti-causal pole-zero system function is stable when all of its poles are in the RHP, i.e., every pole has a positive real part.

Any function $x(t)$ can be written in the form $x(t) \equiv x(t)\theta(t) + x(t)\theta(-t) - x(0)\delta_{0t}$, where the unit pulse δ_{0t} is 1 when $t = 0$ and 0 otherwise (this is not the Dirac delta distribution!); the first term is causal with its Laplace transform denoted by $X_c(s)$, while the second term is anti-causal with its Laplace transform denoted by $X_a(s)$. Thus, we write $X(s) = X_c(s) + X_a(s)$ (note that the time pulse function is non-zero at a single point and does not contribute to the transform integral) and introduce the following notation for extracting the additive causal part $X_c(s)$:

$$\{X(s)\}_c \equiv X_c(s) = \int_0^\infty e^{-st} x(t)\, dt. \tag{3.10}$$

An equivalent way to write the causal extraction implied by the above equation in the time domain is

$$\left\{\mathcal{L}^{-1}\{X(s)\}\right\}_c = \theta(t)\, \mathcal{L}^{-1}\{X(s)\}. \tag{3.11}$$

As an example, consider the causal time function $h_c(t)$ whose Laplace transform is $e^{sT}/(s_0 + s)$

$$h_c(t) = \left\{\mathcal{L}^{-1}\left\{e^{sT}/(s_0+s)\right\}\right\}_c. \tag{3.12}$$

Following the preceding prescription, we have

$$\begin{aligned} h_c(t) &= \left\{\mathcal{L}^{-1}\left\{e^{sT}/(s_0+s)\right\}\right\}_c = \theta(t)\left\{\mathcal{L}^{-1}\left\{e^{sT}/(s_0+s)\right\}\right\} \\ &= \theta(t)\, e^{-s_0 T} e^{-s_0 t} \theta(t+T) = e^{-s_0 T} e^{-s_0 t} \theta(t), \end{aligned} \tag{3.13}$$

where we used the identity $\theta(t)\,\theta(t+T) \equiv \theta(t)$. Taking the Laplace transform of the right hand side also shows that

$$\left\{e^{sT}/(s_0+s)\right\}_c = e^{-s_0 T}/(s_0+s). \tag{3.14}$$

3.3 Phase and Group Delay Response: Continuous Time

If a stable and causal pole-zero system function $H(s)$ has an inverse $1/H(s)$ that is also stable and causal, then the zeros of $H(s)$, i.e., the poles of $1/H(s)$, must also be in the LHP, in which case $H(s)$ is a minimum phase or minimum group delay system. By contrast, a system function whose zeros and poles are in the RHP is known as a maximum phase or maximum group delay system, while a function with zeros and poles in both the LHP and RHP is mixed phase.

System functions with minimum phase (minimum group delay) are of particular importance in signal processing because of the fact that they and their inverses are stable and causal. To understand the minimum group delay property, we consider the Laplace transform evaluated on the imaginary axis $s = i\omega$, i.e., the Fourier transform (see section 3.7), where $\omega \in \mathbb{R}$ is the frequency. The group delay τ_d of a signal is defined to be a measure of the signal phase distortion as a function of frequency

$$\tau_d = -d\phi/d\omega, \quad \text{where } H(\omega) = |H(\omega)|\, e^{i\phi(\omega)}. \tag{3.15}$$

As an example, consider a minimum phase system function with a zero and a pole both in the LHP

$$H(s) = (s - z_0)/(s - p_0), \quad z_0 = -1 + i, \; p_0 = -0.5 + 2\,i. \tag{3.16}$$

The ROC is $\boldsymbol{Re}(s) > -0.5$, which includes the imaginary axis. Moving the zero to the RHP, e.g., to $1 + i$, produces a mixed phase system function whose magnitude is the same as the original minimum phase system. Figure 3.3 shows the phases and group delays for both the minimum phase and mixed phase system functions in this example.

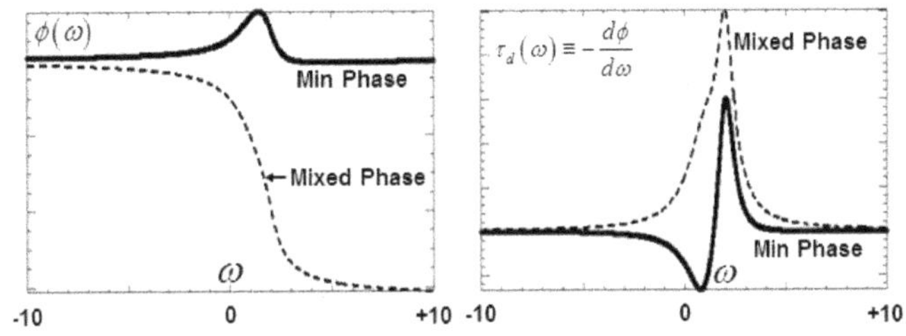

Figure 3.3: The phase $\phi(\omega)$ and time-delay $\tau_d(\omega)$ for a minimum phase and a mixed phase system.

We define the spectrum $S(s)$ and the frequency spectrum $S(\omega)$ of a function $x(t) \in L_2(\mathbb{R})$ by

$$S(s) = X(s)X^*(-s^*), \quad S(\omega) = |X(\omega)|^2, \tag{3.17}$$

where $S(\omega)$ is found by evaluating $S(s)$ on the imaginary axis $s = i\omega$, $\omega \in \mathbb{R}$. Thus, $S(\omega) = |X(i\omega)|^2$ is a positive function of frequency ω, and when multiplied by the infinitesimal quantity $d\omega$, it is the energy of $x(t)$ in an interval $[\omega - d\omega/2, \omega + d\omega/2]$.

An important problem in signal processing is the reconstruction of a system function $h(t)$ from its frequency spectrum, which clearly has no phase information. A unique reconstruction is possible if $h(t)$ is minimum phase because the lost phase can be recovered using a Hilbert transform relation discussed in section 3.14. For a pole-zero system function the reconstruction can be performed by spectral factorization separating zeros and poles in the LHP and the RHP into minimum phase and maximum phase factors. To see this, consider a minimum phase system function of the form (3.7) whose zeros z_m and poles p_n are in the LHP and whose spectrum is

$$S(s) \equiv H(s)H^*(-s^*) = (-1)^{M+N} \frac{\prod_{m=1}^{M} (s - z_m)(s + z_m^*)}{\prod_{n=1}^{N} (s - p_n)(s + p_n^*)}, \quad M < N. \tag{3.18}$$

Clearly $S(s)$ has zeros and poles that come in pairs: $\{z_m, -z_m^*\}$ and $\{p_n, -p_n^*\}$, i.e., every zero/pole in the LHP has an associated zero/pole in the RHP with equal imaginary part and equal magnitude but real part with opposite sign. Hence, a pole-zero system spectrum $S(s)$ is the product of a minimum phase factor and a maximum phase factor: the minimum phase factor $H(s)$ is found by grouping those zeros and poles that lie in the LHP, while the maximum phase factor is just $H^*(-s^*)$.

For instance, consider the spectrum $2a/(s^2 - a^2)$, $a > 0$, which has poles at $\pm a$ with ROC $-a <$ ***Re*** $(s) < +a$, which includes the imaginary axis. Now by removing the pole at $+a$ and taking inverse Laplace transform of $\sqrt{2a}(s+a)^{-1}$, we obtain a causal function, namely, $h^{(c)}(t) = \sqrt{2a}\exp(-at)\ \theta(t)$ corresponding to a stable and causal system function $H^{(c)}(s)$ whose inverse $1/H^c(s)$ is also stable and causal since its zero (the pole of $H^{(c)}(s)$) is in the LHP. Thus, $h^{(c)}(t)$ is the minimum phase function whose spectrum is the given $S(s)$. The anti-causal component corresponding to the pole at $+a$ then gives rise to a maximum-phase function $h^{(a)}(t)$.

In general, a pole-zero system function $H(s)$ with no zeros or poles on the imaginary axis can be written as the product $H_{\text{mnp}}H_{\text{ap}}$ where H_{ap} is an all-pass system function, that is, its spectrum as defined by (3.18) is equal to 1. An example of an all-pass system function is $H(s) = (s + p^*)/(s - p)$ where p is in the LHP. Clearly, $S(s) = H(s)H^*(-s^*) = 1$. Figure 3.4 shows the phase and time-delay response for this all-pass filter with $p = -0.5 + i$. Now consider the mixed phase system function $H(s) = (s - z_0)/(s - p_0)$

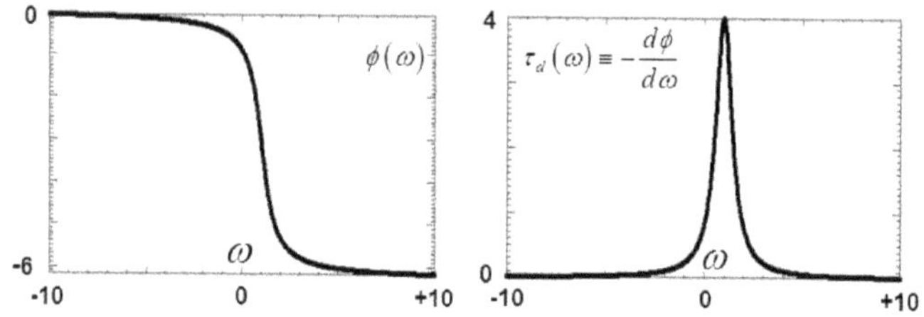

Figure 3.4: Phase and time-delay frequency response for a first order all-pass filter.

where z_0 is the RHP while p_0 is in the LHP. Then $H(s) = H_{\text{mnp}}H_{\text{ap}}$ where $H_{\text{mnp}} = (s + z_0^*)/(s - p_0)$ and $H_{\text{ap}} = (s - z_0)/(s + z_0^*)$. If both z_0 and p_0 are in the RHP, then $H_{\text{mnp}} = (s + z_0^*)/(s + p_0^*)$ and $H_{\text{ap}} = (s - z_0)(s + p_0^*)/(s + z_0^*)(s - p_0)$.

3.4 The Z Transform

The **Z** transform of a discrete time sequence x_n is defined by

$$\mathcal{Z}\{x_n\} \equiv X(z) = \sum_{n=-\infty}^{\infty} x_n z^{-n}, \tag{3.19}$$

and its ROC is, in general, an annular region centered at the origin of the complex z plane. A causal sequence x_n that vanishes for $n < 0$ has a ROC of the form $|z| > c$ with poles inside the circle with radius c, while an anti-causal sequence that vanishes when $n \geq 0$ has its poles outside the circle $|z| = c$ with a ROC given by $|z| < c$. The **Z** transform of a discrete time convolution is the product of the individual **Z** transforms, i.e.,

$$y_n = h * x[n] = \sum_{m=-\infty}^{\infty} x_{n-m}h_m \Leftrightarrow Y(z) = H(z)X(z). \tag{3.20}$$

In addition, the input $x_n = z^n$ in figure 3.1 produces the output

$$\sum_{m=-\infty}^{\infty} z^{n-m} h_m = z^n \, H(z)\,, \tag{3.21}$$

and so z^n, $z \in \mathbb{C}$, $n \in \mathbb{Z}$, is an eigenfunction of the linear discrete time-invariant system of figure 3.1, which is sometimes depicted in the equivalent *direct form* filter realization shown in figure 3.5 (triangles indicate multiplication with the enclosed quantity).

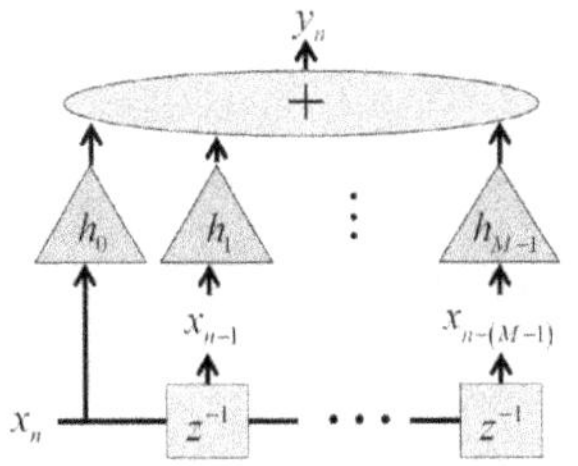

Figure 3.5: Direct form filter realization of a LTI system.

The inverse **Z** transform of $X(z)$ in (3.19) is defined by the sequence $x_n = \mathcal{Z}^{-1}\{X(z)\}$ and is formally given by the contour integral

$$x_n = \frac{1}{2i\pi} \oint_C z^{n-1} X(z)\, dz, \tag{3.22}$$

where the closed contour C is traversed anti-clockwise inside the ROC.

In analogy to the Laplace transform case, we consider the pole-zero system function $H(z)$ with zeros z_m and poles p_n in the complex z-plane

$$H(z) = \prod_{m=1}^{M} \left(1 - z_m z^{-1}\right) \Big/ \prod_{n=1}^{N} \left(1 - p_n z^{-1}\right), \;\; M < N, \tag{3.23}$$

which can be expanded in partial fractions

$$H(z) = \sum_{n=1}^{N} A_n / \left(1 - p_n z^{-1}\right). \tag{3.24}$$

Let us denote the inverse transform of each term in (3.24) by $h_k^{(n)}$ and consider the case $|p_n| < 1$, where the ROC is given by $|z| > |p_n|$; then we have the convergent geometric series

$$\left(1 - p_n z^{-1}\right)^{-1} = \sum_{k=0}^{\infty} p_n^k z^{-k}, \tag{3.25}$$

from which we obtain the result $h_k^{(n)} = A_n \, \theta[k] \; p_n^k$, where $\theta[k]$ is the discrete step function: 0 for $k < 0$ and 1 for $k \geq 0$. If, on the other hand, $|p_n| > 1$, the ROC is $|z| < |p_n|$ and we have the convergent geometric series

$$\frac{1}{1 - p_n z^{-1}} = 1 - \frac{1}{1 - z/p_n} = 1 - \sum_{k=0}^{\infty} z^k / p_n^k = - \sum_{k=-\infty}^{-1} p_n^k z^{-k}, \tag{3.26}$$

from which we obtain $h_k^{(n)} = -A_n\,\theta\,[-k-1]\,p_n^k$. If $H(z)$ has poles inside and outside the unit circle, then its ROC is the annular region between the poles inside and outside the unit circle.

For any $X(z)$, we can write $X(z) = X_c(z) + X_a(z)$, where $X_c(z)$ is the **Z** transform of $\{x_0, x_1, \ldots\}$, i.e., the causal portion of the sequence x_n, while $X_a(z)$ is the **Z** transform of $\{\ldots, x_{-2}, x_{-1}\}$, i.e., the anti-causal portion of the sequence. Thus,

$$\left\{\mathcal{Z}^{-1}X(z)\right\}_c = \theta[n]\,\mathcal{Z}^{-1}\left\{X(z)\right\}. \tag{3.27}$$

For instance, consider the system function

$$H(z) = \frac{1}{(1-pz^{-1})(1-pz)}, \quad p \in \mathbb{R},\ 0 < p < 1,$$

which has one pole p inside the unit circle and one pole $1/p$ outside the unit circle. The causal sequence is found using the residue theorem for the contour integral of equation (3.22) with the circular contour at radius 1,

$$h_n = \frac{1}{2i\pi}\oint_C H(z)\,z^{n-1}dz = \left[\frac{z^n}{1-pz}\right]_{z=p} = \frac{p^n}{1-p^2}.$$

Then

$$H_c(z) \equiv \sum_{n=0}^{\infty} h_n z^{-n} = \frac{1}{1-p^2}\sum_{n=0}^{\infty}\frac{p^n}{z^n} = \frac{1}{1-p^2} \times \frac{1}{1-pz^{-1}}.$$

3.5 Phase and Group Delay Response: Discrete Time

A stable and causal pole-zero system function must have all its poles inside the unit circle, i.e., $|p_n| < 1$. If, in addition, the inverse system function $1/H(z)$ is stable and causal, then the zeros of $H(z)$ (i.e., the poles of the inverse system function) must be inside the unit circle, in which case $H(z)$ is a minimum phase or minimum group delay function. The group delay τ_d for a minimum group delay function is defined by evaluating its **Z** transform on the unit circle $z = e^{i\omega}$ (i.e., the discrete time Fourier transform—see section 3.7), $\omega \in [-\pi, \pi]$,

$$\tau_d = -d\phi/d\omega, \text{ where } H\left(e^{i\omega}\right) = \left|H\left(e^{i\omega}\right)\right| e^{i\phi(\omega)}. \tag{3.28}$$

For instance, consider the minimum phase system function

$$H(z) = \left(1 - z_0 z^{-1}\right)\left(1 - p_0 z^{-1}\right)^{-1}, \quad z_0 = 0.75\,e^{0.2i\pi},\ p_0 = 0.5\,e^{0.3i\pi}, \tag{3.29}$$

corresponding to a system having a zero and a pole at normalized frequencies 0.1 and 0.15 Hz, both inside the unit circle. The ROC is $|z| > 0.75$, which includes the unit circle. Moving the zero outside the unit circle, e.g., to $1.5\,e^{0.2i\pi}$, produces a mixed phase system function which can be normalized by a positive function of frequency so that the new mixed phase spectrum magnitude is equal to the magnitude of the minimum phase system spectrum. Figure 3.6 shows phases and time delays for both the minimum phase and the mixed phase system functions in this example.

Linear phase frequency selective filters of the form $H(e^{i\omega}) = e^{-i\omega N_d}$, $\omega_1 \le \omega \le \omega_2$, and zero otherwise, have a constant group delay N_d. For narrow-band communication signals if the input to a linear phase filter with group delay N_d is $s_n \cos(2\pi f_0 n)$ centered at $-f_0$ and $+f_0$ Hz, then the output phase is linear in $[\pm f_0 - \Delta f, \pm f_0 + \Delta f]$.

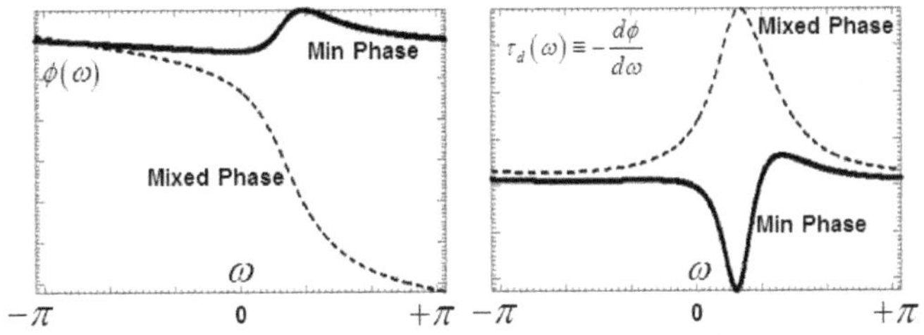

Figure 3.6: Phases $\phi(\omega)$ and time-delays $\tau_d(\omega)$ for a minimum phase and a mixed phase system.

We define the spectrum $S(z)$ of a sequence $x_n \in l_2(\mathbb{Z})$ with **Z** transform $X(z)$ and its frequency spectrum $S(e^{i\omega})$ by

$$S(z) \equiv X(z)X^*(1/z^*), \quad S(e^{i\omega}) = \left|X(e^{i\omega})\right|^2. \tag{3.30}$$

The spectrum $S(z)$ when evaluated on the unit circle $z = e^{i\omega}$, $\pi < \omega \leq +\pi$, becomes the frequency spectrum. $S(e^{i\omega})$ is a periodic and positive definite function of frequency ω, and when multiplied by the infinitesimal quantity $d\omega$ it gives the energy of the sequence in the interval $[\omega - d\omega/2, \omega + d\omega/2]$. The relations (3.30) apply to any discrete time sequence with finite energy.

In analogy to the reconstruction of a minimum phase continuous time function from its frequency spectrum, a minimum phase sequence can be recovered from its frequency spectrum using the Hilbert transform relations in section 3.14. When the sequence represents a pole-zero system function $H(z)$ with poles, zeros, and spectrum $S(z)$, spectral factorization consists of separating the zeros and poles inside and outside the unit circle to produce the minimum phase and maximum phase factors of the spectrum. To see this, consider the system function (3.23) that is minimum phase, i.e., $|z_m| < 1$ and $|p_n| < 1$. Using (3.30), the spectrum of $H(z)$ is

$$S(z) = \frac{\prod_{m=1}^{M} \left(1 - z_m z^{-1}\right)\left(1 - z_m^* z\right)}{\prod_{n=1}^{N} \left(1 - p_n z^{-1}\right)\left(1 - p_n^* z\right)}, \quad M < N. \tag{3.31}$$

Clearly, the zeros and poles of $S(z)$ come in pairs that are *inverse points* through the unit circle, i.e., $\{z_m, 1/z_m^*\}$ and $\{p_n, 1/p_n^*\}$. Hence, given a pole-zero spectrum $S(z)$, we can reconstruct its minimum phase factor $H(z)$ by grouping those zeros and poles that lie inside the unit circle, while the corresponding zeros and poles outside the unit circle form the maximum phase factor $H^*(1/z^*)$.

In analogy to a continuous time all-pass system function we define a discrete time all-pass system function

$$H_{\text{ap}}(z) = \frac{z^{-P} A_P^*(1/z^*)}{A_P(z)}, \quad A_P \equiv 1 + a_1 z^{-1} + \ldots + a_P z^{-P}, \tag{3.32}$$

whose magnitude frequency response is unity

$$\left|H_{\text{ap}}(e^{i\omega})\right| = H_{\text{ap}}(z) H_{ap}^*(1/z^*)_{|z=e^{i\omega}} = 1. \tag{3.33}$$

Clearly poles and zeros of (3.32) come in pairs $(p_k, 1/p_k^*)$ and so the system function can also be expressed as

$$H_{\text{ap}}(z) = \prod_{k=1}^{P} \frac{p_k^* - z^{-1}}{1 - p_k z^{-1}}. \tag{3.34}$$

All-pass filters are used as phase compensators or in transforming low-pass filters into other band-pass forms. As an example consider the $P = 1$ stable all pass system function $H_{ap}(z) = (p^* - z^{-1})/(1 - pz^{-1})$ where $p = re^{i\theta}$ and $0 < r < 1$. The magnitude response is 1 while the continuous phase response and its derivative are

$$\phi(\omega) = -\omega - 2\tan^{-1}\left(r\sin(\omega-\theta)/(1-r\cos(\omega-\theta)\right), \quad \frac{d\phi(\omega)}{d\omega} = (r^2-1)/\left|1-re^{i(\theta-\omega)}\right|^2, \ 0 < r < 1.$$

The derivative of the phase is non-positive and so the continuous phase response decreases monotonically for $\omega \in [0, 2\pi]$. This result can be generalized to the phase response of a stable all-pass system function of any order $P \geq 1$. The group delay is the negative of the phase response derivative and is, therefore, non-negative. Figure 3.7 shows the phase response and group delay of an all-pass system function of order 2 with two poles $p_1 = 0.65\ \exp(0.9\ i)$ and $p_2 = 0.85\ \exp(1.1\ i)$.

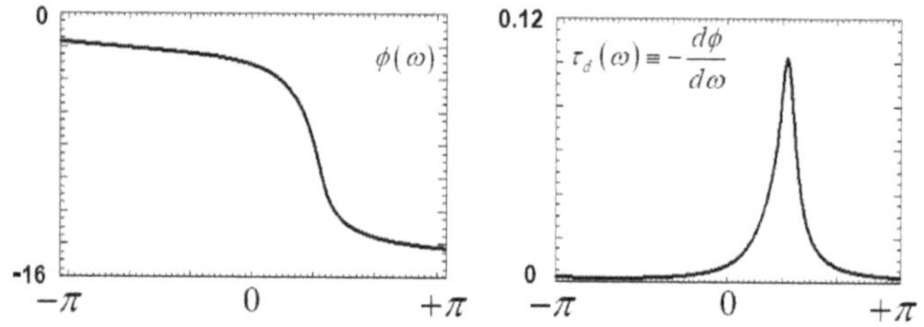

Figure 3.7: Phase and time-delay frequency response for a second order all-pass filter.

Any stable and causal system function can be written as the product of a minimum phase and an all-pass system function. For instance, consider a non-minimum phase system function $H(z)$ with one zero z_0 outside the unit circle, $|z_0| > 1$, with several poles and zeros inside the unit circle, i.e.,

$$H(z) = (z_0^* - z^{-1})H'(z) = H'(z)(1 - z_0 z^{-1}) \times \frac{z_0^* - z^{-1}}{1 - z_0 z^{-1}},$$

where $H'(z)$ is minimum phase. Thus, the phase response of $H(z) = H_{\text{mnp}}H_{\text{ap}}$ is the sum of phase responses of the minimum phase and the all-pass components. Since the phase of the all-pass component is negative, the phase of the minimum phase component is less than the original system function, $\phi_{\text{mnp}}(\omega) < \phi(\omega)$. Equivalently, since the group delay of the all-pass component is positive, the minimum phase component has minimum group delay, i.e., $\tau_{\text{mnp}}(\omega) < \tau(\omega)$.

The relationship between the Laplace transform variable s and **Z** transform variable z is found by a discretization of the time variable $t_n = n\Delta T$ when

$$z = e^{s\Delta T} \quad \Rightarrow \quad \int e^{-st} x(t)\, dt \rightarrow \sum_n x[n\Delta T]\, z^{-n}. \tag{3.35}$$

Under this transformation, the poles and zeros on the imaginary axis $s = e^{i\omega}$ between $-i\pi/\Delta T$ and $+i\pi/\Delta T$ in the s-plane map to the unit circle $|z| = 1$ in the z-plane, while poles and zeros in the left half s-plane between $-i\pi/\Delta T$ and $+i\pi/\Delta T$ map to the interior of the unit circle $|z| \leq 1$ in the z-plane. The Laurent series expansion for the inverse transform $s = \ln z$ for $\boldsymbol{Re}(z) \geq 0$ and $z \neq 0$ is

$$\ln z = 2\left\{\frac{z-1}{z+1} + \frac{1}{3}\left(\frac{z-1}{z+1}\right)^3 + \frac{1}{5}\left(\frac{z-1}{z+1}\right)^5 + \ldots\right\}, \tag{3.36}$$

which upon keeping only the first term in the expansion gives the well-known bilinear transform [1]

$$s = \frac{1}{\Delta T} \ln z \Rightarrow s = \frac{1 + s\Delta T/2}{1 - s\Delta T/2}. \tag{3.37}$$

The bilinear transformation is used in building digital filters from their analogue pole-zero structure. Under the bilinear transformation, the entire left half s-plane maps to the interior of the unit circle in the z-plane, while the imaginary axis in the s-plane maps to the unit circle in the z-plane. The map produces a frequency warping in digital filters designed from their analogue forms in the sense that for $z = e^{i\omega\Delta T}$, we have $s = 2(\Delta T)^{-1}\tan(\omega\Delta T/2)$. Thus, band-edge frequencies must be "prewarped" before the transformation so that they are warped to their correct digital specifications after the transformation.

3.6 Minimum Phase and Front Loading Property

An important property of minimum phase functions known as *front loading* of energy can be illustrated by an inverse filtering (deconvolution) example from seismology.

Deconvolution is an important part of seismic data processing in exploration geophysics and is intended to compress the seismic wavelet (i.e., the seismic waveform—not related to the wavelet transform) to produce the Earth's reflectivity sequence in a horizontally layered earth model with constant propagation velocity in each layer. The main assumption is that the source (dynamite, air gun, etc.) generates a vertically traveling compressional P-wave that is reflected at each horizontal interface at normal incidence [2].

The reflectivity at each interface at depth z is given in terms of the acoustic impedance of the two layers $r(z) = (\rho_2 c_2 - \rho_1 c_1)/(\rho_2 c_2 + \rho_1 c_1)$, where ρ is the layer density and c is the compressional wave velocity in that layer. Further assuming a constant density, we have $r(z) = (c_2 - c_1)/(c_2 + c_1)$. The reflectivity sequence in depth is related to the reflectivity sequence in time through the layer wave velocity c. Thus, we imagine the reflectivity sequence $r(t)$ as a time series; the transformation to depth is made through the layer velocities that are usually available from well logs.

Finally, assuming linear propagation of the source wavelet $w(t)$ that is mostly undistorted while propagating down from the source and back up to the receiver, we arrive at the convolutional model for the reflectivity time series as measured by a receiver (assumed co-located with the source) $x(t) = r * w(t) + n(t)$, where $n(t)$ is the random ambient noise and $x(t)$ is a seismogram. For now, we neglect the ambient noise (we will return to this problem later when we discuss the Wiener filter) and consider the seismogram $x(t) = r * w(t)$. The inverse filtering problem, or deconvolution, is solved by constructing the inverse filter $\tilde{w}(t)$ to the source wavelet so that $\tilde{w} * w(t) = \delta(t)$ if the source wavelet is known.

As an example, consider the discrete time source wavelet $\boldsymbol{w} = [1, -0.5]^T$ whose $\mathbf{Z}$ transform is $W(z) = 1 - 1/2z$; this is a minimum phase function since the zero at 0.5 is inside the unit circle. The inverse $\tilde{\boldsymbol{w}}$ is the causal sequence whose $\mathbf{Z}$ transform is $\tilde{W}(z) = 1 + (1/2)z^{-1} + (1/2)^2 z^{-2} + \ldots$. The more terms we include in the inverse filter, the closer the output of the convolution will be to a unit spike at time 0: two terms will produce $[1, 0, -0.25]^T$, while six terms will produce 1 at time zero with the first non-zero value -0.015625 at time 6. The original wavelet and its inverse are both stable, as expected, since the original wavelet is minimum phase. If the wavelet is not minimum phase, e.g., $\boldsymbol{w} = [-0.5, 1]^T$ (time reversed version of the original minimum phase wavelet), whose zero is now outside the unit circle, the inverse filter does not exist.

[2]The normal incidence assumption for seismic prospecting is impossible in practice since it assumes the same location for the source and receivers. However, it is a reasonable assumption when the depth of the layers is much greater than the distance between the source and receivers. This approximation also leads to the neglect of any shear S-wave generation.

This example highlights an important property of minimum phase signals: minimum phase signals have most of their energy loaded in the front. That is, given a causal sequence x_n, $n = 0, 1, \ldots$, whose total energy is denoted by E_x, and if for all sequences y_n with total energy E_x,

$$\sum_{n=0}^{N} |x_n|^2 \geq \sum_{n=0}^{N} |y_n|^2, \quad \forall N \geq 0, \tag{3.38}$$

then x_n is a minimum phase sequence and will possess a stable and causal inverse. Theorem 14 shows the mathematical characterization of front loaded signals [31].

Theorem 14. *The causal and bounded signal x_n is front loaded if, and only if, its **Z** transform $X(z)$ is an outer function, i.e., it can be written as*

$$X(z) = \alpha \exp\left[\frac{1}{2\pi}\int_0^{2\pi} \frac{e^{i\omega}+z}{e^{i\omega}-z} v\left(e^{i\omega}\right) d\omega\right], \quad z = re^{i\theta}, \quad and\ r < 1, \tag{3.39}$$

where $|\alpha| = 1$ and $v\left(e^{i\omega}\right) = \ln\left|X\left(e^{i\omega}\right)\right|$.

The **Z** transform $X(z)$ evaluated on the unit circle is the Fourier transform of x_n (see equation (3.57) in section 3.7). Theorem 14 can be used to construct a causal minimum phase front loaded sequence if the magnitude of its Fourier transform is specified (see equation (3.82)). To prove the form of the function v, we note that

$$\ln\left|X\left(re^{i\omega}\right)\right| = \boldsymbol{Re}\left[\ln X\left(re^{i\omega}\right)\right] = \frac{1}{2\pi}\int_0^{2\pi} \mathcal{P}_r(\omega-\eta) v\left(e^{i\eta}\right) d\eta\,, \tag{3.40}$$

where $\mathcal{P}_r$ is the *Poisson kernel* [1] defined for $0 < r < 1$,

$$\mathcal{P}_r(\omega) = \sum_{-\infty}^{+\infty} r^{|k|} e^{ik\omega} = \boldsymbol{Re}\left(\frac{1+re^{i\omega}}{1-re^{i\omega}}\right) = \frac{1-r^2}{1+r^2-2r\cos(\omega)}\,. \tag{3.41}$$

The form for v follows by taking the limit $r \to 1$ in $\ln\left|X\left(re^{i\omega}\right)\right|$: the limit is equal to $v\left(e^{i\omega}\right)$ because of properties of the Poisson kernel, and it is also equal to $\ln\left|X\left(e^{i\omega}\right)\right|$ by Fatou's theorem [3] [30]. The imaginary part of the term in parentheses (the middle part of equation (3.41)) is known as the *conjugate Poisson kernel* and denoted by $\mathcal{Q}_r(\omega)$; both $\mathcal{P}_r$ and $\mathcal{Q}_r$ are used in the construction of the analytic signal for periodic functions [see equations (3.90a) and (3.90b)].

3.7 The Fourier Transform

Assuming that the ROC of the Laplace transform includes the imaginary axis of the complex s-plane ($s = i\omega$, $\omega \in \mathbb{R}$), we define the continuous time Fourier transform as the Laplace transform evaluated on the imaginary axis. Thus,

$$X(\omega) = \int_{-\infty}^{\infty} x(t)\, e^{-i\omega t} dt. \tag{3.42}$$

[3] The space of functions $f(z)$, $z = r\exp(i\theta)$, that are analytic inside the unit circle $|z| < 1$ and have finite p-norm (equation (1.3) but with the integral defined over the finite interval $[0, 2\pi]$), is known as the Hardy space $\mathbb{H}^p$. Fatou's theorem, a statement regarding these functions on the unit disk and their point-wise extension to the disc boundary, shows that the radial limit $r \to 1$ exists (almost everywhere in θ) and the resulting function has finite p-norm on the unit circle.

The inverse Fourier transform $x(t) = \mathcal{F}^{-1}\{X(\omega)\}$ is then given by

$$x(t) = (2\pi)^{-1} \int_{-\infty}^{\infty} X(\omega)\, e^{i\omega t} d\omega. \tag{3.43}$$

The product of the numerical constants multiplying the right hand sides of (3.42) and (3.43) must equal $1/2\pi$. In many instances (in order to use orthonormal basis functions) the numerical constants are chosen equal to $1/\sqrt{2\pi}$. In this text, however, we adopt the convention of using 1 for the forward transform (3.42) and $1/2\pi$ for the inverse transform (3.43). With this convention, the equality of the energy in the time and the frequency domains (Parseval's relation)

$$\int_{-\infty}^{\infty} |x(t)|^2\, dt = \frac{1}{2\pi} \int_{-\infty}^{\infty} |X(\omega)|^2\, d\omega, \tag{3.44}$$

follows from the more general result (3.45). Let $x(t), y(t) \in L_2(\mathbb{R})$ be two finite energy signals. Denoting their Fourier transforms by $X(\omega)$ and $Y(\omega)$, we have

$$\int_{-\infty}^{\infty} x^*(t-k)\, y(t-l)\, dt = \frac{1}{2\pi} \int_{-\infty}^{\infty} X^*(\omega)\, e^{i\omega k}\, Y(\omega)\, e^{-i\omega l}\, d\omega. \tag{3.45}$$

An important special case of (3.45) is the convolution theorem for Fourier transforms: the convolution of two continuous time functions is found by inverse transforming the product of their individual Fourier transforms, namely,

$$x * y(t) \equiv \int_{-\infty}^{\infty} x(\tau)\, y(t-\tau)\, d\tau = \frac{1}{2\pi} \int_{-\infty}^{\infty} X(\omega)\, Y(\omega)\, e^{i\omega t} d\omega, \tag{3.46}$$

or equivalently,

$$\mathcal{F}\{x * y(t)\} = X(\omega)\, Y(\omega). \tag{3.47}$$

All finite energy signals as defined by equations (1.1) have finite Fourier transforms [4]. The rate of decay of the Fourier transform $X(\omega)$ as $\omega \to \infty$ is a measure of the smoothness of the function $x(t)$: the faster the decay, the smoother the function. Similarly, the rate of decay of $x(t)$ is determined by the smoothness of $X(\omega)$. The squared magnitude of the Fourier transform, $|X(\omega)|^2$, is the frequency spectrum of $x(t)$.

3.8 The Short-Time Fourier Transform and the Spectrogram

Using a real and symmetric window function with finite support $[-T/2, +T/2]$, we define the short-time Fourier transform (STFT) of $s(t)$ by

$$S(t,\omega) = \int_{-\infty}^{\infty} s(\tau)\, w(t-\tau) e^{-i\omega\tau} d\tau \equiv \left|S(t,\omega)\right| e^{i\phi(t,\omega)}, \tag{3.48}$$

[4] The Fourier transform is actually defined for a wider class of signals than those with finite energy, namely, absolutely integrable, or L_1, functions.

where the effective integration interval is $[t - T/2, t + T/2]$. In this definition of the STFT, the signal is "fixed" in time τ while the window is shifted so as to be centered at t before multiplication and Fourier transformation on the variable τ. An alternative formulation in terms of a shifted signal instead of a shifted window is

$$Y(t,\omega) = \int_{-T/2}^{+T/2} s(t+\tau)\, w(-\tau) e^{-i\omega\tau} d\tau = \left|Y(t,\omega)\right| e^{i\psi(t,\omega)}, \tag{3.49}$$

where windowed sections of the shifted signal are Fourier transformed. Although both formulations have the same magnitude, whose square is the spectrogram, they have unequal phases that are linearly related: transforming the variable $\tau \to \tau' - t$ in equation (3.49) gives

$$\psi(t,\omega) = \omega t + \phi(t,\omega), \text{ while } \left|Y(t,\omega)\right| = \left|S(t,\omega)\right|. \tag{3.50}$$

The original signal $s(t)$ is exactly recoverable from its STFT; the reconstruction formula using the shifted signal transform coefficients $Y(t,\omega)$ is

$$\begin{aligned} s(t) &= \int_{-\infty}^{+\infty}\int_{-\infty}^{+\infty} Y(\tau,\omega) w(\tau - t) e^{-i\omega(\tau - t)}\, d\omega d\tau \\ &= \int_{-\infty}^{+\infty}\int_{-\infty}^{+\infty} \left|Y(\tau,\omega)\right| w(\tau - t)\, e^{+i\left(\psi(\tau,\omega) - \omega\tau + \omega t\right)}\, d\omega d\tau. \end{aligned} \tag{3.51}$$

If the time variation in $\left|Y(t,\omega)\right|$ is slow compared to its phase variation, the stationary phase approximation [13] in the reconstruction formula suggests that the maximum contribution to the integrals comes from the neighborhood of the stationary phase location defined by

$$\frac{\partial}{\partial\omega}\left[\psi(\tau,\omega) - \omega\tau + \omega t\right] = 0, \text{ and } \frac{\partial}{\partial\tau}\left[\psi(\tau,\omega) - \omega\tau + \omega t\right] = 0, \tag{3.52}$$

whose solution is the pair

$$\hat{t} = \tau - \frac{\partial\psi(\tau,\omega)}{\partial\omega} = -\frac{\partial\phi(\tau,\omega)}{\partial\omega}, \quad \hat{\omega} = \omega + \frac{\partial\psi(\tau,\omega)}{\partial\tau} = \frac{\partial\phi(\tau,\omega)}{\partial\tau}. \tag{3.53}$$

For example, the Fourier transform of a periodic signal has a concentrated shape in the frequency domain and its phase varies slowly with respect to time in the vicinity of the oscillation frequency, while elsewhere the phase variation is rapid and leads to destructive interference; thus, most contributions to the reconstructed signal come from areas with stationary phase. Analogously, impulsive signals that have concentrated form in time have slow phase variations with respect to frequencies in the vicinity of the time location where the impulsive signal occurs.

Equation (3.53) is the basis for the concept of *time-frequency reassignment* [32] using spectrograms, which are defined as the squared magnitude of the short-time Fourier transform

$$P_{\text{sp}}(t,\omega) \equiv \left|S(t,\omega)\right|^2 = \left|Y(t,\omega)\right|^2. \tag{3.54}$$

Instead of using the coordinates (t,ω) for the calculated quantity $P_{\text{sp}}(t,\omega)$, the reassignment procedure uses the coordinates $(\hat{t},\hat{\omega})$; denoting the reassigned spectrogram as $\hat{P}_{\text{sp}}(\hat{t},\hat{\omega})$, we have

$$\hat{P}_{\text{sp}}(\hat{t},\hat{\omega}) = P_{\text{sp}}(t,\omega). \tag{3.55}$$

The reassigned coordinates in equations (3.53) have exact representations in terms of three STFT functions $S(t,\omega)$, $S_t(t,\omega)$, and $S_d(t,\omega)$, computed using equation (3.48) and three window functions, namely, the original window $w(t)$, the time-weighted window $tw(t)$, and the time-differentiated window dw/dt, respectively [33]. Changing the variable τ to t in (3.53), the new coordinates are

$$\hat{t} = t - \boldsymbol{Re}\left[\frac{S_t(t,\omega)S^*(t,\omega)}{|S(t,\omega)|^2}\right], \quad \hat{\omega} = \omega + \boldsymbol{Im}\left[\frac{S_d(t,\omega)S^*(t,\omega)}{|S(t,\omega)|^2}\right]. \tag{3.56}$$

The time-differentiated window is usually calculated using the Fourier transform relations $w(t) \leftrightarrow W(\omega)$ and $dw(t)/dt \leftrightarrow i\omega W(\omega)$.

The spectrogram is an example of a time-frequency distribution, a useful analysis tool that delineates the frequency content of the signal as a function of time, similarly to the way in which a musical score denotes an individual tone in a piece as time progresses. The spectrogram is a fixed resolution technique because the time resolution T of the window fixes the frequency resolution at $1/T$. To localize a frequency component in time using a spectrogram, one must choose a very short window, which will inevitably lead to poor frequency resolution. Conversely, to increase the frequency resolution, one must take Fourier transforms of long sections of the data, clearly worsening time localization. Notwithstanding its inherent resolution problems, the spectrogram has been successful in part because most experimental data have temporal variations that are "slow." The reassignment correction to the spectrogram often produces superior results to an otherwise smeared spectrogram as shown in section 3.9.

3.9 The Discrete Time Fourier Transform

The discrete time Fourier transform (DTFT) of a sequence x_n is found by evaluating its **Z** transform $S(z)$ on the unit circle $z = \exp(i\omega)$,

$$X\left(e^{i\omega}\right) = \sum_{n=-\infty}^{\infty} x_n\, e^{-in\omega}, \tag{3.57}$$

with the inverse transform [5],

$$x_n = \frac{1}{2\pi}\int_{-\pi}^{\pi} X\left(e^{i\omega}\right)\, e^{in\omega}\, d\omega. \tag{3.58}$$

Parseval's relation is

$$\sum_{n=-\infty}^{\infty} |x_n|^2 = \frac{1}{2\pi}\int_{-\pi}^{\pi} \left|X(e^{i\omega})\right|^2\, d\omega, \tag{3.59}$$

while the Fourier transform of a convolution equation is

$$\mathcal{F}^{-1}\{x * y[n]\} = X(e^{i\omega})Y(e^{i\omega}). \tag{3.60}$$

The discrete short-time Fourier transform for the shifted (real, symmetric and finite support) window is

$$X_{\mathrm{w}}(n,\omega) = \sum_{m=-\infty}^{\infty} x_m\, w[n-m]\, e^{-im\omega} \equiv \left|X_{\mathrm{w}}(n,\omega)\right|\, e^{i\phi_w(n,\omega)}, \tag{3.61}$$

[5] We denote by $X(\omega)$ the Fourier transform of a continuous time variable and by $X(e^{i\omega})$ the Fourier transform of a discrete time sequence since the latter is a periodic function of ω with period 2π. Consequently, the inversion formula is an integral over any interval of length 2π, which, without loss of generality, we take to be the interval $[-\pi,\pi]$.

while the shifted signal version is

$$X_s(n,\omega) = \sum_{m=-\infty}^{\infty} x_{m+n}\, w[-m]\, e^{-im\omega} \equiv \left|X_s(n,\omega)\right| e^{i\phi_s(n,\omega)}. \tag{3.62}$$

The spectrogram is $P_{sp}(n,\omega) = \left|X_w\right|^2 = \left|X_s\right|^2$.

In practice we use a discrete time (complex) sequence x_n, $0 \le n \le N-1$, representing the values of a continuous time signal $x(t)$ sampled at $f_s = 1/\Delta T$ Hz, i.e., $x_n = x[n\Delta T]$. The Fourier transform (3.42) is then approximated by a sum (excluding the overall factor ΔT)

$$X(\omega) = \sum_{n=0}^{N-1} x_n e^{-in\Delta T\,\omega}. \tag{3.63}$$

Setting $\omega = 2\pi f$ and evaluating (3.63) at a discrete set of N frequency values $f_k = kf_s/N, 0 \le k \le N-1$, we obtain the discrete Fourier transform (DFT),

$$X_k = \sum_{n=0}^{N-1} x_n e^{-2i\pi kn/N}, \quad 0 \le k \le N-1. \tag{3.64}$$

When N is even the elements for $k = 0, \ldots, N/2-1$ are the non-negative frequency components in the range $[0, f_s/2)$ while the elements for $k = N/2, \ldots, N-1$ are the negative frequency components in the range $[-f_s/2, 0)$. If N is odd, then the non-negative range is $k = 0, \ldots, (N-1)/2$ while the negative range is $k = (N+1)/2, \ldots, N-1$. The inverse DFT is

$$x_n = \frac{1}{N}\sum_{k=0}^{N-1} X_k e^{2i\pi kn/N}, \quad 0 \le n \le N-1. \tag{3.65}$$

We can view the inverse DFT as an expansion of $\boldsymbol{x} \in \mathbb{C}^N$ in terms of an orthonormal basis. Let us define $w_N \equiv \exp(2i\pi/N)$ and construct N vectors $\boldsymbol{e}_n \equiv \left[1, w_N^n, \ldots, w_N^{(N-1)n}\right]^T/\sqrt{N}$, $0 \le n \le N-1$; these vectors form an orthonormal basis for $\mathbb{C}^N$. Thus, $\boldsymbol{x} = \left[x_0, \ldots, x_{N-1}\right]^T$ has an expansion:

$$\boldsymbol{x} = \sum_{n=0}^{N-1} \langle \boldsymbol{e}_n, \boldsymbol{x}\rangle\, \boldsymbol{e}_n, \qquad \langle \boldsymbol{e}_m, \boldsymbol{e}_n\rangle = \delta_{mn}, \tag{3.66}$$

which is equivalent to equations (3.64) and (3.65). The DFT and its inverse can be written as matrix equations using an $N \times N$ complex and symmetric (not Hermitian) DFT matrix $\boldsymbol{F}$ for the vectors $\boldsymbol{X} = \left[X_0, \ldots, X_{N-1}\right]^T$ and $\boldsymbol{x} = \left[x_0, \ldots, x_{N-1}\right]^T$,

$$\boldsymbol{X} = \boldsymbol{F}\boldsymbol{x}, \quad \left[\boldsymbol{F}\right]_{kn} = \left[\boldsymbol{F}\right]_{nk} = w_N^{-nk}, \quad \text{and} \quad \boldsymbol{x} = \boldsymbol{F}^{-1}\boldsymbol{X}, \;\; \boldsymbol{F}^{-1} = \boldsymbol{F}^*/N = \boldsymbol{F}^+/N. \tag{3.67}$$

The algorithm to calculate the DFT and its inverse when N is a power of 2 is the fast Fourier transform (FFT). The squared magnitude of the Fourier transform $P_k = \left|X_k\right|^2$ is the frequency spectrum of x_n.

Equations (3.61) and (3.62) become finite sums for discrete values of frequency giving the discrete STFT coefficients X_{nk}, Y_{nk}, and discrete spectrogram coefficients P_{nk}, all of which suffer from the fixed resolution property discussed at the end of section 3.8. Figure 3.8 shows three spectrograms for the Bowhead whale sound of figure 1.1 calculated for three different window sizes. Evidently, increasing the FFT length increases the frequency resolution at the expense of time localization. Figure 3.9 shows improved resolution with three reassigned spectrograms using equations (3.55) and (3.56), for the same parameters as in the spectrograms of figure 3.8.

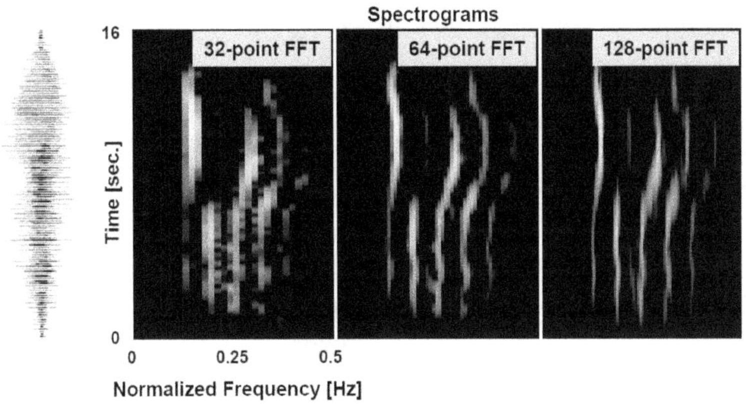

Figure 3.8: Three spectrograms and reassigned versions (Bowhead whale sound).

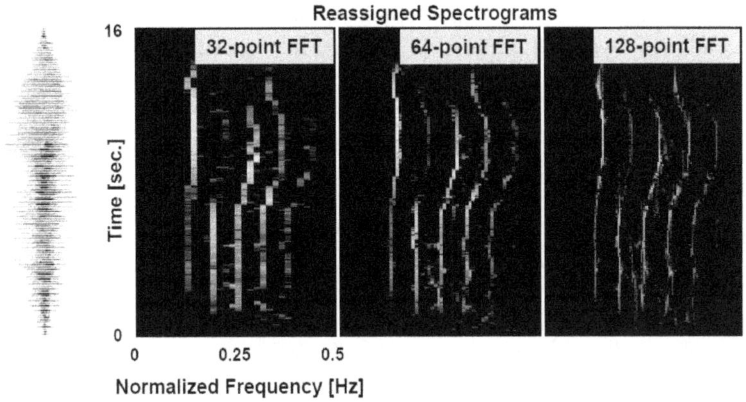

Figure 3.9: Three reassigned spectrograms.

3.10 The Chirp Z Transform

A closely related transform is the chirp **Z** transform (CZT) defined for a finite set of points z_k which can be inside, on, or outside the unit circle. For data with a finite number of samples, we have [34]

$$S_k = \sum_{n=0}^{N-1} s_n z_k^{-n}, \quad z_k = AW^{-k}, \ 1 \leq k \leq K\,, \tag{3.68}$$

where $A = A_0 \exp(2i\pi\theta_0)$ and $W = W_0 \exp(2i\pi\phi_0)$ are arbitrary complex numbers. When $A_0 = 1$, $\theta_0 = 0$, $W_0 = 1$, and $\phi_0 = -2i\pi/N$, we obtain the DFT of the sequence x_n. More generally, (3.68) gives the **Z** transform along a contour different from the unit circle, and that contour can spiral in and out with respect to the unit circle. Clearly, this transform can be calculated for as many points K as desired, where K is not necessarily equal to N. In addition, the angular spacing of the points z_k is arbitrary, and so this transform is sometimes referred to as the "zoom" **Z** transform. Equation (3.68) can be efficiently calculated using the Bluestein method (originally used to calculate the DFT using a convolution), which uses the identity $2nk \equiv n^2 + k^2 - (k-n)^2$ to write

$$X_k = \sum_{n=0}^{N-1} s_n \; A^{-n} \; W^{n^2/2} W^{k^2/2} W^{(k-n)^2/2}, \quad 1 \leq k \leq K. \tag{3.69}$$

Equation (3.69) can then be computed as follows:

- $p_n = s_n A^{-n} W^{n^2/2}$, for $0 \leq n \leq N-1$,
- $q_k = \sum_{n=0}^{N-1} p_n w_{k-n}$, $0 \leq k \leq K-1$, where $w_n = W^{-n^2/2}$,
- $X_k = q_k W^{k^2/2}$.

The second step involves a finite convolution that can be computed quickly as discussed below in section 3.11. The flexibility of the CZT has allowed for applications to digital filter design and spectral analysis of speech. Although the CZT algorithm is quite efficient, the inverse transform did not have a fast implementation until recently (fifty years after the CZT was first introduced) when an algorithm to compute the inverse CZT in $\mathcal{O}(N \log N)$ time was reported [35].

3.11 Finite Convolutions

The result of the convolution theorem for two sequences with finite number of elements x_n, $n = 0, 1, \ldots, N-1$, and h_m, $m = 0, 1, \ldots, M-1$, assuming $N \geq M$, is

$$x * h\,[n] = \sum_{m=0}^{M-1} x_{n-m} h_m, \quad m \leq n, \; n = 0, \ldots, N+M-2, \tag{3.70}$$

which has $N+M-1$ elements. It is often more efficient to use the inverse Fourier transform of the product of two Fourier transforms to compute the convolution sequence for two time series with finite number of elements. When one (or both) time series has (have) a large number of samples, computation of Fourier transforms using the FFT algorithm is a fast way to obtain convolutions between the given time series; this is done by first appending zeros to the end of both sequences to produce two sequences of length $N+M-1$,

$$\hat{x} \leftrightarrow [x_0, \ldots, x_{N-1}, 0, \ldots, 0] \;\text{ and }\; \hat{h} \leftrightarrow [h_0, \ldots, h_{M-1}, 0, \ldots, 0]\,, \tag{3.71}$$

and next computing the $(N+M-1)$-point discrete Fourier transforms, multiplying the Fourier transforms, and finally taking the inverse Fourier transform (all transforms are computed using the FFT). Thus, for $0 \leq l \leq N+M-2$ we have

$$x * h\,[l] = \frac{1}{N+M-1} \sum_{k=0}^{N+M-2} \hat{X}_k \hat{H}_k \; e^{2i\pi kl/(N+M-1)}. \tag{3.72}$$

When $N+M-1$ is not a power of 2, padding with zeros is performed to obtain sequence lengths that are powers of 2 exceeding $N+M-1$, and then using FFTs to perform the convolution, after which the first $N+M-1$ points are extracted.

For some applications, e.g., machine learning using convolutional neural networks (see section 12.11), we pad with zeros on both sides of the sequence to produce outputs that are centered at the middle of each data segment of length M that is assumed to be odd; thus, $(M-1)/2$ zeros are appended to the beginning and end of the sequence x_n, $0 \leq n \leq N-1$, resulting in a data sequence of $N+M-1$ elements. The filter coefficients are now aligned with the first M elements of the new data sequence, and the sum of the products of these M elements produces the first output sample at index 0. The last output sample at index $N-1$ is produced when the filter is aligned with the last M elements of the data; this is illustrated in figure 3.10 for data with 6 elements and a filter with 3 elements.

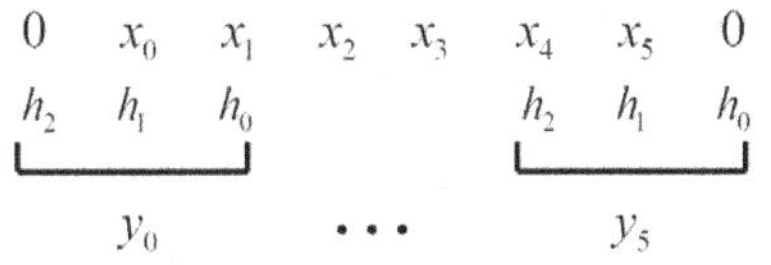

Figure 3.10: Centered convolution: 6×1 data and 3×1 filter.

The finite impulse response (FIR) forms of (3.1) for a filter of even or odd number of elements M are often written (for odd $M \equiv 2M'+1$ and even $M \equiv 2M'$, respectively)

$$x * h[n] = \sum_{m=-M'}^{M'} h_m x_{n-m}, \quad x * h[n] = \sum_{m=-M'}^{M'-1} h_m x_{n-m}. \tag{3.73}$$

For odd $N = 2N'+1$ we have $-N' \leq n \leq N'$, while for even $N = 2N'$ we have $-N' \leq n \leq N'-1$. It is, of course, understood that negative indices are mapped to non-negative indices for computer implementation through the associations $x_{-N'} \ldots \leftrightarrow x_0 \ldots x_{N-1}$, and $h_{-M'} \ldots \leftrightarrow h_0 \ldots h_{M-1}$. The representations in (3.73) are useful when non-causal FIR filters are used. For instance, as we shall discuss in chapter 4, when a non-causal least squares shaping filter h_m on an input data x_n is sought to match a desired time series d_n, the solution is found by minimizing the squared norm of the error sequence that includes data samples before and after the time index n, namely,

$$e_n \equiv d_n - \sum_{m=-M}^{M} h_m x_{n-m} \, . \tag{3.74}$$

On the other hand, solving for a causal FIR filter to estimate or predict a desired time series k steps ahead, $k = 0, 1, 2, \ldots$, only data samples prior to time index n can be used, and so the error sequence is

$$e_{n+k} \equiv d_{n+k} - \sum_{m=0}^{M-1} h_m x_{n-m} \, , \quad k = 0, 1, 2, \ldots \, . \tag{3.75}$$

3.12 The Cepstrum

The cepstrum or complex cepstral coefficients $c[n]$ of a discrete time sequence x_n are defined as the inverse **Z** transform of the natural logarithm of the sequence's **Z** transform $X(z)$, or more commonly, as the inverse Fourier transform of the natural logarithm of the sequence's Fourier transform $\ln X(e^{i\omega})$. The real cepstral coefficients of a sequence are defined as the inverse Fourier transform of the natural logarithm of the magnitude (or the squared magnitude) of the sequence's Fourier transform. Whether the cepstrum is real or complex should be clear from the context, so we use the same notation $c[n]$ and refer to both as the cepstral coefficients.

$$c[n] = \mathcal{Z}^{-1}\{\ln X(z)\}, \quad c[n] = \mathcal{F}^{-1}\{\ln X(e^{i\omega})\}, \tag{3.76a}$$

$$c[n] = \mathcal{F}^{-1}\{|\ln X(e^{i\omega})|\}, \text{ or } c[n] = \mathcal{F}^{-1}\left\{|\ln X(e^{i\omega})|^2\right\}. \tag{3.76b}$$

If $w_n = x_n * y_n$ is a convolution of two sequences, then $\ln W(e^{i\omega}) = \ln X(e^{i\omega}) + \ln Y(e^{i\omega})$, and the (complex or real) cepstral coefficients of w are equal to the sum of the (complex or real) cepstral coefficients of x and y, i.e., $\boldsymbol{c}^{(w)} = \boldsymbol{c}^{(x)} + c^{(y)}$; this property of the cepstral coefficients allows them to have applications to *blind deconvolution* problems when the received signal is the output of the convolution of a source signal with an unknown transmission channel response and the channel response is to be eliminated from the received signal. Another application is to echo cancellation when a received signal is the sum of an original transmitted signal and a possibly attenuated and time shifted version of the original signal, for instance, $x_n = s_n + \beta s_{n-N}$ when the cepstrum of the received sequence is

$$c^{(x)}[n] = c^{(s)}[n] + \mathcal{F}^{-1}\{\ln|1 + \beta^2 + 2\beta\cos(\omega N)|\}. \tag{3.77}$$

The index n is not in the time domain of the original signal but in a pseudo-time domain known as the *quefrency*. To remove the echo, we note that its contribution to the cepstral coefficients of x is the second term on the right hand side of (3.77); this contribution is the sum of a collection of delta distributions of different amplitudes separated by N. The removal of the echo contribution is known as *liftering*—the equivalent of filtering but in the quefrency domain. This method of echo suppression removes all phase information; to preserve the phase information, we should use the complex cepstrum.

Although the main difference between the real and complex cepstrua is the absence of phase information in the former, the real cesptrum is widely used in speech applications that are not phase-sensitive, e.g., vocoders. As we shall see in section 9.9, the most important application of cepstral coefficients is to the representation of speech using a linear predictive model known as linear predictive coding or LPC, which includes a minimum phase system function, rendering the real and complex cepstral coefficients essentially identical.

An important application of the real cepstrum is to the processing of speech. The convolutional model of speech (also known as the source-filter model), namely, an excitation sequence convolved with the system function of the vocal tract $s_n = e_n * v_n$, makes the cepstrum an essential tool in separating the two components. Taking Fourier transforms (in practice we work on short duration frames of speech, and so the frequency spectrum is simply the squared magnitude or just the magnitude of the frame's Fourier transform), we view the speech frequency spectrum $S(e^{i\omega})$ as the product of two components: a rapidly varying part $|E(e^{i\omega})|$ due to the excitation sequence and a slowly varying part $|V(e^{i\omega})|$ due to the vocal tract

$$S(e^{i\omega}) = |E(e^{i\omega})| \times |V(e^{i\omega})|.$$

If we wish to suppress one of the components using linear filtering, we must take logarithms; this method of transforming a convolutional model into an additive model is generally referred to as *homomorphic signal processing* [1]. For the speech signal, we have

$$\ln S(e^{i\omega}) = \ln|E(e^{i\omega})| + \ln|V(e^{i\omega})|,$$

where the original slowly varying vocal tract component v_n corresponds to the slowly varying low-quefrency part c_v, while the rapidly varying excitation sequence corresponds to the rapidly varying high-quefrency portion c_e.

The left panel of figure 3.11 shows a frame of speech 120 milliseconds long sampled at 8 kHz, and the right panel shows its log spectrum with a pitch frequency of approximately 150 Hz (measured between the harmonics). If the signal model in the time domain is that of a convolution between an excitation and the vocal tract response, then this figure shows the sum of the two components: a slowly varying envelope (the vocal tract response) and a rapidly varying excitation component.

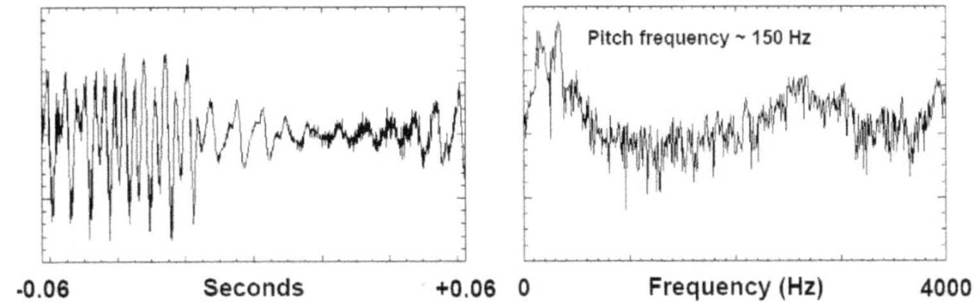

Figure 3.11: A frame of speech and its log spectrum.

Taking an inverse Fourier transform, we obtain the cepstral coefficients that are now in the quefrency domain, shown on the left of figure 3.12. The pitch period is seen as the spike at 0.0065 seconds from the middle and corresponds to the pitch frequency of 153.8 Hz. The two components can now be separated by a liftering operation ("low-time" and "high-time" masking) in the quefrency domain [1]. The right hand side of figure 3.12 shows the low frequency component $\ln |V(e^{i\omega})|$ that results from "low time" liftering followed by Fourier transformation. Figure 3.13 shows the cepstral processing of a frame of a speech sequence.

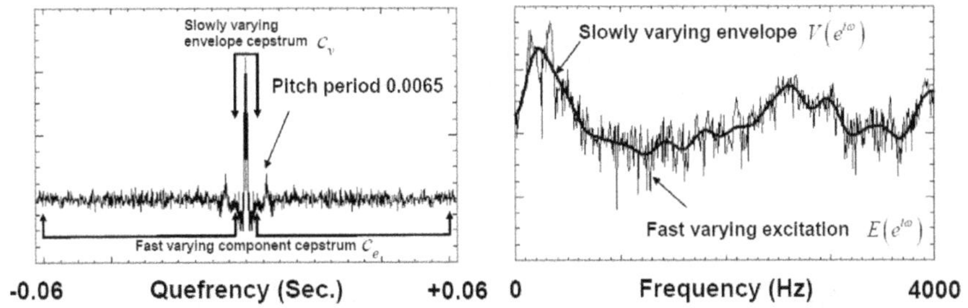

Figure 3.12: The cepstrum of the data in figure 3.11 with the indicated pitch period (left), and the extracted low frequency component (right).

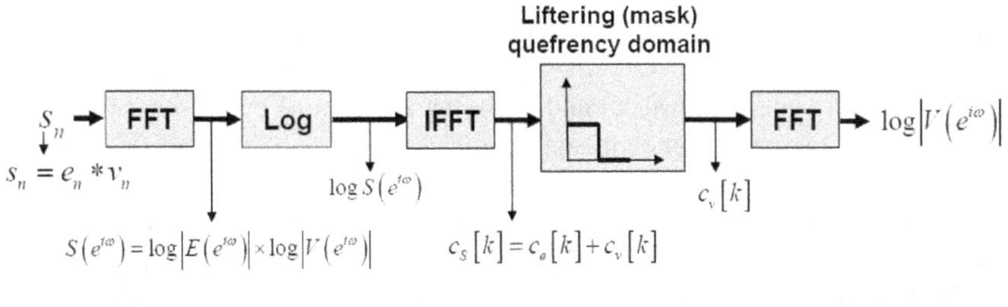

Figure 3.13: Extracting the vocal tract convolutional component.

The STFT is often used in pitch estimation (in single frame cepstral analysis). Human auditory perception of pitch, however, does not follow a linear scale. A *mel* (from the word *melody* to indicate a scale based on comparison of pitch) is a unit measuring perceived pitch or frequency of a tone with an arbitrary frequency of 1000 Hz chosen to represent 1000 mels. Experiments in which physical frequencies were changed until listeners perceived twice, 10 times, half, and one-tenth the reference pitch were conducted, and those physical frequencies were marked as 2000, 10,000, 500, and 100 mels, respectively. Although no single mapping between real frequencies and mel frequencies exists, a popular one is $f_{\text{mel}} = 2595 \, \log_{10}\left(1 + f_{\text{Hz}}/700\right)$. A speech recognition system can then use the *mel cepstrum* (either based on spectral values at a number of mel critical frequencies, or by integrating the spectrum of the frame over a number of mel critical frequency bands, usually $15 - 40$ bands, whose bandwidths change with frequency) for the usual cepstral processing. More recent cepstral processing techniques include quantities known as Δ and $\Delta\Delta$ cepstra, which are polynomial approximations to the first and second derivatives of cepstral coefficients belonging to the adjacent 1 or 2 frames [36]; these quantities contain dynamic information about the changes in the cepstral coefficients.

Figure 3.14 shows the steps in the calculation of mel frequency cepstral features, including Δ and $\Delta\Delta$ features (the Discrete Cosine Transform (DCT) step is sometimes included to decorrelate the log-spectra of frames). Speaker models are then built using the cepstral feature vectors in Gaussian Mixture Models.

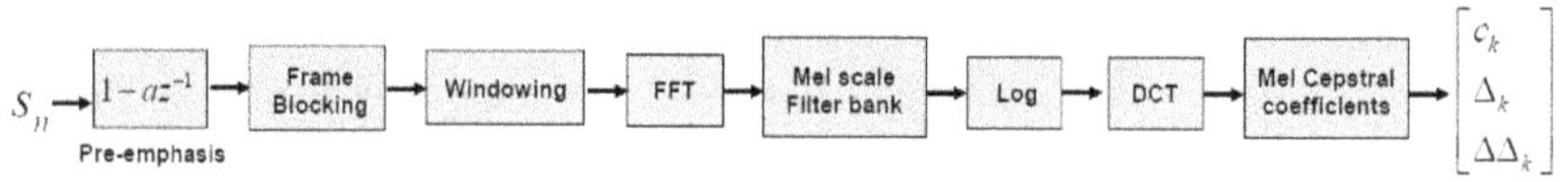

Figure 3.14: Calculation of mel frequency cepstral feature vectors.

Earlier methods of speech processing used LPC (see section 9.9) when the linear prediction (LP) parameters were used as features in speech recognition systems. Cepstral methods (including the mel cepstrum) based on LP parameters have been shown to improve recognition rates and so have replaced the direct use of LP parameters [37]. The relationship between the LP parameters of an all-pole minimum phase system function and its associated cepstral coefficients will be derived in theorem 49 of section 9.9.

3.13 The Orthogonal Discrete Wavelet Transform

In sections 1.12 and 1.13 we introduced multi-resolution analysis (MRA) subspaces and showed the Haar orthonormal basis for $L_2(\mathbb{R})$. The general scaling and wavelet equations (1.76) and (1.84) are applicable to all orthogonal wavelet systems (specialization to the Haar system was shown in equations (1.64) and (1.66)). Each finite order MRA system with orthogonal basis is characterized by a pair of *quadrature mirror filters* (QMF), namely, $h_0[n]$ and $h_1[n]$ with even number of elements; $h_0[n]$ is a half-band low-pass filter, while $h_1[n]$ is a half-band high-pass filter. For instance, the Haar filters define the Daub2 system with low-pass filter coefficients $h_0[n] = [1/\sqrt{2},\ 1/\sqrt{2}]$, and $h_1[n]$ is found by the "alternating flip" of $h_0[n]$, i.e., reversing the order of elements of $h_0[n]$ and changing the sign of every other term so that $h_1[n] = [1/\sqrt{2},\ -1/\sqrt{2}]$. The Daub4 low-pass filter elements are $h_0[n] = \left[1+\sqrt{3},\ 3+\sqrt{3},\ 3-\sqrt{3},\ 1-\sqrt{3}\right]/4\sqrt{2}$ and $h_1[n]$ is found by the alternating flip method.

The key to the fast orthogonal discrete wavelet transform is in the interpretation of equation (1.92) describing recursive relations for the scaling coefficients c_{mn} and the wavelet coefficients d_{mn}. According to equation (1.92) the scaling coefficients c_{mn} at scale m are obtained from the coefficients $c_{m-1,n}$ at the previous finer scale $m-1$ by correlating them with the low-pass filter coefficients $h_0[n]$ (or equivalently, convolving with the time-reversed version $h_0[-n]$), and then keeping all the even numbered samples $2n$, i.e., down sampling by a factor of 2. A similar interpretation applies to the wavelet coefficients, except that the correlation is performed with the coefficients $h_1[n]$ (or equivalently, convolution with the time-reversed version $h_1[-n]$), followed by down sampling by a factor of 2.

The "Analysis box" on the left of figure 3.15 shows the results expressed in equation (1.92) and is applicable to all orthogonal discrete wavelets (including Haar). In addition, the inverse operation to the analysis portion of figure 3.15 is shown on the right hand side, namely, the "Synthesis box." The analysis and synthesis operations show a single level (scale) application of the orthogonal discrete wavelet transform starting with the approximation (scaling) coefficients $c_{m-1,n}$ at scale $m-1$, producing two sets of coefficients at scale m: the coarser approximation c_{mn} and the error or detail (i.e., the difference between the approximations at the two levels) d_{mn}. This one level transform is exactly invertible using the operations indicated in the "Synthesis" box of figure 3.15: both the approximation and error (detail) coefficients at level m are up sampled by a factor of 2 (through single zero insertions between samples), after which each is convolved with $h_0[n]$ and $h_1[n]$, respectively, and the results are added to reproduce (reconstruct) the original input sequence.

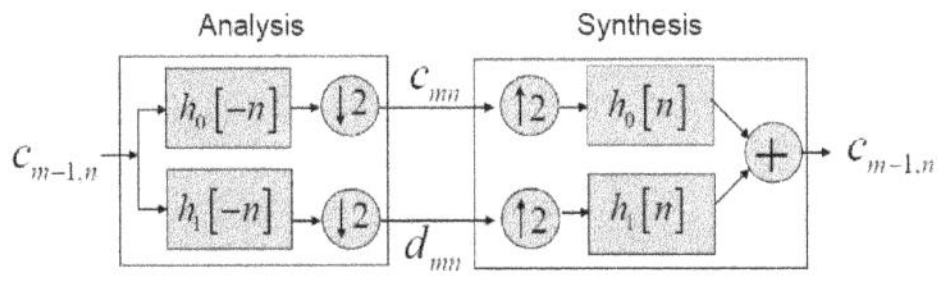

Figure 3.15: Linear filter theory interpretation of equations (1.92).

The orthogonal discrete wavelet transform of x_n, $n = 0, \ldots, N-1$ (assuming N to be a power of 2) is based on repeated application of the analysis portion of figure 3.15 to the scaling c coefficients starting with the data x_n as the first input and keeping the wavelet coefficients d at every stage; at the last stage, we keep the remaining c coefficients (these represent the highest scale approximation to the original data) in addition to the final d coefficients. The number of levels (scales) depends on N and the number of filter coefficients; for instance, for $N = 1024$ and the Daub4 system with filters of length 4, there are 512 d-coefficients at the first level. At the second level, there are 256 d coefficients, and so on, until at level 5 there are 32 d coefficients and the same number of c coefficients. We normally stop at a level for which the number of c or d coefficients is no less than 8 times the number of filter elements, which in this example is level 5, at which there are 992 d coefficients (from 5 scales) and 32 c coefficients from level 5 only. Figure 3.16 shows a three-level finite resolution forward and inverse wavelet transform.

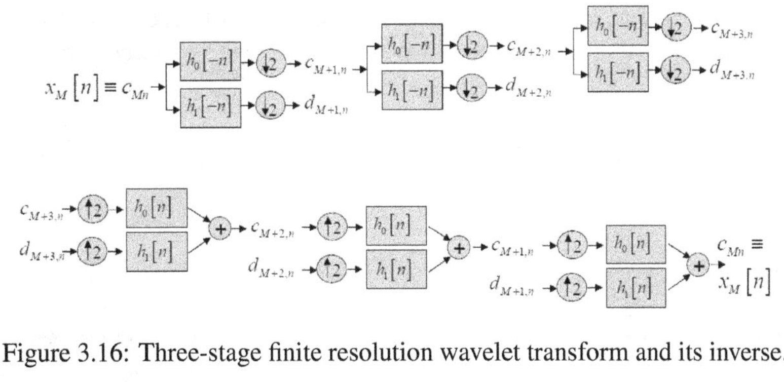

Figure 3.16: Three-stage finite resolution wavelet transform and its inverse.

Consider a denoising application of the Haar wavelet for a 256-point binary signal with added white noise with zero mean and unit variance shown in figure 3.17-(a). We apply a six-level transform using the Haar wavelet; figures 3.17-(b),(c) show the transform coefficients for both the original and the noisy signals. Clearly, the noise coefficients occupy levels $1-4$, while the signal occupies levels $4-7$. In order to

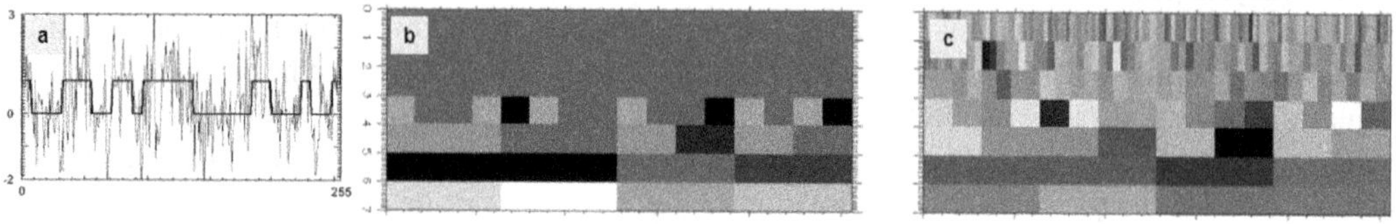

Figure 3.17: Signal, noise, and their six-level Haar wavelet transforms.

leave the signal unaffected, we zero out the coefficients in the first three levels and then perform an inverse Haar transform. Figure 3.18 is a plot of the recovered signal together with the original and shows excellent noise cancellation: the original signal levels are more or less derivable from the reconstructed signal.

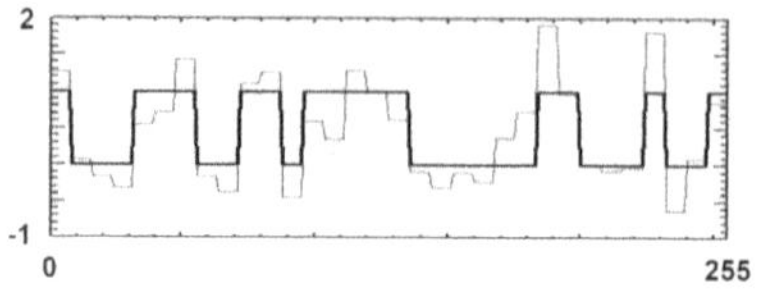

Figure 3.18: Original and denoised signal.

This example illustrates a general concept: additive noise may be thought of as a low-scale (high-frequency) phenomenon. Thus, suppressing lower scale coefficients should, in general, produce a less noisy signal. However, in applications to image data, we must be far more careful when analyzing the lower scale coefficients with large magnitudes because these are also the same locations at which edge information resides.[6] So instead of a simple suppression of all coefficients in lower scales, a *thresholding strategy* must be used. The threshold values in each scale must be chosen carefully, assuming that coefficients with edge information exceed the calculated thresholds. A simple method to calculate a threshold for a scale is to find the variance σ^2 of that scale's wavelet coefficients and then set the threshold for that scale equal to $\alpha\sigma$ for some positive constant α, which may vary for different scales. More sophisticated procedures are based on fitting a statistical distribution function to the observed wavelet coefficients and then choosing thresholds based on percentiles of the fitted distribution. Thresholding techniques can also be used to detect edges in an image, as opposed to local noise at a given scale, using the fact that edge information is present at the same location of the image at all scales. Once a threshold T has been determined, it can be applied either as a hard or a soft policy, i.e.,

$$y(x) = 0, \ \ |x| < T, \quad \text{and for } |x| > T: \ y(x) = x \leftrightarrow \text{hard}, \quad y(x) = x - T\,\mathbf{sign}(x) \leftrightarrow \text{soft}.$$

[6]The disadvantage of using the Fourier transform in detecting edges in an image is that the basis functions are not localized, and so edge information spreads across all spatial frequencies.

3.14 The Hilbert Transform Relations

The problem of inverting a system function h can be simplified by the minimum phase, or minimum group delay, property: h is minimum phase if it is stable and causal and its inverse $\tilde{h}$ is also stable and causal, as described in 3.1. This property can be studied most easily in a pole-zero system function (continuous or discrete time). As noted in section 3.2, a continuous time pole-zero system function with zeros and poles representing a stable and causal system must have all of its poles in the left half complex s plane; if its inverse is also causal and stable, then all of its zeros must be in the left half complex s plane, too. For discrete time system functions, all zeros and poles must be inside the unit circle of the complex z-plane. Examples of such minimum phase system functions were shown in figures 3.3 and 3.6 for continuous time functions and discrete time sequences, respectively. In this section, we consider the Fourier transform $H_{\text{mnp}}(\omega)$ and discuss an important relation between its phase and magnitude if the associated time function is to be minimum phase.

The *Hilbert transform* [7] of a real function $x(t)$, $t \in \mathbb{R}$, is defined by [38]

$$\mathcal{H}\{x(t)\} \equiv \frac{1}{\pi}\int_{-\infty}^{\infty} \frac{x(\tau)}{t-\tau}d\tau. \tag{3.78}$$

The integral on the right hand side of (3.78) is understood as a Cauchy Principal Value, i.e., it is the limit $\varepsilon \to 0$ of the integral over the range $|t-\tau| > \varepsilon > 0$. The integral is a convolution in the time domain, and so its Fourier transform is the product of the Fourier transforms of $x(t)$ and $1/\pi t$. To find the Fourier transform of $1/t$, consider $\omega \geq 0$ and the integral along the closed contour in the complex t plane as shown in figure 3.19; this can be broken up into four integrals:

$$\oint \frac{e^{-i\omega t}}{t}dt = \int_{-R}^{-\varepsilon} \frac{e^{-i\omega t}}{t}dt + \int_{C_\varepsilon} \frac{e^{-i\omega t}}{t}dt + \int_{\varepsilon}^{+R} \frac{e^{-i\omega t}}{t}dt + \int_{C_R} \frac{e^{-i\omega t}}{t}dt. \tag{3.79}$$

The integral along the closed contour vanishes since the integrand has no poles inside the contour; the integral along the large semi-circle C_R tends to 0 as $R \to \infty$ by Jordan's lemma and the assumption $\omega \geq 0$, while the limit $\varepsilon \to 0$ of the integral along the small semi-circle C_ε is found after the substitution $t = \varepsilon e^{i\theta}$, $\theta \in [\pi, 2\pi]$ to be $-i\pi$. When $\omega < 0$, we close the two semicircles in the upper half of the complex plane and obtain the value $+i\pi$ for the integral on the small semi-circle as $\varepsilon \to 0$. Therefore, we write $\mathcal{F}\{1/t\} = -i\pi$ for $\omega \geq 0$ and $\mathcal{F}\{1/t\} = +i\pi$ for $\omega < 0$ and

$$\mathcal{H}\{x(t)\} = \mathcal{F}^{-1}\{X(\omega)(-i)[\theta(\omega) - \theta(-\omega)]\}. \tag{3.80}$$

Thus, the Hilbert transform of a real function $x(t)$ is found by multiplying the positive frequencies of $X(\omega)$ by $-i$, multiplying the negative frequencies by $+i$, and then taking the inverse Fourier transform. For instance, if $x(t) = \cos(\omega_0 t)$, $\omega_0 > 0$, then using $\cos(\omega_0 t) = (e^{+i\omega_0 t} + e^{-i\omega_0 t})/2$, we have $\mathcal{H}\{\cos(\omega_0 t)\} = \sin(\omega_0 t)$. Similarly, we find $\mathcal{H}\{\sin(\omega_0 t)\} = -\cos(\omega_0 t)$. Thus,

$$\mathcal{H}\left\{e^{i\omega t}\right\} = -i\,\textbf{sign}(\omega)\,e^{i\omega t}, \quad \omega \in \mathbb{R}. \tag{3.81}$$

[7] The Hilbert transform is named after the mathematician David Hilbert who first used it in 1905 when working on the Riemann hypothesis; it was proven to be a bounded linear operator on $L_p(\mathbb{R})$ functions by Riesz in 1928. The Hilbert transform arises in many applications including signal processing, optics, circuit theory, and control engineering. In optics, it is often known as the *dispersion relation*, *Kramers-Kronig relation*, or *Cauchy principal value integral*. In circuits and control theory, it is known as the *Bode relation*. The Hilbert transform has important biomedical applications including the extraction of the QRS complex in electrocardiography (ECG).

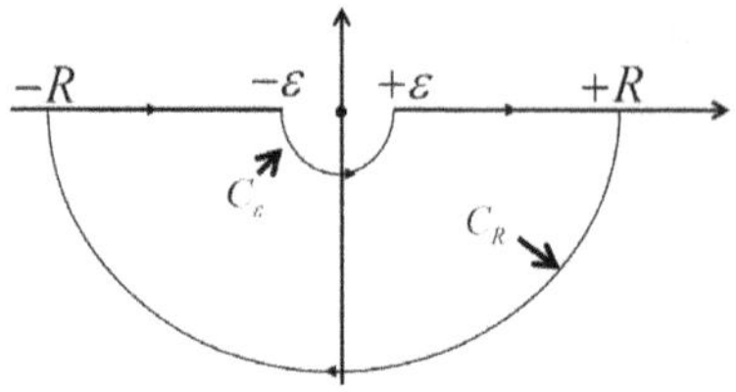

Figure 3.19: Contour for evaluating the Fourier transform of $1/t$ when $\omega > 0$.

The defining property of a minimum phase system function $H_{\text{mnp}}(\omega)$ is that its phase function is the negative of the Hilbert transform of the natural logarithm of its magnitude (this follows from theorem 14 for front loaded causal signals), i.e.,

$$\text{Phase}\left\{H_{\text{mnp}}(\omega)\right\} = -\,\mathcal{H}\left\{\ln\left|H_{\text{mnp}}(\omega)\right|\right\}. \tag{3.82}$$

Minimum phase discrete time functions obey the same Hilbert transform relation. Figure 3.20 refers to the minimum phase system function examples in sections 3.3 and 3.5, figures 3.3 and 3.6: the agreement between the phase of the Fourier transform and the Hilbert transform of the magnitude of the Fourier transform of the system functions is nearly perfect for the Laplace transform example and perfect for the **Z** transform case; the very small discrepancies in the Laplace transform case (difficult to see at this scale) are due to the fact that a discrete Fourier transform is used to compute the Hilbert transform for the continuous frequency variable $\omega \in \mathbb{R}$.

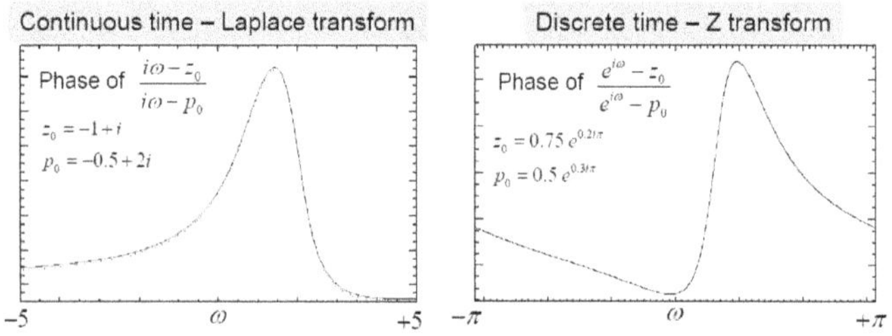

Figure 3.20: Comparison of phase functions.

The Hilbert transform relation between the phase and log-magnitude of a signal frequency spectrum allows us to solve the problem of reconstructing a minimum phase system function H_{mnp} from its magnitude-spectrum $S_{\text{mnp}} = |H_{\text{mnp}}|^2$ alone: the phase of H_{mnp} is the negative of the Hilbert transform of the (natural) logarithm of the square root of the spectrum:

$$\text{Phase}\left\{H_{\text{mnp}}\right\} = -\,\mathcal{H}\left\{\ln|H_{\text{mnp}}|\right\} = -1/2\,\mathcal{H}\left\{\ln S_{\text{mnp}}\right\}. \tag{3.83}$$

The Hilbert transform of a periodic signal $x(t)$ with period $2T$ can be derived using the complex Fourier series representation

$$x(t) = \sum_{n=-\infty}^{+\infty} c_n\, e^{i\pi n t/T}, \quad c_n = \frac{1}{2T}\int_{-T}^{+T} x(t) e^{-i\pi n t/T}\, dt. \tag{3.84}$$

Using (3.81) and $\mathcal{H}\{c\} = 0$ we find,

$$\mathcal{H}\{x(t)\} = -i \sum_{n=1}^{\infty} c_n \left(e^{+\frac{i\pi nt}{T}} - e^{-\frac{i\pi nt}{T}}\right) = \frac{1}{2T} \int_{-T}^{+T} x(t') \sum_{n=1}^{\infty} 2 \sin\left(\frac{n\pi(t-t')}{T}\right) dt'. \tag{3.85}$$

The sum inside the integral on the right hand side of (3.85) can be evaluated using

$$2 \sum_{n=1}^{\infty} \sin\left(\frac{n\pi t}{T}\right) = 2 \, \boldsymbol{Im} \sum_{n=1}^{\infty} e^{\frac{in\pi t}{T}} = \cot\left(\frac{\pi t}{2T}\right), \tag{3.86}$$

which is valid in the sense of distributions (generalized functions), i.e.,

$$2 \, \boldsymbol{Im} \lim_{\varepsilon \to 0} \sum_{n=1}^{\infty} e^{-\varepsilon n + \frac{in\pi t}{T}} = 2 \, \boldsymbol{Im} \lim_{\varepsilon \to 0} \frac{e^{-\varepsilon + \frac{i\pi t}{T}}}{1 - e^{-\varepsilon + \frac{i\pi t}{T}}} = \cot\left(\frac{\pi t}{2T}\right). \tag{3.87}$$

Thus, the Hilbert transform of a periodic function is:

$$\mathcal{H}\{x(t)\} = \frac{1}{2T} \int_{-T}^{+T} x(t') \cot\left[\frac{\pi(t-t')}{2T}\right] dt', \tag{3.88}$$

where the integral is understood as a Cauchy Principal Value [see the discussion following (3.78)].

3.15 The Analytic Signal and Instantaneous Frequency

The Hilbert transform of a real signal $x(t)$ is used to construct the *analytic signal*; a name coined by Gabor who used it to describe a signal with a single-sided Fourier transform [8] and defined by

$$x^{[a]}(t) \equiv x(t) + i\,\mathcal{H}\{x(t)\}. \tag{3.89}$$

For instance, the analytic signal associated with $\cos(\omega_0 t)$ is $e^{+i\omega_0 t}$. Since the signal $x(t)$ is real, its spectrum is an even function of frequency, i.e., $|X(\omega)| = |X(-\omega)|$, and so

- $\left|X^{[a]}(\omega)\right| = 2\left|X(\omega)\right|$ when $\omega > 0$.
- $\left|X^{[a]}(0)\right| = \left|X(0)\right|$, and $\left|X^{[a]}(\omega)\right| = 0$ when $\omega < 0$.
- $\int_{-\infty}^{\infty} \left|X(\omega)\right|^2 d\omega = \frac{1}{2} \int_{-\infty}^{\infty} \left|X^{[a]}(\omega)\right|^2 d\omega.$

[8] The history of single-sideband signals goes back to Carson who invented *single side-band radio* in 1915 to reduce bandwidth of communications over copper transmission lines by filtering one-half of the symmetric spectrum of a real signal. Hartley from Bell laboratories used the Hilbert transform to generate single side-band radio signals in 1925; his method became known as the *phase shift method* [see (3.79)]. Vakman in 1972 proved that the imaginary part of an "analytic signal" must be the Hilbert transform of its real part if the following three conditions are imposed: (1) a small change in the amplitude of the real signal results in a small change in the amplitude, (2) multiplying the real signal by a real number does not affect the phase of the complex signal, (3) a real sinusoid of the form $a\cos(\omega t + \phi)$ corresponds to a complex signal with amplitude a and phase $\omega t + \phi$.

The reason for the term "analytic" is the fact that the analytic signal defined by (3.89) for a periodic function [see (3.88)] can be viewed as the boundary value of a *complex analytic function* of a complex variable z whose real and imaginary parts satisfy the Cauchy-Riemann equations [see the discussion following (3.6)]. Defining $z = r\exp(it)$, $t \in [-\pi, +\pi]$, and the complex function $Y(z) = X(z) + i\,\tilde{X}(z)$, where

$$X(z) = \frac{1}{2\pi}\int_0^{+2\pi} \frac{1-r^2}{1-2r\cos(t-\tau)+r^2}\, x(\tau)\, d\tau = \frac{1}{2\pi}\int_0^{2\pi} \mathcal{P}_r(t-\tau)x(\tau)\, d\tau, \tag{3.90a}$$

$$\tilde{X}(z) = \frac{1}{2\pi}\int_0^{2\pi} \frac{2r\sin(t-\tau)}{1-2r\cos(t-\tau)+r^2}\, x(\tau)\, d\tau = \frac{1}{2\pi}\int_0^{2\pi} \mathcal{Q}_r(t-\tau)x(\tau)\, d\tau, \tag{3.90b}$$

then $Y(z)$ is analytic and its boundary value as $r \to 1$ is the analytic signal $x^{[a]}(t)$. $\mathcal{P}_r$ is the Poisson kernel defined in (3.41) and $\mathcal{Q}_r$ is the conjugate Poisson kernel.

A simple application of the analytic signal is the extraction of the envelope of an amplitude modulated waveform. For instance, consider the modulated signal $x(t) = \cos(2\pi f_0 t)\exp(-100t^2)$ sampled at 1 kHz with $f_0 = 50$ Hz for $-4 \le t \le +4$ seconds. The ideal discrete time Hilbert transform filter h_n, $n \in \mathbb{Z}$, is given by $h_0 = 0$ and

$$h_n = \frac{i}{2\pi}\int_{-\pi}^{0} e^{in\omega} d\omega - \frac{i}{2\pi}\int_0^{\pi} e^{in\omega} d\omega = \frac{2}{\pi n}\sin^2\left(\frac{n\pi}{2}\right),\ n \neq 0, \tag{3.91}$$

which can be implemented as an FIR filter using a window function. The envelope is the magnitude of the analytic signal $\left|x^{[a]}\right|$, as shown in figure 3.21. This example illustrates the concept of instantaneous frequency as defined by the time derivative of the phase of the analytic signal; in this example, the instantaneous frequency is, of course, f_0. An important application of the analytic signal is the construction of the Wigner time-frequency energy distribution function.

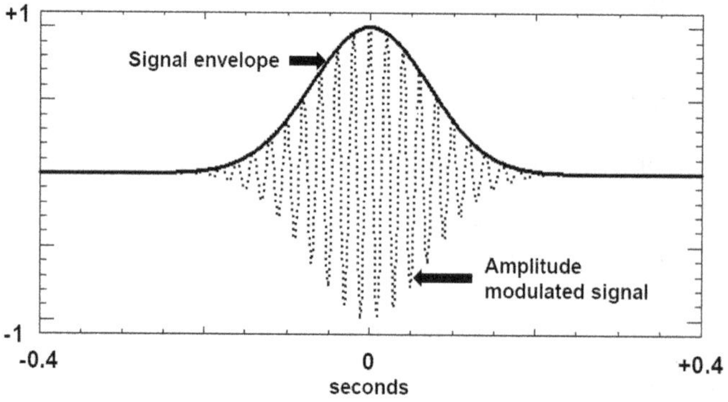

Figure 3.21: Amplitude modulated signal and its extracted envelope.

3.16 Time-Frequency Distribution Functions

Time-frequency distribution functions are powerful tools for analysis of signals whose frequency content may change with time. Capturing this change with high time and frequency resolutions has been the subject of much work in the last several decades. In section 3.8 we described the spectrogram as an example of a time-frequency distribution and noted that it suffers from poor joint resolution when the frequency content changes rapidly with time. Time-varying spectra were studied in the classical works of Gabor, Ville, Page, and Wigner [39, 40]. Their work was not motivated by improving on the spectrogram, but by a desire to construct a joint time and frequency distribution of the energy of a waveform based on general mathematical principles. The Wigner distribution is one in a class of time-frequency functions that have offered a viable approach [41]; Wigner was primarily concerned with constructing such a function for the quantum mechanical wave function [42], which with its probabilistic interpretation led naturally to the concept of a distribution function.

Consider a waveform $s(t)$ whose instantaneous energy per unit time and at time t is given by $|s(t)|^2$, while the energy per unit linear frequency $f = \omega/2\pi$ is given by the squared magnitude of the signal's Fourier transform $|S(f)|^2$, with Parseval's relation

$$\int_{-\infty}^{\infty} |s(t)|^2 dt = \int_{-\infty}^{\infty} |S(f)|^2 df. \tag{3.92}$$

The goal is to devise a joint function of time and frequency $P(t, f)$ that represents the energy of a waveform per unit time and per unit frequency. This joint function must necessarily satisfy the *marginal* conditions

$$\int_{-\infty}^{\infty} P(t, f)dt = |S(f)|^2 \text{ and } \int_{-\infty}^{\infty} P(t, f)df = |s(t)|^2 . \tag{3.93}$$

The Wigner distribution of a real waveform $s(t)$ is defined in terms of the analytic signal (3.89)

$$P_{\text{w}}(t, f) = \int_{-\infty}^{\infty} s^{[a]}(t + \tau/2)\, s^{[a]*}(t - \tau/2)\, e^{-2i\pi f\tau} d\tau, \tag{3.94}$$

and it has the following properties:

- It is a real function, i.e., $P_{\text{w}}(f, t) = P_{\text{w}}^*(f, t)$.
- It satisfies the marginal properties (3.93).
- It is shift-invariant, i.e., if $s^{[a]} \leftrightarrow P_{\text{w}}(t, f)$, then $s^{[a]}(t - t_0)e^{2i\pi f_0 t} \leftrightarrow P_{\text{w}}(t - t_0, f - f_0)$.

As an example of the Wigner distribution, consider the following complex linearly chirped waveform (all signals will be assumed analytic for the rest of this section) and its Wigner distribution:

$$s(t) = e^{-at^2 + i\pi bt^2 + 2i\pi f_0 t}, \quad P_{\text{w}}(t, f) = e^{-2at^2 - \;(2\pi f - bt - 2\pi f_0)^2/2a}, \tag{3.95}$$

which shows the linearly varying frequency as a function of time. Clearly, the Wigner distribution is a quadratic function of the signal, and so it will, in general, exhibit interference between negative and positive frequency components of a real valued signal. However, if the analytic signal is used in the computation, negative frequencies will be absent and no interference will occur. In addition, the analytic signal formulation guarantees that the first moment of the distribution conditioned on t is the instantaneous frequency, namely, the time derivative of the signal phase function

$$\langle f \rangle_t = \left|s(t)\right|^{-2} \int_{-\infty}^{\infty} f \, P_{\mathrm{w}}(t, f) \, df = \frac{d\theta}{dt}, \quad s(t) = \left|s(t)\right| e^{i\theta(t)} . \tag{3.96}$$

An algorithm to compute the Wigner distribution follows from the original definition with the transformation $\tau \to 2\tau$ in the integral:

$$P_{\mathrm{w}}(t, f) = 2 \int_{-\infty}^{\infty} s\,(t + \tau)\, s^*\,(t - \tau)\, e^{-4i\pi f \tau} d\tau, \tag{3.97}$$

where s is the analytic signal. We use the discretizations $t \to n\Delta t$ and $\tau \to m\Delta t$ to write

$$P_{\mathrm{w}}(n\Delta t, f) = 2\Delta t \sum_{m=-\infty}^{+\infty} s[n + m] s^*[n - m] e^{-4i\pi f m \Delta t}, \tag{3.98}$$

and evaluate (3.98) for any value of the frequency $0 \le f \le f_s/2$ where $f_s = 1/\Delta t$ is the sampling frequency of the analytic signal s. Figure 3.22 shows the Wigner distribution of the Bowhead whale sound of figure 3.8: there are four distinct chirps that are highly resolved in both time and frequency, and there are cross-term artifacts in between all chirp components.

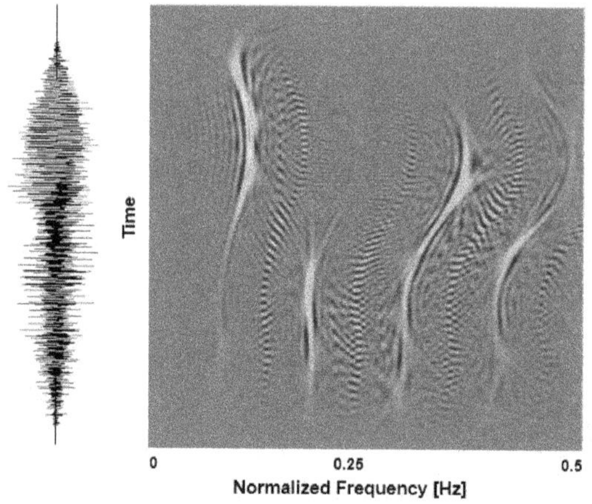

Figure 3.22: The Wigner distribution for the Bowhead whale sound.

The cross-term artifacts in figure 3.22 can be reduced by a more general formulation of the Wigner time-frequency distribution in terms of the signal ambiguity function

$$A(\tau, \nu) = \int_{-\infty}^{\infty} s\,(t + \tau/2)\, s^*\,(t - \tau/2)\, e^{+2i\pi\nu t}\, dt, \tag{3.99}$$

where ν is Doppler frequency and τ is the lag. The ambiguity function is commonly used to evaluate the performance of a radar signal waveform; it reveals the range-Doppler position of ambiguous responses, defines the range and Doppler resolutions, and satisfies the conditions $A(-\nu,-\tau) = A^*(\nu,\tau)$ and $A(-\nu,\tau) = A^*(\nu,-\tau)$. The Wigner distribution and the ambiguity function of an analytic signal s are Fourier transform pairs [43], i.e.,

$$P_{\mathrm{w}}(t,f) = \int_{-\infty}^{\infty}\int_{-\infty}^{\infty} A(\tau,\nu)e^{-2i\pi\nu t-2i\pi f\tau}\, d\nu d\tau. \tag{3.100}$$

A general class of joint time-frequency distributions is defined by introducing a multiplicative kernel $K(\nu,\tau)$ in the Doppler-lag domain [41], which can be used to reduce cross-term artifacts. The general Cohen class of time-frequency distributions is defined by

$$C(t,f) = \int_{-\infty}^{\infty}\int_{-\infty}^{\infty} A(\tau,\nu)K(\tau,\nu)e^{-2i\pi\nu t-2i\pi f\tau}\, d\nu d\tau. \tag{3.101}$$

Many desirable properties of a time-frequency distribution can be described in terms of the kernel; these properties (not exhaustive) cannot all hold at the same time, so the kernel function must be chosen judiciously.

- The time marginal property is satisfied if $K(0,\nu) = 1$.
- The frequency marginal property is satisfied if $K(\tau,0) = 1$.
- Parseval's relation for total energy is satisfied if $K(0,0) = 1$.
- The distribution is a real function if $K(-\tau,-\nu) = K^*(\tau,\nu)$.
- The distribution is scale-invariant $C(t/\eta,\eta f) = C(t,f)$ if $K(\tau,\nu) = K(\tau\nu)$, i.e., if K is a function of the product $\tau\nu$ and not a function of τ and ν independently.
- The first moment of the distribution conditioned on t is the instantaneous frequency if $\partial K/\partial\tau = 0$ at $\tau = 0$.
- The first moment of the distribution conditioned on f is the group delay if $\partial K/\partial\nu = 0$ at $\nu = 0$, where the group delay is found from the phase of the analytic signal Fourier transform $S(f) = |S(f)|\exp(2i\pi\chi)$ to be $-d\chi/d(2\pi f)$.
- For a finite duration signal, the distribution is zero before the signal begins and after the signal has ended if $\int K(\tau,\nu)\exp(-2i\pi\nu t)\, d\nu = 0$ for $\tau \leq 2\,|t|$.
- For a band limited signal, the distribution is zero outside the signal band if $\int K(\tau,\nu)\exp(-2i\pi\nu t)\, d\tau = 0$ for $\nu \leq 4\pi\,|f|$.

An experimental approach to kernel design is to study the magnitude of the (analytical) signal ambiguity function and design a kernel as a masking function $M(\tau,\nu)$ in the Doppler-lag domain [44]. The kernel is designed such that the product $K(\tau,\nu)M(\tau,\nu)$ satisfies the symmetries of the ambiguity function. Then, a two dimensional inverse Fourier transform of the masked ambiguity function will produce a time-frequency distribution which may or may not satisfy the properties mentioned above.

To compute the ambiguity function, we use an analytic signal s and discretized variables $t = n\Delta t$, $\tau = 2m\Delta t$ (we use 2τ in (3.99) to avoid having to interpolate between sample values), and $\nu = l\Delta f$ to obtain

$$A(2m\Delta t, l\Delta f) = \Delta t \sum_{n=-\infty}^{+\infty} s[n+m]s^*[n-m]\, e^{+2i\pi ln\, \Delta t\, \Delta f}, \tag{3.102}$$

where $\Delta t = 1/f_s$ and Δf is some appropriate fraction of the sampling frequency f_s.

The method of reassignment discussed in section 3.8 is also applicable to the general Cohen class of time-frequency distributions defined by (3.101) through a generalization of the reassigned coordinates (3.56)

$$\hat{t}(t,\omega) = t - \frac{\int_{-\infty}^{\infty}\int_{-\infty}^{\infty} \tau P_{\mathrm{w}}(t-\tau,\omega-\nu)K(\tau,\nu)\, d\tau d\nu}{\int_{-\infty}^{\infty}\int_{-\infty}^{\infty} P_{\mathrm{w}}(t-\tau,\omega-\nu)K(\tau,\nu)\, d\tau d\nu}, \tag{3.103a}$$

$$\hat{\omega}(t,\omega) = \omega - \frac{\int_{-\infty}^{\infty}\int_{-\infty}^{\infty} \nu P_{\mathrm{w}}(t-\tau,\omega-\nu)K(\tau,\nu)\, d\tau d\nu}{\int_{-\infty}^{\infty}\int_{-\infty}^{\infty} P_{\mathrm{w}}(t-\tau,\omega-\nu)K(\tau,\nu)\, d\tau d\nu}, \tag{3.103b}$$

where $P_{\mathrm{w}}(t,\omega)$ is the Wigner distribution of the signal and K is the Kernel defining that distribution [33].

CHAPTER 4

Least Squares Filters

4.1 Introduction

FIR filters are at the core of signal processing. Although most applications involve random components (signal, clutter, noise) which require specific assumptions regarding their statistics in order for them to be amenable to linear signal processing methods, we often use least squares techniques to deal with them. Our discussion in this chapter involves several examples to illustrate the application of least squares filtering techniques to problems that involve both non-random and random signals.

In section 4.2 we describe general solutions to constrained quadratic minimization problems. In sections 4.3 and 4.4 we discuss two special applications that illustrate least squares filter design (possibly with constraints). Both applications solve for a least squares filter by minimizing a non-negative definite quadratic function of the filter coefficients whose quadratic term is of the form $\boldsymbol{h}^{+}\boldsymbol{R}\boldsymbol{h}$ where $\boldsymbol{R}$ is a Hermitian and positive semi-definite matrix. The design is based on the minimization of the L_2 norm of the error between the output of the filter and a desired output as shown in figure 4.1: $h(t)$ is the filter to be determined by minimizing $\|e\|^2 = \|y - d\|^2$, i.e., the total energy of the difference output, together with some constraints to exclude the trivial (zero) solution.

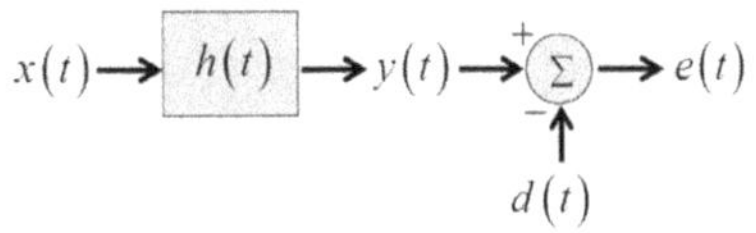

Figure 4.1: Design of an optimal filter $h(t)$ to match a known signal $d(t)$.

In section 4.3 we design a discrete time low-pass filter: referring to figure 4.1 the input is assumed to be a unit delta pulse sequence δ_{0n}, while the desired function is the discrete time representation of an ideal low-pass filter

$$h_d[n] = \frac{1}{2\pi}\int_{-\omega_p}^{\omega_p} e^{in\omega}d\omega = \frac{\sin(n\omega_p)}{\pi n},\ n \in \mathbb{Z}, \tag{4.1}$$

where ω_p is the pass frequency. Using Parseval's relation (3.44), the minimization problem is solved in the frequency domain. In section 4.4 we work in the time domain; we find a least squares filter by orthogonal projection and show applications to enhancing a signal in noisy measurements. In section 4.6 we show an example of time-delay estimation and a solution using time domain least squares filtering.

4.2 Quadratic Minimization Problems

Consider minimizing a quadratic form $\boldsymbol{h}^+\boldsymbol{R}\boldsymbol{h}$, the objective function, with $\boldsymbol{h} \in \mathbb{C}^M$, where $\boldsymbol{R}$ is an $M \times M$ positive semi-definite matrix. Without a constraint, the solution is the trivial zero solution, so we consider a quadratic constraint $\boldsymbol{h}^+\boldsymbol{h} = \|\boldsymbol{h}\|^2 = 1$. Thus, the constrained minimization problem is

$$\arg\min_{\boldsymbol{h}} \left(\boldsymbol{h}^+\boldsymbol{R}\boldsymbol{h}\right), \text{ subject to } \boldsymbol{h}^+\boldsymbol{h} = 1. \tag{4.2}$$

This problem is transformed to an unconstrained minimization problem using a Lagrange multiplier λ:

$$\arg\min_{\boldsymbol{h}} \left[\boldsymbol{h}^+\boldsymbol{R}\boldsymbol{h} + \lambda\left(1 - \boldsymbol{h}^+\boldsymbol{h}\right)\right]. \tag{4.3}$$

Differentiating with respect to $\boldsymbol{h}^+$ (treating $\boldsymbol{h}$ and its Hermitian conjugate as independent variables), we obtain an eigenvector and eigenvalue equation for $\boldsymbol{R}$,

$$\boldsymbol{R}\boldsymbol{h} = \lambda\boldsymbol{h} \;\Rightarrow\; \boldsymbol{h}^+\boldsymbol{R}\boldsymbol{h} = \lambda|\boldsymbol{h}|^2 = \lambda, \tag{4.4}$$

which shows the eigenvalue to equal the original quadratic objective function. Thus, the solution minimizing the quadratic form and satisfying the unit norm condition is the normalized eigenvector $\boldsymbol{h}_{\min}$ associated with the smallest eigenvalue of $\boldsymbol{R}$, namely, $\lambda_{\min}$.

A modified version of this problem often encountered in computer vision involves minimization carried out on an ellipsoid instead of a sphere, i.e., the constraint is $\boldsymbol{h}^+\boldsymbol{S}\boldsymbol{h} = 1$ for a Hermitian positive semi-definite matrix $\boldsymbol{S}$,

$$\arg\min_{\boldsymbol{h}} \left(\boldsymbol{h}^+\boldsymbol{R}\boldsymbol{h}\right) \text{ subject to } \boldsymbol{h}^+\boldsymbol{S}\boldsymbol{h} = 1. \tag{4.5}$$

To solve (4.5), we use the SVD of $\boldsymbol{S}$, which, in view of its Hermitian and positive definite property, is

$$\boldsymbol{S} = \boldsymbol{Q}\boldsymbol{D}\boldsymbol{Q}^+, \;\; \boldsymbol{Q}^+\boldsymbol{Q} = \boldsymbol{I}. \tag{4.6}$$

Thus, $\boldsymbol{Q}$ is a unitary matrix and $\boldsymbol{D}$ is a diagonal matrix of (real and positive) eigenvalues s_k of $\boldsymbol{S}$. Defining the square root matrix

$$\boldsymbol{S}^{1/2} = \boldsymbol{Q}\boldsymbol{D}^{1/2}\boldsymbol{Q}^+, \tag{4.7}$$

and changing variables $\boldsymbol{g} = \boldsymbol{S}^{1/2}\boldsymbol{h}$, we have the minimization problem for $\boldsymbol{g}$ on a sphere $|\boldsymbol{g}|^2 = 1$, i.e., equation (4.2), with a new Hermitian positive semi-definite matrix $\boldsymbol{R}' \equiv \boldsymbol{S}^{-1/2}\boldsymbol{R}\boldsymbol{S}^{1/2}$ instead of $\boldsymbol{R}$.

Another variant involves inclusion of a linear term (one for real and symmetric $\boldsymbol{R}$ and two complex conjugate terms for complex Hermitian $\boldsymbol{R}$) in the objective function

$$\arg\min_{\boldsymbol{h}} \left(\boldsymbol{h}^+\boldsymbol{R}\boldsymbol{h} - \boldsymbol{d}^+\boldsymbol{h} - \boldsymbol{h}^+\boldsymbol{d}\right), \text{ subject to } \boldsymbol{h}^+\boldsymbol{h} = 1. \tag{4.8}$$

Setting the derivative with respect to $\boldsymbol{h}$ to zero, we find the solution that includes the Lagrange multiplier λ; λ is found by using the constraint equation, so the final solution is given by

$$\boldsymbol{h} = \left(\boldsymbol{R} - \lambda\boldsymbol{I}\right)^{-1}\boldsymbol{d}, \;\; \boldsymbol{d}^+\left(\boldsymbol{R} - \lambda\boldsymbol{I}\right)^{-2}\boldsymbol{d} = 1. \tag{4.9}$$

The normalization constraint in the original minimization problem (4.2) is unnecessary if other linear constraints are present. For instance, consider the linear constraint $\boldsymbol{A}^+\boldsymbol{h} = \boldsymbol{b}$, where $\boldsymbol{A}^+$ is a rectangular

$K \times M$ full rank matrix, $K < M$, and $\boldsymbol{b}$ is a $K \times 1$ vector. Introducing a $K \times 1$ vector Lagrange multiplier $\boldsymbol{\lambda}^+$, the equivalent unconstrained minimization problem is [1]

$$\arg\min_{\boldsymbol{h}} \left[\boldsymbol{h}^+\boldsymbol{R}\boldsymbol{h} + \boldsymbol{\lambda}^+\left(\boldsymbol{b} - \boldsymbol{A}^+\boldsymbol{h}\right)\right]. \tag{4.10}$$

Differentiating with respect to $\boldsymbol{h}$ and setting the derivative to zero gives [2]

$$\boldsymbol{h}^+\boldsymbol{R} = \boldsymbol{\lambda}^+\boldsymbol{A}^+ \;\Rightarrow\; \boldsymbol{h} = \boldsymbol{R}^{-1}\boldsymbol{A}\boldsymbol{\lambda}, \tag{4.11}$$

which upon multiplication on the left by $\boldsymbol{A}^+$ and use of the constraint gives

$$\boldsymbol{\lambda} = \left(\boldsymbol{A}^+\boldsymbol{R}^{-1}\boldsymbol{A}\right)^{-1}\boldsymbol{b} \;\Rightarrow\; \boldsymbol{h} = \boldsymbol{R}^{-1}\boldsymbol{A}\left(\boldsymbol{A}^+\boldsymbol{R}^{-1}\boldsymbol{A}\right)^{-1}\boldsymbol{b}. \tag{4.12}$$

This solution has applications to minimum variance distortionless spectral estimation in section 8.9 and the minimum variance distortionless response beamformer (and its linearly constrained from) in chapter 11. The solution is the sum of two orthogonal projections: $\boldsymbol{P}$, given in (4.13), is the projection onto the range space of $\boldsymbol{A}^+$ and $\boldsymbol{I} - \boldsymbol{P}$ is the projection onto the null space of $\boldsymbol{A}^+$,

$$\boldsymbol{P} = \boldsymbol{A}\left(\boldsymbol{A}^+\boldsymbol{A}\right)^{-1}\boldsymbol{A}^+. \tag{4.13}$$

Projecting the solution vector $\boldsymbol{h}$ onto the range space of $\boldsymbol{A}^+$ gives

$$\boldsymbol{P}\boldsymbol{h} = \boldsymbol{A}\left(\boldsymbol{A}^+\boldsymbol{A}\right)^{-1}\boldsymbol{b}, \tag{4.14}$$

which is independent of $\boldsymbol{R}$, since the right hand side of (4.14) is the generalized inverse (an example of the minimum norm solution (1.24)) of the constraint equation, which does not involve $\boldsymbol{R}$.

If, in addition to the linear set of constraints, there is a quadratic constraint $\boldsymbol{h}^+\boldsymbol{h} = 1$, then the optimization problem is

$$\arg\min_{\boldsymbol{h}} \left(\boldsymbol{h}^+\boldsymbol{R}\boldsymbol{h}\right), \quad \text{subject to } \boldsymbol{A}^+\boldsymbol{h} = \boldsymbol{b} \text{ and } \boldsymbol{h}^+\boldsymbol{h} = 1. \tag{4.15}$$

Introducing a vector Lagrange multiplier $\boldsymbol{\lambda}$ for the linear constraint and a scalar multiplier μ for the quadratic constraint, we have the unconstrained minimization problem

$$\arg\min_{\boldsymbol{h}} \left[\boldsymbol{h}^+\boldsymbol{R}\boldsymbol{h} + \boldsymbol{\lambda}^+\left(\boldsymbol{b} - \boldsymbol{A}^+\boldsymbol{h}\right) + \mu\left(\boldsymbol{h}^+\boldsymbol{h} - 1\right)\right]. \tag{4.16}$$

Setting the derivative with respect to $\boldsymbol{h}$ to zero, we find

$$\boldsymbol{h} = \boldsymbol{R}'^{-1}\boldsymbol{A}\boldsymbol{\lambda}, \quad \boldsymbol{R}' \equiv \boldsymbol{R} + \mu\boldsymbol{I}, \tag{4.17}$$

which upon multiplication on the left by $\boldsymbol{A}^+$ and use of the linear constraint gives

$$\boldsymbol{\lambda} = \left(\boldsymbol{A}^+\boldsymbol{R}'^{-1}\boldsymbol{A}\right)^{-1}\boldsymbol{b} \;\Rightarrow\; \boldsymbol{h} = \boldsymbol{R}'^{-1}\boldsymbol{A}\left(\boldsymbol{A}^+\boldsymbol{R}'^{-1}\boldsymbol{A}\right)^{-1}\boldsymbol{b}. \tag{4.18}$$

$\boldsymbol{R}'$, defined in equation (4.17), is known as the *diagonally loaded* version of $\boldsymbol{R}$. The Lagrange multiplier μ is the solution to the quadratic constraint equation using the solution $\boldsymbol{h}$ in (4.18), namely,

$$\boldsymbol{h}^+\boldsymbol{h} = 1 = \boldsymbol{b}^+\left(\boldsymbol{A}^+\boldsymbol{R}'^{-1}\boldsymbol{A}\right)^{-1}\boldsymbol{A}^+\boldsymbol{R}'^{-2}\boldsymbol{A}\left(\boldsymbol{A}^+\boldsymbol{R}'^{-1}\boldsymbol{A}\right)^{-1}\boldsymbol{b}, \tag{4.19}$$

---

[1] $\boldsymbol{A}^+$ and $\boldsymbol{\lambda}^+$ are the Hermitian conjugates of $\boldsymbol{A}$ and $\boldsymbol{\lambda}$, respectively; this choice instead of using $\boldsymbol{A}^T$ and $\boldsymbol{\lambda}^T$ is motivated by the notation for the optimal beamforming weight vector of section 11.4.

[2] If we choose to differentiate with respect to $\boldsymbol{h}^+$, then in keeping with our assumption of independence, we would write the Hermitian conjugated constraint in terms of $\boldsymbol{h}^+$ and introduce the Lagrange multiplier vector in the form $\left(\boldsymbol{h}^+\boldsymbol{A} - \boldsymbol{b}^+\right)\boldsymbol{\lambda}$.

which does not have a closed form solution. A numerical solution is based on elimination of the linear constraint beginning with the SVD of the linear constraint matrix [45], after which the problem is transformed into the simpler problem (4.8). The linearly constrained minimization problem with an additional quadratic constraint has applications to minimum variance beamformers presented in chapter 11 in the context of signal mismatch.

4.3 Frequency Domain Least Squares Filters

Consider the design of a low-pass filter based on the minimization of a quadratic function of the filter subject to a constraint. Although an ideal low-pass filter frequency response has a jump discontinuity at the pass frequency (from 1 to 0), when limited to a finite number of coefficients (which we take to be odd so as to have a symmetric filter) it will necessarily have a transition band that must be specified in the design. Thus, a low-pass filter h_n with odd number of coefficients $N+1$ (even N) is defined by the following response of its Fourier transform

$$H_d\left(e^{i\omega}\right)=\begin{cases}1, & -\omega_p\leq\omega\leq\omega_p,\\ 0, & \omega_s\leq\omega\leq\pi,\quad -\pi\leq\omega\leq-\omega_s,\end{cases} \tag{4.20}$$

where ω_p is the pass frequency and ω_s is the stop frequency. The form of the response in the transition band is not specified. In addition, the frequency domain is where the minimization problem of figure 4.1 should be performed (Parseval's relation guarantees the equivalence of the total energy in both domains), where the transition band can be excluded from the minimization problem (an impossible task in the time domain).

Clearly, no FIR filter will achieve the strict requirements of the desired response, i.e., zero error, and so the criterion for a successful design is minimization of a combination of the energy of the difference between the actual filter and the desired response in the pass band $[0,\omega_p]$ and the stop band $[\omega_s,\pi]$, where we use the symmetry of the response (since the filter coefficients are real) to limit the energies to positive frequencies. If we set the further requirement that the filter should have linear phase, then the filter coefficients in the discrete time domain satisfy a symmetry requirement (N is even) $h_n=h_{N-n}$, $0\leq n\leq N$. To see this, consider

$$H\left(e^{i\omega}\right)=\sum_{n=0}^{N}h_n e^{-in\omega}=\left|H\left(e^{i\omega}\right)\right|e^{-i\phi(\omega)}, \tag{4.21}$$

which in view of the linear phase requirement, i.e., $\phi(\omega)=-\tau_p\omega$ with τ_p being the phase delay, leads to the equation

$$\tan\left(\tau_p\omega\right)\sum_{n=0}^{N}h_n\cos\left(n\omega\right)=\sum_{n=0}^{N}h_n\sin\left(n\omega\right)\Rightarrow\sum_{n=0}^{N}h_n\sin\left(\tau_p\omega-n\omega\right)=0.$$

The solution to this equation is $\tau_p=N/2$ and $h_n=h_{N-n}$, $0\leq n\leq N$. Thus,

$$H\left(e^{i\omega}\right)=\sum_{n=0}^{N}h_n e^{-in\omega}=e^{iN\omega/2}\sum_{n=0}^{N/2}\tilde{h}_n\cos\left(n\omega\right)\equiv e^{iN\omega/2}\,\tilde{\boldsymbol{h}}^T\boldsymbol{c}_\omega, \tag{4.22}$$

where we have introduced the vectors $\tilde{\boldsymbol{h}}$ and $\boldsymbol{c}_\omega$,

$$\begin{aligned}\tilde{\boldsymbol{h}}&=\left[\tilde{h}_0,\ldots,\tilde{h}_{N/2}\right]^T=2\left[h_{N/2}/2,\ h_{N/2-1},\ldots,h_0\right]^T,\\ \boldsymbol{c}_\omega&\equiv\left[1,\cos\left(\omega\right),\ldots,\cos\left(N\omega/2\right)\right]^T.\end{aligned} \tag{4.23}$$

We use the vector $\tilde{\boldsymbol{h}}$ in the rest of our calculations with the understanding that the actual low-pass filter $\boldsymbol{h}$ coefficients can be found from $\tilde{\boldsymbol{h}}$ by the defining equation (4.23) and the symmetry requirement $h_n = h_{N-n}, 0 \le n \le N$.

The pass and the stop band error energies E_p and E_s are the integrals of the squared magnitude of the error over the corresponding bands. A linear combination of the two in the form $\alpha E_p + (1-\alpha) E_s$ will be minimized subject to a constraint $\tilde{\boldsymbol{h}}^T \tilde{\boldsymbol{h}} = 1$ (to exclude the zero solution). In the pass band we drop the linear phase term and calculate the energy of the squared magnitude of the deviation from the desired response as

$$E_p = \tilde{\boldsymbol{h}}^T \left[\int_0^{\omega_p} (\mathbf{1} - \boldsymbol{c}_\omega)(\mathbf{1} - \boldsymbol{c}_\omega)^T d\omega \right] \tilde{\boldsymbol{h}}, \quad \mathbf{1} \equiv [1, \ldots, 1]^T, \tag{4.24}$$

where we have used the fact that the DC response of the filter (value of the Fourier transform at $\omega = 0$) is 1, i.e., $H\left(e^{i0}\right) = 1 = \tilde{\boldsymbol{h}}^T \mathbf{1}$, where $\mathbf{1} = [1, \ldots, 1]^T$. The stop band deviation energy is simply the energy of the filter in that band (since the desired response is zero there):

$$E_s = \tilde{\boldsymbol{h}}^T \left[\int_{\omega_s}^{\pi} \boldsymbol{c}_\omega \boldsymbol{c}_\omega{}^T d\omega \right] \tilde{\boldsymbol{h}}. \tag{4.25}$$

The integrals in (4.24) and (4.25) can be exactly calculated since the integrands are either constant, a single cosine, or the product of two cosines. The problem is then to minimize the quadratic function $\tilde{\boldsymbol{h}}^T \boldsymbol{R} \tilde{\boldsymbol{h}}$, where $\boldsymbol{R}$ is an $(N/2+1) \times (N/2+1)$ symmetric matrix equal to the appropriate linear combination of the integrals in equations (4.24) and (4.25) and subject to the constraint $\tilde{\boldsymbol{h}}^T \tilde{\boldsymbol{h}} = 1$. This is the same as in problem (4.2), and the solution is found from (4.3) and (4.4): $\tilde{\boldsymbol{h}}$ is the eigenvector corresponding to the smallest eigenvalue of the symmetric matrix $\boldsymbol{R}$.

Once $\tilde{\boldsymbol{h}}$ is found, we use the first line of equation (4.23) to find the filter coefficients $[h_0, \ldots, h_{N/2}]$ and then construct the full filter vector by the symmetry requirement $h_n = h_{N-n}$. A final normalization of the low-pass filter $\boldsymbol{h}$ ensures that the DC response is equal to 1, i.e., the $N+1$ filter coefficients sum to 1. Figure 4.2 shows the frequency response of three low-pass filters with lengths 51, 101, and 201 in the transition band and portions of the pass and the stop bands: as the number of filter points increases, the response more closely approximates the ideal response. This is an example of an *eigenfilter* [46].

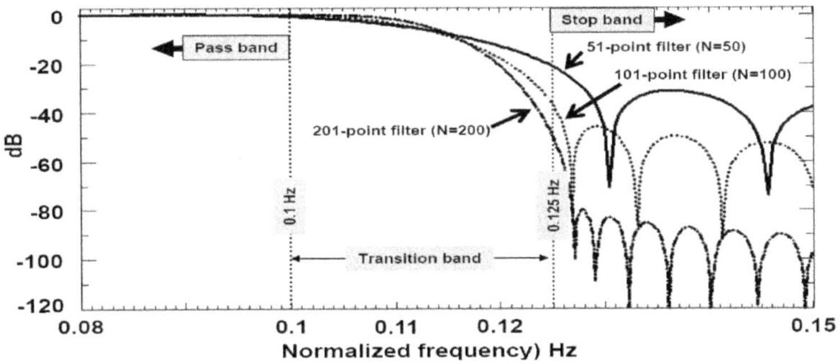

Figure 4.2: Low-pass eigenfilters with lengths 51, 101, and 201, and $\alpha = 0.5$.

4.4 Time Domain Least Squares Shaping Filters

As illustrated in figure 4.1, we seek to match a specified (desired) data $d(t)$ by linearly filtering an input data $x(t)$ and minimizing

$$\|e\|_2^2 = \langle d-y, d-y\rangle = \|y\|_2^2 + \|d\|_2^2 - 2\,\mathbf{Re}\,\langle y, d\rangle, \tag{4.26}$$

where y is given by the convolution of the filter with the input. If the norms of the functions e, y, and d represent three sides of a triangle then the norm of the error e is minimized when the triangle is a right triangle with y as the hypotenuse, i.e., Pythagoras's theorem is satisfied

$$\|e\|_2^2 + \|d\|_2^2 = \|y\|_2^2. \tag{4.27}$$

This happens when the error e is orthogonal to the desired data d, i.e., $\langle e, d\rangle = 0$. To see this we use Parseval's relations to rewrite (4.26) in the frequency domain, namely,

$$2\pi\,\|e\|_2^2 = \langle Y, Y\rangle + \langle D, D\rangle - \langle Y, D\rangle - \langle D, Y\rangle. \tag{4.28}$$

The inner products on the right hand side are calculated in the frequency domain, for instance,

$$\langle Y, D\rangle \equiv \int_{-\infty}^{\infty} Y_\omega^* D_\omega d\omega, \tag{4.29}$$

where the subscript ω in the integrand indicate the frequency dependence of the Fourier transform functions. Now the convolution relation is $Y_\omega = H_\omega X_\omega$, and so (4.28) becomes

$$2\pi\,\|e\|_2^2 = \int_{-\infty}^{\infty} H_\omega^* |X_\omega|^2 H_\omega d\omega + 2\pi\,\|d\|_2^2 - 2\,\mathbf{Re}\int_{-\infty}^{\infty} H_\omega^* X_\omega^* D_\omega D\omega. \tag{4.30}$$

Treating H_ω and its complex conjugate as independent functions, the right hand side is minimized by choosing the filter to satisfy the equation

$$H_\omega^* \left|X_\omega^*\right|^2 - X_\omega D_\omega^* = 0. \tag{4.31}$$

Multiplying both sides with D_ω and dividing by X_ω we find $\langle Y, D\rangle - \langle D, D\rangle$, which is the frequency domain form of the time domain orthogonality relation

$$\langle e, d\rangle = 0, e = d - y.$$

Now we consider discrete time desired data with dimension $N \times 1$, and a discrete time filter with M elements, $M \ll N$. Using the finite convolution equation, we have

$$y_{n+l} = \sum_{m=0}^{M-1} h_m x_{n-m}. \tag{4.32}$$

For a finite number of observations, (4.32) can be written in matrix form $\boldsymbol{y} = \boldsymbol{A}\boldsymbol{h}$ with $\boldsymbol{h} = [h_0, \ldots, h_{M-1}]^T$ representing the filter. When $l = 0$, we are solving an *estimation* least squares filter, while $l < 0$ and $l > 0$ correspond to *interpolation* and *prediction* least squares filters, respectively.

The error sequence $e_{n+k} \equiv d_{n+k} - y_{n+k}$ and the form of the data matrix $\boldsymbol{A}$ depend on the index n in the finite convolution equation (4.32). For instance, we may begin the index n at the value $M-1$, the first output index for which exactly M input data values starting at x_0 are used in the sum; that is, we make no assumptions about what the data prior to $n = 0$ might be, and so we take $n \in [M-1, N-1]$ without ever referencing values of x outside the given $[0, N-1]$ range. This form of computing the data matrix is called the *covariance* method and is defined by the finite convolution equation for $M-1 \le n \le N-1$. The corresponding $(N-M+1) \times M$ data matrix $\boldsymbol{A}_{\text{cov}}$ is

$$\boldsymbol{A}_{\text{cov}} = \begin{bmatrix} x_{M-1} & x_{M-2} & \cdots & x_0 \\ x_M & x_{M-1} & \cdots & x_1 \\ \vdots & \vdots & \vdots & \vdots \\ x_{N-1} & x_{N-2} & \cdots & x_{N-M} \end{bmatrix}. \tag{4.33}$$

If we denote the columns of the Hermitian conjugate matrix $\boldsymbol{A}_{\text{cov}}^+$ by $\boldsymbol{c}_k$, $0 \le k \le N-M$, then the rows of $\boldsymbol{A}_{\text{cov}}$ are given by $\boldsymbol{c}_k^+$ and we have [3]

$$\boldsymbol{A}_{\text{cov}}^+ \boldsymbol{A}_{\text{cov}} = \sum_{k=0}^{N-M} \boldsymbol{c}_k \boldsymbol{c}_k^+. \tag{4.34}$$

Since the rows of $\boldsymbol{A}_{\text{cov}}$ are given by $[x_{M-1+m}, \ldots, x_m]$ for $0 \le m \le N-M$, the columns of $\boldsymbol{A}_{\text{cov}}^+$ are given by $[x_{M-1+m}^*, \ldots, x_m^*]^T$, and we find that the *Gramian matrix*

$$[\boldsymbol{A}_{\text{cov}}^+ \boldsymbol{A}_{\text{cov}}]_{ij} = \sum_{m=0}^{N-M} x_{N-1+m-i}^* x_{N-1+m-j} = \sum_{m=M-1}^{N-1} x_{m-i}^* x_{m-j}, \tag{4.35}$$

is, in general, not Toeplitz.

If, on the other hand, we assume that the data x_n with indices n outside the range $[0, N-1]$ are actually zero, then the indices to use in equation (4.32) are $0 \le n \le M+N-2$; for instance, the first output term is $y_l = h_0 x_0$ and the last is $y_{N+M-2+l} = h_{M-1} x_{N-1}$; this is known as the *auto-correlation* method and is defined by the finite convolution equation for indices $0 \le n \le M+N-2$. The $(N+M-1) \times M$ data matrix $\boldsymbol{A}_{\text{ac}}$ is now Toeplitz, i.e., $[\boldsymbol{A}_{\text{ac}}]_{ij} = [\boldsymbol{A}_{\text{ac}}]_{i-1,j-1}$, $1 \le i, j$, $i \le N+M-2$, and $j \le M-1$,

$$\boldsymbol{A}_{\text{ac}} = \begin{bmatrix} x_0 & 0 & \cdots & \cdots & \cdots & 0 \\ x_1 & x_0 & 0 & \cdots & \cdots & \vdots \\ x_2 & x_1 & x_0 & 0 & \cdots & 0 \\ \vdots & x_2 & x_1 & x_0 & \ddots & \vdots \\ \vdots & \vdots & \vdots & \vdots & \vdots & 0 \\ x_{M-1} & x_{M-2} & x_{M-3} & \cdots & x_1 & x_0 \\ x_M & x_{M-1} & x_{M-2} & \cdots & x_2 & x_1 \\ \vdots & \vdots & \vdots & \vdots & \vdots & \vdots \\ x_{N-1} & x_{N-2} & \cdots & \cdots & \cdots & x_{N-M} \\ 0 & x_{N-1} & \cdots & \cdots & \cdots & x_{N-M+1} \\ 0 & 0 & x_{N-1} & \cdots & \cdots & x_{N-M+2} \\ \vdots & \vdots & 0 & \ddots & \ddots & \vdots \\ \vdots & \vdots & \vdots & \ddots & \ddots & \vdots \\ 0 & \vdots & \vdots & 0 & 0 & x_{N-1} \end{bmatrix}. \tag{4.36}$$

[3] If the columns of an $N \times N$ matrix $\boldsymbol{A}$ are denoted by $\boldsymbol{a}_n$ and the rows of an $N \times N$ matrix $\boldsymbol{B}$ are denoted by $\boldsymbol{b}_n^T$, then we have $[\boldsymbol{AB}]_{ij} = \sum_n [\boldsymbol{A}]_{in} [\boldsymbol{B}]_{nj} = \sum_n [\boldsymbol{a}_n]_i [\boldsymbol{b}_n^T]_j$, and so $\boldsymbol{AB} = \sum_n \boldsymbol{a}_n \boldsymbol{b}_n^T$.

Note that the middle $N-M+1$ rows (excluding the top $M-1$ and the bottom $M-1$ rows) form the data matrix $\boldsymbol{A}_{\text{cov}}$. The auto-correlation method leads to a Toeplitz Gramian matrix, for if we denote the columns of $\boldsymbol{A}_{\text{ac}}^{+}$ by $\boldsymbol{c}_k$, $0 \le k \le N+M-2$, then the rows of the data matrix $\boldsymbol{A}_{\text{ac}}$ are given by $\boldsymbol{c}_k^{+}$ and we have

$$\boldsymbol{A}_{\text{ac}}^{+}\boldsymbol{A}_{\text{ac}} = \sum_{k=0}^{N+M-2} \boldsymbol{c}_k \boldsymbol{c}_k^{+}. \tag{4.37}$$

Now $\boldsymbol{c}_i$, i.e., column i of $\boldsymbol{A}_{\text{ac}}^{+}$, is $[0,\ldots,0,\boldsymbol{x}^*,0,\ldots,0]^T$, where there are i zeros on the front and $M-1-i$ zeros on the end. Using the above formula and assuming $j \le i$, we have

$$[\boldsymbol{A}_{\text{ac}}^{+}\boldsymbol{A}_{\text{ac}}]_{ij} = \sum_{m=0}^{N-1-(i-j)} x_{m+i-j}\, x_m^*, \quad j \le i, \tag{4.38}$$

which proves the Toeplitz property for $j \le i$; when $i < j$, we use the Hermitian property $[\boldsymbol{A}_{\text{ac}}^{+}\boldsymbol{A}_{\text{ac}}]_{ij} = [\boldsymbol{A}_{\text{ac}}^{+}\boldsymbol{A}_{\text{ac}}]_{ji}^*$. Thus, the auto-correlation method always leads to a Toeplitz Gramian matrix.

Two other methods closely related to the auto-correlation method are the *pre-window* and the *post-window* methods. The $N \times M$ pre-window data matrix is constructed from $\boldsymbol{A}_{\text{ac}}$ by taking the first N rows with $0 \le n \le N-1$, while the $N \times M$ post-window data matrix uses the last N rows of $\boldsymbol{A}_{\text{ac}}$ with $M-1 \le n \le N+M-2$; both matrices $\boldsymbol{A}_{\text{pre}}$ and $\boldsymbol{A}_{\text{pos}}$ are

$$\mathbf{A}_{\text{pre}} = \begin{bmatrix} x_0 & 0 & \cdots & \cdots & 0 \\ x_1 & x_0 & 0 & \ddots & \vdots \\ \vdots & \ddots & x_0 & \ddots & \vdots \\ \vdots & \vdots & \vdots & \ddots & 0 \\ x_{M-1} & \cdots & \cdots & x_1 & x_0 \\ \vdots & \vdots & \vdots & \vdots & \vdots \\ x_{N-1} & \cdots & \cdots & \cdots & x_{N-M} \end{bmatrix} \quad \mathbf{A}_{\text{pos}} = \begin{bmatrix} x_{M-1} & x_{M-2} & \cdots & \cdots & x_0 \\ \vdots & \vdots & \vdots & \vdots & \vdots \\ x_{N-1} & x_{N-2} & \cdots & \cdots & x_{N-M} \\ 0 & x_{N-1} & x_{N-2} & \cdots & x_{N-M+1} \\ 0 & 0 & x_{N-1} & \cdots & x_{N-M+2} \\ \vdots & \vdots & \vdots & \vdots & \vdots \\ 0 & \cdots & \cdots & 0 & x_{N-1} \end{bmatrix}$$

As an example, consider approximately twenty seconds of a pulse signal of unit amplitude sampled at 25 Hz and corrupted with zero-mean white noise with variance 0.25 and four narrow-band interferences of unit amplitude with frequencies 3.75, 5.25, 6.75, and 7.75 Hz and phases 2.68683, 1.07419, 0.563822, and 2.60462 as shown at the top of figure 4.3. The output of the least squares shaping filter with 91 elements using the covariance method at the bottom of figure 4.3 shows a PSNR improvement of nearly 20 dB (PSNR here is defined as $20\log_{10}$ of maximum signal amplitude divided by the standard deviation of the noise).

As another example, consider approximately 4 minutes of two noisy observations of a 0.008 Hz sinusoid sampled at 1 Hz, as shown in figure 4.4. Noises on the two channels are independent and white with zero-mean and equal variance 0.1. This is a two-channel generalization when the outputs of two least squares filters are combined and then compared to the desired signal.

Denoting the data matrix for $\boldsymbol{x}_1$ by $\boldsymbol{A}_1$ and the data matrix for $\boldsymbol{x}_2$ by $\boldsymbol{A}_2$, we have the following augmented matrix equation for equal number of elements in both filters

$$\boldsymbol{A}\,\boldsymbol{h} = \boldsymbol{d}, \quad \boldsymbol{A} \equiv [\boldsymbol{A}_1 \,|\, \boldsymbol{A}_2], \quad \boldsymbol{h} \equiv [\boldsymbol{h}_1 \,|\boldsymbol{h}_2]^T. \tag{4.39}$$

The two data matrices can be constructed using any of the four methods outlined above and appended as indicated. The augmented data matrix is then used to calculate the Gramian matrix and the least squares

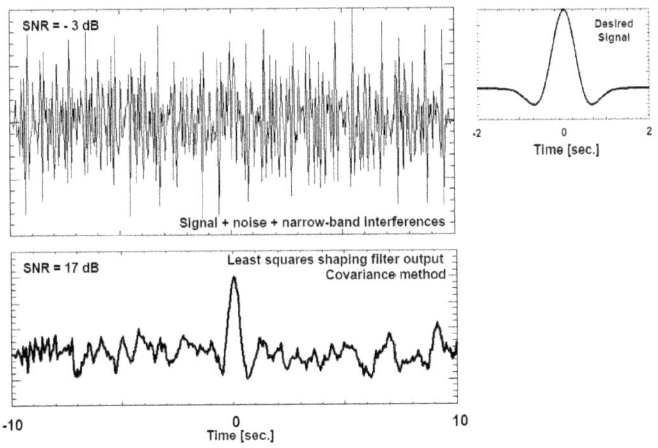

Figure 4.3: Pulse signal before and after least squares filtering.

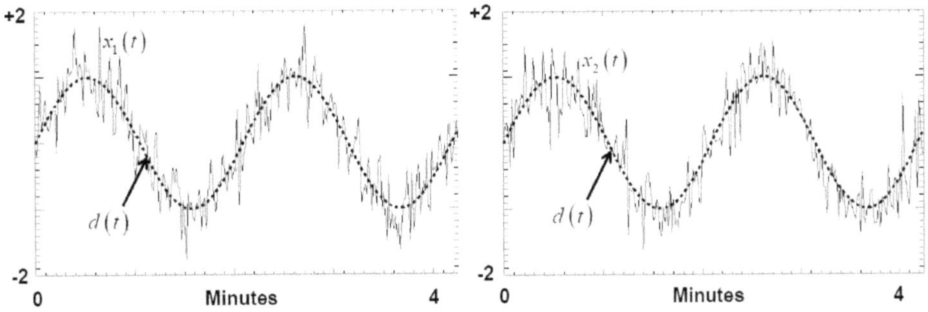

Figure 4.4: Desired signal and two noisy observations.

filter $\boldsymbol{h}$ whose first half will be $\boldsymbol{h}_1$ and whose second half will be $\boldsymbol{h}_2$. Using the covariance method and filters with 30 elements, the final filtered output is shown in figure 4.5; it has a much smaller noise variance of 0.008 which implies approximately 11 dB of noise reduction.

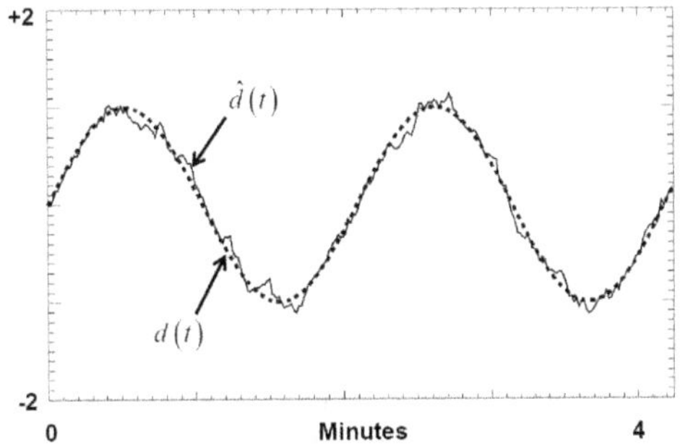

Figure 4.5: Two-channel least squares filter output using the covariance method.

As a final example consider a desired signal model $d_n = -0.3x_n + 0.2x_{n-1} + \nu_n$ for $0 \le n \le 99$, where x and ν are independent Gaussian random noise sequences with zero mean and unit variance (we will discuss random variables and stochastic processes in chapters 5 and 6) as illustrated in figure 4.6. The

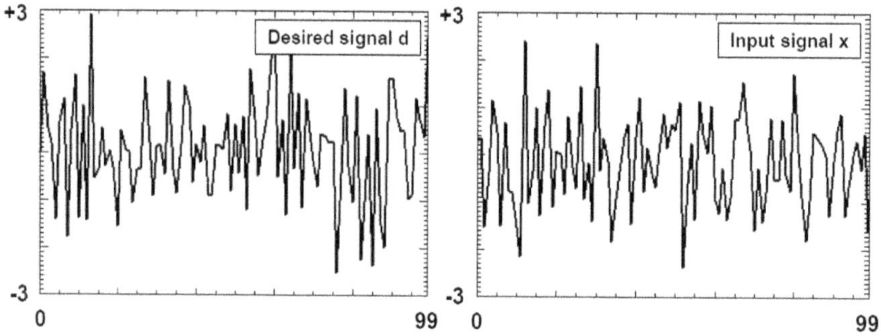

Figure 4.6: Desired signal d_n and input signal x_n.

problem here is to estimate the two-point filter coefficients $[-0.3, 0.2]$ from the observed data x_n and the desired sequence values d_n. Using the covariance method and 100 realizations, the least squares two-point filters have a mean of $[-0.3, 0.2]$ with a standard deviation of 0.1. Reducing the variance of ν to 0.09 produces a mean of $[-0.3, 0.19]$ with a standard deviation of 0.03.

The terms "auto-correlation" and "covariance" were originally introduced in linear prediction of speech where the correlation matrix of a data sequence is estimated from the speech data vector by a process that is similar to the formation of the Gramian matrix $\boldsymbol{A}^+\boldsymbol{A}$ from the data matrix $\boldsymbol{A}$. This procedure, as we have seen above, leads to the elements of the Gramian matrix being equal to sums of products of two elements of the data vector at different positions. As we shall see in our discussion of optimal filtering of random processes, the sums of products are equivalent to averaging procedures that provide estimates of the data correlation matrix elements; thus, least squares time domain filters applied to noisy observations are approximations to filters whose statistical optimality criteria are the minimization of noise power.

4.5 Gradient Descent Iterative Solution to Least Squares Filtering

An iterative solution to the least squares filtering problem is based on *gradient descent*. Consider filter vector $\boldsymbol{h}$ and the convolution equation

$$y_n = \sum_m h_m x_{n-m}, \quad \boldsymbol{h} = [\ldots, h_m, \ldots]^T. \tag{4.40}$$

The average squared error for a given desired signal vector $\boldsymbol{d} = [\ldots, d_n, \ldots]^T$ is

$$\mathcal{E} = \frac{1}{N} \sum_n |y_n - d_n|^2, \tag{4.41}$$

where the range of values in the sum and corresponding assumptions about the data values in that range follow the discussion in section 4.4 and the associated four methods, namely, auto-correlation, covariance, pre-window, and post-window. This is a quadratic function of the filter vector, namely,

$$\mathcal{E} = \boldsymbol{h}^+ \boldsymbol{R} \boldsymbol{h} - 2\,\boldsymbol{Re}\,(\boldsymbol{h}^+ \boldsymbol{r}) + \|\boldsymbol{d}\|^2, \tag{4.42}$$

where

$$[\boldsymbol{R}]_{km} \equiv \frac{1}{N}\sum_n x^*_{n-k}x_{n-m}, \; [\boldsymbol{r}]_m \equiv \frac{1}{N}\sum_n d_n x^*_{n-m}, \; \|\boldsymbol{d}\|^2 \equiv \frac{1}{N}\sum_n |d_n|^2. \tag{4.43}$$

Thus, the error $\mathcal{E}$ is a real quadratic function of the filter vector $\boldsymbol{h}$, which together with the observation that the matrix $\boldsymbol{R}$ is positive definite, means that $\mathcal{E}$ has a global minimum.

To solve the minimization equation $\partial\mathcal{E}/\partial\boldsymbol{h} = \boldsymbol{0}$ using iterative gradient descent, we note that the solution at iteration $k+1$, denoted by $\boldsymbol{h}^{(k+1)}$, is found from the equation $\boldsymbol{h}^{(k+1)} = \boldsymbol{h}^{(k)} + \Delta\boldsymbol{h}^{(k)}$, where the update term $\Delta\boldsymbol{h}^{(k)}$ must be chosen so as to reach the minimum value as $k \to \infty$; convergence to the minimum is assured if $\mathcal{E}^{(k+1)} < \mathcal{E}^{(k)}$. Consider the simplest case of a real filter with 1 element, $M = 1$. Let us assume that $|\Delta h/h| \ll 1$ so that $\mathcal{E}_h + \Delta h\, \partial\mathcal{E}_h/\partial h \leq \mathcal{E}_h$. The inequality is satisfied if we choose

$$\Delta h = -\,\mu\, \partial\mathcal{E}_h/\partial h, \tag{4.44}$$

for a positive constant $\mu > 0$ that must be sufficiently small so as to justify the Taylor approximation. Equation (4.44) is known as the gradient descent method and is illustrated in figure 4.7.

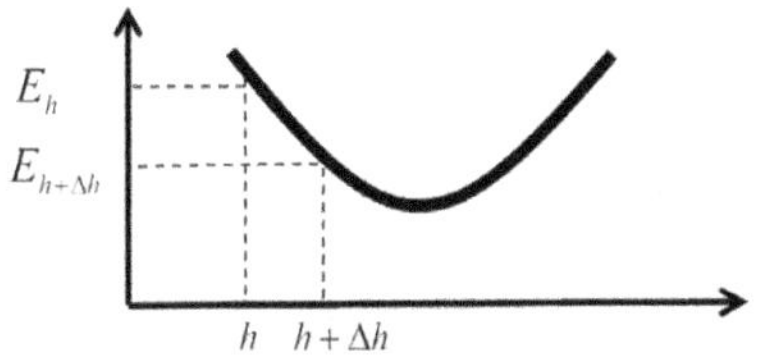

Figure 4.7: Gradient descent towards the global minimum.

Returning to the general case of a complex filter with $M > 1$ elements, we use the Taylor approximation (treating $\boldsymbol{h}$ and its complex conjugate $\boldsymbol{h}^*$ as independent variables)

$$\begin{aligned}\mathcal{E}\left(\boldsymbol{h}+\Delta\boldsymbol{h}, \boldsymbol{h}^*+\Delta\boldsymbol{h}^*\right) &\approx \mathcal{E}\left(\boldsymbol{h}, \boldsymbol{h}^*\right) + \sum_m \left(\Delta h_m\, \partial\mathcal{E}/\partial h_m + \Delta h^*_m\, \partial\mathcal{E}/\partial h^*_m\right) \\ &\equiv \mathcal{E}\left(\boldsymbol{h}, \boldsymbol{h}^*\right) + \Delta\boldsymbol{h}^T\partial\mathcal{E}/\partial\boldsymbol{h} + \Delta\boldsymbol{h}^+\partial\mathcal{E}/\partial\boldsymbol{h}^*,\end{aligned} \tag{4.45}$$

where h_m denotes the M components of the filter vector $\boldsymbol{h}$. The gradient descent choice is $\Delta h_m = -2\mu\; \partial\mathcal{E}/\partial h^*_m$, $\mu > 0$, i.e., $\Delta\boldsymbol{h} = -2\mu\nabla\mathcal{E}$ where the gradient is with respect to the elements of the complex conjugate vector $\boldsymbol{h}^*$ (for real filter vectors, the gradient descent choice is $\Delta\boldsymbol{h} = -\mu\nabla\mathcal{E}$). Using the gradient descent formula we have

$$\mathcal{E}\left(\boldsymbol{h}+\Delta\boldsymbol{h}, \boldsymbol{h}^*+\Delta\boldsymbol{h}^*\right) \approx \mathcal{E}\left(\boldsymbol{h}, \boldsymbol{h}^*\right) - 4\mu\left|\partial\mathcal{E}/\partial\boldsymbol{h}\right|^2 < \mathcal{E}\left(\boldsymbol{h}, \boldsymbol{h}^*\right). \tag{4.46}$$

The gradient descent formula is also known as the *steepest descent* algorithm since the gradient defines the direction of fastest change; it is orthogonal to the surface $\mathcal{E}$ = constant (this follows from the relation $d\boldsymbol{h}.\nabla\mathcal{E} = d\mathcal{E} = 0$ on the surface). The derivative of $\mathcal{E}$ with respect to a filter coefficient h^*_k is

$$\frac{\partial\mathcal{E}}{\partial h^*_k} = \frac{2}{N}\sum_n x^*_{n-k}\Big(\sum_m h_m x_{n-m} - d_n\Big) \;\leftrightarrow\; \frac{\partial\mathcal{E}}{\partial\boldsymbol{h}} = 2(\boldsymbol{Rh} - \boldsymbol{r}), \tag{4.47}$$

with $\boldsymbol{R}$ and $\boldsymbol{r}$ defined in equation (4.43). Equation (4.47) can be used in an iterative scheme to reach the minimum by choosing

$$\Delta\boldsymbol{h} = -\,2\mu\,(\boldsymbol{Rh} - \boldsymbol{r}).$$

Figure 4.8 illustrates the gradient descent method for a real filter with 21 elements. The desired output is $d_n = \cos(0.02\,\pi n)$ for $0 \le n \le 999$, and the data to be filtered is a noisy version of the desired output. The summation range for n corresponds to the covariance method of section 4.4. Convergence to the minimum is achieved in 15 iterations for $\mu = 0.01$ starting with initial random filter values (zero mean and standard deviation of 0.01).

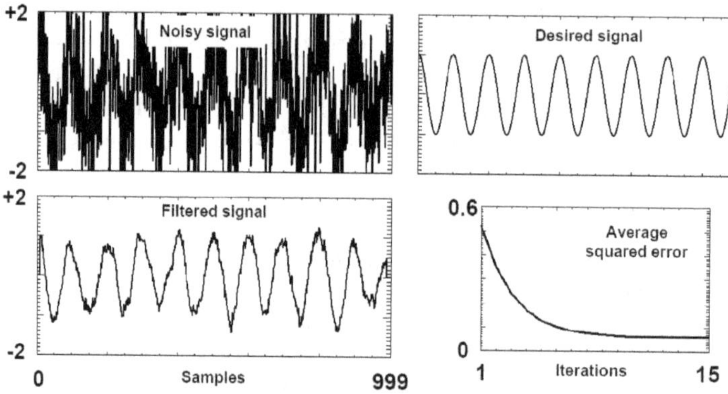

Figure 4.8: Iterative solution to the least squares filtering example.

The gradient descent method will be discussed in two major applications later in this text: adaptive filter methods in chapter 10 and neural networks in chapter 12.

4.6 Time Delay Estimation

An interesting example of least squares filtering is the problem of time delay estimation [47]. Consider a wide-band signal $s(t)$ and two spatially separated receivers with data $p(t)$ and $q(t)$ that follow the model equations

$$p(t) = s(t) + \eta(t), \quad q(t) = \alpha s(t - T) + \eta(t). \tag{4.48}$$

In this model, $\eta(t)$ represents background noise or clutter recorded on both sensors, and T is an unknown time delay whose value is to be found. The wide-band signal is assumed to be random, while the nature of $\eta(t)$ defines two classes of problems of interest: the class discussed in this section comprises a wide-band noise function $\eta(t)$, while the second class (discussed in section 7.9) includes a narrow-band sinusoidal clutter function $\eta(t)$. Although random signals will be discussed in chapters 5 and 6, our approach will proceed without the required concepts, using only least squares filtering ideas.

According to the sampling theorem (theorem 6 in section 1.9), we can write

$$s(t - T) = \sum_{k=-\infty}^{\infty} s(k\,\Delta t)\,\text{sinc}[2B(t - T - k\,\Delta t)], \tag{4.49}$$

where B defines the signal band and $\text{sinc}(t) \equiv \sin \pi t/\pi t$. Evaluating (4.49) for $t = n\,\Delta t$, $T = d\,\Delta t$, and further choosing normalized frequencies with $\Delta t = 1$ and assuming that the signal is sampled at exactly the Nyquist rate, i.e., $B\,\Delta T = 1/2$, we find

$$s[n - d] = \sum_{k=-\infty}^{\infty} s[k]\,\text{sinc}[n - d - k] \equiv \sum_{k=-\infty}^{\infty} s[k]\,h_d[n - k], \tag{4.50}$$

which defines the interpolation filter $h_d[k]$

$$h_d[k] \equiv \mathrm{sinc}(k-d) = \frac{\sin \pi(k-d)}{\pi(k-d)}. \tag{4.51}$$

Consider a filter with $2M+1$ elements and define the error sequence e_n associated with the least squares filter w_k by

$$e_n \equiv q_n - \sum_{k=-M}^{M} w_k p_{n-k} \ , -N \le n \le N, \tag{4.52}$$

whose average squared norm

$$\mathcal{E} \equiv \frac{1}{2N+1} \sum_{n=-N}^{N} |e_n|^2 = \frac{1}{2N+1} \sum_{n=-N}^{N} \left| q_n - \sum_{k=-M}^{M} w_k p_{n-k} \right|^2 \tag{4.53}$$

is to be minimized. Although the least squares filter can be calculated using any of the methods described in section 4.4, we use the auto-correlation method, which leads to a Toeplitz Gramian matrix. Before calculating the derivative of $\mathcal{E}$ with respect to the filter coefficients, we make the following assumptions:

$$\sum_{n=-N}^{N} \eta_{n-l} s^*_{n-m} = 0, \quad \text{and} \quad \frac{1}{2N+1} \sum_{n=-N}^{N} \eta_{n-l} \eta^*_{n-m} = \sigma^2_\eta \delta_{ml}, \tag{4.54}$$

and we define the signal "correlation" matrix elements (this is consistent with later discussions in section 6.6)

$$R_{ml} = R^*_{lm} \equiv \frac{1}{2N+1} \sum_{n=-N}^{N} s_{n-l} s^*_{n-m}. \tag{4.55}$$

As will be seen in chapter 6, equations (4.54) and (4.55) are justified by the statistical requirements that the signal and noise are uncorrelated, that the noise is white with power σ^2_η, and that all random signals of interest are (wide sense) stationary. Thus, we have

$$\begin{aligned} \mathcal{E} = \ & \sigma^2_\eta (1 + \sum_l |w_l|^2) + \sum_{l,m} w^*_l w_m R_{lm} - \alpha \sum_{m,k} w^*_m h_d[k] R_{mk} \\ & - \alpha^* \sum_{m,k} w_m h_d[k] R_{km} + |\alpha|^2 \sum_{m,k} h_d[k] h_d[m] R_{km}. \end{aligned} \tag{4.56}$$

Differentiating both sides of (4.56) with respect to w^*_j and setting the result equal to 0, we find

$$\sigma^2_\eta w_k + \sum_{m=-M}^{M} w_m R_{km} = \alpha \sum_{m=-M}^{M} h_d[m] R_{km}, \quad -M \le k \le M\ , \tag{4.57}$$

which can be written as

$$(\sigma^2_\eta + R_{kk}) w_k = \alpha h_d[k] R_{kk} + \sum_{m \ne k} (\alpha h_d[m] - w_m) R_{km}, \quad -M \le k \le M\ . \tag{4.58}$$

Equations (4.58) relate the least squares filter coefficients $w_m^{(\mathrm{ls})}$ to the delayed *sinc* function $h_d[m]$, the noise power σ_η^2, and the amplitude α of the delayed signal. An approximate least squares solution for the time-delay d can be found by assuming *stationarity* of the signal, the Toeplitz property of the matrix $\boldsymbol{R}$, and the wide-band property of the signal that together with stationarity implies

$$\|s\|^2 = R_{00} = R_{kk} \gg |R_{k-m}|\,, \quad k \neq m. \tag{4.59}$$

Thus, neglecting the sum on the right hand side of equation (4.58), we find

$$w_k^{(\mathrm{ls})} \approx \alpha \, \frac{\mathrm{SNR}}{1+\mathrm{SNR}} \, h_d[k]\,, \quad \mathrm{SNR} \equiv \frac{\|s\|^2}{\sigma_\eta^2}\,. \tag{4.60}$$

The *signal to noise ratio* (SNR) is defined as the ratio of signal power to noise power. The above approximation then allows for the calculation of the time delay by finding an appropriate d and scale factor λ so that the least squares filter (derived from the two data channels) best matches the scaled *sinc* function $\lambda h_d[k]$.

CHAPTER 5

Random Variables and Estimation Theory

5.1 Real Random Variables and Random Vectors

A *continuous real random variable* X is a quantity whose possible real values $x \in \mathbb{R}$ are outcomes of a random phenomenon. Associated with X is a probability density $f_X(x)$ defined on $\mathbb{R}$; the density function is non-negative and its integral over the entire real line is equal to 1. The density function $f_X(x)$ defines the probabilities for ranges of values of the random variable X:

$$\Pr\{a \le X \le b\} = \int_a^b f_X(x)\, dx. \tag{5.1}$$

The expectation value of any function $g(X)$ of a random variable X is defined by

$$\mathrm{E}\big[g(X)\big] = \int_{-\infty}^{\infty} g(x) f_X(x)\, dx, \tag{5.2}$$

which shows the expectation operator "E" to be linear. The *moments* of a random variable X are defined as the expectation values $\mathrm{E}\big[X^n\big]$, $n = 1, 2, \ldots$. The first moment $\mathrm{E}[X]$ is the *mean* and is denoted by μ_X while the *variance*, denoted by σ_X^2, is defined as the second moment about the mean, i.e.,

$$\sigma_X^2 = \mathrm{E}\Big[\big(X - \mu_X\big)^2\Big] = \mathrm{E}\Big[X^2\Big] - \mu_X^2, \tag{5.3}$$

where the last equality follows from the linearity of the expectation operator.

Two random variables X and Y have a *joint density* function $f_{XY}(x, y)$. The marginal densities of X and Y are defined by

$$f_X(x) = \int_{-\infty}^{\infty} f_{XY}(x, y)\, dy, \quad f_Y(y) = \int_{-\infty}^{\infty} f_{XY}(x, y)\, dx. \tag{5.4}$$

The space of all real-valued and continuous random variables defined on the real line $\mathbb{R}$ forms a vector space:

- The additive zero is the random variable which takes the value 0 with probability 1.

- If X is a real-valued random variable with density function $f_X(x)$, then so is $X' = aX$ for all $a \in \mathbb{R}$ and the density function of X' is $f_X(x'/a)/|a|$.
- If X and Y are two real random variables, then so is their sum $U = X + Y$. The density function for U is

$$f_U(u) = \int_{-\infty}^{\infty} f_{XY}(x, u - x)\, dx. \tag{5.5}$$

Two random variables X and Y are independent if their joint density function is the product of their individual marginal densities, i.e.,

$$X \text{ and } Y \text{ independent } \Leftrightarrow\ f_{XY}(x, y) = f_X(x) f_Y(y). \tag{5.6}$$

The density function of the sum of two independent random variables $U = X + Y$ is the convolution of the individual marginal density functions,

$$X \text{ and } Y \text{ independent } \Rightarrow f_U(u) = \int_{-\infty}^{\infty} f_X(x) f_Y(u - x) dx, \quad U = X + Y. \tag{5.7}$$

The *correlation* (or *cross-correlation*) of two random variables X and Y is defined by

$$\mathbf{Corr}(X, Y) \equiv R_{XY} \equiv \mathrm{E}[XY] = \int_{-\infty}^{\infty}\int_{-\infty}^{\infty} xy f_{XY}(x, y)\, dxdy \tag{5.8}$$

and their *covariance* (or *cross-covariance*) is

$$\mathbf{Cov}(X, Y) \equiv C_{XY} \equiv \mathrm{E}[(x - \mu_X)(y - \mu_Y)] = R_{XY} - \mu_X \mu_Y. \tag{5.9}$$

If X and Y are independent, then $R_{XY} = \mu_X \mu_Y$ and $C_{XY} = 0$, in which case either of the latter two equations defines two uncorrelated random variables. Thus, independent random variables are uncorrelated, but the converse is not true in general (it is true only if the two random variables are jointly Gaussian—see section 5.4). The vector space of real random variables is endowed with an inner-product, namely, the covariance function C_{XY},

$$\langle X, Y \rangle = C_{XY}. \tag{5.10}$$

The induced norm is $\|X\|^2 = C_{XX} = \sigma_X^2$. The equation $\|X - Y\| = 0$ is understood to mean that the two random variables are equal with probability 1, in which case we say they are equal *almost surely* [1].

[1] A rigorous foundation for the study of random variables starts with the concept of a *probability space* that consists of a sample space Ω (set of all possible outcomes), a collection of events $\mathcal{F}$ that are subsets of the sample space (called a σ-algebra), and a non-negative function P called a *measure*, that assigns to each event a real number between 0 and 1. Certain requirements must be satisfied, e.g., the empty set ϕ and the full space must be members of $\mathcal{F}$ and their measures must be 0 and 1, respectively [48]. A simple example is provided by the coin toss. The sample space Ω has two elements: H (for head) and T (for tail). The σ-algebra can be the set of all subsets of the sample space, i.e., $\{\phi, \{H\}, \{T\}, \{H, T\}\}$, and it has to satisfy certain conditions. The probability measure P gives a weight to each member of the σ-algebra. The collection of the probability space, the associated σ-algebra, and the probability measure define a *measurable probability space*. All random variables that differ from one another on sets of probability measure 0 are considered to be the same—this means that we deal with an inner-product space of equivalence classes of random variables. This is analogous to the equivalence class of functions in $L_2(\mathbb{R})$ that differ from one another only on sets of measure 0.

The *correlation coefficient* between two real valued random variables X and Y is defined by

$$\rho_{XY} \equiv \left\langle \frac{X-\mu_X}{\sigma_X}, \frac{Y-\mu_Y}{\sigma_Y} \right\rangle = \frac{C_{XY}}{\sqrt{C_{XX}C_{YY}}}, \quad -1 \leq \rho_{XY} \leq 1, \tag{5.11}$$

where the bounds on the correlation coefficient follow from the Cauchy-Schwarz inequality (1.9). Two random variables are *orthogonal* if they have zero covariance or zero correlation coefficient.

Given two random variables X and Y with joint density function $f_{XY}(x,y)$, the density of X *conditioned* on Y (i.e., for a specific value y) is defined by

$$f_{X|Y}(x|y) \equiv f_{XY}(x,y)/f_Y(y), \tag{5.12}$$

which leads to

$$f_{X|Y}(x|y)\, f_Y(y) = f_{Y|X}(y|x)\, f_X(x). \tag{5.13}$$

An alternative form of this equation is *Bayes' formula*:

$$f_{X|Y}(x|y) = \frac{f_{Y|X}(y|x)\, f_X(x)}{f_Y(y)}. \tag{5.14}$$

If X and Y are independent then $f_{X|Y}(x|y) = f_X(x)$, i.e., the observation of Y (any observed value y) has no effect on the density function of X and provides no information that can be used to predict X.

The conventional notation of using a capital letter for a random variable and the lower case version of the same letter to refer to the values of that random variable can become confusing once we define random signals when we often use capital letters to denote Fourier and other transforms of signals; henceforth we abandon this notation and use lower case letters to refer to both the random variable and its value. For instance, we will now use $f_x(x)$ to denote the density function of the random variable x (the subscript) whose values are denoted by x (in parentheses).

A real valued random vector $\boldsymbol{x}$ of finite dimension has a joint density function defined by

$$f_{\boldsymbol{x}}(\boldsymbol{x}) = f_{x_1 x_2 \ldots}(x_1, x_2, \ldots). \tag{5.15}$$

The expectation value of a function $g(\boldsymbol{x})$ of the random vector $\boldsymbol{x}$, denoted by $\mathrm{E}[g(\boldsymbol{x})]$, is now given by the following multiple integral:

$$\int_{-\infty}^{\infty} \cdots \int_{-\infty}^{\infty} g(x_1, x_2, \ldots)\, f_{x_1 x_2 \cdots}(x_1, x_2, \ldots)\, dx_1 dx_2 \cdots \equiv \int g(\boldsymbol{x})\, f_{\boldsymbol{x}}(\boldsymbol{x})\, d\boldsymbol{x}. \tag{5.16}$$

The marginal density function of each random variable x_k is defined by the integral of the joint density over all the remaining random variables, i.e.,

$$f_{x_k}(x_k) = \int_{-\infty}^{\infty} \cdots \int_{-\infty}^{\infty} f_{x_1 x_2 \cdots}(x_1, x_2, \ldots)\, dx_1 \cdots dx_{k-1} dx_{k+1} \cdots, \tag{5.17}$$

which implies that

$$\mathrm{E}[x_k] = \int_{-\infty}^{\infty} x_k\, f_{x_k}(x_k)\, dx_k. \tag{5.18}$$

The following two results for jointly distributed random variables are useful when working with conditional expectation values:

$$\mathrm{E}[y] = \int y f_y(y)\, dy = \int\int y f_{y|x}(y|x) f_x(x)\, dx dy \equiv \mathrm{E}_x\big[\mathrm{E}[y|x]\big], \tag{5.19a}$$

$$\mathbf{Cov}(x,y) = \mathrm{E}_z\big[\mathbf{Cov}(x,y|z)\big] + \mathbf{Cov}\big(\mathrm{E}[x|z], \mathrm{E}[y|z]\big). \tag{5.19b}$$

The *correlation matrix* (or *cross-correlation matrix*) of two real random vectors $\boldsymbol{x}$ and $\boldsymbol{y}$ of equal dimension is defined by

$$\boldsymbol{R_{xy}} \equiv \mathrm{E}\left[\boldsymbol{x}\boldsymbol{y}^T\right], \tag{5.20}$$

and their *covariance matrix* (or *cross-covariance matrix*) is

$$\boldsymbol{C}_{xy} \equiv \mathrm{E}\left[(\boldsymbol{x}-\boldsymbol{\mu}_x)(\boldsymbol{y}-\boldsymbol{\mu}_y)^T\right] = \boldsymbol{R}_{xy} - \boldsymbol{\mu}_x\,\boldsymbol{\mu}_y^T, \tag{5.21}$$

where

$$\boldsymbol{\mu}_x = \mathrm{E}[\boldsymbol{x}] = \left[\mu_{x_1}, \mu_{x_2}, \ldots\right]^T, \quad \boldsymbol{\mu}_y = \mathrm{E}[\boldsymbol{y}] = \left[\mu_{y_1}, \mu_{y_2}, \ldots\right]^T \tag{5.22}$$

are the vectors of the mean values of the individual elements x_k and y_k, respectively, each of which is defined with respect to its marginal density function.

The inner product between two real random vectors with the same dimension is the trace of the corresponding cross-covariance matrix,

$$\langle \boldsymbol{x}, \boldsymbol{y}\rangle = \textbf{Trace}\,(\boldsymbol{C}_{xy}) = \mathrm{E}\left[(\boldsymbol{y}-\boldsymbol{\mu}_y)^T(\boldsymbol{x}-\boldsymbol{\mu}_x)\right]. \tag{5.23}$$

Thus, two random vectors are uncorrelated or orthogonal if the trace of their cross-covariance matrix vanishes. We often use the terms *auto-correlation* and *auto-covariance* matrix for the quantities $\boldsymbol{R}_{xx}$ and $\boldsymbol{C}_{xx}$. The auto-covariance matrix of a real random vector is symmetric and non-negative definite. We will prove a more general result for complex random vectors in section 5.2.

Finally, if $\boldsymbol{u} = h(\boldsymbol{x})$ represents an invertible transformation between the random variables $\boldsymbol{x}$ and $\boldsymbol{u}$, then their probability density functions are related by the requirement $f_{\boldsymbol{x}}(\boldsymbol{x})d\boldsymbol{x} = f_{\boldsymbol{u}}(\boldsymbol{u})d\boldsymbol{u}$, or

$$f_{\boldsymbol{u}}(\boldsymbol{u}) = \left|\partial\boldsymbol{u}/\partial\boldsymbol{x}\right|^{-1} f_{\boldsymbol{x}}(\boldsymbol{x}), \tag{5.24}$$

where the first term on the right hand side is the inverse of the Jacobian of the transformation, and the right hand side is evaluated at $\boldsymbol{x} = h^{-1}(\boldsymbol{u})$. For instance, consider the random variables x, y and the transformed random variable $u = xy$. To find the probability density function of u, we define a second transformed random variable $w = y$; thus, the inverse transformation is given by $x = u/w$ and $y = w$. Using (5.24) we have

$$f_{uw}(u,w) = \begin{vmatrix} y & x \\ 0 & 1 \end{vmatrix}^{-1} f_{xy}(x,y) = w^{-1} f_{xy}(u/w, w),$$

which can be used to calculate the probability density function of u by integrating the right hand side over w. For example, let us find the probability density function of the sum of two independent random variables $u = x + y$. Defining $w = y$, the Jacobian of the transformation $(x,y) \rightarrow (u,w)$ is 1 and we have $f_{uw}(u,w) = f_{xy}(u-w,w) = f_x(u-w)f_y(w)$, where we used the independence of x and y. Integrating the right hand side over w gives the density function of u as the convolution of the density functions of x and y:

$$f_u(u) = \int_{-\infty}^{\infty} f_{uw}(u,w)\, dw = \int_{-\infty}^{\infty} f_{xy}(u-w,w)\, dw = \int_{-\infty}^{\infty} f_x(u-w) f_y(w)\, dw. \tag{5.25}$$

5.2 Complex Random Variables and Random Vectors

A complex random variable $z = x + iy$ is defined in terms of its real and imaginary parts, which are assumed to be real random variables with a joint probability density function $f_{xy}(x, y)$. We define the probability density function $f_z(z)$ and integration measure dz by the relation

$$f_z(z)\,dz \equiv f_{xy}(x,y)\,dxdy. \tag{5.26}$$

The expectation of the product of two random variables is calculated using the corresponding joint probability density function. For instance, if $z = x + iy$ and $w = u + iv$, then

$$\mathrm{E}[zw] = \mathrm{E}[xu] - \mathrm{E}[yv] + i\left(\mathrm{E}[xv] + \mathrm{E}[yu]\right), \tag{5.27}$$

where all joint probability density functions on the right hand side are found as marginals of the probability density function $f_{xyuv}(x, y, u, v)$; for instance,

$$f_{xu}(x,u) = \int\int f_{xyuv}(x,y,u,v)\,dydv, \quad \text{and} \quad \mathrm{E}[xu] = \int\int xuf_{xu}(x,u)\,dxdu. \tag{5.28}$$

The correlation and covariance between complex random variables z and w are

$$R_{zw} = \mathrm{E}[zw^*],\ C_{zw} = \mathrm{E}[(z - \mathrm{E}[z])(w - \mathrm{E}[w])^*] = R_{zw} - \mathrm{E}[z]\left(\mathrm{E}[w]\right)^*, \tag{5.29}$$

and the covariance between z and w defines their inner product

$$\langle z, w\rangle \equiv C_{zw}. \tag{5.30}$$

The inner product (5.30) defines a convention different from (1.7): we now have $\langle z, aw\rangle = a^* \langle z, w\rangle$ for $a \in \mathbb{C}$. As in the case of real random variables, in order for C_{zw} to be a valid inner product, all complex random variables that differ from one another on sets of probability measure zero must be identified as the same random variable. Given the covariance C_{zw}, two complex random variables, unlike real ones, have a *pseudo-covariance* defined by $\tilde{C}_{zw} = \mathrm{E}[(z - \mathrm{E}[z])(w - \mathrm{E}[w])]$, which we assume to be zero in all signal processing applications of interest; we will discuss this property in more detail in the context of complex random vectors.

If $\boldsymbol{z} = [z_1, \ldots, z_N]^T$ is a complex random vector whose complex elements are $z_n = x_n + iy_n$, $1 \le n \le N$, then its probability density function (multiplied by the volume measure) is defined by

$$f_{\boldsymbol{z}}(\boldsymbol{z})\,dz \equiv f_{x_1 y_1 \ldots x_N y_N}(x_1, y_1, \ldots, x_N, y_N)\,dx_1 \ldots dx_N dy_1 \ldots dy_N. \tag{5.31}$$

The cross-correlation and cross-covariance matrices for two $N \times 1$ complex random vectors $\boldsymbol{z}$ and $\boldsymbol{w}$ are:

$$\boldsymbol{R}_{zw} = \mathrm{E}[\boldsymbol{z}\boldsymbol{w}^+],\quad \boldsymbol{C}_{zw} = \mathrm{E}[(\boldsymbol{z} - \mathrm{E}[\boldsymbol{z}])(\boldsymbol{w} - \mathrm{E}[\boldsymbol{w}])^+] = \boldsymbol{R}_{zw} - \mathrm{E}[\boldsymbol{z}]\left(\mathrm{E}[\boldsymbol{w}]\right)^+. \tag{5.32}$$

Writing $\boldsymbol{z} = \boldsymbol{x} + i\boldsymbol{y}$ and $\boldsymbol{w} = \boldsymbol{u} + i\boldsymbol{v}$ we have

$$\boldsymbol{C}_{zw} = \boldsymbol{C}_{xx} + \boldsymbol{C}_{yy} + i\left(\boldsymbol{C}_{yx} - \boldsymbol{C}_{xy}\right), \tag{5.33}$$

and the inner product between $\boldsymbol{z}$ and $\boldsymbol{w}$ is the trace of their cross-covariance matrix, namely,

$$\langle \boldsymbol{z}, \boldsymbol{w}\rangle \equiv \textbf{Trace}\left(\boldsymbol{C}_{zw}\right). \tag{5.34}$$

Unlike real random vectors, two complex random vectors $\boldsymbol{z}$ and $\boldsymbol{w}$ possess a *pseudo-covariance matrix*

$$\tilde{\boldsymbol{C}}_{zw} = \mathrm{E}\Big[\big(\boldsymbol{z} - \mathrm{E}[\boldsymbol{z}]\big)\big(\boldsymbol{w} - \mathrm{E}[\boldsymbol{w}]\big)^T\Big]. \tag{5.35}$$

Two complex random vectors are uncorrelated provided both $\boldsymbol{C}_{zw}$ and $\tilde{\boldsymbol{C}}_{zw}$ vanish. In most signal processing applications, however, we demand that the pseudo-covariance matrix be zero. Complex random vectors with zero pseudo-covariance are known as *proper* [2]. For any complex random vector $\boldsymbol{z} = \boldsymbol{x} + i\boldsymbol{y}$ we have

$$\boldsymbol{C}_{zz} = \boldsymbol{C}_{xx} + \boldsymbol{C}_{yy} + i\ (\boldsymbol{C}_{yx} - \boldsymbol{C}_{xy}),\quad \tilde{\boldsymbol{C}}_{zz} = \boldsymbol{C}_{xx} - \boldsymbol{C}_{yy} + i\ (\boldsymbol{C}_{xy} + \boldsymbol{C}_{yx}). \tag{5.36}$$

Thus, for a proper complex random vector the following results hold:

$$\boldsymbol{C}_{xx} = \boldsymbol{C}_{yy},\quad \boldsymbol{C}_{yx} = -\boldsymbol{C}_{xy} = -\boldsymbol{C}_{yx}^T. \tag{5.37}$$

Defining $\mathrm{E}[z_j] = \mu_j$, $\sigma_j^2 \equiv \mathrm{E}\big[|z_j - \mu_j|^2\big]$, and $\sigma_{jk} = \sigma_{kj}^* \equiv \mathrm{E}\big[(z_j - \mu_j)(z_k - \mu_k)^*\big]$, the complex correlation coefficient between z_j and z_k is:

$$\rho_{jk} = \frac{\sigma_{jk}}{\sigma_j \sigma_k} = \rho_{kj}^*,\qquad |\rho_{jk}| \le 1. \tag{5.38}$$

The auto-covariance matrix of an $N \times 1$ complex random vector $\boldsymbol{z}$ with mean vector $\boldsymbol{\mu}$ is

$$\boldsymbol{C}_{zz} = \mathrm{E}\Big[(\boldsymbol{z} - \boldsymbol{\mu})(\boldsymbol{z} - \boldsymbol{\mu})^+\Big] = \begin{bmatrix} \sigma_1^2 & \sigma_{12} & \cdots & \sigma_{1N} \\ \sigma_{12}^* & \sigma_2^2 & \cdots & \vdots \\ \vdots & \vdots & \ddots & \sigma_{1,N-1} \\ \sigma_{1N}^* & \cdots & \sigma_{1,N-1}^* & \sigma_N^2 \end{bmatrix}, \tag{5.39}$$

which is Hermitian; it is positive definite if the elements of $\boldsymbol{z}$ are linearly independent. Some important properties of the auto-covariance matrix are summarized in theorem 15.

Theorem 15. *Let $\boldsymbol{z} = [z_1, \ldots, z_N]^T$ denote a complex random vector with mean $\boldsymbol{\mu}$ and auto-covariance matrix $\boldsymbol{C}_{zz}$ as defined in equation* (5.39).

- *$\boldsymbol{C}_{zz}$ is Hermitian and non-negative definite; it is positive definite if the elements of $\boldsymbol{z}$ are linearly independent.*

- *$\boldsymbol{C}_{zz}$ has real and non-negative eigenvalues, and eigenvectors corresponding to distinct eigenvalues are orthogonal. If an eigenvalue is degenerate, i.e., it belongs to several distinct eigenvectors, then the Gram-Schimdt process can be used to produce a set of orthonormal eigenvectors from the degenerate set. Thus, we have the decomposition $\boldsymbol{C}_{zz} = \boldsymbol{U}\boldsymbol{\Lambda}\boldsymbol{U}^+$ where $\boldsymbol{U} = [\boldsymbol{u}_1, \ldots, \boldsymbol{u}_N]$ is the unitary matrix of orthonormal eigenvectors and $\boldsymbol{\Lambda}$ is the diagonal matrix of eigenvalues. The decomposition can also be written as $\boldsymbol{C}_{zz} = \sum_{n=1}^{N} \lambda_n \boldsymbol{u}_n \boldsymbol{u}_n^+$ with the inverse matrix $\boldsymbol{C}_{zz}^{-1} = \sum_{n=1}^{N} \lambda_n^{-1} \boldsymbol{u}_n \boldsymbol{u}_n^+$.*

[2] A proper complex random vector possesses the maximal entropy property, i.e., the entropy $h(\boldsymbol{z}) = -\int f_{\boldsymbol{z}}(\boldsymbol{z}) \ln f_{\boldsymbol{z}}(\boldsymbol{z})\, d\boldsymbol{z}$ [see equation (5.213) in section 5.18 for a discussion of entropy] of a complex zero-mean random vector $\boldsymbol{z}$ with correlation matrix $\boldsymbol{R}$ satisfies the inequality $h(\boldsymbol{z}) \le \ln\big[(\pi e)^N |\boldsymbol{R}|\big]$. Equality is achieved when the pseudo-correlation matrix vanishes, i.e., when $\boldsymbol{z}$ is a proper complex random vector [49].

- *In addition, $\boldsymbol{C}_{zz}$ has a "square root" decomposition $\boldsymbol{C}_{zz} = \boldsymbol{T}\boldsymbol{T}^{+}$, where $\boldsymbol{T} \equiv \boldsymbol{U}\Lambda^{1/2}$.*

For an arbitrary complex vector $\boldsymbol{a}$ we have

$$\boldsymbol{a}^{+}\boldsymbol{C}_{zz}\boldsymbol{a} = E\left[\left|\boldsymbol{a}^{+}(\boldsymbol{z}-\boldsymbol{\mu})\right|^{2}\right] \geq 0, \tag{5.40}$$

with equality only when the elements of $\boldsymbol{z}$ are linearly dependent. The eigenvalue decomposition of the auto-covariance matrix can be written in the equivalent form

$$\Lambda = \boldsymbol{U}^{+}\boldsymbol{C}_{zz}\boldsymbol{U} = \mathrm{E}\left[\boldsymbol{U}^{+}(\boldsymbol{z}-\boldsymbol{\mu})(\boldsymbol{z}-\boldsymbol{\mu})^{+}\boldsymbol{U}\right], \tag{5.41}$$

which is the auto-covariance matrix $\boldsymbol{C}_{\boldsymbol{vv}}$ of the random vector $\boldsymbol{v} = \boldsymbol{U}^{+}(\boldsymbol{z}-\boldsymbol{\mu})$. Thus, a zero eigenvalue of the auto-covariance matrix at diagonal location (k,k) corresponds to an equation $\mathrm{E}[v_k v_k^*] = \|v_k\|^2 = 0$ where $v_k = \boldsymbol{u}_k^{+}(\boldsymbol{z}-\boldsymbol{\mu})$ and $\boldsymbol{u}_k$ is the k-th column of $\boldsymbol{U}$. The solution to this equation is, of course, $\boldsymbol{u}_k^{+}(\boldsymbol{z}-\boldsymbol{\mu}) = 0$; thus, a zero eigenvalue implies that the elements of $\boldsymbol{z}$ are linearly dependent. The number of non-zero eigenvalues is the number of linearly independent random variables that make up the random vector $\boldsymbol{z}$. We will always assume that elements of a random vector are linearly independent and so the auto-covariance matrix is positive definite, in which case the matrix $\boldsymbol{T}$ is invertible; it can be used to define a "whitened" random vector

$$\boldsymbol{w} \equiv \boldsymbol{T}^{-1}(\boldsymbol{z}-\boldsymbol{\mu}) \tag{5.42}$$

whose mean is zero and whose auto-covariance matrix (same as its auto-correlation matrix) is the identity:

$$\mathrm{E}[\boldsymbol{w}] = \boldsymbol{0}, \text{ and } \boldsymbol{C}_{ww} = \boldsymbol{R}_{ww} \equiv \mathrm{E}[\boldsymbol{w}\boldsymbol{w}^{+}] = \boldsymbol{T}^{-1}\boldsymbol{C}_{zz}(\boldsymbol{T}^{+})^{-1} = \boldsymbol{I}. \tag{5.43}$$

5.3 Random Processes

We can extend the definition of a random vector from a finitely indexed set $n \in [N, N']$ of random variables x_n to a continuously indexed set $t \in [T, T']$ denoted by x_t or $x(t)$ and defining a *random process* or a *stochastic function* [50]; t often represents time but can be replaced by a continuous variable representing a spatial dimension. Thus, a *random process* or *stochastic function* x_t depends on a continuous parameter $t \in [T, T']$, T and $T' \in \mathbb{R}$, is square integrable over a continuous probability space $(\Omega, \mathcal{F}, P)$ (see footnote 1), and takes on values associated with a probability density function $f(x_t)$. A (complex) random process is fully defined if, in addition to $f(x_t)$, all its joint density functions $f(x_{t_1}, x_{t_2})$, $f(x_{t_1}, x_{t_2}^*)$, ..., $f(x_{t_1}, \ldots, x_{t_k})$, etc. for all time instances t, t_k are known. If the index set is $n \in \mathbb{Z}$, then x_n is a discrete time random process.

The mean of a random process $\mu(t) = \mathrm{E}[x_t]$ may be a function of time t. The auto-correlation function of a (complex) random process $R_{xx}(t,t') \equiv \mathrm{E}\left[x_t x_{t'}^*\right]$ and the auto-covariance function $C_{xx}(t,t') \equiv \mathrm{E}\left[\left(x_t - \mu_t\right)\left(x_{t'} - \mu_{t'}\right)^*\right]$ may be functions of both t and t'. The auto-covariance and the auto-correlation function are related by

$$C_{xx}(t,t') = R_{xx}(t,t') - \mu(t)\mu^*(t'). \tag{5.44}$$

Similarly, we define the cross-covariance function between two (complex) random processes x_t, y_t with means μ_x, μ_y (which are, in general, functions of time)

$$C_{xy}(t,t') \equiv \mathrm{E}\left[\left(x_t - \mu_x(t)\right)\left(y_{t'} - \mu_y(t')\right)^*\right], \quad C_{xy}(t,t') = R_{xy}(t,t') - \mu_x(t)\mu_y^*(t'). \tag{5.45}$$

A class of useful random processes for which many theoretical results exist is the set of *strict sense stationary* random processes (SSS), all of whose individual and joint density functions are invariant under time translation; i.e., $f(x_t)$ is a constant, $f(x_t, x^*_{t-\tau})$ is a function of τ, and similarly for all higher order density functions. An example of an SSS random process is $x_t = A\cos(\omega t + \phi)$, where ϕ is uniformly distributed in $[-\pi, \pi]$. The density function is $f(x_t) = 1/\pi\sqrt{A^2 - x^2}$, $-A < x < +A$, and $f(x_t, x_{t-\tau}) = f(x_t)f(x_t|x_{t-\tau}) = f(x)\delta(x_t - x_{t-\tau})$ which depends on τ alone; once the process values at two times are known, the shape of the function is known and all higher order density functions are independent of the time origin, and so the process is SSS; the auto-correlation function of this process is $R(\tau) = A^2\cos(\omega\tau)/2$.

If multiple processes are *independent and identically distributed* (IID), they form an SSS random process. Random processes whose means are independent of t and whose auto-correlation and auto-covariance functions are functions of the time difference alone and have finite values at $t = 0$ (i.e., they have finite average power) are known as *wide sense stationary* (WSS) and will be discussed in chapter 6. An SSS random process is clearly WSS, but the converse is not necessarily true (unless the process is Gaussian [see section 5.4]).

A random process (discrete or continuous time) x_t is said to be a *Markov* process if the future and past of the process are conditionally independent given the process present value. That is, $f\left(x_{t_{n+1}}|x_{t_n}, x_{t_{n-1}}, \ldots\right) = f\left(x_{t_{n+1}}|x_{t_n}\right)$, $0 \le t_1 < t_2 \ldots < t_{n+1}$. x_t is said to have the *independent increment* property if x_{t_1}, $x_{t_2} - x_{t_1}, x_{t_3} - x_{t_2}, \ldots$, are independent. An independent increment process is Markov but the converse is not necessarily true; for example, iid processes are Markov but not independent increment. The Poisson process N_t with rate λ is defined by an independent increment arrival (or counting) process where the probability of the number of arrivals (or counts) k in any interval of time t is $P[N_t = k] = (\lambda t)^k \exp(-\lambda t)/k!$; its independent increments property can be used to show that its mean is equal to λt, its auto-correlation function $R(t, t') = \lambda \min(t, t') + \lambda^2 tt'$, and its auto-covariance function $C(t, t') = \lambda \min(t, t')$.

Dynamic linear systems with random inputs can be described by differential equations, for instance, the Wiener process w_t, $0 \le t \le T$, describing the position of a particle undergoing Brownian motion in a fluid [3]. The particle can be modeled as being subject to a frictional force $-\dot{w}_t/\tau_0$ (for some constant τ_0 with dimension of time) and a force due to random collisions with the fluid molecules. The equation of motion is a linear second order stochastic differential equation

$$\ddot{w}_t + \dot{w}_t/\tau_0 = \nu_t, \tag{5.46}$$

where the collisional force $\nu_t \sim \mathcal{N}\left(0, \sigma_\nu^2\right)$ and σ_ν has dimension of acceleration. The sample functions of the Wiener process (i.e., specific paths of a single particle) are fractal in nature and not differentiable in the usual sense. Definitions of continuity and differentiability are meaningful in "mean square" sense.

A random process x_t is continuous at t if $\mathrm{E}\left[|x_{t+\delta} - x_t|^2\right] \to 0$ as $\delta \to 0$. Now

$$\mathrm{E}\left[|x_{t+\delta} - x_t|^2\right] = R_{xx}(t+\delta, t+\delta) - R_{xx}(t+\delta, t) + R_{xx}(t, t) - R_{xx}(t, t+\delta), \tag{5.47}$$

and so the left hand side of (5.47) will tend to 0, as $\delta \to 0$, provided that $R_{xx}(t, t')$ is a continuous function in both t and t'. The continuity of a random process at t also implies that its mean is continuous at the same point; this follows from

$$\mathrm{E}\left[|x_{t+\delta} - x_t|^2\right] \ge |\mu_{t+\delta} - \mu_t|^2, \tag{5.48}$$

[3]The Wiener process is the theoretical name for the physical process first observed by Robert Brown in 1827 while studying erratic motion of pollen in water, and later described by Albert Einstein as the result of continuous collision with water molecules; it is defined to be zero at $t = 0$ and has independent Gaussian distributed increments, i.e., $w_{t'+t} - w_{t'} \sim \mathcal{N}(0, \sigma_w^2 t)$. The Wiener process can also be defined as the limit of a symmetric random walk when the jumps and times between them $\to 0$ simultaneously. Modern applications of the Wiener process include modeling the logarithm of a stock price with the process maximum over an interval being closely related to the value of an option on the stock.

and the fact that if the left hand side $\to 0$, as $\delta \to 0$, so does the right hand side. A random process x_t has a derivative denoted by $\dot{x}_t$ if

$$\mathrm{E}\left[\left|\frac{x_{t+\delta} - x_t}{\delta} - \dot{x}_t\right|^2\right] \to 0 \quad \text{as} \quad \delta \to 0. \tag{5.49}$$

Results for differentiability of a random process [50] are summarized in theorem 16.

Theorem 16. *A random process x_t is differentiable at t if $\partial^2 R_{xx}(t,t')/\partial t \partial t'$ exists at both t and $t' = t$. When the process $\dot{x}_t$ exists, then*

$$\mu_{\dot{x}} = d\mu_x/dt, \quad R_{\dot{x}\dot{x}}(t,t') = \partial^2 R_{xx}(t,t')/\partial t \partial t'. \tag{5.50}$$

If the process x_t is WSS, then $\dot{x}_t$ exists at all t provided that the second derivative $d^2 R_{xx}(\tau)/d\tau^2$ exists at $\tau = 0$. If a WSS process is differentiable, then

$$R_{\dot{x}\dot{x}}(\tau) = -\, d^2 R_{xx}(\tau)/d\tau^2. \tag{5.51}$$

As an example, consider the random process $x_t = A\cos(\omega t)$, where $A \sim \mathcal{N}(0,\sigma^2)$. The mean and auto-correlation function are $\mu = 0$ and $R_{xx}(t,t') = \sigma^2 \cos(\omega t)\cos(\omega t')$, respectively. Since the second mixed partial derivative of $R_{xx}(t,t')$ is equal to $\omega^2\sigma^2 \sin^2(\omega t)$ when evaluated at $t' = t$, we conclude that the process is differentiable at all t. Now consider the Wiener process w_t whose auto-correlation function is $R_{ww}(t,t') = \sigma_w \min(t,t')$. We have,

$$\frac{\partial R_{ww}(t,t')}{\partial t'} = \sigma_w\, \theta(t-t'), \quad \frac{\partial^2 R_{ww}(t,t')}{\partial t \partial t'} = \sigma_w\, \delta(t-t'). \tag{5.52}$$

Using equation (5.50) the auto-correlation function of the process $\nu_t = \dot{w}_t$ is proportional to the Dirac delta distribution; this process is known as *White Gaussian Noise* (WGN).

5.4 Gaussian Random Variables and Random Vectors

A *Gaussian* real random variable x with mean $\mu_x = \mathrm{E}[x]$ and variance $\sigma_x^2 = \mathrm{E}[(x-\mu_x)^2] = \mathrm{E}[x^2] - \mu_x^2$ is defined by the probability density function [4]

$$f_x(x) = \frac{1}{\sqrt{2\pi}\,\sigma_x} \exp\left[-\frac{(x-\mu_x)^2}{2\sigma_x^2}\right], \quad x \in \mathbb{R}. \tag{5.53}$$

The Gaussian density function (5.53) is also known as the *normal density* and we write $x \sim \mathcal{N}(\mu_x, \sigma_x^2)$. The importance of the normal distribution is embodied in the *central limit theorem* [5].

[4] The Gaussian distribution is followed by many naturally occurring phenomena, and can provide the best model approximation to other processes. It is, however, wise to remember that "Gaussianity, like the queen of England, reigns but does not govern."

[5] Theorem 17 is an extension of the classical result [51]; it is attributed to the Russian mathematician Lyapunov, whose generalization requires only independent random variables X_n, which need not have the same density (in the classical form of the theorem they are assumed to have identical distributions) [48].

Theorem 17. *If X_n, $n = 1, \ldots, N$, are independently distributed real random variables with means μ_n and variances σ_n^2, and if we define the real random variable Y_N by*

$$Y_N = \sum_{n=1}^{N} \frac{X_n - \mu_n}{\sigma_N}, \quad \sigma_N^2 \equiv \sum_{n=1}^{N} \sigma_n^2, \tag{5.54}$$

then, symbolically, $Y_\infty \sim \mathscr{N}(0,1)$, i.e., the density function of $Y_N \to$ a Gaussian distribution with zero mean and unit variance as $N \to \infty$, provided that the Lyapunov condition holds for some $\delta > 0$:

$$\frac{1}{\sigma_N^{2+\delta}} \sum_{n=1}^{N} E\left[\left|X_n - \mu_n\right|^{2+\delta}\right] \to 0 \quad as \quad N \to \infty.$$

A real Gaussian random vector $\boldsymbol{x} = [x_1, \ldots, x_N]^T$ with mean $\mathrm{E}[\boldsymbol{x}] \equiv \boldsymbol{\mu}_x = [\mu_1, \ldots, \mu_N]^T$ and covariance matrix $\boldsymbol{C}_{xx} = \mathrm{E}\left[(\boldsymbol{x} - \boldsymbol{\mu}_x)(\boldsymbol{x} - \boldsymbol{\mu}_x)^T\right]$ has the density function

$$f_{\boldsymbol{x}}(\boldsymbol{x}) = \frac{1}{\sqrt{(2\pi)^N |\boldsymbol{C}_{xx}|}} \exp\left[-\frac{1}{2}(\boldsymbol{x} - \boldsymbol{\mu}_x)^T \boldsymbol{C}_{xx}^{-1} (\boldsymbol{x} - \boldsymbol{\mu}_x)\right], \tag{5.55}$$

and denoted by $\boldsymbol{x} \sim \mathscr{N}(\boldsymbol{\mu}_x, \boldsymbol{C}_{xx})$.

In order to derive the probability density function of a complex Gaussian random vector $\boldsymbol{z} = \boldsymbol{x} + i\boldsymbol{y}$, we first introduce the "augmented" $2N \times 1$ random vector $\boldsymbol{z}'$, and calculate its mean and covariance,

$$\boldsymbol{z}' \equiv \begin{bmatrix} \boldsymbol{z} \\ \boldsymbol{z}^* \end{bmatrix}, \quad \boldsymbol{\mu}_{z'} = \mathrm{E}[\boldsymbol{z}'] = \begin{bmatrix} \boldsymbol{\mu}_z \\ \boldsymbol{\mu}_z^* \end{bmatrix}, \quad \boldsymbol{C}_{z'z'} = \mathrm{E}\left[(\boldsymbol{z}' - \boldsymbol{\mu}_{z'})(\boldsymbol{z}' - \boldsymbol{\mu}_{z'})^+\right] = \begin{bmatrix} \boldsymbol{C}_{zz} & \tilde{\boldsymbol{C}}_{zz} \\ \tilde{\boldsymbol{C}}_{zz}^* & \boldsymbol{C}_{zz}^* \end{bmatrix}. \tag{5.56}$$

If $\boldsymbol{z}$ is a proper complex random vector then $\tilde{\boldsymbol{C}}_{zz} = \boldsymbol{0}$ and $\boldsymbol{C}_{z'z'}$ is block-diagonal. Next we define the $2N \times 1$ augmented real random vector $\boldsymbol{u} = \left[\boldsymbol{x}^T, \boldsymbol{y}^T\right]^T$ and find its mean and covariance,

$$\boldsymbol{u} \equiv \begin{bmatrix} \boldsymbol{x} \\ \boldsymbol{y} \end{bmatrix}, \quad \boldsymbol{\mu}_u = \mathrm{E}[\boldsymbol{u}] = \begin{bmatrix} \boldsymbol{\mu}_x \\ \boldsymbol{\mu}_y \end{bmatrix}, \quad \boldsymbol{C}_{uu} = \mathrm{E}\left[(\boldsymbol{u} - \boldsymbol{\mu}_u)(\boldsymbol{u} - \boldsymbol{\mu}_u)^T\right] = \begin{bmatrix} \boldsymbol{C}_{xx} & \boldsymbol{C}_{xy} \\ \boldsymbol{C}_{yx} & \boldsymbol{C}_{yy} \end{bmatrix}. \tag{5.57}$$

In order to find the probability density function of the complex random vector $\boldsymbol{z} = \boldsymbol{x} + i\boldsymbol{y}$ according to (5.31), we must find the probability density function of $\boldsymbol{u}$. The augmented vector $\boldsymbol{z}'$ can be expressed in terms of $\boldsymbol{u}$,

$$\boldsymbol{z}' \equiv \begin{bmatrix} \boldsymbol{x} + i\boldsymbol{y} \\ \boldsymbol{x} - i\boldsymbol{x} \end{bmatrix} = \begin{bmatrix} \boldsymbol{I}_N & +i\boldsymbol{I}_N \\ \boldsymbol{I}_N & -i\boldsymbol{I}_N \end{bmatrix} \begin{bmatrix} \boldsymbol{x} \\ \boldsymbol{y} \end{bmatrix} \equiv \boldsymbol{K}\boldsymbol{u}, \tag{5.58}$$

where $\boldsymbol{I}_N$ is the $N \times N$ identity matrix and $\boldsymbol{K}^+\boldsymbol{K} = \boldsymbol{K}\boldsymbol{K}^+ = 2\boldsymbol{I}_{2N}$. The $2N \times 2N$ matrix $\boldsymbol{K}$ is, therefore, invertible and $\boldsymbol{K}^{-1} = 1/2\, \boldsymbol{K}^+$. Multiplying both sides of (5.58) with $\boldsymbol{K}^+$ and then calculating the mean and covariance matrix of $\boldsymbol{u}$ we obtain

$$\boldsymbol{u} = \frac{1}{2}\boldsymbol{K}^+\boldsymbol{z}', \quad \boldsymbol{\mu}_u = \frac{1}{2}\boldsymbol{K}^+\boldsymbol{z}', \quad \boldsymbol{C}_{uu} = \frac{1}{4}\boldsymbol{K}^+ \boldsymbol{C}_{z'z'} \boldsymbol{K}, \quad \boldsymbol{C}_{uu}^{-1} = \boldsymbol{K}^+ \boldsymbol{C}_{z'z'}^{-1} \boldsymbol{K}, \tag{5.59}$$

where we used the relations $\boldsymbol{K}^{-1} = 1/2\boldsymbol{K}^+$ and $\left[\boldsymbol{K}^+\right]^{-1} = 1/2\boldsymbol{K}$. Thus, we have

$$\begin{aligned}(\boldsymbol{u} - \boldsymbol{\mu}_u)^T \boldsymbol{C}_{uu}^{-1} (\boldsymbol{u} - \boldsymbol{\mu}_u) &= \frac{1}{4} (\boldsymbol{u} - \boldsymbol{\mu}_u)^T \boldsymbol{K}^+\boldsymbol{K}\boldsymbol{C}_{uu}^{-1}\boldsymbol{K}^+\boldsymbol{K} (\boldsymbol{u} - \boldsymbol{\mu}_u) \\ &= (\boldsymbol{z}' - \boldsymbol{\mu}_{z'})^+ \boldsymbol{C}_{z'z'}^{-1} (\boldsymbol{z}' - \boldsymbol{\mu}_{z'}). \end{aligned} \tag{5.60}$$

Next we use

$$2^{2N}\left|\boldsymbol{C}_{uu}\right| = \left|2\boldsymbol{I}_{2N}\boldsymbol{C}_{uu}\right| = \left|\boldsymbol{K}\boldsymbol{K}^{+}\boldsymbol{C}_{uu}\right| = \left|\boldsymbol{K}\boldsymbol{C}_{uu}\boldsymbol{K}^{+}\right| = \left|\boldsymbol{C}_{z'z'}\right|, \tag{5.61}$$

and (5.60) in the Gaussian probability density function (5.55) for the $2N \times 1$ real random vector $\boldsymbol{u}$, namely,

$$f_{\boldsymbol{u}}(\boldsymbol{u}) = \frac{1}{\sqrt{(2\pi)^{2N}\left|\boldsymbol{C}_{uu}\right|}} \exp\left[-\frac{1}{2}(\boldsymbol{u}-\boldsymbol{\mu}_u)^T \boldsymbol{C}_{uu}^{-1}(\boldsymbol{u}-\boldsymbol{\mu}_u)\right], \tag{5.62}$$

to find the probability density function of the complex Gaussian random vector $\boldsymbol{z}$ according to to (5.31) (the volume measure is unaffected by the $\boldsymbol{K}$ matrix):

$$f_{\boldsymbol{z}}(\boldsymbol{z}) = \frac{1}{\pi^N\sqrt{\left|\boldsymbol{C}_{z'z'}\right|}} \exp\left[-\frac{1}{2}(\boldsymbol{z}'-\boldsymbol{\mu}_{z'})^{+}\boldsymbol{C}_{z'z'}^{-1}(\boldsymbol{z}'-\boldsymbol{\mu}_{z'})\right], \quad \boldsymbol{z}' = \begin{bmatrix} \boldsymbol{z} \\ \boldsymbol{z}^* \end{bmatrix}. \tag{5.63}$$

The probability density function (5.63) is the most general form for a complex Gaussian random vector $\boldsymbol{z} = \boldsymbol{x} + i\boldsymbol{y}$ which has a covariance matrix $\boldsymbol{C}$ and a pseudo-covariance matrix $\tilde{\boldsymbol{C}}$, both of which are nonzero. As mentioned earlier, almost all signal processing applications require a proper complex random vector satisfying equations (5.36) and (5.37). Assuming that $\boldsymbol{z}$ is a proper complex Gaussian random vector, we have

$$\boldsymbol{C}_{z'z'} = \begin{bmatrix} \boldsymbol{C}_{zz} & \boldsymbol{0} \\ \boldsymbol{0} & \boldsymbol{C}_{zz}^* \end{bmatrix}, \quad \left|\boldsymbol{C}_{z'z'}\right| = \left|\boldsymbol{C}_{zz}\right|^2, \tag{5.64}$$

and the final form of the probability density function of a proper complex Gaussian random vector $\boldsymbol{z} = \boldsymbol{x} + i\boldsymbol{y} \sim \mathscr{CN}(\boldsymbol{\mu}, \boldsymbol{C}_{zz})$ whose pseudo-covariance matrix vanishes:

$$f_{\boldsymbol{z}}(\boldsymbol{z}) = \frac{1}{\pi^N\left|\boldsymbol{C}_{zz}\right|} \exp\left[-(\boldsymbol{z}-\boldsymbol{\mu}_z)^{+}\boldsymbol{C}_{zz}^{-1}(\boldsymbol{z}-\boldsymbol{\mu}_z)\right]. \tag{5.65}$$

The two probability density functions (5.55) and (5.65), when viewed as functions of random parameters on which the random vectors $\boldsymbol{x}$ or $\boldsymbol{z}$ depend, are known as likelihood functions in estimation theory [see section 5.9]. Maximization of the likelihood as a function of parameters to be estimated is often performed by maximizing the log-likelihood function, which in the two cases (5.55) and (5.65) takes the following forms:

$$\ln f_{\boldsymbol{x}}(\boldsymbol{x}) = -\frac{N}{2}\ln(2\pi) - \frac{1}{2}\ln\left|\boldsymbol{C}_{xx}\right| - \frac{1}{2}(\boldsymbol{x}-\boldsymbol{\mu}_x)^T\boldsymbol{C}_{xx}^{-1}(\boldsymbol{x}-\boldsymbol{\mu}_x), \tag{5.66a}$$

$$\ln f_{\boldsymbol{z}}(\boldsymbol{z}) = -N\ln\pi - \ln\left|\boldsymbol{C}_{zz}\right| - (\boldsymbol{z}-\boldsymbol{\mu}_z)^{+}\boldsymbol{C}_{zz}^{-1}(\boldsymbol{z}-\boldsymbol{\mu}_z). \tag{5.66b}$$

It is useful to consider the case $\boldsymbol{C} = \sigma^2\boldsymbol{I}$; for instance, when estimating signal parameters from observed data $x_n = s_n + \nu_n$, $0 \le n \le N-1$, the noise vector $\boldsymbol{\nu}$ is often assumed to be zero-mean white Gaussian noise (WGN) $\sim \mathscr{N}(\boldsymbol{0}, \sigma^2\boldsymbol{I})$ or $\sim \mathscr{CN}(\boldsymbol{0}, \sigma^2\boldsymbol{I})$. Substituting $\sigma^2\boldsymbol{I}$ for the covariance matrix in equations (5.66) we find

$$\ln f_{\boldsymbol{x}}(\boldsymbol{x}) = -\frac{N}{2}\ln(2\pi) - \frac{N}{2}\ln\sigma^2 - \frac{1}{2\sigma^2}(\boldsymbol{x}-\boldsymbol{\mu}_x)^T(\boldsymbol{x}-\boldsymbol{\mu}_x), \tag{5.67a}$$

$$\ln f_{\boldsymbol{z}}(\boldsymbol{z}) = -N\ln\pi - N\ln\sigma^2 - \frac{1}{\sigma^2}(\boldsymbol{z}-\boldsymbol{\mu}_z)^{+}(\boldsymbol{z}-\boldsymbol{\mu}_z). \tag{5.67b}$$

Two independent random variables are uncorrelated. Two uncorrelated random variables, however, are not necessarily independent unless they are jointly Gaussian. To see this, consider $N = 2$, $\boldsymbol{\mu} = \boldsymbol{0}$,

$\sigma_k^2 = \mathrm{E}\big[|z_k|^2\big]$, $k = 1, 2$, and $\sigma_{12} = \sigma_{21}^* = \mathrm{E}\big[z_1 z_2^*\big]$. The covariance matrix (same as the correlation matrix since $\boldsymbol{\mu} = \mathbf{0}$) is:

$$\boldsymbol{C}_{zz} = \mathrm{E}\big[\boldsymbol{z}\boldsymbol{z}^+\big] = \begin{bmatrix} \sigma_1^2 & \sigma_{12} \\ \sigma_{21} & \sigma_2^2 \end{bmatrix}, \tag{5.68}$$

and its inverse is:

$$\boldsymbol{C}_{zz}^{-1} = \frac{1}{1 - |\rho|^2} \begin{bmatrix} 1/\sigma_1^2 & -\rho/\sigma_1\sigma_2 \\ -\rho^*/\sigma_1\sigma_2 & 1/\sigma_2^2 \end{bmatrix}, \quad \rho \equiv \frac{\sigma_{12}}{\sigma_1\sigma_2}. \tag{5.69}$$

Thus, two jointly Gaussian random variables are independent if, and only if, they are uncorrelated, i.e. $\rho = 0$, in which case the covariance matrix and its inverse are 2×2 diagonal matrices and the joint density becomes the product of individual Gaussian distributions; theorem 18 is the generalization of this result to an $N \times 1$ random vector (real or complex).

Theorem 18. *Let $\boldsymbol{z} = \big[z_1, \ldots, z_N\big]^T$ be a complex Gaussian random vector with mean $\boldsymbol{\mu} = \big[\mu_1, \ldots, \mu_N\big]^T$ and covariance matrix $\boldsymbol{C}_{zz} = E\big[(\boldsymbol{z}-\boldsymbol{\mu})(\boldsymbol{z}-\boldsymbol{\mu})^+\big]$ whose diagonal elements are $\sigma_n^2 = E\big[|z_n - \mu_n|^2\big]$, $1 \leq n \leq N$. If the correlation coefficients $\rho_{mn} \equiv E\big[(z_m - \mu_m)(z_n - \mu_n)^*\big]/\sigma_m\sigma_n = \delta_{mn}$, then the individual elements z_n of the random vector are independent from each other and each z_n has a complex Gaussian distribution with mean μ_n and variance σ_n^2, i.e., $z_n \sim \mathscr{CN}(\mu_n, \sigma_n^2)$ and $\boldsymbol{z} \sim \mathscr{CN}(\boldsymbol{\mu}, \boldsymbol{diag}\big[\sigma_1^2, \ldots, \sigma_N^2\big])$. If $\boldsymbol{z}$ is a real random vector, then $z_n \sim \mathscr{N}(\mu_n, \sigma_n^2)$ and $\boldsymbol{z} \sim \mathscr{N}(\boldsymbol{\mu}, \boldsymbol{diag}\big[\sigma_1^2, \ldots, \sigma_N^2\big])$.*

The linear transformation $\boldsymbol{T} \equiv \boldsymbol{U}\boldsymbol{\Lambda}^{1/2}$ in theorem 15 can be used to transform a real Gaussian random vector with density (5.55) to a real Gaussian random vector $\sim \mathscr{N}(\mathbf{0}, \boldsymbol{I})$, and a complex Gaussian random vector with density (5.65) to a complex Gaussian random vector $\sim \mathscr{CN}(\mathbf{0}, \boldsymbol{I})$. Another possible factorization of the covariance matrix is the Cholesky decomposition using triangular matrices; in the context of WSS random processes, the Cholesky factorization can be used for causal filtering and signal modeling [see section 5.5].

The usual statistical measure of non-Gaussianity of a random variable z with mean μ and variance σ^2 is the (normalized) *excess-kurtosis*

$$k_z = \mathrm{E}\left[\left|\frac{z-\mu}{\sigma}\right|^4\right] - 3, \tag{5.70}$$

which is zero if $z \sim \mathscr{N}(\mu, \sigma^2)$ or $z \sim \mathscr{CN}(\mu, \sigma^2)$; it measures the degree of peakedness (or spikiness) of a distribution. A distribution whose excess-kurtosis is positive is called *super-Gaussian* (spikier than Gaussian), while a distribution whose excess-kurtosis is negative is called *sub-Gaussian* (flatter than Gaussian).

5.5 Gram-Schmidt Decorrelation

In transform coding and compression applications, we seek a linear transformation of the form $\boldsymbol{y} = \boldsymbol{U}^+\boldsymbol{x}$ to remove self-correlations of a (complex) zero-mean random vector $\boldsymbol{x} = [x_1, \ldots, x_N]^T$, $\mathrm{E}[\boldsymbol{x}] = \mathbf{0}$, whose $N \times N$ auto-correlation matrix $\boldsymbol{R}_{xx} = \mathrm{E}[\boldsymbol{x}\boldsymbol{x}^+]$ is not diagonal, i.e., there are non-zero mutual correlations. The two vectors $\boldsymbol{y}$ and $\boldsymbol{x}$ are *informationally equivalent*, but since components of $\boldsymbol{y}$ are uncorrelated, each component adds new information that is absent in the rest of the components; for this reason $\boldsymbol{y}$ is known as the *innovations representation* of random vector $\boldsymbol{x}$. We will look at three such linear transformations in this and the next two sections.

We may use the Gram-Schmidt procedure with the expectation operator as the inner product, i.e., $\langle x_k, x_m \rangle = E\big[x_k x_m^*\big] \equiv R_{km}$, to construct the random vector $\boldsymbol{\nu} = [\nu_1, \ldots \nu_N]^T$ whose elements have zero mutual correlations, i.e., the auto-correlation matrix $\boldsymbol{R}_{\nu\nu} = \mathrm{E}[\boldsymbol{\nu}\boldsymbol{\nu}^{+}]$ is diagonal. Defining the first element $\nu_1 = x_1$ (note that we are using a variant of the Gram-Schmidt algorithm that produces orthogonal, i.e., uncorrelated, random variables that are not normalized and whose squared norms are the individual variances), and for $2 \leq n \leq N$ we have

$$\nu_n = x_n - \sum_{k=1}^{n-1} \mathrm{E}\big[x_n x_k^*\big]\, \nu_k / \|\nu_k\|^2, \quad \|\nu_k\|^2 \equiv \mathrm{E}\big[\nu_k \nu_k^*\big] = \mathrm{E}\big[\,|\nu_k|^2\,\big]. \tag{5.71}$$

The new random variables $\{\nu_1, \ldots \nu_N\}$, denoted by the random vector $\boldsymbol{\nu}$, are now mutually uncorrelated zero-mean random variables whose correlation matrix $\boldsymbol{R}_{\nu\nu}$ is diagonal with diagonal entries $\sigma_k^2 \equiv \mathrm{E}\big[\nu_k \nu_k^*\big]$.

The Gram-Schmidt algorithm (5.71) can be written in matrix form $\boldsymbol{x} = \boldsymbol{L}\boldsymbol{\nu}$ where

$$\boldsymbol{L} = \begin{bmatrix} 1 & 0 & 0 & \ldots & 0 \\ L_{21} & 1 & 0 & \ldots & 0 \\ L_{31} & L_{32} & 1 & \ldots & \vdots \\ \vdots & \vdots & \vdots & \ddots & 0 \\ L_{N1} & L_{N2} & L_{N3} & \ldots & 1 \end{bmatrix}, \tag{5.72}$$

$L_{nk} = \mathrm{E}\big[x_n \nu_k^*\big] / \|\nu_k\|^2, 1 \leq k \leq n-1$, and $L_{nn} = 1$. The matrix $\boldsymbol{L}$ is non-singular (since the determinant of a triangular matrix is the product of the diagonal elements, which in this case is equal to 1), is invertible, and its inverse is a lower triangular matrix whose diagonal elements are equal to 1. Thus, we can write $\boldsymbol{\nu} = \boldsymbol{L}^{-1}\boldsymbol{x}$. The random vector $\boldsymbol{\nu}$ is the *innovations representation* of $\boldsymbol{x}$. The elements of $\boldsymbol{\nu}$ are mutually uncorrelated and each one represents new information. The two vectors $\boldsymbol{x}$ and $\boldsymbol{\nu}$ are informationally equivalent. Thus, the auto-correlation matrix of $\boldsymbol{\nu}$ is diagonal and the auto-correlation matrix of $\boldsymbol{x} = \boldsymbol{L}\boldsymbol{\nu}$ is

$$\boldsymbol{R}_{xx} = \mathrm{E}\big[\boldsymbol{x}\boldsymbol{x}^{+}\big] = \boldsymbol{L}\,\mathrm{E}\big[\boldsymbol{\nu}\boldsymbol{\nu}^{+}\big]\,\boldsymbol{L}^{+} = \boldsymbol{L}\boldsymbol{R}_{\nu\nu}\boldsymbol{L}^{+}, \tag{5.73}$$

which can be used to construct its Cholesky factors; $\boldsymbol{R}_{xx} = \boldsymbol{L}'\boldsymbol{L}'^{+}$ with $\boldsymbol{L}' \equiv \boldsymbol{L}\sqrt{\boldsymbol{R}_{\nu\nu}}$, where the square root of the diagonal matrix $\boldsymbol{R}_{\nu\nu}$ is a diagonal matrix whose diagonal elements are σ_k.

As an example consider $\boldsymbol{x}_N \equiv [x_1, \ldots, x_N]^T$ and $\boldsymbol{x}_n \equiv [x_1, \ldots, x_n]^T$, $2 \leq n \leq N$, for which the Gram-Schmidt algorithm gives $\nu_1 = x_1$ and

$$\nu_n = x_n - \mathrm{E}\big[x_n \boldsymbol{x}_{n-1}^{+}\big]\mathrm{E}\big[\boldsymbol{x}_{n-1}\boldsymbol{x}_{n-1}^{+}\big]^{-1}\boldsymbol{x}_{n-1} \equiv x_n - \hat{x}_n^{(n-1)}, \quad 2 \leq n \leq N. \tag{5.74}$$

The projection $\hat{x}_n^{(n-1)}$ is the best linear approximation to x_n given the values $x_1, \ldots, x_{n-1}$

$$\hat{x}_n^{(n-1)} = \begin{bmatrix} R_{n1} & \cdots & R_{n,n-1} \end{bmatrix} \begin{bmatrix} R_{11} & \cdots & R_{1,n-1} \\ \vdots & \ddots & \vdots \\ R_{n-1,1} & \cdots & R_{n-1,n-1} \end{bmatrix}^{-1} \begin{bmatrix} x_1 \\ \vdots \\ x_{n-1} \end{bmatrix}, \tag{5.75}$$

for $2 \leq n \leq N$. If the index n is interpreted as a time step, then the projection term is the one-step ahead linear predictor of x_n given the vector $\boldsymbol{x}_{n-1}$ with prediction error ν_n. Had we started the Gram-Schmidt process with x_N and worked backward through the indices to 1, we would have obtained the one-step backward linear predictor. Linear prediction (LP) of stationary random processes is a major topic in optimal signal processing, and the Gram-Schmidt orthogonalization is its mathematical foundation.

As another example, consider two $N \times 1$ correlated zero-mean complex random vectors $\boldsymbol{y}$ and $\boldsymbol{x}$. The correlation between them can be canceled by means of a linear transformation matrix $\boldsymbol{H}$ where the residual vector $\mathbf{e} = \boldsymbol{y} - \boldsymbol{H}\boldsymbol{x}$ is no longer correlated with $\boldsymbol{x}$, i.e., $\boldsymbol{R}_{ex} = \mathrm{E}[\boldsymbol{e}\boldsymbol{x}^+] = \mathbf{0}$. Thus,

$$\boldsymbol{R}_{ex} = \mathrm{E}\big[(\boldsymbol{y} - \boldsymbol{H}\boldsymbol{x})\boldsymbol{x}^+\big] = \boldsymbol{R}_{yx} - \boldsymbol{H}\boldsymbol{R}_{xx} = \mathbf{0}, \tag{5.76}$$

whose solution is $\boldsymbol{H} = \boldsymbol{R}_{yx}\boldsymbol{R}_{xx}^{-1}$. The vector $\hat{\boldsymbol{y}} = \boldsymbol{H}\boldsymbol{x}$ is the orthogonal projection of $\boldsymbol{y}$ onto $\boldsymbol{x}$; it is the minimum mean square approximation to $\boldsymbol{y}$ since it minimizes $\|\boldsymbol{y} - \hat{\boldsymbol{y}}\|_2^2$.

In the context of zero-mean WSS random processes [see section 5.3] when $\mathbf{e}$ consists of time samples of a white noise process, the Gram-Schmidt decorrelation by a lower triangular matrix is equivalent to synthesizing the signal $\boldsymbol{z}$ through causal filtering of a white noise process; an autoregressive (AR) process is defined by exactly this causal filtering scheme. Two closely related linear transformations of random vectors, principal components analysis and the Karhunen-Loéve transformation, with applications to data reduction and image compression will be discussed in sections 5.6 and 5.7.

5.6 Principal Components Analysis

Principal components analysis (PCA) is used to perform data reduction or discover relationships that are otherwise unknown, e.g., pattern recognition. Consider a random vector $\boldsymbol{x}$ with mean $\boldsymbol{\mu}_x$ and dimension $d \times 1$. The $d \times d$ auto-covariance matrix $\boldsymbol{C}_{xx} = \mathrm{E}[(\boldsymbol{x} - \boldsymbol{\mu}_x)(\boldsymbol{x} - \boldsymbol{\mu}_x)^+]$ is Hermitian and positive definite but not necessarily Toeplitz. $\boldsymbol{x}$ can be a $d \times 1$ set of consecutive values of a random time series, or d features in a pattern classification problem where it is referred to as a *feature vector*, or one instant of recordings from a uniform line array of d sensors where it is referred to as a *snapshot*; it can also be an image with d pixels that have been put in the form of a $d \times 1$ vector. The idea behind PCA is to find a linear combination of the elements of $\boldsymbol{x}$ with maximum variance subject to a constraint that the coefficients have unit norm (otherwise any multiple of the coefficient vector can change the maximum). This linear combination would then exhibit the largest variation in the random vector $\boldsymbol{x}$ and that could replace it. The variation in the random vector $\boldsymbol{x}$ is encoded in the d eigenvalues of its covariance matrix and in many circumstances we find that a number $L \ll d$ may be sufficient to accurately represent $\boldsymbol{x}$, where L is the number of the largest eigenvalues of the auto-covariance matrix we choose to keep while discarding the remaining $d - L$. Let us define

$$y_1 = \boldsymbol{w}_1^+ \boldsymbol{x} = \sum_{m=1}^{d} w_{1m}^* x_m, \qquad \|\boldsymbol{w}\| = 1, \tag{5.77}$$

whose mean is $\mu_1 = \boldsymbol{w}_1^+ \boldsymbol{\mu}_x$. Using $y_1^* = \boldsymbol{w}_1^T \boldsymbol{x}^* = \boldsymbol{x}^+ \boldsymbol{w}_1$ we find the variance

$$\mathrm{E}\Big[|y_1 - \mu_1|^2\Big] = \mathrm{E}\Big[\boldsymbol{w}_1^+ (\boldsymbol{x} - \boldsymbol{\mu}_x)(\boldsymbol{x} - \boldsymbol{\mu}_x)^+ \boldsymbol{w}_1\Big] = \boldsymbol{w}_1^+ \boldsymbol{C}_{xx} \boldsymbol{w}_1. \tag{5.78}$$

The first *principal direction* is found by maximizing the *Rayleigh quotient*

$$\underset{\boldsymbol{w}_1}{\arg\max} \Big[(\boldsymbol{w}_1^+ \boldsymbol{C}_{xx} \boldsymbol{w}_1) / \|\boldsymbol{w}_1\|^2\Big], \tag{5.79}$$

which is equivalent to maximizing the numerator subject to the constraint $\|\boldsymbol{w}_1\|^2 = 1$. The solution is the eigenvector of the covariance matrix corresponding to the largest eigenvalue λ_1, i.e., $\boldsymbol{C}_{xx}\boldsymbol{w}_1 = \lambda_1 \boldsymbol{w}_1$. The projection of the data along the first principal direction, i.e., y_1, is the first *principal component*. Since the covariance matrix is Hermitian, all eigenvectors corresponding to distinct eigenvalues are orthogonal (if there are any degenerate eigenvectors, then we can use the Gram-Schmidt method to orthogonalize them). Assuming the eigenvectors $\boldsymbol{w}_m$ to be orthonormal and in descending order of associated eigenvalues, we choose the second principal component as the linear combination $y_2 = \boldsymbol{w}_2^+\boldsymbol{x}$, and so on. Thus, defining $\boldsymbol{W}$ as the unitary matrix whose columns are the principal directions (eigenvectors) $\boldsymbol{w}_m$, $1 \le m \le d$, the principal components are the projections of the data along the principal directions, i.e., elements of the random vector $\boldsymbol{y} = [y_1, \ldots, y_d]^T = \boldsymbol{W}^+\boldsymbol{x}$, where $\boldsymbol{w}_m^+$ is the m-th row of $\boldsymbol{W}^+$. Data reduction is achieved by taking the first $L < d$ elements of $\boldsymbol{y}$ corresponding to the approximation $\hat{\boldsymbol{x}} = \boldsymbol{W}[y_1, \ldots, y_L, 0, \ldots, 0]^T$.

In practice we have K realizations $\boldsymbol{x}_k$, $1 \le k \le K$, of the random vector $\boldsymbol{x}$ to estimate the $d \times d$ covariance matrix

$$\hat{\boldsymbol{C}}_{xx} = \frac{1}{K}\sum_{k=1}^{K}(\boldsymbol{x}_k - \hat{\boldsymbol{\mu}}_x)(\boldsymbol{x}_k - \hat{\boldsymbol{\mu}}_x)^+, \quad \hat{\boldsymbol{\mu}}_x = \frac{1}{K}\sum_{k=1}^{K}\boldsymbol{x}_k. \tag{5.80}$$

The factor $1/K$ in the covariance matrix estimate can be replaced by $1/(K-1)$ to ensure that the sample covariance matrix using the sample mean is an unbiased estimator, i.e., its expectation value is the true covariance matrix (see section 5.13); the factor, however, does not affect the calculation of eigenvalues and eigenvectors. The estimation of the $d \times d$ sample covariance matrix can be performed by a single matrix multiplication if we use the $d \times K$ data matrix $\boldsymbol{X}$ whose columns are the mean-subtracted data realizations

$$\hat{\boldsymbol{C}}_{xx} = \frac{1}{K}\boldsymbol{X}\boldsymbol{X}^+, \quad \boldsymbol{X} \equiv [\boldsymbol{x}_1, \ldots, \boldsymbol{x}_K] - \hat{\boldsymbol{\mu}}_x. \tag{5.81}$$

Let $\boldsymbol{W} = [\boldsymbol{w}_1, \ldots, \boldsymbol{w}_d]$ denote the $d \times d$ unitary matrix of eigenvectors of the covariance matrix (or its estimate) and assume that the eigenvectors $\boldsymbol{w}_1, \ldots, \boldsymbol{w}_d$ are in descending order of the associated eigenvalues, i.e., $\lambda_1 \ge \ldots \ge \lambda_d \ge 0$. The *principal components* are the rows of the $d \times K$ matrix $\boldsymbol{Y} = \boldsymbol{W}^+\boldsymbol{X}$. Note that

$$E[\boldsymbol{Y}\boldsymbol{Y}^+] = \boldsymbol{W}^+\boldsymbol{C}_{xx}\boldsymbol{W} = \Lambda = \textbf{diag}[\lambda_1, \ldots, \lambda_d]. \tag{5.82}$$

Thus, each principal component $\boldsymbol{y}_m = \boldsymbol{w}_m^+\boldsymbol{X}$ is uncorrelated with the other components and contributes a portion $\lambda_m/\textbf{Trace}(\Lambda)$ of the total variation. Data reduction is achieved by keeping the principal components corresponding to the largest $1 \le L < d$ eigenvalues, i.e., the first L rows of $\boldsymbol{Y}$. Setting $\boldsymbol{w}_{L+1} = \ldots = \boldsymbol{w}_d = \boldsymbol{0}$ and denoting the resulting matrix of eigenvectors by $\boldsymbol{W}_L$, the L principal components are $\boldsymbol{Y}_L = \boldsymbol{W}_L^+\boldsymbol{X}$ and the reconstructed data matrix is

$$\boldsymbol{X}_L = \boldsymbol{W}_L\boldsymbol{Y}_L + \hat{\boldsymbol{\mu}}_x, \tag{5.83}$$

corresponding to a portion $(\lambda_1 + \ldots + \lambda_L)/\textbf{Trace}(\Lambda)$ of the total variation and a reduction rate of L/d.

As an example, consider a zero-mean random vector $\boldsymbol{x} = [x_1, x_2]^T$, $d = 2$, with variances $[\boldsymbol{C}]_{11} = \mathrm{E}[x_1^2] = 2.4$, $[\boldsymbol{C}]_{22} = \mathrm{E}[x_2^2] = 5.6$, and correlation $[\boldsymbol{C}]_{12} = [\boldsymbol{C}]_{21} = \mathrm{E}[x_1x_2] = 1.2$, where $\boldsymbol{C}$ is the 2×2 covariance matrix. The scatter plot on the left hand side of figure 5.1 shows $K = 10,000$ samples (realizations) of $\boldsymbol{x}$ that were generated by first producing K samples of a two dimensional zero-mean uncorrelated random Gaussian noise with unit variance, then calculating the Cholesky factorization $\boldsymbol{C} = \boldsymbol{L}\boldsymbol{L}^H$, and finally applying the 2×2 matrix $\boldsymbol{L}$ to the random noise samples. The covariance matrix eigenvalues are 6 and 2 corresponding to normalized eigenvectors $\alpha[1, 3]^T$ and $\alpha[-3, 1]^T$, $\alpha = 1/\sqrt{10}$, respectively. The two solid arrows through the origin of the left hand side of figure 5.1 indicate the two orthogonal principal directions given by the normalized eigenvectors of the covariance matrix, while the scatter plot of the transformed data $\boldsymbol{y} = \boldsymbol{W}^+\boldsymbol{x}$ is shown on the right hand side. The reduced data vector

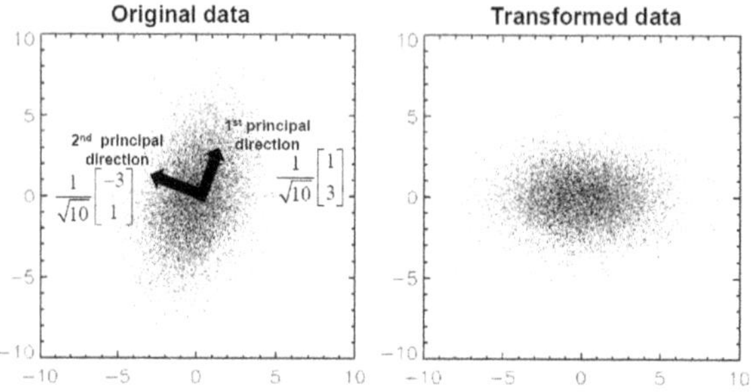

Figure 5.1: Principal directions and transformed data.

is found by keeping the first eigenvector $\boldsymbol{w}_1$ (the first column of $\boldsymbol{W}$) and neglecting the second to find the $1 \times 10,000$ first principal component $\boldsymbol{y}_1^T = \boldsymbol{w}_1^+ [\boldsymbol{x}_1, \ldots, \boldsymbol{x}_{10000}]$, which is responsible for 75% of the variation in the data; the corresponding reduction rate is 50%.

As a second example, consider a 256×256 color image of a baboon whose three RGB channels and their pairwise scatter diagrams are shown in figure 5.2. The individual pixel correlations evident in figure

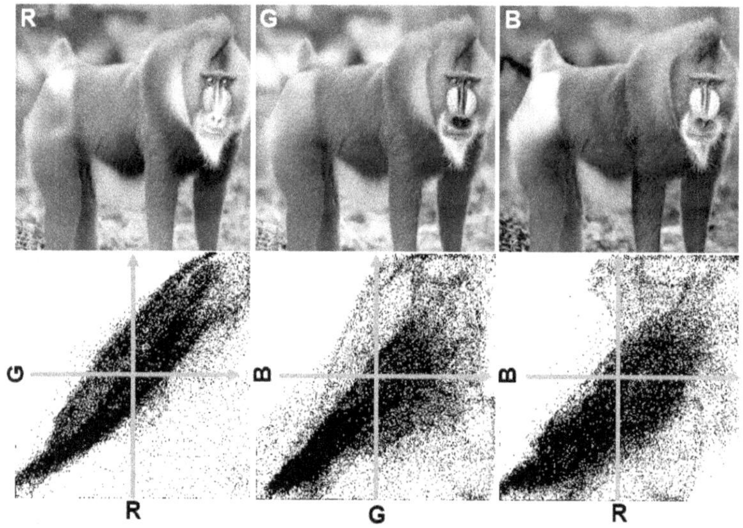

Figure 5.2: Three RGB channels and pairwise scatter diagrams.

5.2 indicate that a linear combination of the three channels can be found to include most of the variance in the three images and hence used as a grayscale version of the original color image; this can be compared with an ideal grayscale luminance image from γ-corrected RGB channels $0.2126R + 0.7152G + 0.0722B$ derived on the basis of human visual sensitivity which peaks in the Green-Yellow region. To this end we construct a 3×256^2 data matrix whose 3×3 correlation matrix eigenvalues represent 86.3%, 9.7%, and 4.0% of the total variation; the corresponding principal components are shown in figure 5.3. We may now use the first principal component as the grayscale luminance representation of the baboon color image.

We can apply PCA to compress a single $N \times N$ image $\boldsymbol{Z}$ by interpreting the columns as features and rows as samples of those features. Using equation (5.80) where $\boldsymbol{x}_k$ are columns of $\boldsymbol{Z}$, the $N \times N$ covariance

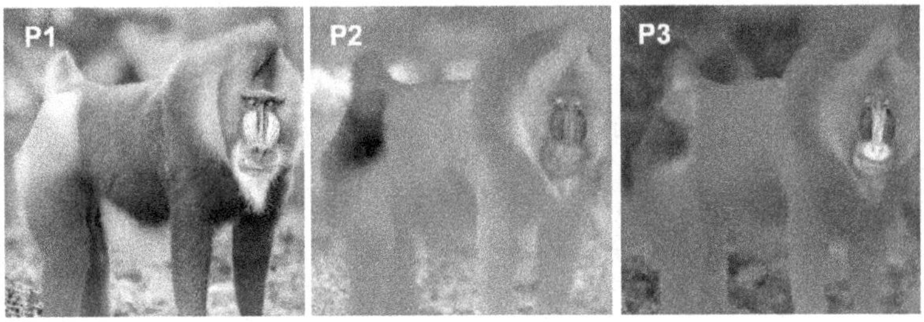

Figure 5.3: Three principal components of the baboon RGB images.

matrix $\boldsymbol{C}_r$ is calculated. The compressed image is given by (5.83) keeping $L \ll N$ eigenvectors. If we use the transpose of the image $\boldsymbol{Z}^T$ instead, then the covariance matrix $\boldsymbol{C}_c$ will have the same eigenvalues as $\boldsymbol{C}_r$ while its eigenvectors are related to the corresponding eigenvectors of $\boldsymbol{C}_r$ by $\boldsymbol{w}_c = \boldsymbol{Z}\boldsymbol{w}_r$ and the principal components are related by $\boldsymbol{Y}_r = \boldsymbol{Y}_c\boldsymbol{Z}$. A third method that takes into account local row and column correlations while producing a smaller covariance matrix is to first divide the image into $K \equiv N^2/d^2$ subimages of size $d \times d$, $d \ll N$ and then to construct a $d^2 \times K$ data matrix $\boldsymbol{X}$ whose columns are the d^2 pixels of the subimages and whose $d \times d$ covariance matrix is calculated using equation (5.80). Once $L \ll d^2$ principal components are retained, the reconstructed data matrix is calculated using (5.83) and reshaped into the original $N \times N$ size. Figure 5.4(a) shows an 8-bit encoded 256×256 image of two parrots. Figures 5.4(b) and 5.4(c) show reconstructions using row-wise and column-wise covariance matrix calculations and retaining 10 eigenvectors with 8-bit encoding with equal compression ratios 12.8:1 (8-bit encoding is applied to the retained eigenvectors, the principal components, and the mean and standard deviations of rows or columns before reconstruction). Figure 5.4(d) is the reconstruction using 16×16 subimages to calculate the covariance matrix and retaining 10 eigenvectors with 8-bit encoding with a compression ratio of 13.4:1. In this example the method using subimages gives a qualitatively better result at even a slightly higher compression ratio.

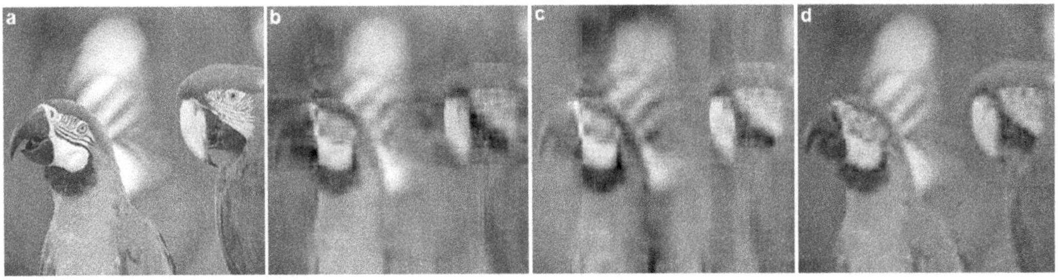

Figure 5.4: PCA compression of a 256×256 image of two parrots.

Equation (5.82) clearly shows the SVD of the covariance matrix, namely, $\boldsymbol{C}_{xx} = \boldsymbol{W}\boldsymbol{\Lambda}\boldsymbol{W}^+$. Had we begun our discussion with this equation, we would have arrived at the principal components and their properties by inspection. When the random vector $\boldsymbol{x}$ is zero-mean, the SVD of its auto-correlation matrix defines the Karhunen-Loéve transformation $\boldsymbol{W}^+\boldsymbol{x}$. Although identical to the PCA for a random vector, the Karhunen-Loéve transformation has a more theoretical development in terms of random processes and will be discussed in section 5.7.

5.7 The Karhunen-Loéve Transformation

The Karhunen-Loéve expansion (or transformation) can be seen as an analogue of the Fourier series expansion but defined for a stochastic function x_t (or random process) that depends on a continuous time parameter $t \in [T, T']$, T and $T' \in \mathbb{R}$, and is square integrable over a continuous probability space $(\Omega, \mathcal{F}, P)$ [see footnote 1] [6]. We assume that the random process x_t has zero mean and a Hermitian positive definite covariance (or correlation) function $C_{xx}(t,t') = \mathrm{E}\big[x_t x_{t'}^*\big]$. The covariance function satisfies the properties of the kernel of a Hilbert-Schmidt operator of theorem 7; denoting the orthonormal eigenfunctions by $\phi_n(t)$ (in descending order of the associated real and positive eigenvalues σ_n^2) we have

$$x_t = \sum_{n=1}^{\infty} X_n \phi_n(t), \quad X_n = \langle \phi_n, x\rangle, \qquad \langle \phi_m, \phi_n\rangle = \delta_{mn}, \tag{5.84}$$

where the eigenfunctions satisfy the Hilbert-Schmidt operator eigenfunction (integral) equation

$$\int_T^{T'} C_{xx}(t,t')\phi_k(t')\, dt' = \sigma_n^2\, \phi_k(t). \tag{5.85}$$

The series in (5.84) converges in mean square and uniformly for all $t \in [0, T]$. Clearly, the transform coefficients have zero mean, i.e., $\mathrm{E}\big[X_n\big] = 0$. In addition, the expansion coefficients are uncorrelated and the variance of each coefficient is given by the corresponding eigenvalue,

$$\begin{aligned}\mathrm{E}\Big[X_n X_m^*\Big] &= \int_T^{T'}\int_T^{T'} \phi_n(t)\, \mathrm{E}\Big[x_t x_{t'}^*\Big]\, \phi_m^*(t')\, dt dt' = \int_T^{T'}\int_T^{T'} \phi_n(t)\, C_{xx}(t,t')\, \phi_m^*(t')\, dt dt' \\ &= \sigma_m^2\, \langle \phi_n \phi_m\rangle = \sigma_m^2\, \delta_{mn}.\end{aligned} \tag{5.86}$$

If we start with the expansion 5.84 and require that the coefficients of the orthonormal functions be uncorrelated (instead of using Mercer's theorem), then we would have

$$\mathrm{E}\Big[x_t x_{t'}^*\Big] = \sum_{n=1}^{\infty}\sum_{m=1}^{\infty} \mathrm{E}\Big[X_n X_m^*\Big]\phi_n(t)\phi_m^*(t') = \sum_{n=1}^{\infty} \sigma_n^2\, \phi_n(t)\phi_n^*(t'). \tag{5.87}$$

Multiplying on the right by $\phi_k(t')$ and integrating over t' gives us the integral equation (5.85).

The integrated variance of the process is given by the sum of the eigenvalues, namely,

$$\int_T^{T'} \mathrm{E}\Big[|x_t|^2\Big]\, dt = \sum_{n=1}^{\infty} \sigma_n^2\, \|\phi_n\|^2 = \sum_{n=1}^{\infty} \sigma_n^2. \tag{5.88}$$

Thus, the truncated Karhunen-Loéve expansion containing the first L terms (the L largest eigenvalues) will represent the fraction $\sum_{n=1}^{L} \sigma_n^2 / \sum_{n=1}^{\infty} \sigma_n^2$ of the total integrated variance. In addition, we have the following result regarding the optimality of the Karhunen-Loéve expansion.

[6] The ideas underlying the Karhunen-Loéve expansion go back to Hotteling (1933), Karhunen (1947), and Loéve (1946) investigating the convergence conditions of the expansion of a stochastic function in terms of an orthonormal basis of a Hilbert space and conditions under which the expansion coefficients would be uncorrelated. There is also a discussion of the idea by Kosambi in 1943.

Theorem 19. *Consider an orthonormal basis ψ_n other than the eigenfunctions ϕ_n of (5.85) and the truncated expansion*

$$x'_L(t) = \sum_{n=1}^{L} X'_n \psi_n(t), \quad X'_n = \langle \psi_n, x' \rangle, \qquad \langle \psi_m, \psi_n \rangle = \delta_{mn}. \tag{5.89}$$

Defining the square error $e_L^2[\psi] = \left| x_t - x'_L(t) \right|^2$ and the mean square error (MSE) $E\left[e_L^2[\psi]\right]$, we have

$$E\left[e_L^2[\psi]\right] = \int_T^{T'} E\left[\left|x_t\right|^2\right] dt - \sum_{n=1}^{L} \int_T^{T'} \int_T^{T'} \psi_k^*(t')\, C_{xx}(t', t'')\, \psi_k(t'')\, dt' dt'', \tag{5.90}$$

and the MSE is at its minimum if, and only if, ψ_n, $n = 1, \ldots, L$, are eigenfunctions of the Hilbert-Schmidt operator in (5.85) *corresponding to the L largest eigenvalues, in which case the second term on the right hand side of* (5.90) *is maximized with the maximum value being the sum of the first L largest eigenvalues and*

$$E\left[e_L^2[\psi]\right]_{min} = \sum_{n=L+1}^{\infty} \sigma_n^2. \tag{5.91}$$

In addition, defining $p_n \equiv E\left[|X_n|^2\right]/E\left[|x_t|^2\right]$ and $p'_n \equiv E\left[|X'_n|^2\right]/E\left[|x_t|^2\right]$, we have $\sum_n p_n = \sum_n p'_n = 1$ and

$$-\sum_{n=1}^{\infty} p_n \ln p_n \geq -\sum_{n=1}^{\infty} p'_n \ln p'_n. \tag{5.92}$$

Thus, the Karhunen-Loéve basis functions corresponding to the L largest eigenvalues in (5.85) minimize the MSE of the truncated expansion and are, therefore, optimal. Equation (5.92) shows that the Karhunen-Loéve expansion has maximum representation entropy (see section 5.18 on entropy and information theory).

As an example, consider the Wiener process w_t, $t \in [0, T]$, with auto-covariance function $C_{ww}(t, t') = \sigma_w^2 \min(t, t')$. Equation (5.85) becomes (using σ^2 to denote the eigenvalue)

$$\sigma_w^2 \int_0^t t' \phi(t')\, dt' + \sigma_w^2 \int_t^T t \phi(t')\, dt' = \sigma^2\, \phi(t). \tag{5.93}$$

Differentiating (5.93) with respect to t twice we find $\sigma^2\, d^2\phi(t)/dt^2 + \sigma_w^2\, \phi(t) = 0$, whose general solution is $\phi(t) = A\cos(\sigma_w t/\sigma) + B\sin(\sigma_w t/\sigma)$. Substituting this into the integral equation (5.93) and letting $t \to 0$ gives $A = 0$. Differentiating (5.93) once with respect to t and letting $t \to T$ we find $d\phi(T)/dt = 0$ which gives the "quantized" eigenvalues $\sigma_n = \sigma_w T/\pi\,(n - 1/2)$. Thus, the Karhunen-Loéve expansion of the Wiener process is:

$$w_t = \sqrt{\frac{2}{T}} \sum_{n=1}^{\infty} w_n \sin\left[\frac{(n-1/2)\pi t}{T}\right], \quad w_n \sim \mathcal{N}\left(0, \sigma_n^2\right). \tag{5.94}$$

We defined WGN as the first derivative of the Wiener process, i.e., $\nu_t = \dot{w}_t$ [see equation (5.51)]. Using the eigenvalue relation $(n-1/2)\pi/T = \sigma_w/\sigma_n$, the Karhunen-Loéve expansion of the WGN noise process is:

$$\nu_t = \sqrt{\frac{2}{T}} \sum_{n=1}^{\infty} \nu_n \cos\left[\frac{(n-1/2)\pi t}{T}\right], \quad \nu_n \sim \mathcal{N}\left(0, \sigma_w^2\right). \tag{5.95}$$

When applied to an $M \times 1$ zero-mean random vector $\boldsymbol{x}$ with an $M \times M$ covariance matrix $\boldsymbol{C}_{xx} = \mathrm{E}\left[\boldsymbol{x}\boldsymbol{x}^+\right]$, the Karhunen-Loéve expansion is identical to PCA discussed in section 5.6: the transform coefficients X_n are now denoted by the principal components y_m and each eigenfunction $\phi_n(t)$ is replaced by a column of the $M \times M$ unitary matrix $\boldsymbol{W}$. Thus, the decorrelated transformed vector (the vector of principal components) is:

$$\boldsymbol{y} = \boldsymbol{W}^+\boldsymbol{x}, \quad \boldsymbol{C}_{xx} = \boldsymbol{W}\Lambda\boldsymbol{W}^+ = \sum_{m=1}^{M} \lambda_m \boldsymbol{w}_m \boldsymbol{w}_m^+, \quad \boldsymbol{W}^+\boldsymbol{W} = \boldsymbol{I}, \tag{5.96}$$

where the diagonal covariance matrix of the transformed vector and the Karhunen-Loéve expansion (the inverse Karhunen-Loéve transformation) are

$$\mathrm{E}\left[\boldsymbol{y}\boldsymbol{y}^+\right] = \boldsymbol{C}_{yy} = \Lambda, \qquad \boldsymbol{x} = \boldsymbol{W}\boldsymbol{y}. \tag{5.97}$$

Clearly, keeping $L < M$ columns of $\boldsymbol{W}$ and discarding the rest will result in an approximation $\hat{\boldsymbol{x}}$ to $\boldsymbol{x}$ that is responsible for the fraction $\sum_{m=1}^{L} \sigma_m^2 / \sum_{m=1}^{N} \sigma_m^2$ of the total variance in $\boldsymbol{x}$ and an MSE:

$$\|\boldsymbol{e}\|^2 = \mathrm{E}\left[|\boldsymbol{x} - \hat{\boldsymbol{x}}|^2\right] = \sum_{m=L+1}^{M} \sigma_m^2. \tag{5.98}$$

The Karhunen-Loéve transformation is not as efficient a method of compression and transmission through communications channels as the discrete cosine transform or the discrete wavelet transform because it is data dependent, i.e., new principal directions and components must be computed and transmitted every time the auto-correlation matrix of the data changes. Interestingly, the Karhunen-Loéve transformation has applications to radio signals from relativistic spaceships in outer space [52]; it has even been suggested as the correct method of eavesdropping on extraterrestrial life (SETI) if the advanced alien civilizations have switched their modes of communication from single carrier wave to spread spectrum transmission (as humans have done here on earth) for which the FFT becomes a useless detection tool. Although they might still transmit a narrow-band frequency to us, it is likely that their cell phone conversations are via spread spectrum!

5.8 Statistical Properties of the Least Squares Filter

Consider the least squares filter (section 4.4), i.e., the least squares solution to the over-determined matrix equation $\boldsymbol{A}\boldsymbol{h} = \boldsymbol{y}$ where $\boldsymbol{A}$ is $N \times M$, $\boldsymbol{y}$ is $N \times 1$, and $\boldsymbol{h}$ is $M \times 1$. Let us assume a linear statistical model for the data:

$$\boldsymbol{A}\boldsymbol{h}_0 + \boldsymbol{e} = \boldsymbol{y}, \tag{5.99}$$

where $\boldsymbol{e}$ is assumed to be a vector of iid random variables $\sim \mathscr{CN}\left(0, \sigma^2\right)$. Thus, $\mathrm{E}\left[\boldsymbol{e}\right] = \boldsymbol{0}$ and $\mathrm{E}\left[\boldsymbol{e}\boldsymbol{e}^+\right] = \sigma^2\boldsymbol{I}$. There are two different possible data matrices $\boldsymbol{A}$: if the data matrix has no observation errors, then it is considered to be non-random. On the other hand, if it includes fluctuations (noise), then it must be considered to be a random data matrix. The least squares solution in either case is:

$$\boldsymbol{h}_{ls} = \left(\boldsymbol{A}^+\boldsymbol{A}\right)^{-1}\boldsymbol{A}^+\boldsymbol{y} = \boldsymbol{h}_0 + \left(\boldsymbol{A}^+\boldsymbol{A}\right)^{-1}\boldsymbol{A}^+\boldsymbol{e}. \tag{5.100}$$

Consider first the case of a non-random data matrix. The mean is $\mathrm{E}[\boldsymbol{h}_{ls}] = \boldsymbol{h}_0$, and so the least squares solution is an unbiased estimator of the true underlying model $\boldsymbol{h}_0$. The residuals are

$$\boldsymbol{\epsilon} = \boldsymbol{y} - \boldsymbol{A}\boldsymbol{h}_{ls} = \boldsymbol{A}\boldsymbol{h}_0 + \boldsymbol{e} - \boldsymbol{A}(\boldsymbol{A}^+\boldsymbol{A})^{-1}\boldsymbol{A}^+(\boldsymbol{A}\boldsymbol{h}_0 + \boldsymbol{e}) = (\boldsymbol{I} - \boldsymbol{P})\boldsymbol{e}, \tag{5.101}$$

where $\boldsymbol{P} \equiv \boldsymbol{A}(\boldsymbol{A}^+\boldsymbol{A})^{-1}\boldsymbol{A}^+$ is the projection matrix onto the columns of the data matrix $\boldsymbol{A}$. The square of the norm of the residuals is:

$$\begin{aligned} \|\boldsymbol{\epsilon}\|^2 &= \mathrm{E}\left[\boldsymbol{\epsilon}^+\boldsymbol{\epsilon}\right] = \mathbf{Trace}\left(\mathrm{E}\left[\boldsymbol{e}^+(\boldsymbol{I} - \boldsymbol{P})\boldsymbol{e}\right]\right) = \mathbf{Trace}\left(\mathrm{E}\left[\boldsymbol{e}\boldsymbol{e}^+(\boldsymbol{I} - \boldsymbol{P})\right]\right) \\ &= \sigma^2\mathbf{Trace}(\boldsymbol{I} - \boldsymbol{P}) = \sigma^2\left(N - \mathbf{Trace}(\boldsymbol{P})\right) = (N - M)\sigma^2, \end{aligned} \tag{5.102}$$

where $N - M$ is the number of degrees of freedom. The covariance of the least squares solution is:

$$\mathbf{Cov}(\boldsymbol{h}_{ls}) = \mathrm{E}\left[(\boldsymbol{h}_{ls} - \boldsymbol{h}_0)(\boldsymbol{h}_{ls} - \boldsymbol{h}_0)^+\right] = \sigma^2\left(\boldsymbol{A}^+\boldsymbol{A}\right)^{-1}, \tag{5.103}$$

which must be compared with that of any other unbiased linear estimate of the form $\hat{\boldsymbol{h}} = \boldsymbol{B}\boldsymbol{y}$. Since $\mathrm{E}\left[\hat{\boldsymbol{h}}\right] = \boldsymbol{B}\boldsymbol{A}\boldsymbol{h}$, we must have $\boldsymbol{B}\boldsymbol{A} = \boldsymbol{I}$ for $\hat{\boldsymbol{h}}$ to be unbiased and the covariance of the linear estimate $\hat{\boldsymbol{h}}$ is:

$$\mathbf{Cov}\left(\hat{\boldsymbol{h}}\right) = \sigma^2\boldsymbol{B}\boldsymbol{B}^+. \tag{5.104}$$

The difference $\boldsymbol{E}$ between the covariances of the two estimates is:

$$\boldsymbol{E} \equiv \mathbf{Cov}(\hat{\boldsymbol{h}}) - \mathbf{Cov}(\boldsymbol{h}_{ls}) = \sigma^2\boldsymbol{B}\boldsymbol{B}^+ - \sigma^2\left(\boldsymbol{A}^+\boldsymbol{A}\right)^{-1}. \tag{5.105}$$

Defining $\boldsymbol{W} \equiv \sigma\left[\boldsymbol{B} - \left(\boldsymbol{A}^+\boldsymbol{A}\right)^{-1}\boldsymbol{A}^+\right]$ and using the requirement $\boldsymbol{B}\boldsymbol{A} = \boldsymbol{I}$ we find

$$\boldsymbol{E} = \sigma^2\left[\boldsymbol{B}\boldsymbol{B}^+ - \left(\boldsymbol{A}^+\boldsymbol{A}\right)^{-1}\right] = \boldsymbol{W}\boldsymbol{W}^+. \tag{5.106}$$

Thus, for $\boldsymbol{c} \neq 0$, $\langle \boldsymbol{c}, \boldsymbol{W}\boldsymbol{W}^+\boldsymbol{c}\rangle = \langle \boldsymbol{W}^+\boldsymbol{c}, \boldsymbol{W}^+\boldsymbol{c}\rangle > 0$, which shows that $\boldsymbol{E}$ is a positive definite matrix. Hence, the least squares solution has minimum covariance among all unbiased linear estimates.

When the data matrix includes random fluctuations and $\boldsymbol{A}$ and $\mathbf{e}$ are independent (not merely uncorrelated), then using conditional expectation values (5.19a) and $\mathrm{E}\left[\boldsymbol{e}|\boldsymbol{A}\right] = \mathbf{0}$, we find

$$\mathrm{E}\left[\boldsymbol{h}_{ls}\right] = \mathrm{E}_{\boldsymbol{A}}\left[\mathrm{E}[\boldsymbol{h}_{ls}|\boldsymbol{A}]\right] = \boldsymbol{h}_0 + \mathrm{E}\left[(\boldsymbol{A}^+\boldsymbol{A})^{-1}\boldsymbol{A}^+\,\mathrm{E}[\boldsymbol{e}|\boldsymbol{A}]\right] = \boldsymbol{h}_0, \tag{5.107}$$

and so the least squares solution is unbiased. Similarly, using (5.19b) we have

$$\begin{aligned} \mathrm{E}\left[(\boldsymbol{h}_{ls} - \boldsymbol{h}_0)(\boldsymbol{h}_{ls} - \boldsymbol{h}_0)^+\right] &= \mathrm{E}_{\boldsymbol{A}}\left[\mathrm{E}[(\boldsymbol{h}_{ls} - \boldsymbol{h}_0)(\boldsymbol{h}_{ls} - \boldsymbol{h}_0)^+|\boldsymbol{A}]\right] + \\ \mathrm{E}_{\boldsymbol{A}}\left[(\mathrm{E}[\boldsymbol{h}_{ls}|\boldsymbol{A}] - \boldsymbol{h}_0)(\mathrm{E}[\boldsymbol{h}_{ls}|\boldsymbol{A}] - \boldsymbol{h}_0)^+\right] &= \sigma^2\,\mathrm{E}\left[\left(\boldsymbol{A}^+\boldsymbol{A}\right)^{-1}\right]. \end{aligned} \tag{5.108}$$

Thus, the least squares solution (5.100) for random data matrices has the same statistical properties as the least squares solution for non-random data matrices.

5.9 Estimation of Random Variables

The problem of estimating a random variable or random vector from the observation of a correlated random variable or vector is of great importance in signal processing. A desired signal is seldom observed in noise-free or clutter-free environments; for instance, an acoustic signature of a submerged vessel is recorded with background noise and clutter, or side-lobe leakage from jamming can corrupt an array output looking in a particular direction even when jamming directions are different from the look direction of the array, or a communication channel can adversely affect the amplitude and/or phase of a transmitted signal.

The approximation $\hat{\boldsymbol{y}}$ to a desired random vector $\boldsymbol{y}$ using a correlated random vector $\boldsymbol{x}$, both of dimension $N \times 1$, is known as signal estimation. Thus, we wish to find the functional relation $\hat{y}_n = g(\boldsymbol{x})$, $0 \le n \le N-1$, which, in general, is nonlinear depending on the optimality criterion. The two most important estimates are:

- The minimum mean square (MMS) estimate: this is found by minimizing the MSE $\mathrm{E}\big[\,|y_n - \hat{y}_n|^2\,\big]$ where $\hat{y}_n = g(\boldsymbol{x})$. The MMS estimate is, in general, nonlinear but if the minimization is performed for a linear functional relationship of the form $\hat{\boldsymbol{y}} = \boldsymbol{\mu}_y + \boldsymbol{H}(\boldsymbol{x} - \boldsymbol{\mu}_x)$ for an $N \times N$ matrix $\boldsymbol{H}$, and $\boldsymbol{\mu}_y \equiv \mathrm{E}[\boldsymbol{y}]$ and $\boldsymbol{\mu}_x \equiv \mathrm{E}[\boldsymbol{x}]$, then the estimate is known as the linear minimum mean square estimate (LMMS). For zero-mean stationary random processes this linear estimate becomes an output of an FIR filter, i.e., $\hat{y}_n = \sum_m h_m x_{n-m}$.

- The maximum a posteriori (MAP) estimate: this is found by maximizing the a posteriori conditional probability density $f(y_n\,|\boldsymbol{x})$; given the observed vector $\boldsymbol{x}$, this is the choice with the highest probability and it is, in general, nonlinear.

As an example consider two real or complex zero-mean jointly Gaussian random variables x and y with (complex) correlation coefficient $\rho = \sigma_{yx}/\sigma_x\sigma_y \neq 0$, $\sigma_{yx} = \mathrm{E}\big[yx^*\big] = \sigma_{xy}^*$, $\sigma_x^2 = \mathrm{E}\big[|x|^2\big]$, and $\sigma_y^2 = \mathrm{E}\big[|y|^2\big]$. The conditional mean is [see section 5.2 for definitions of density function of a complex random variable and expectation values of functions of a complex random variable]

$$\mathrm{E}\big[y\,|x\,\big] = \int y\, f_{y|x}\,(y\,|x)\;dy. \tag{5.109}$$

We now show that the conditional mean is an MMS estimate by first solving for the LMMS estimate. Let $\hat{y} = Hx$; the scalar quantity H is found by minimizing the MSE

$$\mathrm{E}\big[\,|e|^2\,\big] = \mathrm{E}\big[\,|y - Hx|^2\,\big] = \sigma_y^2 - H\sigma_{xy} - H^*\sigma_{yx} + |H|^2\,\sigma_x^2, \tag{5.110}$$

where $e = y - \hat{y}$, $\sigma_{yx} \equiv \mathrm{E}\big[yx^*\big]$, $\sigma_y^2 \equiv \mathrm{E}\big[\,|y|^2\,\big]$, etc. Differentiating with respect to H^* and setting the result to zero gives the solution $H = \sigma_{yx}/\sigma_x^2$ and the LMMS estimate $\big(\sigma_{yx}/\sigma_x^2\big)x$. The same solution could have been found by using the orthogonality principle, which in this case corresponds to $\mathrm{E}\big[ex^*\big] = 0$, i.e., e and x are uncorrelated zero-mean random variables. A linear transformation from the pair (x, y) to the pair (x, e) implies that the latter pair also has a joint Gaussian distribution but the fact that x and e are uncorrelated means that x and e are independent Gaussian random variables. Since $e = y - \hat{y}$ and $\hat{y} = Hx$, then $y = Hx + e$ and we have

$$\mathrm{E}\big[y\,|x\,\big] = \mathrm{E}\big[(Hx + e)\,|x\,\big] = Hx + \mathrm{E}\big[e\,|x\,\big] = Hx + \mathrm{E}[e] = Hx, \tag{5.111}$$

where we used the independence of e and x and the zero mean property of e. Thus, the conditional mean is the linear minimum MSE (LMMS) estimator for zero-mean jointly Gaussian random variables x and y. If the random variables have non-zero means then the conditional mean is

$$\mathrm{E}\big[y\,|x\,\big] = \mu_y + H(x - \mu_x). \tag{5.112}$$

Next we show that if we drop the joint Gaussian assumption, the conditional mean $\mathrm{E}[y\,|x]$ remains the minimum MSE of y. Consider the functional relation $\hat{y} = g(x)$ and minimize the MSE:

$$\mathrm{E}\big[\,|e|^2\,\big] = \mathrm{E}\big[\,|y - g(x)|^2\,\big] = \int\!\!\int |y - g(x)|^2\, f_{xy}\,(x, y)\; dxdy. \tag{5.113}$$

Using the identity $f_{xy}\,(x, y) = f_{y|x}\,(y\,|x\,)\, f_x\,(x)$ we have

$$\mathrm{E}\big[\,|e|^2\,\big] = \int f_x\,(x)\; dx \int |y - g(x)|^2\, f_{y|x}\,(y\,|x\,)\; dy. \tag{5.114}$$

Since $f_x(x) \geq 0$ we can neglect the integral over x in the minimization of the MSE and minimize the integral over y alone

$$\int |y - g(x)|^2\, f_{y|x}\,(y\,|x\,)\, dy, \tag{5.115}$$

with respect to the function g. We expand the integrand, take a functional derivative with respect to $g(x')$, and set the result equal to zero to find

$$g(x) = \int y\, f_{y|x}\,(y\,|x\,)\; dy \equiv \mathrm{E}\big[y\,|x\,\big]. \tag{5.116}$$

Thus, the conditional mean $g(x) = \mathrm{E}[y\,|x]$ is the unrestricted solution to the minimization of the MSE.

5.10 Jointly Gaussian Random Vectors, the Conditional Mean and Covariance

As seen in (5.116), the conditional mean is the unrestricted (i.e., not necessarily linear) minimum mean square estimate of a random variable y using a correlated random variable x. To simplify notation, for now we drop subscripts for density functions; for instance, $f(y\,|x\,)$ is used to describe the density of y conditioned on x, while $f(y)$ refers to the density of y. When x and y have a joint Gaussian distribution [(5.55) for real random variables or (5.65) for complex random variables], the conditional mean is also the MAP estimate; to see this we find the conditional distribution of y given x,

$$f(y\,|x\,) = \frac{f(y, x)}{f(x)} = \alpha_{_G} \exp\left[-\beta_{_G} \frac{|y - \mu_{y|x}\,|^2}{\sigma^2_{_{y|x}}}\right], \tag{5.117}$$

where $\alpha_{_G} = 1/\sqrt{2\pi}\sigma_{_{y|x}}\,,\, \beta_{_G} = 1/2$, for real random variables, and $\alpha_{_G} = 1/\pi\sigma^2_{_{y|x}}\,,\, \beta_{_G} = 1$, for complex random variables [see (5.55) and (5.65)], and

$$\mu_{y|x} \equiv \mathrm{E}[y\,|x\,] = \mu_y + (x - \mu_x)\,\sigma_{_y}\rho_{_{yx}}/\sigma_{_x}, \tag{5.118a}$$

$$\sigma^2_{_{y|x}} \equiv \mathrm{E}\left[|y - \mu_{y|x}\,|^2\,|x\right] = \sigma^2_{_y}\left(1 - |\rho_{_{yx}}|^2\right), \qquad \rho_{_{yx}} = \sigma_{_{yx}}/\sigma_x\sigma_y. \tag{5.118b}$$

Thus, the conditional distribution $f(y\,|x\,)$ is a Gaussian whose mean and variance are $\mu_{y|x}$ and $\sigma^2_{_{y|x}}$, respectively, i.e., $f(y\,|x\,) \sim \mathscr{N}\left(\mu_{y|x}\,, \sigma^2_{_{y|x}}\right)$ or $f(y\,|x\,) \sim \mathscr{CN}\left(\mu_{y|x}\,, \sigma^2_{_{y|x}}\right)$ and in either case $f(y\,|x\,)$ attains its maximum at the conditional mean $\mu_{y|x}$.

The conditional mean $\mu_{y|x}$ improves the estimate of y by incorporating the information gained by the observed value of x. The improvement appears as a reduction in the variance: in the absence of information provided by x, the best estimate of y is its mean μ_y and the variance of the estimate is σ_y^2. When x is taken into account, the estimate is the conditional mean $\mu_{y|x}$ and the variance of this estimate is $\sigma_y^2(1-|\rho_{yx}|^2)$, which is smaller than σ_y^2 because $|\rho_{yx}| \leq 1$. Figure 5.5 shows the joint Gaussian density (two dimensional image) of two real random variables x and y, their (Gaussian) marginal densities, and the (Gaussian) density for y conditioned on x, with $\mu_y = 0.2$, $\mu_x = 0.3$, $\sigma_y = 0.5$, $\sigma_x = 0.75$, and $\rho_{yx} = 0.625$.

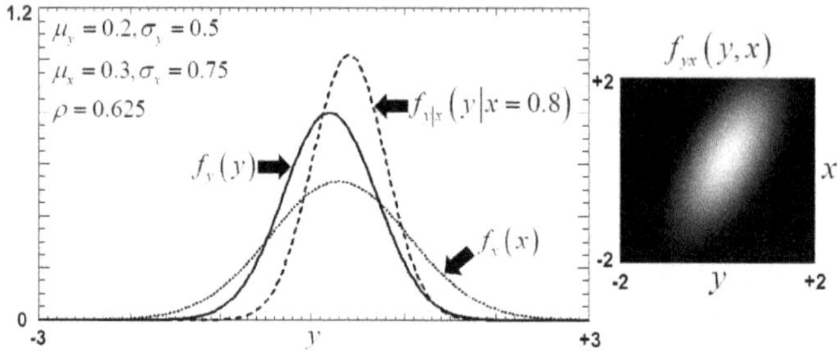

Figure 5.5: Various density functions for two correlated random variables.

This two dimensional example can be generalized to an $N \times 1$ random vector $\boldsymbol{y}$ and an $M \times 1$ random vector $\boldsymbol{x}$, whose joint Gaussian density is [7]:

$$f(\boldsymbol{w}) = \frac{1}{\sqrt{(2\pi)^{N+M}|\boldsymbol{C}_{ww}|}} \exp\left\{-\frac{1}{2}(\boldsymbol{w}-\boldsymbol{\mu}_w)^+ \boldsymbol{C}_{ww}^{-1}(\boldsymbol{w}-\boldsymbol{\mu}_w)\right\}, \tag{5.119}$$

where $\boldsymbol{w} = [\boldsymbol{y}, \boldsymbol{x}]$, $\boldsymbol{\mu}_w = [\boldsymbol{\mu}_y, \boldsymbol{\mu}_x]$, and

$$\boldsymbol{C}_{ww} \equiv \mathrm{E}\big[(\boldsymbol{w}-\boldsymbol{\mu}_w)(\boldsymbol{w}-\boldsymbol{\mu}_w)^+\big] = \begin{bmatrix} \boldsymbol{C}_{yy} & \boldsymbol{C}_{yx} \\ \boldsymbol{C}_{xy} & \boldsymbol{C}_{xx} \end{bmatrix}, \tag{5.120}$$

with $\boldsymbol{C}_{yx} \equiv \mathrm{E}\big[(\boldsymbol{y}-\boldsymbol{\mu}_y)(\boldsymbol{x}-\boldsymbol{\mu}_x)^+\big]$, etc. The conditional density $f(\boldsymbol{y}\,|\boldsymbol{x})$ is found using Bayes' formula $f(\boldsymbol{y},\boldsymbol{x})/f(\boldsymbol{x})$ and the fact that the marginal density $f(\boldsymbol{x})$ is a Gaussian with mean $\boldsymbol{\mu}_x$ and $M \times M$ covariance matrix $\boldsymbol{C}_{xx}$:

$$f(\boldsymbol{y}\,|\boldsymbol{x}) = \sqrt{\frac{|\boldsymbol{C}_{xx}|}{(2\pi)^N|\boldsymbol{C}_{ww}|}}\, e^{-\frac{1}{2}(\boldsymbol{w}-\boldsymbol{\mu}_w)^+\boldsymbol{C}_{ww}^{-1}(\boldsymbol{w}-\boldsymbol{\mu}_w)+\frac{1}{2}(\boldsymbol{x}-\boldsymbol{\mu}_x)^+\boldsymbol{C}_{xx}^{-1}(\boldsymbol{x}-\boldsymbol{\mu}_x)}. \tag{5.121}$$

The density (5.121) is a Gaussian whose mean vector $\boldsymbol{\mu}_{y|x}$ and covariance matrix $\boldsymbol{C}_{yy|x}$ can be expressed in terms of the mean vectors and covariance matrices of $\boldsymbol{x}$ and $\boldsymbol{y}$ using the inverse of a block partitioned matrix.

Theorem 20. *Given a square matrix $\boldsymbol{X}$ of the form*

$$\boldsymbol{X} = \begin{bmatrix} \boldsymbol{A} & \boldsymbol{B} \\ \boldsymbol{C} & \boldsymbol{D} \end{bmatrix} \tag{5.122}$$

[7] We use the probability density function for real jointly Gaussian random vectors here without loss of generality. The final results (5.126) are valid for both real and complex Gaussian random vectors and probability density functions (5.55) and (5.65); therefore, we use the Hermitian conjugate instead of transposition.

whose upper left block is a square non-singular matrix $\boldsymbol{A}$, the inverse of $\boldsymbol{X}$ has the following block partitioned form:

$$\boldsymbol{X}^{-1} = \begin{bmatrix} \boldsymbol{A}^{-1} + \boldsymbol{A}^{-1}\boldsymbol{B}\boldsymbol{\Delta}^{-1}\boldsymbol{C}\boldsymbol{A}^{-1} & -\boldsymbol{A}^{-1}\boldsymbol{B}\boldsymbol{\Delta}^{-1} \\ -\boldsymbol{\Delta}^{-1}\boldsymbol{C}\boldsymbol{A}^{-1} & \boldsymbol{\Delta}^{-1} \end{bmatrix}, \quad \boldsymbol{\Delta} \equiv \boldsymbol{D} - \boldsymbol{C}\boldsymbol{A}^{-1}\boldsymbol{B}. \tag{5.123}$$

$\boldsymbol{\Delta}$ is known as the Schur complement of $\boldsymbol{X}$.

Equation (5.123) together with the identity (valid for any complex vector $\boldsymbol{z}$ and Hermitian non-singular matrix $\boldsymbol{H}$)

$$\boldsymbol{z}^+\boldsymbol{H}\boldsymbol{z} + \frac{1}{2}(\boldsymbol{z}^+\boldsymbol{u} + \boldsymbol{u}^+\boldsymbol{z}) + a \equiv \left(\boldsymbol{z} + \frac{1}{2}\boldsymbol{H}^{-1}\boldsymbol{u}\right)^+ \boldsymbol{H}\left(\boldsymbol{z} + \frac{1}{2}\boldsymbol{H}^{-1}\boldsymbol{u}\right) + \left(a - \frac{1}{4}\boldsymbol{u}^+\boldsymbol{H}^{-1}\boldsymbol{u}\right), \tag{5.124}$$

when used in the exponent of the density $f(\boldsymbol{y}\,|\boldsymbol{x}\,)$ of equation (5.121), lead to

$$f(\boldsymbol{y}\,|\boldsymbol{x}\,) = \frac{1}{\sqrt{(2\pi)^N \left|\boldsymbol{C}_{yy|x}\right|}} \exp\left\{-\frac{1}{2}(\boldsymbol{y} - \boldsymbol{\mu}_{y|x})^+ \boldsymbol{C}_{yy|x}^{-1}\,(\boldsymbol{y} - \boldsymbol{\mu}_{y|x})\right\}, \tag{5.125}$$

$$\boldsymbol{\mu}_{y|x} \equiv E\left[\boldsymbol{y}\,|\boldsymbol{x}\,\right] = \boldsymbol{\mu}_y + \boldsymbol{C}_{yx}\boldsymbol{C}_{xx}^{-1}\,(\boldsymbol{x} - \boldsymbol{\mu}_x)\,, \quad \boldsymbol{C}_{yy|\boldsymbol{x}} = \boldsymbol{C}_{yy} - \boldsymbol{C}_{yx}\boldsymbol{C}_{xx}^{-1}\boldsymbol{C}_{xy}. \tag{5.126}$$

When $N = M = 1$ we recover the earlier results for two jointly Gaussian scalar random variables y and x. Thus, in the absence of information provided by the observation of $\boldsymbol{x}$, the prior density function $f(\boldsymbol{y})$ is used to find the mean vector $\boldsymbol{\mu}_y$ as the best (MMS) estimate of $\boldsymbol{y}$. The MMS estimate is improved once the correlated vector $\boldsymbol{x}$ has been observed: the conditional, or the posterior, density $f(\boldsymbol{y}\,|\boldsymbol{x}\,)$ is now used to calculate the conditional mean $\boldsymbol{\mu}_{\boldsymbol{y}|\boldsymbol{x}}$, which is the MMS estimate of $\boldsymbol{y}$ incorporating the information contained in $\boldsymbol{x}$. The prior and the posterior densities are, of course, related by Bayes' formula

$$f(\boldsymbol{y}\,|\boldsymbol{x}\,) = \frac{f(\boldsymbol{x}\,|\boldsymbol{y}\,)}{f(\boldsymbol{x})} f(\boldsymbol{y}), \tag{5.127}$$

and the conditional mean is also known as the Bayes' estimate, which is equal to the mean, mode, and the median of the posterior distribution.

The Bayes' improvement on the original estimate, represented by the additional term $\boldsymbol{C}_{yx}\boldsymbol{C}_{xx}^{-1}(\boldsymbol{x} - \boldsymbol{\mu}_x)$, depends on the deviation between the measurement $\boldsymbol{x}$ and its expected value $\boldsymbol{\mu}_x$, and on the correlation between $\boldsymbol{x}$ and $\boldsymbol{y}$ through the term $\boldsymbol{C}_{yx}$. In addition, high measurement covariance is penalized by an inverse dependence on that covariance through the term $\boldsymbol{C}_{xx}^{-1}$. Finally, using the relation $\boldsymbol{C}_{yx} = \boldsymbol{C}_{xy}^+$, and given an arbitrary complex vector $\boldsymbol{a}$ (if the random vectors are real then the covariance matrices $\boldsymbol{C}_{yy}$ and $\boldsymbol{C}_{xx}$ are real, symmetric, and positive definite and we would use an arbitrary real vector $\boldsymbol{a}$)

$$\boldsymbol{a}^+\boldsymbol{C}_{yy|\boldsymbol{x}}\boldsymbol{a} = \boldsymbol{a}^+\boldsymbol{C}_{yy}\boldsymbol{a} - \boldsymbol{b}^+\boldsymbol{C}_{xx}\boldsymbol{b} < \boldsymbol{a}^+\boldsymbol{C}_{yy}\boldsymbol{a}, \quad \boldsymbol{b} \equiv \boldsymbol{C}_{xy}\boldsymbol{a}, \tag{5.128}$$

where we have used the fact that $\boldsymbol{C}_{yy}$ and $\boldsymbol{C}_{xx}$ are positive definite. Thus, the prior covariance has been "reduced" by the correlated observation.

5.11 The Conditional Mean and the Linear Model

The results of section 5.10 can be used to illustrate the MMS solution to the linear model estimation problem: estimating the complex Gaussian random vector $\boldsymbol{y}$ with mean vector $\boldsymbol{\mu}_y$ and covariance matrix $\boldsymbol{C}_{yy}$ from a linear observation model including noise, i.e., an observed complex random vector

$$\boldsymbol{x} = \boldsymbol{A}\boldsymbol{y} + \boldsymbol{\nu}, \tag{5.129}$$

where $\boldsymbol{\nu}$ is a zero-mean complex Gaussian random noise vector with correlation matrix $\boldsymbol{R}_{\nu\nu}$ and $\boldsymbol{A}$ is a known matrix that must satisfy the relation $\boldsymbol{\mu}_x = \boldsymbol{A}\boldsymbol{\mu}_y$. The MMS estimate in the presence of the observed vector $\boldsymbol{x}$ is given by the conditional mean (5.126)

$$\boldsymbol{\mu}_{y|x} \equiv E\left[\boldsymbol{y}\,|\boldsymbol{x}\,\right] = \boldsymbol{\mu}_y + \boldsymbol{C}_{yx}\boldsymbol{C}_{xx}^{-1}\left(\boldsymbol{x} - \boldsymbol{\mu}_x\right). \tag{5.130}$$

Using the linear model equation, the covariance matrices are

$$\boldsymbol{C}_{xx} = \boldsymbol{A}\boldsymbol{C}_{yy}\boldsymbol{A}^{+} + \boldsymbol{R}_{\nu\nu}, \quad \text{and} \quad \boldsymbol{C}_{yx} = \boldsymbol{C}_{yy}\boldsymbol{A}^{+}, \tag{5.131}$$

and the conditional mean estimate (5.130) becomes

$$\boldsymbol{\mu}_{y|x} = \boldsymbol{\mu}_y + \boldsymbol{C}_{yy}\boldsymbol{A}^{+}\left(\boldsymbol{A}\boldsymbol{C}_{yy}\boldsymbol{A}^{+} + \boldsymbol{R}_{\nu\nu}\right)^{-1}\left(\boldsymbol{x} - \boldsymbol{A}\boldsymbol{\mu}_y\right). \tag{5.132}$$

This equation is often written in terms of the *Kalman gain* $\boldsymbol{K}$

$$\boldsymbol{\mu}_{y|x} = \boldsymbol{\mu}_y + \boldsymbol{K}\left(\boldsymbol{x} - \boldsymbol{A}\boldsymbol{\mu}_y\right), \quad \boldsymbol{K} \equiv \boldsymbol{C}_{yy}\boldsymbol{A}^{+}\left(\boldsymbol{A}\boldsymbol{C}_{yy}\boldsymbol{A}^{+} + \boldsymbol{R}_{\nu\nu}\right)^{-1}, \tag{5.133}$$

while the covariance matrix conditioned on the observation vector $\boldsymbol{x}$ is

$$\boldsymbol{C}_{yy|x} = (\boldsymbol{I} - \boldsymbol{K}\boldsymbol{A})\boldsymbol{C}_{yy}. \tag{5.134}$$

As an example, consider estimating a zero-mean complex scalar random variable y from a zero-mean complex observed vector $\boldsymbol{x} = \boldsymbol{a}y + \boldsymbol{\nu}$, where $\boldsymbol{a}$ is a known vector. The zero-mean complex vector $\boldsymbol{\nu}$ is assumed to be white noise, i.e., its elements are zero-mean, independent, and identically distributed complex random variables ν_n with $\mathrm{E}[\nu_n\nu_m^*] = \sigma_\nu^2\delta_{nm}$, while y is assumed to be a zero-mean complex random variable uncorrelated with $\boldsymbol{\nu}$, $\mathrm{E}\left[y\nu_n^*\right] = 0$ and with variance $\mathrm{E}\left[|y|^2\right] = \sigma_y^2$. The conditional mean (5.132) now becomes:

$$\boldsymbol{\mu}_{y|x} = \frac{\sigma_y^2}{\sigma_\nu^2}\boldsymbol{a}^{+}\left(\boldsymbol{I} + \frac{\sigma_y^2}{\sigma_\nu^2}\boldsymbol{a}\boldsymbol{a}^{+}\right)^{-1}\boldsymbol{x}, \tag{5.135}$$

which when set equal to the linear estimate $\hat{y} = \boldsymbol{h}^{+}\boldsymbol{x}$ (we use the Hermitian conjugate instead of transpose to define the linear estimate) gives the optimal vector

$$\boldsymbol{h}_{\text{opt}} = \frac{\sigma_y^2}{\sigma_\nu^2}\left(\boldsymbol{I} + \frac{\sigma_y^2}{\sigma_\nu^2}\boldsymbol{a}\boldsymbol{a}^{+}\right)^{-1}\boldsymbol{a} = \left(\frac{\sigma_\nu^2}{\sigma_y^2} + |\boldsymbol{a}|^2\right)^{-1}\boldsymbol{a}, \tag{5.136}$$

where we used the identity $\left(\boldsymbol{I} + \boldsymbol{b}\boldsymbol{b}^{+}\right)^{-1} \equiv \boldsymbol{I} - \left(1 + |\boldsymbol{b}|^2\right)^{-1}\boldsymbol{b}\boldsymbol{b}^{+}$ to get the final form on the right hand side of (5.136). The LMMS estimate is $\hat{y} = \boldsymbol{h}_{\text{opt}}^{+}\boldsymbol{x}$.

We expect that solving for the LMMS vector $\boldsymbol{h}_{\text{opt}}$ by minimizing $\mathrm{E}\left[|e|^2\right]$ for $e = y - \hat{y}$ should reproduce the conditional mean. The MMS solution for the linear estimate $\hat{y} = \boldsymbol{h}^{+}\boldsymbol{x}$ is found through the orthogonality principle $\mathrm{E}\left[e\boldsymbol{x}^{+}\right] = \boldsymbol{0}$. Thus,

$$\mathrm{E}\left[y\boldsymbol{x}^{+}\right] = \boldsymbol{h}^{+}\mathrm{E}\left[\boldsymbol{x}\boldsymbol{x}^{+}\right] = \boldsymbol{h}^{+}\boldsymbol{R}_{xx}, \tag{5.137}$$

where $\boldsymbol{R}_{xx} \equiv \mathrm{E}\left[\boldsymbol{x}\boldsymbol{x}^{+}\right]$. The solution to this equation is $\boldsymbol{h}_{\text{opt}}^{+} = \mathrm{E}\left[y\boldsymbol{x}^{+}\right]\boldsymbol{R}_{xx}^{-1}$. Now, $\mathrm{E}\left[y\boldsymbol{x}^{+}\right] = \sigma_y^2\boldsymbol{a}^{+}$, $\boldsymbol{R}_{xx} = \sigma_\nu^2 + \sigma_y^2\boldsymbol{a}\boldsymbol{a}^{+}$, and we recover the conditional mean solution (5.136); this is not a coincidence and we have just verified the general result that the conditional mean is the LMMS solution for jointly Gaussian random vectors.

Theorem 21. *If $\boldsymbol{x}$ and $\boldsymbol{y}$ are two complex random vectors, then the conditional mean $E[\boldsymbol{y}\,|\boldsymbol{x}]$ is the unrestricted MMS estimate of $\boldsymbol{y}$ in the presence of the observed vector $\boldsymbol{x}$; the generally nonlinear functional form $\hat{\boldsymbol{y}} = \boldsymbol{g}(\boldsymbol{x})$ of the conditional mean is found through the minimization of*

$$E\left[\left|\boldsymbol{y} - \boldsymbol{g}(\boldsymbol{x})\right|^2\right]. \tag{5.138}$$

If we restrict the functional form $\boldsymbol{g}(\boldsymbol{x})$ to be linear, namely, $\hat{\boldsymbol{y}} = \boldsymbol{\mu}_y + \boldsymbol{H}(\boldsymbol{x} - \boldsymbol{\mu}_x)$, then the LMMS estimate corresponds to

$$\boldsymbol{H}_{opt} = \boldsymbol{C}_{yx}\boldsymbol{C}_{xx}^{-1}, \tag{5.139}$$

where $\boldsymbol{C}_{xx} \equiv E\big[(\boldsymbol{x} - E[\boldsymbol{x}])(\boldsymbol{x} - E[\boldsymbol{x}])^+\big]$ and $\boldsymbol{C}_{yx} \equiv E\big[(\boldsymbol{y} - E[\boldsymbol{y}])(\boldsymbol{x} - E[\boldsymbol{x}])^+\big]$. When the two random vectors have a joint Gaussian density, then the MMS estimate (the conditional mean) becomes the LMMS estimate.

The last statement is proven by minimizing the MSE

$$\begin{aligned} \mathrm{E}\big[\,|\boldsymbol{e}|^2\big] &= \mathrm{E}\big[|\boldsymbol{y} - \boldsymbol{\mu}_y - \boldsymbol{H}(\boldsymbol{x} - \boldsymbol{\mu}_x)|^2\big] \\ &= \sigma_y^2 - \textbf{Trace}(\boldsymbol{H}\boldsymbol{C}_{xy}) - \textbf{Trace}(\boldsymbol{H}^+\boldsymbol{C}_{yx}) + \textbf{Trace}(\boldsymbol{H}^+\boldsymbol{H}\boldsymbol{C}_{xx}), \end{aligned} \tag{5.140}$$

with respect to the elements of the matrix $\boldsymbol{H}$, which leads to the solution (5.139); the corresponding optimal linear estimate is

$$\hat{\boldsymbol{y}}_{\text{opt}} = \boldsymbol{\mu}_y + \boldsymbol{C}_{yx}\boldsymbol{C}_{xx}^{-1}(\boldsymbol{x} - \boldsymbol{\mu}_x). \tag{5.141}$$

The LMMS estimate is the basis for the Wiener filter in sections 7.5 and 7.6. Our discussion here has centered on the estimation of one random vector using an observed correlated random vector. In section 5.13 we will discuss the problem of estimating parameters of a probability density function given the observation of a single random vector; our parameter estimation methods have applications to estimation of signal correlation matrix and signal power spectral density.

5.12 The Kalman Filter

The conditional mean and covariance matrix estimates for the linear observation model $\boldsymbol{x} = \boldsymbol{A}\boldsymbol{y} + \boldsymbol{\nu}$ of the previous section 5.11 form the basis for the Bayesian description of the Kalman filtering problem, i.e., the recursive estimation of a linear time-varying state vector $\boldsymbol{y}$. The state vector $\boldsymbol{y}$ includes all time-dependent variables of a system; for instance, the state vector could be composed of three Euler angles of a rigid body in three dimensional space motion. We denote the state vector's dependence on discrete-time by $n \in \mathbb{Z}$. The Kalman filtering problem is then summarized as the recursive estimation of the state vector $\boldsymbol{y}_{n+1}$ for a linear evolution model [8]

$$\boldsymbol{y}_{n+1} = \boldsymbol{B}_n\boldsymbol{y}_n + \boldsymbol{\epsilon}_n, \quad \boldsymbol{\epsilon}_n \sim \mathcal{N}(\boldsymbol{0}, \sigma_\epsilon^2\boldsymbol{I}), \tag{5.142}$$

with known evolution matrix $\boldsymbol{B}_n$, from the observation vector $\boldsymbol{x}_n$ which is the output of a linear observation model

$$\boldsymbol{x}_n = \boldsymbol{A}_n\boldsymbol{y}_n + \boldsymbol{\nu}_n, \quad \boldsymbol{\nu}_n \sim \mathcal{N}(\boldsymbol{0}, \sigma_\nu^2\boldsymbol{I}), \tag{5.143}$$

with known observation matrix $\boldsymbol{A}_n$, and a knowledge of the state vector estimate at step n. We make the simplifying assumption that both noise processes, $\boldsymbol{\epsilon}_n$ and $\boldsymbol{\nu}_n$, are white, Gaussian with constant variances,

[8] We use real jointly Gaussian random vectors with probability density function (5.55) but continue to use the superscript "+" instead of "T". Extensions to complex random vectors that are not proper have been discussed in recent years [53].

and uncorrelated with each other. In addition, we assume that both $\boldsymbol{y}_n$ and $\boldsymbol{x}_n$ satisfy the *Markov property*, i.e., any density function at step $n+1$ conditioned on random variables at steps $\leq n$, depend on those at step n and not on any previous steps. Also, we assume the simplest initial condition for the a priori density function $f(\boldsymbol{y}_0)$, namely that is Gaussian with zero-mean and covariance $\boldsymbol{C}_0$. Thus, given the conditional mean and the error covariance at step n,

$$\boldsymbol{\mu}_{n|n} = \mathrm{E}\big[\boldsymbol{y}_n \,|\boldsymbol{x}_n\big], \;\; \boldsymbol{C}_{n|n} = \mathrm{E}\big[(\boldsymbol{y}_n - \boldsymbol{\mu}_{n|n})(\boldsymbol{y}_n - \boldsymbol{\mu}_{n|n})^+ \,|\boldsymbol{x}_n\big], \tag{5.144}$$

we will show how to find $\boldsymbol{\mu}_{n+1|n+1}$ and $\boldsymbol{C}_{n+1|n+1}$ at step $n+1$. The latter two quantities are defined with respect to the conditional density function $f(\boldsymbol{y}_{n+1} \,|\boldsymbol{x}_{n+1})$, also known as the posterior density. Using Bayes' formula we have

$$f(\boldsymbol{y}_{n+1} \,|\boldsymbol{x}_{n+1}) = \frac{f(\boldsymbol{x}_{n+1} \,|\boldsymbol{y}_{n+1})}{f(\boldsymbol{x}_{n+1} \,|\boldsymbol{x}_n)} f(\boldsymbol{y}_{n+1} \,|\boldsymbol{x}_n), \tag{5.145}$$

where $f(\boldsymbol{y}_{n+1} \,|\boldsymbol{x}_n)$ is the a priori density. The process to find the posterior density from $f(\boldsymbol{y}_n \,|\boldsymbol{x}_n)$ is comprised of two steps: propagation of the conditional density at step n to obtain the a priori density, and updating to the posterior density, as shown in figure 5.6. The conditional density $f(\boldsymbol{y}_n \,|\boldsymbol{x}_n)$ is $\sim$

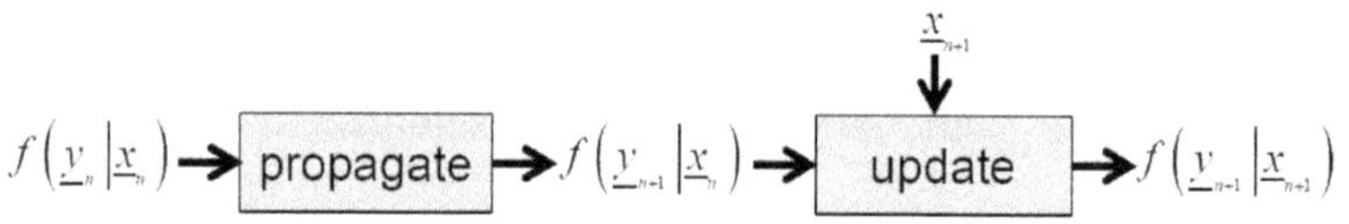

Figure 5.6: The Kalman estimation process.

$\mathcal{N}(\boldsymbol{\mu}_{n|n}, \boldsymbol{C}_{n|n})$ while the evolution equation shows that

$$f(\boldsymbol{y}_{n+1} \,|\boldsymbol{y}_n) \sim \mathcal{N}(\boldsymbol{B}_n \boldsymbol{y}_n, \sigma_\epsilon^2 \boldsymbol{I}). \tag{5.146}$$

The propagated conditional density is found from

$$f(\boldsymbol{y}_{n+1} \,|\boldsymbol{x}_n) = \int f(\boldsymbol{y}_{n+1} \,|\boldsymbol{y}_n) f(\boldsymbol{y}_n \,|\boldsymbol{x}_n)\, d\boldsymbol{y}_n, \tag{5.147}$$

which is an infinite integral of the product of two Gaussian functions and can be calculated to be a Gaussian whose mean and covariance are

$$\boldsymbol{\mu}_{n+1|n} \equiv \mathrm{E}[\boldsymbol{y}_{n+1} \,|\boldsymbol{x}_n] = \boldsymbol{B}_n \boldsymbol{\mu}_{n|n}, \tag{5.148a}$$

$$\boldsymbol{C}_{n+1|n} \equiv \mathrm{E}[(\boldsymbol{y}_{n+1} - \boldsymbol{\mu}_{n+1|n})(\boldsymbol{y}_{n+1} - \boldsymbol{\mu}_{n+1|n})^+ \,|\boldsymbol{x}_n] = \boldsymbol{B}_n \boldsymbol{C}_{n|n} \boldsymbol{B}_n^+ + \sigma_\epsilon^2. \tag{5.148b}$$

This completes the propagation step. Defining the Kalman gain

$$\boldsymbol{K}_{n+1} \equiv \boldsymbol{C}_{n+1|n} \boldsymbol{A}_{n+1}^+ \Big(\boldsymbol{A}_{n+1} \boldsymbol{C}_{n+1|n} \boldsymbol{A}_{n+1}^+ + \sigma_\nu^2\Big)^{-1}, \tag{5.149}$$

we use the results of section 5.11 (appropriately modified with the time indices $n+1$ and n for the current discussion) to update to the required conditional mean and error covariance at step $n+1$,

$$\boldsymbol{\mu}_{n+1|n+1} = \boldsymbol{\mu}_{n+1|n} + \boldsymbol{K}_{n+1}(\boldsymbol{x}_{n+1} - \boldsymbol{A}_{n+1} \boldsymbol{\mu}_{n+1|n}), \tag{5.150a}$$

$$\boldsymbol{C}_{n+1|n+1} = (\boldsymbol{I} - \boldsymbol{K}_{n+1} \boldsymbol{A}_{n+1}) \boldsymbol{C}_{n+1|n}. \tag{5.150b}$$

Thus, starting with the initial density function $f(\boldsymbol{y}_0) \sim \mathcal{N}(\boldsymbol{0}, \boldsymbol{C}_0)$, the Kalman recursive algorithm is:

- $\boldsymbol{\mu}_{n+1|n} = \boldsymbol{B}_n \boldsymbol{\mu}_{n|n}$
- $\boldsymbol{C}_{n+1|n} = \boldsymbol{B}_n \boldsymbol{C}_{n|n} \boldsymbol{B}_n^+ + \sigma_\epsilon^2$
- $\boldsymbol{K}_{n+1} \equiv \boldsymbol{C}_{n+1|n} \boldsymbol{A}_{n+1}^+ \left(\boldsymbol{A}_{n+1} \boldsymbol{C}_{n+1|n} \boldsymbol{A}_{n+1}^+ + \sigma_\nu^2 \right)^{-1}$
- $\boldsymbol{\mu}_{n+1|n+1} = \boldsymbol{\mu}_{n+1|n} + \boldsymbol{K}_{n+1} (\boldsymbol{x}_{n+1} - \boldsymbol{A}_{n+1} \boldsymbol{\mu}_{n+1|n})$
- $\boldsymbol{C}_{n+1|n+1} = (\boldsymbol{I} - \boldsymbol{K}_{n+1} \boldsymbol{A}_{n+1}) \boldsymbol{C}_{n+1|n}$

Although the Bayesian description of the Kalman filter is the most straightforward, other approaches, most notably the innovations approach, exist. The Gaussianity assumption makes the Kalman filter both the maximum likelihood and the minimum variance estimator. The Bayesian algorithm, although simple in nature, is fraught with numerical issues; existing robust numerical methods include square-root filters. We will show in section 7.7 that the Wiener filter together with an autoregressive state model is equivalent to a Kalman filter.

5.13 Parameter Estimation and the Cramer-Rao Lower Bound

The probability density function for a random variable x or a random vector $\boldsymbol{x}$ often depends on one or more parameters that we assume to be non-random. For instance, the Gaussian density function (5.53) for a random variable x is defined in terms of two parameters: the mean μ and the variance σ^2, while the Gaussian density function for a random vector (5.55) or (5.65) depends on the mean vector and the covariance matrix of the random vector. In general we consider a probability density function to depend on a vector of parameters $\boldsymbol{\theta}$, which should be estimated from observed values of the random variable x or the random vector $\boldsymbol{x}$. For instance, when $x \sim \mathscr{N}(\mu, \sigma^2)$ we set the parameter vector $\boldsymbol{\theta} \equiv [\mu, \sigma^2]$ and attempt to estimate the true values of the parameters using N observed data samples $x_1, \ldots, x_N$, and so the estimator for $\boldsymbol{\theta}$ is a function of the observed data samples, which are considered random variables. Thus, any estimate produced from them is also a random variable whose probability density function depends on the density function of the individual data samples x_n, $1 \le n \le N$, namely, $f(x; \boldsymbol{\theta})$.

Consider a scalar parameter θ whose estimator $\hat{\theta}_N$ is some function of the N observed values of a random variable $x \sim f_x(x; \theta)$. For example, an estimator for the parameter μ in (5.53) is the sample mean (the sample mean is actually an estimator for the mean of any density function),

$$\hat{\mu}_N = \frac{1}{N} \sum_{n=1}^{N} x_n . \tag{5.151}$$

This is one of many possible estimators of the mean; other estimators include any one of the N observed samples x_n. We limit the number of possible estimators to those that satisfy certain standards of quality. In order to discuss appropriate measures of quality we denote the mean of an estimator $\hat{\theta}$ of the parameter θ by μ_θ and its variance by σ_θ^2 and consider the MSE of the estimator:

$$\mathrm{E}\left[|\theta - \hat{\theta}|^2 \right] = \mathrm{E}\left[|\theta - \mu_\theta - \hat{\theta} + \mu_\theta|^2 \right] = \mathrm{E}\left[|\hat{\theta} - \mu_\theta|^2 \right] + |\theta - E[\hat{\theta}]|^2 . \tag{5.152}$$

The first desirable feature of an estimator $\hat{\theta}$ is that it should be unbiased; the estimator bias is defined as the difference between the true value of the parameter and the mean value of its estimator, i.e., $\theta - \mathrm{E}[\hat{\theta}]$.

An estimator is unbiased if $\mathrm{E}[\hat{\theta}] = \theta$. If an unbiased estimator does not exist, then we attempt to find an *asymptotically unbiased estimator* for which $E[\hat{\theta}] \to \theta$ as $N \to \infty$ (the estimate $\hat{\theta}$ always depends on the number of samples N). For the normal distribution (5.55), the sample mean and all the estimators mentioned above are unbiased, since $\mathrm{E}[x_n] = \mu$, and

$$\mathrm{E}\left[\frac{1}{N}\sum_{n=1}^{N} x_n\right] = \frac{N}{N}\,\mathrm{E}[x_n] = \mu. \tag{5.153}$$

Now consider estimating the variance σ^2 of the normal distribution when the mean is unknown; an estimator is the sample variance

$$\hat{\sigma}^2 = \frac{1}{N}\sum_{n=1}^{N} |x_n - \hat{\mu}|^2, \tag{5.154}$$

which is biased; an unbiased estimator is found by multiplying (5.154) by $N/(N-1)$ [see the discussion after equation (5.80)]. The biased estimate (5.154), however, is asymptotically unbiased:

$$\mathrm{E}[\hat{\sigma}^2] = (1 - 1/N)\sigma^2 \to \sigma^2 \quad \text{as} \quad N \to \infty. \tag{5.155}$$

The next measure of quality within the class of unbiased estimators is that of smallest variance; we seek an unbiased estimator that has minimum variance and is known as the *minimum variance unbiased* (MVU) estimator. Theorem 22 provides a lower bound on the variance of all unbiased estimators [54].

Theorem 22. *If the real random vector $\boldsymbol{x}$ is described by the joint probability density function $f(\boldsymbol{x};\theta)$ where θ is a deterministic (non-random) real scalar parameter, and if $f(\boldsymbol{x};\theta)$ satisfies certain regularity conditions including the existence of $\partial \ln f(\boldsymbol{x};\theta)/\partial\theta$ for all $\boldsymbol{x}$ and θ, then the variance of all unbiased estimators of θ, denoted by σ_θ^2, is bounded below by*

$$\sigma_\theta^2 \geq \left\{E\left[\left(\partial \ln f(\boldsymbol{x};\theta)/\partial\theta\right)^2\right]\right\}^{-1}. \tag{5.156}$$

The right hand side of (5.156) *is known as the* Cramer-Rao lower bound *(CRLB) and it is is often written in the equivalent form*

$$\sigma_\theta^2 \geq -\left\{E\left[\partial^2 \ln f(\boldsymbol{x};\theta)/\partial\theta^2\right]\right\}^{-1}. \tag{5.157}$$

The existence of an unbiased estimator that achieves the CRLB is not guaranteed; if such an estimator exists it is called an efficient estimator. *If the first derivative of the log-likelihood function has the form*

$$\frac{\partial}{\partial\theta}\ln f(\boldsymbol{x};\theta) = I(\theta)\,\big(g(\boldsymbol{x}) - \theta\big), \tag{5.158}$$

for two functions I and g, then the estimator is efficient.

The equivalent form of the CRLB (5.157) follows from differentiating the equation $\int f(\boldsymbol{x};\theta)\,d\boldsymbol{x} = 1$ with respect to θ twice and using the identity $f\,\partial \ln f/\partial\theta = \partial f/\partial\theta$ to write

$$0 = \frac{\partial}{\partial\theta}\int \frac{\partial \ln f}{\partial\theta} f\,d\boldsymbol{x} = \int \left(\frac{\partial^2 \ln f}{\partial\theta^2} f + \left(\frac{\partial \ln f}{\partial\theta}\right)^2 f\right) d\boldsymbol{x}, \tag{5.159}$$

which implies the equality of the right hand sides of (5.156) and (5.157).

As an example, consider estimating an unknown frequency ω of a real signal $s_n = A\cos(n\omega)$ in white Gaussian noise using N samples $x_n = s_n + \nu_n$ where $\nu_n \sim \mathcal{N}(0, \sigma_\nu^2)$, $0 \le n \le N-1$. Since $\mathrm{E}[x_n] = s_n$ and $\sigma_x^2 = \sigma_\nu^2$, the log-likelihood function (5.67a) is

$$\ln f(\boldsymbol{x};\omega) = -\frac{N}{2}\ln(2\pi) - \frac{N}{2}\ln\sigma_\nu^2 - \frac{1}{2\sigma_\nu^2}\sum_{n=0}^{N-1}(x_n - s_n)^2, \quad s_n = A\cos(n\omega). \tag{5.160}$$

Taking the expectation value of the second derivative of (5.160) with respect to ω gives

$$-\mathrm{E}\left[\frac{\partial^2 \ln f}{\partial\omega^2}\right] = \frac{1}{\sigma_\nu^2}\,\mathrm{E}\left[\sum_{n=0}^{N-1}(x_n - s_n)\frac{\partial^2 s_n}{\partial\omega^2} - \left(\frac{\partial s_n}{\partial\omega}\right)^2\right] = \frac{1}{\sigma_\nu^2}\sum_{n=0}^{N-1}\left(\frac{\partial s_n}{\partial\omega}\right)^2, \tag{5.161}$$

where we have used $\mathrm{E}[x_n - s_n] = 0$. Using $\partial s_n/\partial\omega = nA\sin(n\omega)$ and (5.157) we find

$$\sigma_\omega^2 \ge \frac{\sigma_\nu^2}{A^2}\left(\sum_{n=0}^{N-1} n^2\sin^2(n\omega)\right)^{-1}. \tag{5.162}$$

Figure 5.7 shows the CRLB as a function of frequency in the interval $[0,\pi]$ for $A/\sigma_\nu = 1$ and $N = 10$. Note that the variances on two specific frequencies $\omega \approx 0.2$ and $\omega \approx 2.94$ are smaller than the variances at all other frequencies for this particular example. Also, the CRLB diverges at $\omega = 0$ and $\omega = \pi$.

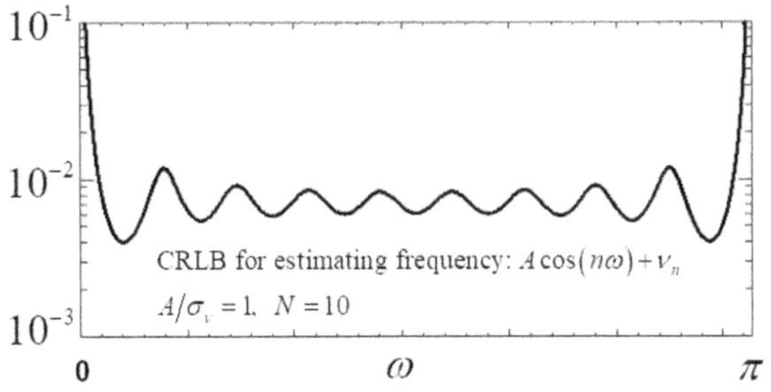

Figure 5.7: CRLB for estimating an unknown frequency.

Theorem 23 generalizes theorem 22 from a real scalar parameter to a differentiable function of that parameter and a complex vector of parameters [54, 55].

Theorem 23. *If instead of θ we consider an unbiased real scalar estimator of a differentiable function of θ, denoted by $g(\theta)$, then the CRLB is*

$$\sigma_{g(\theta)}^2 \ge \frac{\left(\partial g/\partial\theta\right)^2}{E\left[\left(\partial \ln f(\boldsymbol{x};\theta)/\partial\theta\right)^2\right]}. \tag{5.163}$$

If $\boldsymbol{\theta}$ is a complex $M \times 1$ vector parameter, then the $M \times M$ Hermitian matrix $\boldsymbol{C}_{\theta\theta} - \mathbf{I}_{\theta\theta}^{-1}$ is positive semi-definite, where $\boldsymbol{C}_{\theta\theta}$ is the Hermitian covariance matrix of the vector estimate $\hat{\boldsymbol{\theta}}$ and $\mathbf{I}_{\theta\theta}$ is the $M \times M$ Hermitian Fisher information matrix

$$\mathbf{I}_{\theta\theta} \equiv E\left[\frac{\partial \ln f_{\boldsymbol{x}}(\boldsymbol{x};\boldsymbol{\theta})}{\partial\boldsymbol{\theta}}\,\frac{\partial \ln f_{\boldsymbol{x}}(\boldsymbol{x};\boldsymbol{\theta})}{\partial\boldsymbol{\theta}+}\right], \tag{5.164}$$

whose (j,k) element is

$$\left[\mathbf{I}_{\theta\theta}\right]_{jk} = E\left[\frac{\partial \ln f(\boldsymbol{x};\boldsymbol{\theta})}{\partial \theta_j}\,\frac{\partial \ln f(\boldsymbol{x};\boldsymbol{\theta})}{\partial \theta_k^*}\right] = -E\left[\frac{\partial^2 \ln f(\boldsymbol{x};\boldsymbol{\theta})}{\partial \theta_j \partial \theta_k^*}\right]. \tag{5.165}$$

Since the covariance matrix of the parameter vector and the Fisher information matrix (and its inverse) are Hermitian, the positive semi-definite property of $\boldsymbol{C}_{\theta\theta} - \mathbf{I}_{\theta\theta}^{-1}$ implies the following CRLBs for the diagonal elements of the covariance matrix:

$$\sigma_{\theta_m}^2 \geq \left[\mathbf{I}_{\theta\theta}^{-1}\right]_{mm}, \quad 1 \leq m \leq M. \tag{5.166}$$

If $\boldsymbol{\eta}$ is a $K \times 1$ vector function of the $M \times 1$ parameter vector $\boldsymbol{\theta}$ whose $K \times M$ Jacobian transformation matrix is $\mathbf{J} \equiv \partial\boldsymbol{\eta}/\partial\boldsymbol{\theta}$, then $\boldsymbol{C}_{\eta\eta} - \mathbf{J}\mathbf{I}_{\theta\theta}^{-1}\mathbf{J}^+$ is positive semi-definite, where $\boldsymbol{C}_{\eta\eta}$ is the $K \times K$ Hermitian covariance matrix of the vector estimate $\hat{\boldsymbol{\eta}}$ and $\mathbf{I}_{\theta\theta}$ is given by (5.164). In this case the CRLBs for the diagonal elements of $\boldsymbol{C}_{\eta\eta}$ are

$$\sigma_{\eta_k}^2 \geq \left[\mathbf{J}\mathbf{I}_{\theta\theta}^{-1}\mathbf{J}^+\right]_{kk}, \quad 1 \leq k \leq K. \tag{5.167}$$

As an example, consider estimating the amplitude, frequency, and phase of a real sinusoid in white Gaussian noise described by N samples $x_n = A\cos(n\omega+\phi)+\nu_n$, where $\nu_n \sim \mathcal{N}(0,\sigma^2), 0 \leq n \leq N-1$; the parameter vector is now $\boldsymbol{\theta} \equiv [A,\omega,\phi]^T$ with $A > 0$, $0 < \omega < \pi$. Defining $\gamma_n \equiv n\omega + \phi$, the log-likelihood function is the same as (5.160) with $s_n = A\cos\gamma_n$. The (j,k) element of the Fisher information matrix (5.165) is

$$[\mathbf{I}_{\theta\theta}]_{jk} = -\mathrm{E}\left[\frac{\partial^2 \ln f}{\partial\theta_j \partial\theta_k}\right] = \frac{1}{\sigma^2}\,\mathrm{E}\left[\sum_{n=0}^{N-1}(x_n - s_n)\frac{\partial^2 s_n}{\partial\theta_j \partial\theta_k} - \frac{\partial s_n}{\partial\theta_j}\frac{\partial s_n}{\partial\theta_k}\right] = \frac{1}{\sigma^2}\sum_{n=0}^{N-1}\frac{\partial s_n}{\partial\theta_j}\frac{\partial s_n}{\partial\theta_k}, \tag{5.168}$$

where we have used $\mathrm{E}[x_n - s_n] = \mathrm{E}[\nu_n] = 0$. Using $\partial s_n/\partial A = \cos\gamma_n$, $\partial s_n/\partial\omega_n = nA\sin\gamma_n$, and $\partial s_n/\partial\phi = A\sin\gamma_n$ in (5.168) we find

$$\mathbf{I}_{\theta\theta} = \frac{1}{\sigma^2}\sum_{n=0}^{N-1}\begin{bmatrix} \cos^2\gamma_n & -nA\sin\gamma_n\cos\gamma_n & -A\sin\gamma_n\cos\gamma_n \\ nA\sin\gamma_n\cos\gamma_n & n^2A^2\sin^2\gamma_n & nA^2\sin^2\gamma_n \\ -A\sin\gamma_n\cos\gamma_n & nA^2\sin^2\gamma_n & A^2\sin^2\gamma_n \end{bmatrix}. \tag{5.169}$$

Using the identities $\cos^2\gamma_n = 1/2(1+\cos 2\gamma_n)$, $\sin^2\gamma_n = 1/2(1-\cos 2\gamma_n)$, $2\sin\gamma_n\cos\gamma_n = \sin 2\gamma_n$, and assuming that ω is not too close to 0 or 2π, we use the results [56]

$$\begin{aligned}
\frac{1}{N^{k+1}}\sum_{n=0}^{N-1} n^k\cos 2\gamma_n &= \frac{1}{N^{k+1}}\,\boldsymbol{Re}\left[e^{i\phi}\left(\frac{-i}{2}\frac{d}{d\omega}\right)^k\sum_{n=0}^{N-1}e^{2in\omega}\right] \\
&= \frac{1}{N^{k+1}}\,\boldsymbol{Re}\left[\left(\frac{-i}{2}\frac{d}{d\omega}\right)^k e^{i(N-1)\omega+i\phi}\frac{\sin N\omega}{\sin\omega}\right] \to 0 \text{ as } N\to\infty, \quad k=0,1,2, \\
\frac{1}{N^{k+1}}\sum_{n=0}^{N-1} n^k\sin 2\gamma_n &\to 0 \text{ as } N\to\infty, \quad k=0,1,
\end{aligned} \tag{5.170}$$

and the sums $\sum_{n=0}^{N-1} n = N(N-1)/2$ and $\sum_{n=0}^{N-1} n^2 = N(N-1)(2N-1)/6$, to find

$$\mathbf{I}_{\theta\theta} \approx \begin{bmatrix} N/2\sigma^2 & 0 & 0 \\ 0 & N(N-1)(2N-1)A^2/12\sigma^2 & N(N-1)A^2/4\sigma^2 \\ 0 & N(N-1)A^2/4\sigma^2 & NA^2/2\sigma^2 \end{bmatrix}, \tag{5.171a}$$

$$\mathbf{I}_{\theta\theta}^{-1} \approx \begin{bmatrix} 2\sigma^2/N & 0 & 0 \\ 0 & 4\sigma^2/A^2N(N^2-1) & -2\sigma^2/A^2N(N+1) \\ 0 & -2\sigma^2/A^2N(N+1) & 2\sigma^2(2N-1)/3A^2N(N+1) \end{bmatrix}. \tag{5.171b}$$

According to (5.166) the CRLBs on the variances of the ML estimates of the elements of $\hat{\boldsymbol{\theta}}$ are given by the diagonal entries of (5.171b), namely, $\left[2\sigma^2/N, 4\sigma^2/A^2N(N^2-1), 2\sigma^2(2N-1)/3A^2N(N+1)\right]^T$. Thus, the lower bound on the variance of the amplitude estimate decreases as $1/N$ and is proportional to the noise variance (independent of the signal amplitude), while the lower bounds on the variances of the frequency and phase estimates fall as $1/N^3$ and $1/N$, respectively, and both are inversely proportional to SNR. If the signal is complex with unknown frequency and complex amplitude and described by N samples $x_n = A\exp(in\omega) + \nu_n$, where $\nu_n \sim \mathscr{CN}(0, \sigma^2)$, $0 \le n \le N-1$, then the complex parameter vector is $\boldsymbol{\theta} \equiv [A, \omega]^T$ with $A \in \mathbb{C}$ and $0 < \omega < \pi$. Using (5.168) and the equations $s_n = A\exp(in\omega)$, $\partial s_n/\partial A = s_n/A$, and $\partial s_n/\partial\omega = ins_n$, the Fisher information matrix and its inverse are

$$\mathbf{I}_{\theta\theta} = \begin{bmatrix} \frac{N}{\sigma^2} & \frac{-iN(N-1)A}{2\sigma^2} \\ \frac{iN(N-1)A^*}{2\sigma^2} & \frac{N(N-1)(2N-1)|A|^2}{3\sigma^2} \end{bmatrix}, \; \mathbf{I}_{\theta\theta}^{-1} = \begin{bmatrix} \frac{4(2N-1)\sigma^2}{3N(5N-1)} & \frac{6i\sigma^2}{N(5N-1)A^*} \\ \frac{-6i\sigma^2}{N(5N-1)A} & \frac{12\sigma^2}{N(N-1)(5N-1)|A|^2} \end{bmatrix}. \tag{5.172}$$

Thus, the CRLB for the variance of the complex amplitude estimate $\hat{A}$ falls as $1/N$ and the CRLB for the variance of the frequency estimate $\hat{\omega}$ is inversely proportional to $|A|^2/\sigma^2$ (SNR) and falls as $1/N^3$.

As an example of (5.167) for transformed variables consider estimating the SNR from observed samples $x_n = A\cos(n\omega) + \nu_n$, $\nu_n \sim \mathscr{N}(0, \sigma^2)$, assuming a known frequency ω. Using the likelihood function (5.160) with the parameter vector $\boldsymbol{\theta} = \left[A, \sigma^2\right]^T$, the first of (5.170), and $\mathrm{E}\left[x_n - A\cos n\omega\right] = 0$, we find

$$\mathrm{E}\left[-\frac{\partial^2 \ln f}{\partial A^2}\right] = \frac{1}{\sigma^2}\sum_{n=0}^{N-1}\cos^2(n\omega) = \frac{1}{\sigma^2}\sum_{n=0}^{N-1}\left(\frac{1}{2} + \frac{1}{2}\cos(2n\omega)\right) \approx \frac{N}{2\sigma^2},$$

$$\mathrm{E}\left[-\frac{\partial^2 \ln f}{\partial[\sigma^2]^2}\right] = -\frac{N}{2\sigma^4} + \frac{1}{\sigma^6}\sum_{n=0}^{N-1}\mathrm{E}\left[(x_n - A\cos n\omega)^2\right] = -\frac{N}{2\sigma^4} + \frac{N\sigma^2}{\sigma^6} = \frac{N}{2\sigma^4},$$

and $\mathrm{E}\left[\partial^2 \ln f/\partial A\partial\sigma^2\right] = 0$. Thus, the inverse Fisher information matrix is:

$$\mathbf{I}_{\theta\theta}^{-1} = \frac{2\sigma^2}{N}\begin{bmatrix} 1 & 0 \\ 0 & \sigma^2 \end{bmatrix}. \tag{5.173}$$

The SNR for this model is $\eta(\boldsymbol{\theta}) = A^2/2\sigma^2$ with the Jacobian $\mathbf{J} = \partial\eta/\partial\boldsymbol{\theta} = [A/\sigma^2, -A^2/2\sigma^4]$, then using (5.167) we find the following CRLB inequality for the variance of the SNR estimate

$$\sigma^2_{\mathrm{SNR}} \ge \mathbf{J}\mathbf{I}_{\theta\theta}^{-1}\mathbf{J}^+ = \frac{2}{N}\left(2\,\mathrm{SNR} + \mathrm{SNR}^2\right). \tag{5.174}$$

5.14 Linear MVU and Maximum Likelihood Estimators

For a random variable $x \sim \mathcal{N}(\mu, \sigma^2)$ or $\sim \mathcal{CN}(\mu, \sigma^2)$, the sample mean was shown to be an unbiased estimator of the true mean. The variance of this estimate can be calculated using the fact that the variance of a sum of independent random variables x_n is the sum of their individual variances and that the variance of ax for a constant a is $|a|^2$ times the variance of x; thus, the variance of the sample mean with sample size N is σ^2/N. We will now show that the sample mean is an efficient estimator, i.e., it achieves the CRLB. Using the log-likelihood functions (5.67a) and (5.67b), we find $\partial^2 \ln f(\boldsymbol{x};\mu)/\partial\mu^2 = \partial^2 \ln f(\boldsymbol{x};\mu)/\partial\mu^*\partial\mu = -N/\sigma^2$ and so the CRLB is σ^2/N. Thus, the sample mean achieves the CRLB on the variance of its estimate and is, therefore, an efficient estimator. The efficiency of the sample mean estimator can also be proven using the last condition of theorem 22, since the first derivative of the log-likelihood function in this example is the product of two terms as in (5.158),

$$\frac{\partial}{\partial\mu}\ln f(\boldsymbol{x};\mu) = N(\hat{\mu}-\mu)/\sigma^2, \quad \text{or} \quad \frac{\partial}{\partial\mu}\ln f(\boldsymbol{x};\mu) = N(\hat{\mu}^*-\mu^*)/\sigma^2. \tag{5.175}$$

An MVU estimator may not exist or it may be difficult to find, so it is often easier to construct sub-optimal estimators that are linear in the data samples. When using sub-optimal estimators, such as linear estimators, an important feature to look for is *consistency*: an estimate $\hat{\theta}_N$ of the parameter θ that depends on N observed data samples is *consistent* when it converges to its true value θ with probability 1, i.e.,

$$\Pr\left(\left|\theta - \hat{\theta}_N\right| = 0\right) \to 1 \quad \text{as} \quad N \to \infty\,. \tag{5.176}$$

In other words, the probability density function of a consistent estimate of a parameter gets more and more concentrated around the true parameter value as the number of data points increases. For an estimator to be consistent it is sufficient for it to be consistent in mean square, namely,

$$\mathrm{E}\left[\left|\theta - \hat{\theta}_N\right|^2\right] \to 0 \text{ as } N \to \infty. \tag{5.177}$$

This can be written as

$$\mathrm{E}\left[\left|\hat{\theta}_N - E[\hat{\theta}_N]\right|^2\right] + \left|\mathrm{E}[\hat{\theta}_N] - \theta\right|^2 \to 0 \text{ as } N \to \infty, \tag{5.178}$$

which shows that an asymptotically unbiased estimator is consistent in mean square if its variance $\to 0$ as $N \to \infty$; the converse to this is that an estimator that is consistent in mean square is asymptotically unbiased with zero variance. As we have seen the estimator of the mean of a Gaussian random variable is efficient, i.e., it achieves its CRLB; thus, the sample mean of $x \sim \mathcal{N}(\mu, \sigma^2)$ or $\sim \mathcal{CN}(\mu, \sigma^2)$ is a consistent estimator in addition to being efficient.

Minimum variance estimators that are restricted to be linear functions of the data are very useful. For instance, consider a scalar parameter that is a linear combination of the components of some random vector $\boldsymbol{x}$, i.e., $\hat{\theta} = \boldsymbol{h}^+\boldsymbol{x}$. The MSE of the estimator is

$$\mathcal{E} \equiv \mathrm{E}\left[\left|\theta - \hat{\theta}\right|^2\right] = \mathrm{E}\left[|\theta|^2\right] - \boldsymbol{h}^+\mathrm{E}\left[\theta^*\boldsymbol{x}\right] - \mathrm{E}\left[\theta\boldsymbol{x}^+\right]\boldsymbol{h} + \boldsymbol{h}^+\boldsymbol{C}_{\boldsymbol{xx}}\boldsymbol{h}, \tag{5.179}$$

where $\boldsymbol{C}_{\boldsymbol{xx}}$ is the covariance matrix of the random vector $\boldsymbol{x}$. Differentiating (5.179) with respect to $\boldsymbol{h}^+$ and setting the result equal to zero, we find the optimal minimum variance estimator

$$\boldsymbol{h}_{\text{opt}} = \boldsymbol{C}_{\boldsymbol{xx}}^{-1}\,\mathrm{E}\left[\theta^*\boldsymbol{x}\right], \tag{5.180}$$

with the MSE

$$\mathcal{E}_{\min} = \mathrm{E}\big[|\theta|^2\big] - \mathrm{E}\big[\theta \boldsymbol{x}^+\big]\, \boldsymbol{C}_{\boldsymbol{x}\boldsymbol{x}}^{-1}\, \mathrm{E}\big[\theta^* \boldsymbol{x}\big] = \mathrm{E}\big[|\theta|^2\big] - \mathrm{E}\big[\theta \boldsymbol{x}^+\big]\, \boldsymbol{h}_{\text{opt}}\,. \tag{5.181}$$

Two methods are usually used for estimating parameters. The first is the moment method which is easy to compute but does not generally yield efficient estimators. In this method all theoretical moments are estimated using the sample moments:

$$\theta_k = \mathrm{E}\big[x^k\big], \quad \hat{\theta}_k = \frac{1}{N}\sum_{n=1}^{N} x_n^k, \quad k = 1, 2, \ldots\,. \tag{5.182}$$

The second is the maximum likelihood (ML) method [55]; it is more difficult to compute but in the absence of an MVU estimator, it generally yields asymptotically efficient estimators. In this method the likelihood function, namely, the probability density function $f(\boldsymbol{x};\boldsymbol{\theta})$ evaluated for the observed values $\boldsymbol{x}$ (or its logarithm), is maximized as a function of the parameter vector $\boldsymbol{\theta}$.

Theorem 24. *The maximum likelihood estimator (MLE) $\hat{\boldsymbol{\theta}}_{ML}$ of a vector parameter $\boldsymbol{\theta}$ is found by maximizing the likelihood function $f(\boldsymbol{x};\boldsymbol{\theta})$ (or its logarithm) with respect to the parameter vector $\boldsymbol{\theta}$; thus, $\hat{\boldsymbol{\theta}}_{ML}$ is the solution of $\partial f(\boldsymbol{x};\boldsymbol{\theta})/\partial\boldsymbol{\theta} = \mathbf{0}$. Furthermore, any unbiased and efficient estimator is an MLE. Maximum likelihood estimators are consistent, asymptotically efficient, and asymptotically normal $\sim \mathcal{N}(\boldsymbol{\theta}, \mathbf{I}_{\theta\theta}^{-1})$, where $\mathbf{I}_{\theta\theta}$ is the Fisher information matrix* (5.164).

The covariance matrix inverse in the exponent of a multi-dimensional Gaussian likelihood function is often difficult to calculate, so the MLE for these likelihood functions are nearly impossible to obtain. In signal processing applications involving long data records of zero-mean WSS random processes (section 5.3), an approximate likelihood function involving the process power spectral density (PSD) can be defined [see equation (8.175) in section 8.22]. We will discuss maximum likelihood estimation of autoregressive (AR) parameters in sections 8.22 and 8.23.

The sample mean can be used to estimate the mean of a random variable x for which N observed values $x_1, \ldots, x_N$ exist, i.e., a random sample of size N. An important question in statistics is that of how close the sample mean is to the true mean of the random variable; the precise mathematical question is then how to choose a positive number ϵ so that the probability that $|\hat{\mu}_N - \mu| \le \epsilon$ is close to 1, say 0.95, in which case we say that the interval $[\hat{\mu}_N - \epsilon, \hat{\mu}_N + \epsilon]$ is a 95% confidence interval for the mean. If we had chosen the probability as 0.99 then the corresponding ϵ would define a 99% confidence interval for the mean. Let us define the normalized random variable:

$$z_N = \frac{\hat{\mu}_N - \mu}{\sigma/\sqrt{N}}. \tag{5.183}$$

For a large random sample $N \gg 1$ theorem 17 states that $z_N \sim \mathcal{N}(0,1)$. Then,

$$\Pr\big(|\hat{\mu}_N - \mu|\big) \le \epsilon \;\Rightarrow\; \Pr\big(|z_N|\big) \le \epsilon\,\sqrt{N}/\sigma. \tag{5.184}$$

The standard normal distribution tables show $\epsilon_{0.95} = 1.96\,\sigma/\sqrt{N}$ and $\epsilon_{0.99} = 2.58\,\sigma/\sqrt{N}$.

5.15 Maximum Likelihood Estimate of the Parameter Vector of a Linear Model

Consider the linear model of a complex $N \times 1$ vector $\boldsymbol{x}$

$$\boldsymbol{x} = \boldsymbol{A}\boldsymbol{\theta} + \boldsymbol{\nu}, \tag{5.185}$$

where $\boldsymbol{A}$ is a known complex $N \times M$ full rank constant matrix, $\boldsymbol{\theta}$ is a non-random $M \times 1$ parameter vector, and $\boldsymbol{\nu}$ is a zero-mean complex Gaussian random vector with correlation matrix $R_{\boldsymbol{\nu}\boldsymbol{\nu}}$. Linear models are particularly useful in signal and data analysis. For instance, consider fitting a polynomial of given degree $M-1$ to N measured real data values $\boldsymbol{x} = \left[x_0, \ldots, x_{N-1}\right]^T$ at times $t_0, \ldots, t_{N-1}$, using the linear model $x_n = c_0 + c_1 t_n + \ldots + c_{M-1} t_n^{M-1} + \nu_n$, where $\nu_n \sim \mathcal{N}(0, \sigma_\nu^2)$. The model can be written in the form of (5.185) with $\boldsymbol{\theta} = \left[c_0, \ldots, c_{M-1}\right]^T$ and $\left[\boldsymbol{A}\right]_{nm} = t_n^m$, $m = 0, \ldots, M-1$, $n = 0, \ldots, N-1$, is a Vandermonde matrix. Another example is provided by the model $x_n = h_0 u_n + \ldots + h_{M-1} u_{n-M+1} + \nu_n$, where the parameter vector $\boldsymbol{\theta} = \left[h_0, \ldots, h_{M-1}\right]^T$ and the form of the (data) matrix $\boldsymbol{A}$ is determined according to the methods described in section 4.4. A final example of the linear model is provided by $N = M$ and $\boldsymbol{A} = \boldsymbol{F}^{-1} = \boldsymbol{F}^*/N$ where $\boldsymbol{F}$ is the DFT matrix defined in (3.67); the ML estimate of the parameter vector is $\hat{\boldsymbol{\theta}} = \boldsymbol{X}$ where $\boldsymbol{X} = \boldsymbol{F}\boldsymbol{x}$ is the DFT of $\boldsymbol{x}$.

Since $\mathrm{E}\left[\boldsymbol{x}\right] = \boldsymbol{A}\boldsymbol{\theta}$ and $\mathrm{E}\left[(\boldsymbol{x} - \boldsymbol{A}\boldsymbol{\theta})(\boldsymbol{x} - \boldsymbol{A}\boldsymbol{\theta})^+\right] = \boldsymbol{R}_{\boldsymbol{\nu}\boldsymbol{\nu}}$, the joint density function of $\boldsymbol{x}$ is (using (5.65) for a complex random vector)

$$f_{\boldsymbol{x}}\left(\boldsymbol{x}; \boldsymbol{\theta}\right) = (\pi)^{-N} |\boldsymbol{R}_{\boldsymbol{\nu}\boldsymbol{\nu}}|^{-1} \exp\left[-\left(\boldsymbol{x} - \boldsymbol{A}\boldsymbol{\theta}\right)^+ \boldsymbol{R}_{\boldsymbol{\nu}\boldsymbol{\nu}}^{-1} \left(\boldsymbol{x} - \boldsymbol{A}\boldsymbol{\theta}\right)\right]. \tag{5.186}$$

Thus, the ML estimation of $\boldsymbol{\theta}$ is equivalent to the minimization problem

$$\arg\min_{\boldsymbol{\theta}} \lambda(\boldsymbol{\theta}), \quad \lambda(\boldsymbol{\theta}) \equiv \left(\boldsymbol{x} - \boldsymbol{A}\boldsymbol{\theta}\right)^+ \boldsymbol{R}_{\boldsymbol{\nu}\boldsymbol{\nu}}^{-1} \left(\boldsymbol{x} - \boldsymbol{A}\boldsymbol{\theta}\right). \tag{5.187}$$

Setting the derivative of $\lambda(\boldsymbol{\theta})$ with respect to $\boldsymbol{\theta}^+$ to zero we find the ML estimate

$$\hat{\boldsymbol{\theta}}_{\mathrm{ML}} = \left(\boldsymbol{A}^+ \boldsymbol{R}_{\boldsymbol{\nu}\boldsymbol{\nu}}^{-1} \boldsymbol{A}\right)^{-1} \boldsymbol{A}^+ \boldsymbol{R}_{\boldsymbol{\nu}\boldsymbol{\nu}}^{-1} \boldsymbol{x}. \tag{5.188}$$

The parameter vector estimate (5.188) is unbiased

$$\mathrm{E}\left[\hat{\boldsymbol{\theta}}_{\mathrm{ML}}\right] = \left(\boldsymbol{A}^+ \boldsymbol{R}_{\boldsymbol{\nu}\boldsymbol{\nu}}^{-1} \boldsymbol{A}\right)^{-1} \boldsymbol{A}^+ \boldsymbol{R}_{\boldsymbol{\nu}\boldsymbol{\nu}}^{-1}\, \mathrm{E}\left[\boldsymbol{A}\boldsymbol{\theta}\right] = \left(\boldsymbol{A}^+ \boldsymbol{R}_{\boldsymbol{\nu}\boldsymbol{\nu}}^{-1} \boldsymbol{A}\right)^{-1} \boldsymbol{A}^+ \boldsymbol{R}_{\boldsymbol{\nu}\boldsymbol{\nu}}^{-1} \boldsymbol{A}\, \boldsymbol{\theta} = \boldsymbol{\theta}, \tag{5.189}$$

and its covariance is

$$\mathrm{E}\left[\left(\hat{\boldsymbol{\theta}} - \boldsymbol{\theta}\right)\left(\hat{\boldsymbol{\theta}}^+ - \boldsymbol{\theta}^+\right)\right] = \left(\boldsymbol{A}^+ \boldsymbol{R}_{\boldsymbol{\nu}\boldsymbol{\nu}}^{-1} \boldsymbol{A}\right)^{-1}. \tag{5.190}$$

In addition, we have

$$\lambda_{\min} \equiv \lambda\left(\hat{\boldsymbol{\theta}}_{\mathrm{ML}}\right) = \boldsymbol{x}^+ \left(\boldsymbol{R}_{\boldsymbol{\nu}\boldsymbol{\nu}}^{-1} - \boldsymbol{R}_{\boldsymbol{\nu}\boldsymbol{\nu}}^{-1} \boldsymbol{A}\left(\boldsymbol{A}^+ \boldsymbol{R}_{\boldsymbol{\nu}\boldsymbol{\nu}}^{-1} \boldsymbol{A}\right)^{-1} \boldsymbol{A}^+ \boldsymbol{R}_{\boldsymbol{\nu}\boldsymbol{\nu}}^{-1}\right) \boldsymbol{x}. \tag{5.191}$$

If the noise vector is white with unknown variance σ_ν^2, then the noise variance is considered an additional parameter whose ML estimate is found by maximizing the log-likelihood function evaluated for $\boldsymbol{\theta} = \hat{\boldsymbol{\theta}}_{\mathrm{ML}}$, namely,

$$\ln f_{\boldsymbol{x}}\left(\boldsymbol{x}; \hat{\boldsymbol{\theta}}_{\mathrm{ML}}\right) = -N \ln \pi - N \ln \sigma_\nu^2 - \lambda_{\min}, \tag{5.192}$$

which upon using $\boldsymbol{R}_{\boldsymbol{\nu}\boldsymbol{\nu}} = \sigma_\nu^2 \boldsymbol{I}$ in (5.191) becomes

$$\ln f_{\boldsymbol{x}}\left(\boldsymbol{x}; \hat{\boldsymbol{\theta}}_{\mathrm{ML}}\right) = -N \ln \pi - N \ln \sigma_\nu^2 - \frac{1}{\sigma_\nu^2} \boldsymbol{x}^+ \left(\boldsymbol{I} - \boldsymbol{A}\left(\boldsymbol{A}^+ \boldsymbol{A}\right)^{-1} \boldsymbol{A}^+\right) \boldsymbol{x}. \tag{5.193}$$

Setting the derivative of (5.193) to zero gives the ML estimate for the unknown noise variance:

$$\hat{\sigma}_\nu^2 = \frac{1}{N} \boldsymbol{x}^+ \left(\boldsymbol{I} - \boldsymbol{A}\left(\boldsymbol{A}^+ \boldsymbol{A}\right)^{-1} \boldsymbol{A}^+\right) \boldsymbol{x}, \tag{5.194}$$

which is also valid if $\boldsymbol{x}$ is a real random variable and $\boldsymbol{\nu} \sim \mathcal{N}\left(\boldsymbol{0}, \sigma_\nu^2 \boldsymbol{I}\right)$.

5.16 Maximum Likelihood Estimate of Complex Amplitude of a Complex Sinusoid in Gaussian Noise

In section 5.13 we gave an example calculation for the CRLB when estimating the complex amplitude of a complex sinusoid with known frequency ω in complex WGN [see the example relating to (5.172)]. Here we wish to present the MLE estimate in the more general case when the noise vector is not white and $\boldsymbol{\nu} \sim \mathscr{CN}(\mathbf{0}, \boldsymbol{R}_{\nu\nu})$. The observed data with the known frequency is $x_n = Ae^{in\omega} + \nu_n, 0 \le n \le N-1$, which we write as $\boldsymbol{x} = A\boldsymbol{e}_\omega + \boldsymbol{\nu}$,

$$\boldsymbol{e}_\omega \equiv [1, e^{i\omega}, \ldots, e^{(N-1)i\omega}]^T, \quad \boldsymbol{x} = [x_0, \ldots, x_{N-1}]^T, \; \boldsymbol{\nu} = [\nu_0, \ldots, \nu_{N-1}]^T,$$

and the log-likelihood function is given by (5.67b), which now takes the form

$$\ln f_{\boldsymbol{x}}(\boldsymbol{x}; A) = -N \ln \pi - \ln |\boldsymbol{R}_{\nu\nu}| - (\boldsymbol{x} - A\boldsymbol{e}_\omega)^+ \boldsymbol{R}_{\nu\nu}^{-1} (\boldsymbol{x} - A\boldsymbol{e}_\omega). \tag{5.195}$$

Now $\partial \ln f / \partial A^* = -\boldsymbol{e}_\omega^+ \boldsymbol{R}_{\nu\nu}^{-1} \boldsymbol{x} + A\, \boldsymbol{e}_\omega^+ \boldsymbol{R}_{\nu\nu}^{-1} \boldsymbol{e}_\omega$, and so the MLE of the complex amplitude A is

$$\hat{A}_{ML} = \boldsymbol{e}_\omega^+ \boldsymbol{R}_{\nu\nu}^{-1} \boldsymbol{x} / \boldsymbol{e}_\omega^+ \boldsymbol{R}_{\nu\nu}^{-1} \boldsymbol{e}_\omega. \tag{5.196}$$

The MLE $\hat{A}_{ML}$ is unbiased since

$$\mathrm{E}\left[\hat{A}_{ML}\right] = \boldsymbol{e}_\omega^+ \boldsymbol{R}_{\nu\nu}^{-1} \mathrm{E}\left[\boldsymbol{x}\right] / \boldsymbol{e}_\omega^+ \boldsymbol{R}_{\nu\nu}^{-1} \boldsymbol{e}_\omega = A,$$

where we used $\mathrm{E}\left[\boldsymbol{x}\right] = A\boldsymbol{e}_\omega$. This MLE is also the MVU linear estimator of the complex amplitude; i.e., it is an unbiased estimator of the form $\hat{A} = \boldsymbol{h}^+\boldsymbol{x}$ with minimum variance. This can be seen by minimizing the variance

$$\sigma_{\hat{A}}^2 \equiv \mathrm{E}\left[\|A - \hat{A}\|^2\right] = \boldsymbol{h}^+ \boldsymbol{R}_{\nu\nu} \boldsymbol{h}, \tag{5.197}$$

subject to the constraint that the estimator must be unbiased, i.e., $A = \mathrm{E}\left[\hat{A}\right] = \boldsymbol{h}^+ \mathrm{E}\left[\boldsymbol{x}\right] = A\boldsymbol{h}^+\boldsymbol{e}_\omega$. Thus, the constraint is $\boldsymbol{h}^+\boldsymbol{e}_\omega = 1$; this is known as the *distortionless* property of the minimum variance estimator. If we consider the linear estimation equation as a "filtering" operation $y = \boldsymbol{h}^+\boldsymbol{x}$, then the output is $y = 1$ when the input is a complex sinusoid at the known frequency ω [9]. Introducing a Lagrange multiplier λ the minimum variance estimation problem is:

$$\arg\min_{\boldsymbol{h}} \boldsymbol{h}^+ \boldsymbol{R}_{\nu\nu} \boldsymbol{h} + \lambda \left(1 - \boldsymbol{h}^+ \boldsymbol{e}_\omega\right). \tag{5.198}$$

Differentiating with respect to $\boldsymbol{h}^+$ and setting the result to zero gives

$$\boldsymbol{R}_{\nu\nu} \boldsymbol{h} = \lambda \boldsymbol{e}_\omega. \tag{5.199}$$

Multiplying (5.199) on the left by $\boldsymbol{h}^+$ and using the constraint $\boldsymbol{h}^+\boldsymbol{e}_\omega = 1$ we find $\lambda = \boldsymbol{h}^+ \boldsymbol{R}_{\nu\nu} \boldsymbol{h}$. The solution to (5.199) is $\boldsymbol{h} = \lambda \boldsymbol{R}_{\nu\nu}^{-1} \boldsymbol{e}_\omega$, which when substituted into the Hermitian conjugate of the constraint

[9] The filtering operation interpretation follows from considering $\boldsymbol{x}$ as a "section" of a vector of more elements $\left[x_0, \ldots, x_{N-1}, x_N, x_{N+1}, \ldots\right]^T$. Defining an $N \times 1$ FIR "filter" $\boldsymbol{h} \equiv \left[h_{N-1}, \ldots, h_0\right]^T$, the "filtering operation" is the convolution equation $y_n = \boldsymbol{h}^+ \boldsymbol{x}^{(n)}$, where $\boldsymbol{x}^{(n)} = \left[x_n, \ldots, x_{n+N-1}\right]^T, n = 0, 1, 2, \ldots$. Our estimation problem then corresponds to the output $y_0 = \boldsymbol{h}^+ \boldsymbol{x}^{(0)}$.

equation $e_{\omega}^{+}h = 1$ gives $\lambda = 1/e_{\omega}^{+}R_{\nu\nu}^{-1}e_{\omega}$. Using this value for λ in (5.199) leads to the linear MVU solution

$$h_{\text{MV}} = R_{\nu\nu}^{-1}e_{\omega}/e_{\omega}^{+}R_{\nu\nu}^{-1}e_{\omega}. \tag{5.200}$$

When used in the linear estimation equation $h^{+}x$, the MVU solution produces the same estimate as (5.196); thus, the linear MVU solution to the estimation problem is also the ML solution. Using h_{MV} on the right hand side of (5.197) gives the variance of the optimal (MVU or ML) estimator as $\sigma_{\text{min}}^2 = 1/e_{\omega}^{+}R_{\nu\nu}^{-1}e_{\omega}$.

The filtering interpretation of the linear MVU estimator (or the equivalent ML estimator) and the associated variance of the estimator σ_{min}^2 form the basis for the minimum variance distortionless spectral estimator (MVDSE) in section 8.9.

5.17 Maximum Likelihood Estimate of a First Order Gaussian Markov Process

Consider a sequence of zero-mean complex Gaussian random variables x_n, $n \in \mathbb{Z}$, whose joint density is a complex Gaussian, satisfying the order 1 autoregressive (AR) relation

$$x_n = -ax_{n-1} + \nu_n, \ a \in \mathbb{C}, \ \boldsymbol{\nu} = \left[\nu_0, \cdots, \nu_{N-1}\right]^T \sim \mathscr{CN}\left(\mathbf{0}, \sigma_\nu^2 I\right). \tag{5.201}$$

If n is interpreted as a time step, then x_n defines a first order Gaussian Markov process [or a Gaussian AR(1) process; see chapter 6]. We further assume that $\mathrm{E}\left[x_n x_{n-l}^*\right]$ is a function of l only and denote this auto-correlation value by R_l. Clearly $R_l = R_{-l}^*$. Multiplying both sides of the AR equation (5.201) by x_{n-l}^*, $l \geq 1$, and taking expectation values, we find $R_l = -aR_{l-1}$ whose solution is $R_l = (-a)^l R_0$, $l \geq 1$. In addition, we have

$$R_0 = \mathrm{E}\left[x_n x_n^*\right] = \mathrm{E}\left[\left(-ax_{n-1} + \nu_n\right)\left(-a^* x_{n-1}^* + \nu_n^*\right)\right] = |a|^2 R_0 + \sigma_\nu^2 \ \Rightarrow \ R_0 = \frac{\sigma_\nu^2}{1-|a|^2}. \tag{5.202}$$

Thus, $x = \left[x_0, \ldots, x_{N-1}\right]^T \sim \mathscr{CN}\left(\mathbf{0}, R\right)$, where R is a Hermitian Toeplitz matrix whose (j,k) element is $R_{j-k} = (-a)^{j-k}R_0 = R_{k-j}^*$ and we can use (5.66a) to write its log-likelihood function to be maximized to find the MLE of the parameter vector. An equivalent but simplified formula can be found by using the probability density function

$$f(x_n;\boldsymbol{\theta}) = \frac{1}{\pi R_0}\exp\left(-\frac{|x_n|^2}{R_0}\right), \quad \boldsymbol{\theta} \equiv \left[a, \sigma_\nu^2\right]^T, \ R_0 = \frac{\sigma_\nu^2}{1-|a|^2}, \quad 0 \leq n \leq N-1, \tag{5.203}$$

to find the joint probability density $f(x;\boldsymbol{\theta})$ by repeatedly using Bayes' formula starting with

$$f(x_1, x_0;\boldsymbol{\theta}) = f(x_1 | x_0;\boldsymbol{\theta}) f(x_0;\boldsymbol{\theta}). \tag{5.204}$$

Since $x_1 = -ax_0 + \nu_1$, the conditional mean $\mu_{1|0} = \mathrm{E}[x_1 | x_0] = -ax_0$, while the conditional variance is $\sigma_{1|0}^2 = \sigma_\nu^2$. Thus, the density of x_1 conditioned on x_0 is

$$f(x_1 | x_0;\boldsymbol{\theta}) = \frac{1}{\pi\sigma_\nu^2}\exp\left(-\frac{|x_1 + ax_0|^2}{\sigma_\nu^2}\right), \tag{5.205}$$

which leads to the following joint density between x_1 and x_0:

$$f(x_1, x_0; \boldsymbol{\theta}) = \frac{1}{\pi R_0} \frac{1}{\pi \sigma_\nu^2} \exp\left(- \frac{|x_0|^2}{R_0} - \frac{|x_1 + a x_0|^2}{\sigma_\nu^2} \right). \tag{5.206}$$

Next consider the joint density $f(x_2, x_1, x_0; \boldsymbol{\theta})$, which using Bayes' formula is $f(x_2 | x_1, x_0; \boldsymbol{\theta}) f(x_1, x_0; \boldsymbol{\theta})$. Since $x_2 = -a x_1 + \nu_2$, $f(x_2 | x_1, x_0; \boldsymbol{\theta})$ is independent of the observed value x_0 and is equal to $f(x_2 | x_1; \boldsymbol{\theta})$ (this is the first order *Markov* property, namely, that the density of x_n conditioned on all previous values is only a function of the preceding value x_{n-1}). Thus, the likelihood function is:

$$f(\boldsymbol{x}; \boldsymbol{\theta}) = f(x_0; \boldsymbol{\theta}) \prod_{n=1}^{N-1} f\left(x_n | x_{n-1}; \boldsymbol{\theta}\right). \tag{5.207}$$

Using the relation $x_n = -a x_{n-1} + \nu_n$, we find the conditional mean $\mu_{n|n-1} = \mathrm{E}[x_n | x_{n-1}] = -a x_{n-1}$ and the conditional variance $\sigma^2_{n|n-1} = \sigma_\nu^2$, and so the conditional density is

$$f(x_n | x_{n-1}; \boldsymbol{\theta}) = \frac{1}{\pi \sigma_\nu^2} \exp\left[- \frac{|x_n + a x_{n-1}|^2}{\sigma_\nu^2} \right], \quad 1 \le n \le N-1. \tag{5.208}$$

Finally, we find the log-likelihood function

$$\ln f(\boldsymbol{x}; \boldsymbol{\theta}) = -N \ln(\pi \sigma_\nu^2) + \ln\left(1 - |a|^2\right) - \frac{\left(1 - |a|^2\right) |x_0|^2}{\sigma_\nu^2} - \frac{1}{\sigma_\nu^2} \sum_{n=1}^{N-1} |x_n + a x_{n-1}|^2. \tag{5.209}$$

Unfortunately, differentiating (5.209) with respect to a^* will produce a nonlinear equation

$$\frac{a}{2}\left[\frac{|x_0|^2}{\sigma_\nu^2} - \frac{1}{1 - |a|^2} \right] - \frac{1}{\sigma_\nu^2} \sum_{n=1}^{N-1} \left(x_n + a x_{n-1} \right) x_{n-1}^* = 0, \tag{5.210}$$

which together with its complex conjugate will not produce a closed form result. Therefore, we maximize the log-likelihood function conditioned on x_0

$$\ln f(\boldsymbol{x} | x_0; \boldsymbol{\theta}) = -N \ln\left(\pi \sigma_\nu^2\right) - \frac{1}{\sigma_\nu^2} \sum_{n=1}^{N-1} |x_n + a x_{n-1}|^2, \tag{5.211}$$

to obtain the MLE of the parameter a,

$$\hat{a}_{\mathrm{ML}} = - \frac{\sum\limits_{n=1}^{N-1} x_n x_{n-1}^*}{\sum\limits_{n=1}^{N-1} |x_{n-1}|^2}. \tag{5.212}$$

Although the numerator of the solution depends on x_{N-1}, the denominator does not. It is then possible for the magnitude of the MLE to exceed 1 for a sufficiently large $|x_{N-1}|$. As we shall see later, this first order Gaussian Markov process is an example of an autoregressive (AR) model of a stationary random process described by a single pole filter driven by white noise, and as such will be unstable should the filter parameter a have a magnitude larger than 1. Thus, the MLE does not always guarantee physically realizable or useful linear system models. The maximum likelihood estimation in this case is equivalent to the covariance method of AR model parameter estimation. Other methods of parameter estimation that guarantee the stability of the AR filter model exist and will be discussed later.

5.18 Information Theory: Entropy and Mutual Information

The most basic concept of information theory is that of *entropy* [10] of a random variable x with density function $f_x(x)$: it is a measure of uncertainty associated with the random variable; the more random and unpredictable x is, the larger is its entropy, which, for a continuous random variable, is defined by [57]

$$H_x = -\int f_x(x) \log f_x(x)\, dx = -\mathrm{E}\big[\log f_x(x)\big], \tag{5.213}$$

and measured in *bits* when the base of logarithm is 2, and *nats* for the natural logarithm. This definition can be extended to a vector random variable $\boldsymbol{x}$ with joint density function $f_{\boldsymbol{x}}(\boldsymbol{x})$. Entropy is not invariant under linear transformation of variables, and in particular it is not scale-invariant,

$$H_{\boldsymbol{y}} = H_{\boldsymbol{x}} - \log|\boldsymbol{W}|, \;\; \boldsymbol{y} = \boldsymbol{W}\boldsymbol{x}, \;\; |\boldsymbol{W}| \neq 0. \tag{5.214}$$

A Gaussian random variable (section 5.4) has the largest entropy among all continuous random variables of equal mean and variance whose support is the entire real line (except for a set of measure 0) [58]; if the support is limited to a closed interval of the real line, then that distinction belongs to a uniformly distributed random variable. If the requirement of variance is changed to the mean deviation from the mean, i.e., instead of a constant variance $\mathrm{E}[(x-\mu)^2]$ we require $\mathrm{E}[|x-\mu|]$ to be a constant, then the distribution with maximum entropy is the Laplace density function [59] with mean μ and variance $2b^2$

$$f(x;\mu,b) = (2b)^{-1}\, \exp\big(|x-\mu|/b\big). \tag{5.215}$$

The Laplace density is a "fat-tailed" non-Gaussian density (with an excess-kurtosis of 3) and illustrates the fact that non-Gaussian density functions are more spiky around their mean value (this is an example of a super-Gaussian density discussed at the end of section 5.4). A Gaussian density represents the most random of all density functions with a prescribed mean and variance; hence, entropy can be used as a measure of Gaussianity. An information theoretic measure of non-Gaussianity of a random variable x is *negentropy* defined by

$$J_x = H_x^{(\mathrm{G})} - H_x, \tag{5.216}$$

where $H_x^{(\mathrm{G})}$ is the entropy of a Gaussian random variable with the same mean and variance as x. Negentropy is non-negative and is zero if, and only if, x is Gaussian.

The conditional entropy of x given y is

$$H_{x|y} = -\mathrm{E}\big[\log f_{x|y}(x|y)\big] = H_{xy} - H_y, \tag{5.217}$$

[10] *Entropy* is generally used for discrete random variables while *differential entropy* is used for continuous random variables. We will use *entropy* to refer to both.

where H_{xy} is the entropy of the joint density

$$H_{xy} = -\int f_{xy}(x,y) \log f_{xy}(x,y)\, dxdy. \tag{5.218}$$

Individual and joint entropies satisfy a number of relations [60] listed below:

- $H_x \geq 0$; $H_x = 0$ if, and only if, $x =$ constant almost surely;
- $H_x + H_y \geq H_{xy}$, with equality if, and only if, x and y are independent;
- $H_{x|y} \leq H_x$, with equality if, and only if, x is independent of y; i.e., entropy is reduced by conditioning;
- $H_{y|x} = H_{yx} - H_x$;
- $H_{xy\ldots zw} = H_x + H_{y|x} + \cdots + H_{w|xy\ldots z}$.

Note that the fourth property is a special case of the fifth, which is known as the *chain rule of entropy* of multiple random variables.

Given two random variables x and y with densities $f_x(x)$ and $f_y(y)$, respectively, the entropy of x is always less than or equal to its cross entropy with y [61], i.e.,

$$-\int f_x(x) \log f_x(x)\, dx \leq -\int f_x(x) \log f_y(x)\, dx \;\Leftrightarrow\; \int f_x(x) \log \frac{f_x(x)}{f_y(x)}\, dx \geq 0.$$

This result leads to a measure of difference between two densities f and g known as the *Kullback-Leibler divergence* (also known as *relative entropy*) [62],

$$D_{\mathrm{KL}}(f||g) \equiv \int f(x)\ \log \frac{f(x)}{g(x)}\, dx = \mathrm{E}_f\left[\log \frac{f(x)}{g(x)}\right]. \tag{5.219}$$

Since the "log" function is concave, we use *Jensen's inequality* [63], $\mathrm{E}[\log(x)] \leq \log\left(\mathrm{E}[x]\right)$, to obtain

$$-D_{\mathrm{KL}}(f||g) \leq \log \int f(x)\, \frac{g(x)}{f(x)}\, dx \leq \log 1 = 0.$$

But "log" is strictly concave and equality holds only when f is a constant density, or if g/f is a constant for all x, i.e., $g(x) = cf(x)$. The last inequality above is an equality only when $c = 1$. Thus, the Kullback-Leibler divergence is non-negative and vanishes if, and only if, $f = g$ (this is also known as the *Gibbs information inequality*). The Kullback-Leibler divergence is extensively used in selection of statistical models, including the selection of order in autoregressive models of wide sense stationary random processes (see section 8.24).

The *mutual information* I_{xy} between two random variables x, y that are not independent is a symmetric measure of reduction in uncertainty in one of the random variables when the second becomes known; it is denoted by I_{xy} and is defined as the Kullback-Leibler divergence between their joint density and the product of their marginal densities,

$$I_{xy} = D_{\mathrm{KL}}(f_{xy}(x,y)||f_x(x)f_y(y)). \tag{5.220}$$

Thus, mutual information is non-negative and vanishes if, and only if, the two random variables are independent. We can express mutual information in terms of entropies,

$$\begin{aligned} I_{xy} &= \int \int f_{xy}(x,y) \log \frac{f_{xy}(x,y)}{f_x(x) f_y(y)} \, dx dy \\ &= H_x + H_y - H_{xy} = H_x - H_{x|y} = H_y - H_{y|x}. \end{aligned} \tag{5.221}$$

The entropy of a random variable is often referred to as its *self-information*, i.e., $H_x = I_{xx}$. When x and y have a joint Gaussian density with correlation function ρ, their mutual information is given by $-0.5 \ln \left(1 - \rho^2\right)$. Mutual information is a complete non-parametric measure of linear and nonlinear dependence between two random variables.

We can extend the concept of mutual information to a set of $N > 2$ random variables $\boldsymbol{x} = [x_1, \ldots, x_N]^T$ by

$$I_{\boldsymbol{x}} = \sum_{n=1}^{N} H_n - H_{\boldsymbol{x}}, \quad H_{\boldsymbol{x}} = H_1 + H_{2|1} + H_{3|1,2} + \ldots + H_{N|1,\ldots,N-1}, \tag{5.222}$$

where H_n denotes the entropy of x_n, and $H_{3|1,2}$ denotes the entropy of x_3 conditioned on x_1 and x_2, and so on. The mutual information of a set of N random variables, forming a vector $\boldsymbol{y}$, that are related to a set with an equal number of random variables and denoted by $\boldsymbol{x}$, through an invertible linear transformation $\boldsymbol{y} = \boldsymbol{W}\boldsymbol{x}$, where $\boldsymbol{W}$ has non-zero determinant $|\boldsymbol{W}|$, satisfies the following relation [60]:

$$I_{\boldsymbol{y}} = \sum_{n=1}^{N} H_n^{(y)} - H_{\boldsymbol{x}} - \log |\boldsymbol{W}|. \tag{5.223}$$

In addition, since the entropy of an $N \times 1$ Gaussian random vector with mean μ and covariance matrix $\boldsymbol{C}$ is given by $1/2 \ln \boldsymbol{C} + N/2(1 + \ln 2\pi)$, the negentropy of another $N \times 1$ random vector, with the same mean and covariance of the Gaussian random vector, is invariant under that transformation, i.e.,

$$J_{\boldsymbol{y}} = J_{\boldsymbol{W}\boldsymbol{x}} = J_{\boldsymbol{x}}. \tag{5.224}$$

Let us now assume that y_n are uncorrelated, have zero mean and unit variance, i.e.,

$$\mathrm{E}\left[\boldsymbol{y}\right] = \boldsymbol{0}, \quad \mathrm{E}\left[\boldsymbol{y}\boldsymbol{y}^{+}\right] = \boldsymbol{W}\,\mathrm{E}\left[\boldsymbol{x}\boldsymbol{x}^{+}\right] \boldsymbol{W}^{+} = \boldsymbol{I}. \tag{5.225}$$

Taking determinants gives $\left|\boldsymbol{W}\right|^2 \left|\mathrm{E}\left[\boldsymbol{x}\boldsymbol{x}^{+}\right]\right| = 1$; therefore, $|\boldsymbol{W}|$ is a constant and the last two terms in (5.223) are independent of $\boldsymbol{W}$. Thus, minimization of mutual information of $\boldsymbol{y}$, i.e., the left hand side of (5.223), is equivalent to minimizing each individual entropy $H_n^{(y)}$. Equation (5.213) shows that minimization of $H_n^{(y)}$ is equivalent to the minimization of each density function $f_n(y_n)$, and since the Gaussian distribution has maximal entropy for a fixed mean and variance, minimization of mutual information is equivalent to minimizing the Gaussianity of each of the random variables y_n. Furthermore, since y_n are uncorrelated, have zero mean, and unit variance, their negentropy is $J_n = H_n^{(\mathrm{G})} - H_n$ (with subscript n referring to y_n), which is equivalent to $H_n = C - J_n$, where $C = H_n^{(\mathrm{G})}$ is the same entropy of a zero-mean and unit variance Gaussian that is independent of n. Substituting this relation into (5.223) and combining the constant last two terms with C into a single constant C', we have [64]

$$I_{\boldsymbol{y}} = C' - \sum_{n=1}^{N} J_n^{(y)}. \tag{5.226}$$

Equation (5.226) shows that reducing the inter-dependence of random variables y_n by minimizing their mutual information is equivalent to increasing their negentropy and hence decreasing their Gaussianity; this fundamental relation is the basis for an information theoretic approach to independent component analysis (ICA) discussed in section 5.19.

5.19 Independent Components Analysis

Independent components analysis (ICA) is also known as *blind source separation.* Consider N source signals $\boldsymbol{s}_t = [s_1(t), \ldots, s_M(t)]^T$ that have been linearly mixed by an $M \times M$ mixing matrix $\boldsymbol{A}$ with outputs $\boldsymbol{x}_t = [x_1(t), \ldots, x_M(t)]^T$, where $\boldsymbol{x}_t = \boldsymbol{A}\boldsymbol{s}_t$, as illustrated in figure 5.8. The number of discrete observation times t_n for the signals is N where we assume $N \gg M$.

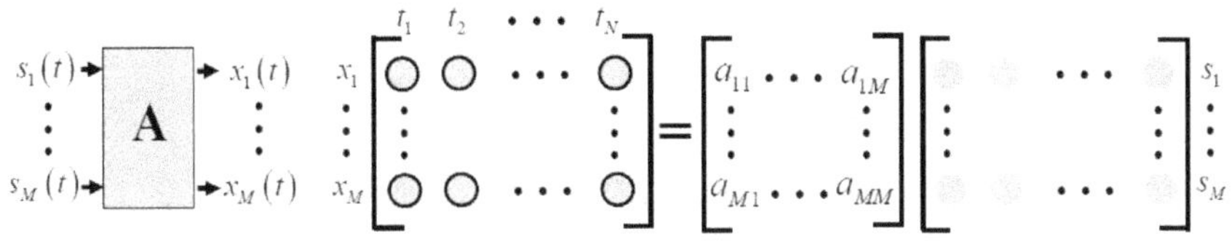

Figure 5.8: Linearly mixed signals for a mixing matrix $\boldsymbol{A}$.

A well-known example is the *cocktail party problem* when the sources are speech signals from different speakers in a room, with microphones placed at different locations in the room providing the mixed signals. Ideally, we would find an un-mixing $M \times M$ matrix $\boldsymbol{W}$ that is the inverse of $\boldsymbol{A}$ such that

$$\boldsymbol{s}_t = \boldsymbol{W}\boldsymbol{x}_t = \Lambda \boldsymbol{P}\boldsymbol{x}_t, \tag{5.227}$$

where Λ is a diagonal matrix of amplitudes and $\boldsymbol{P}$ is a permutation matrix. In practice and in the absence of any environmental noise and other complications such as multi-path propagation, we are content with determining the input signals up to a multiplicative constant.

Under ideal circumstances we must assume that the source signals are statistically independent. In practice, of course, ICA algorithms recover mutually uncorrelated components. The left panel of figure 5.9 shows 500 realizations of two uniformly distributed and independent real random signals s_1 and s_2 with range of values in $[0, 1]$, mean of 0.5, variance $1/12$, and excess-kurtosis $-6/5$, while the right panel shows a linearly mixed set $x_1 = 0.8s_1 + 0.5s_2$ and $x_2 = 0.75s_1 + 0.15s_2$.

In addition to statistical independence of the source signals we must assume that each source has a non-Gaussian distribution; for if each distribution is Gaussian, then their joint density is a Gaussian with a diagonal covariance matrix and a linear transformation will produce outputs that have the same Gaussian density function. All ICA algorithms based on non-Gaussianity use whitened data [11]. Since the observed data and source signals are related by the mixing matrix $\boldsymbol{A}$, removing the mean $\mathrm{E}[\boldsymbol{x}_t]$ also removes the mean from the source signals. Whitening is performed by SVD of the covariance matrix of $\boldsymbol{x}_t$ and then transforming to $\boldsymbol{x'}_t$

$$\mathrm{E}\left[\boldsymbol{x}_t\boldsymbol{x}_t^+\right] = \boldsymbol{U}\Lambda\boldsymbol{U}^+, \quad \boldsymbol{x'}_t \equiv \Lambda^{-1/2}\boldsymbol{U}^+\boldsymbol{x}_t, \tag{5.228}$$

so that $\mathrm{E}\left[\boldsymbol{x'}_t\boldsymbol{x'}_t^+\right] = \boldsymbol{I}$. Figure 5.10 shows the whitened form of the mixed real data of the right panel of figure 5.9.

[11] If the source signals are real, only one Gaussian source can be present for a successful separation. For complex source signals, multiple Gaussian sources can be separated so long as they have distinct "circularity" coefficients [65, 66].

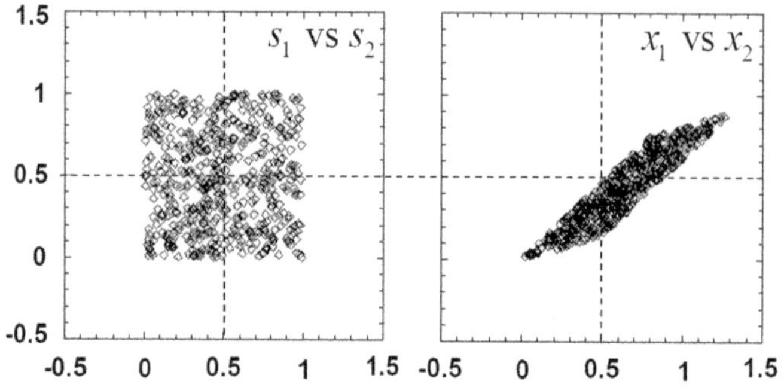

Figure 5.9: Two uniformly distributed independent random variables and their linearly mixed version.

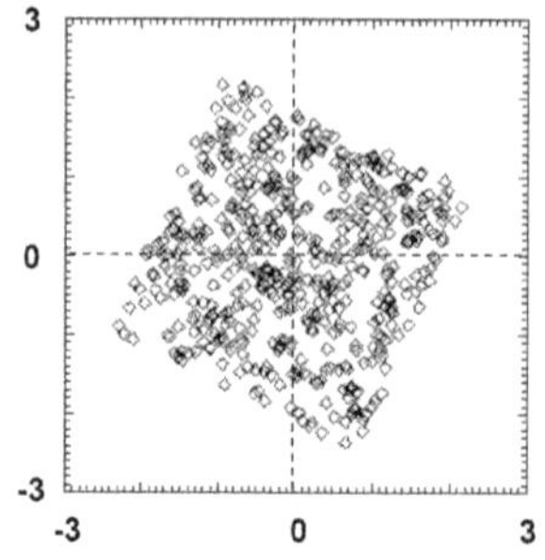

Figure 5.10: Whitened mixed real data of figure 5.9.

The transformed mixed model after whitening the observed data becomes

$$\boldsymbol{x}'_t = \Lambda^{-1/2}\boldsymbol{U}^+\boldsymbol{A}\boldsymbol{s}_t \equiv \boldsymbol{A}'\boldsymbol{s}_t, \tag{5.229}$$

where the new mixing matrix $\boldsymbol{A}'$ is unitary provided that the original signals are uncorrelated and have unit variance, i.e., $\mathrm{E}[\boldsymbol{s}_t\boldsymbol{s}_t^+] = \boldsymbol{I}$,

$$\boldsymbol{A}'\boldsymbol{A}'^+ = \boldsymbol{A}'\mathrm{E}[\boldsymbol{s}_t\boldsymbol{s}_t^+]\boldsymbol{A}'^+ = \mathrm{E}[\boldsymbol{x}'_t\boldsymbol{x}'^+_t] = \boldsymbol{I}. \tag{5.230}$$

The assumption that the covariance matrix of the original signals is the identity matrix also implies the unitarity of the un-mixing matrix $\boldsymbol{W}$,

$$\boldsymbol{W}\boldsymbol{W}^+ = \boldsymbol{W}\,\mathrm{E}[\boldsymbol{x}'_t\boldsymbol{x}'^+_t]\,\boldsymbol{W}^+ = \mathrm{E}[\boldsymbol{s}_t\boldsymbol{s}_t^+] = \boldsymbol{I}, \tag{5.231}$$

which reduces the number of independent (complex) components of the un-mixing matrix from M^2 to $M(M-1)/2$.

A statistical approach to recovering the independent components is based on the recognition that linearly mixing of signals leads to more Gaussianity, as suggested by the central limit theorem (theorem 17). Since a non-zero value of excess-kurtosis (5.70) is a measure of non-Gaussianity of a signal, a simple ICA method is based on maximizing the magnitude excess-kurtosis of the (whitened) mixed signal. Let $\boldsymbol{x}_t$ denote the whitened mixed signal with $\mathrm{E}[\boldsymbol{x}_t] = 0$ and $\mathrm{E}[\boldsymbol{x}_t\boldsymbol{x}_t^+] = \boldsymbol{I}$. An ICA algorithm based on excess-kurtosis attempts to find the first independent component by solving the following constrained maximization problem for $\boldsymbol{w}_1$ (i.e., the first row of the unitary un-mixing matrix $\boldsymbol{W}$),

$$\arg\max_{\boldsymbol{w}_1} \left| \mathrm{E}\left[\left| \boldsymbol{w}_1^T \boldsymbol{x}_t \right|^4 \right] - 3 \left(\mathrm{E}\left[\left| \boldsymbol{w}_1^T \boldsymbol{x}_t \right|^2 \right] \right)^2 \right|, \quad \text{subject to } \boldsymbol{w}_1^+ \boldsymbol{w}_1 = 1, \tag{5.232}$$

where the quantity to be maximized is the un-normalized excess-kurtosis.

The graph on the left hand side of figure 5.11 shows the objective function of equation (5.232) for the whitened mixed data example of figure 5.10 as a function of the angle θ defining the unit direction vector $\boldsymbol{w}_1 = [\cos\theta, \sin\theta]$; the maximum occurs at $\theta = 52.5°$ and the corresponding direction vector is shown on the right hand side of figure 5.11. The second direction vector, for the second independent component, is obviously given by $\boldsymbol{w}_2 = [-\cos 52.5°, \sin 52.5°]$. Note that both direction vectors are defined up to a sign. We should note PCA is of little use here since it looks for directions of maximum variation in the data, which are the two diagonals of the square.

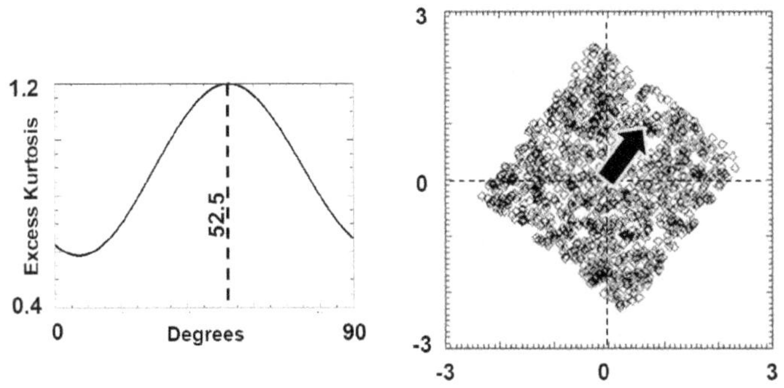

Figure 5.11: Excess-kurtosis and the direction vector for the first independent component of data depicted in figure 5.10.

For problems with more than two independent components, once the first row of the un-mixing matrix is found, the second row is found by solving (5.232) with the additional orthogonality constraint $\boldsymbol{w}_2^+ \boldsymbol{w}_1 = 0$. In general, once $\boldsymbol{w}_k$, $1 \le k \le K < M$, have been found, the next one is found by solving (5.232) with the additional K orthogonality constraints $\boldsymbol{w}_{K+1}^+ \boldsymbol{w}_k = 0$, $1 \le k \le K < M$.

An iterative gradient algorithm to solve the constrained maximization problem, starting with a randomly chosen unit-norm vector, can be found using the derivative of the objective function in (5.232) with respect to $\boldsymbol{w}_1$ (treating it and its complex conjugate as independent variables),

$$2\,k_s \left(\mathrm{E}\left[(\boldsymbol{w}_1^+ \boldsymbol{x'}_t^*) \left| \boldsymbol{w}_1^+ \boldsymbol{x'}_t \right|^2 \boldsymbol{x'}_t \right] - 3\,\mathrm{E}\left[\left| \boldsymbol{w}_1^T \boldsymbol{x'}_t \right|^2 \right] \mathrm{E}\left[(\boldsymbol{w}_1^+ \boldsymbol{x'}_t^*) \boldsymbol{x'}_t \right] \right),$$

where k_s is the sign of the excess-kurtosis of $\boldsymbol{w}_1^T \boldsymbol{x'}_t$. Since the mixed data have been whitened (equation (5.230)), and $|\boldsymbol{w}_1| = 1$ is the constraint, we use the relations

$$\mathrm{E}\left[\left| \boldsymbol{w}_1^T \boldsymbol{x'}_t \right|^2 \right] = \boldsymbol{w}_1^+ \left(\mathrm{E}\left[\boldsymbol{x'}_t \boldsymbol{x'}_t^+ \right] \right)^* \boldsymbol{w}_1 = 1, \quad \text{and } \mathrm{E}\left[(\boldsymbol{w}_1^+ \boldsymbol{x'}_t^*) \boldsymbol{x'}_t) \right] = \boldsymbol{w}_1^*,$$

to obtain

$$2\,k_s \left(\mathrm{E}\left[(\boldsymbol{w}_1^+ \boldsymbol{x'}_t^*) \left| \boldsymbol{w}_1^+ \boldsymbol{x'}_t \right|^2 \boldsymbol{x'}_t \right] - 3\,\boldsymbol{w}_1^* \right). \tag{5.233}$$

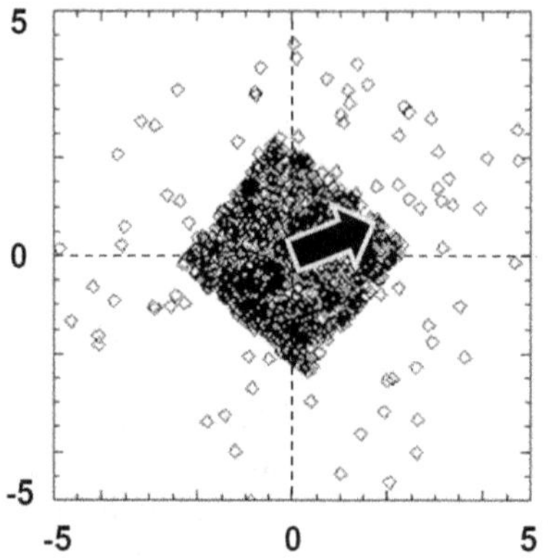

Figure 5.12: Effect of additional outliers (see figure 5.11).

The excess-kurtosis maximization algorithm is not robust; it breaks down in the presence of outliers. Figure 5.12 shows the same example of figure 5.10 with some additional outliers; the algorithm fails to pick the correct direction vector for the first independent component.

Since zero mutual information between two random variables guarantees their independence, a more robust algorithm is based on the information theoretic approach to ICA that minimizes pairwise mutual information among the mixed signals. Equation (5.226) and the discussion following it showed that the optimal un-mixing transformation matrix $\boldsymbol{W}$ should minimize pair-wise mutual information of, or equivalently, maximize the individual negentropies. Thus, the objective function to be maximized is the negentropy given by equation (5.216) whose calculation requires difficult estimation of probability density functions. In practice, approximations to negentropy are used; one such approximation is [64]

$$J_y \approx k \left|\mathrm{E}[G(y)] - \mathrm{E}[G(\nu)]\right|^2, \quad y \sim \mathcal{N}(0,1), \quad \nu \sim \mathcal{N}(0,1), \tag{5.234}$$

where k is a constant and $G(.)$ is a non-quadratic function; the following choices lead to robust negentropy estimates

$$G(y) = -\exp(-y^2/2), \quad \text{or} \quad G(y) = \left[\ln\cosh(ay)\right]/a, \quad 1 \le a \le 2. \tag{5.235}$$

Thus, the information based ICA algorithm finds a unit vector $\boldsymbol{w}$ that maximizes the non-Gaussianity of $\boldsymbol{w}^T\boldsymbol{x}$ using the approximation to the negentropy (5.234); it can be implemented as an approximate Newton iteration, starting with an initial random weight vector of unit norm [64]:

- $\boldsymbol{v} = \mathrm{E}\left[\boldsymbol{x}\, G'(\boldsymbol{w}_n^T\boldsymbol{x})\right] - \mathrm{E}\left[G''(\boldsymbol{w}_n^T\, \boldsymbol{x})\right]\boldsymbol{w}_n$
- $\boldsymbol{w}_{n+1} = \boldsymbol{v}/\left\|\boldsymbol{v}\right\|$
- stop when $\left|\left\langle \boldsymbol{w}_{n+1}, \boldsymbol{w}_n \right\rangle\right| \approx 1$.

In practice the expectation operations are replaced by their sample mean values, and the stopping criterion is a reflection of the fact that the un-mixing matrix is determined up to a sign. The rest of the independent components are found by using additional orthonormality constraints discussed earlier.

The left hand side of figure 5.13 shows two independent signals that have been mixed together, using the indicated mixing matrix, to produce the mixed signals shown on the right hand side of the same figure. Figure 5.14 shows the scatter diagram for the mixed components (left) and the whitened mixed components

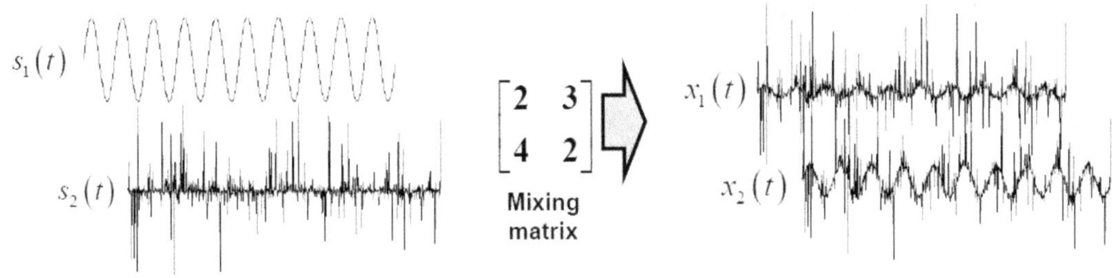

Figure 5.13: Two independent signals (left) and their mixed versions (right).

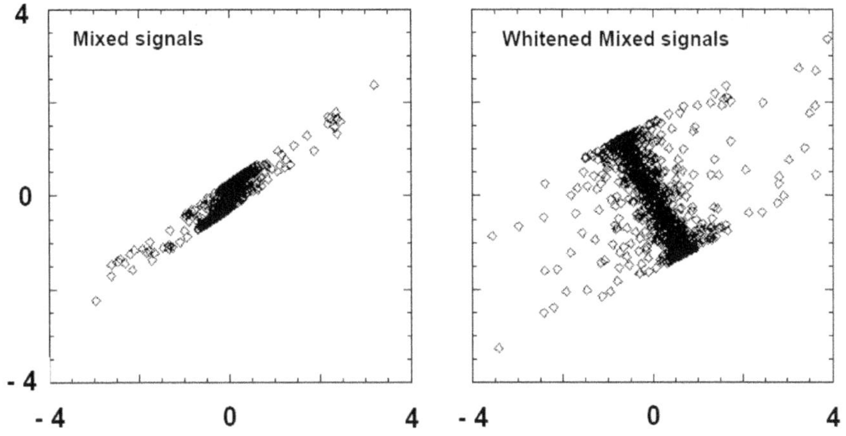

Figure 5.14: Mixed signals (left) and whitened mixed signals (right).

(right). The presence of the outliers render the kurtosis based method useless.

The negentropy based ICA algorithm is known as FastICA; its output, shown on the left hand side of figure 5.15, clearly separates the two independent components. The PCA result, right hand side of figure 5.15, shows no separation of the two independent components.

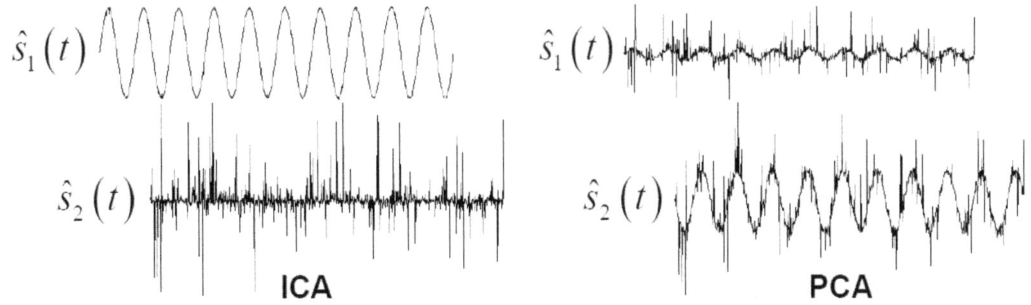

Figure 5.15: Independent components using ICA (left) and components using PCA (right).

5.20 Maximum Likelihood ICA

Another approach to identifying independent components of observed mixed signals is based on the maximum likelihood principle [64]. Using the mixing equation (5.227) we can relate the joint density of the independent components, denoted by g, and the joint density of the mixed signals, denoted by f, at any single observation time t, namely,

$$f(\boldsymbol{x}_t) = \left|\boldsymbol{W}\right| g(\boldsymbol{s}_t) = \left|\boldsymbol{W}\right| \prod_{m=1}^{M} g_m(s_m), \;\; s_m = s_m(t) = \boldsymbol{w}_m^T \boldsymbol{x}_t, \tag{5.236}$$

where g_m is the marginal density of the independent component s_m, and $\boldsymbol{w}_m^T$ is the mth row of the unmixing matrix $\boldsymbol{W}$. Thus, the likelihood function for observations at all times t_n, $1 \leq n \leq N$, is

$$L(\boldsymbol{W}|\boldsymbol{x}) = \left|\boldsymbol{W}\right|^N \prod_{t=1}^{N}\prod_{m=1}^{M} g_m(\boldsymbol{w}_m^T \boldsymbol{x}_t), \tag{5.237}$$

and the log-likelihood is

$$\ln L(\boldsymbol{W}|\boldsymbol{x}) = N \ln\left|\boldsymbol{W}\right| + \sum_{t=1}^{N}\sum_{m=1}^{M} \ln g_m(\boldsymbol{w}_m^T \boldsymbol{x}_t). \tag{5.238}$$

The log-likelihood is often written as an average over all observations when the right hand side is also written as an expectation value,

$$\frac{1}{N} \ln L(\boldsymbol{W}|\boldsymbol{x}) = \ln\left|\boldsymbol{W}\right| + \mathrm{E}\Big[\sum_{t=1}^{N}\sum_{m=1}^{M} \ln g_m(\boldsymbol{w}_m^T \boldsymbol{x}_t)\Big], \tag{5.239}$$

with the maximum likelihood problem statement

$$\underset{\boldsymbol{W}}{\arg\max} \left(\ln\left|\boldsymbol{W}\right| + \mathrm{E}\Big[\sum_{m=1}^{M} \ln g_m(\boldsymbol{w}_m^T \boldsymbol{x}_t)\Big]\right). \tag{5.240}$$

An iterative gradient ascent algorithm to solve (5.240) is

$$\boldsymbol{W}^{[t+1]} = \boldsymbol{W}^{[t]} + \mu \frac{\partial}{\partial \boldsymbol{W}} \left(\ln\left|\boldsymbol{W}\right| + \mathrm{E}\Big[\sum_{m=1}^{M} \ln g_m(\boldsymbol{w}_m^T \boldsymbol{x}_t)\Big]\right), \tag{5.241}$$

for an appropriate step-size $\mu > 0$ that will result in increasing the log-likelihood after each step. In practice, we use each observed data $\boldsymbol{x}_m(t)$, $1 \leq m \leq M$, in succession. It is easy to show

$$\frac{\partial}{\partial \boldsymbol{W}} \left(\mathrm{E}\Big[\sum_{m=1}^{M} \ln g_m(\boldsymbol{w}_m^T \boldsymbol{x}_t)\Big]\right) = \mathrm{E}\Big[\boldsymbol{h}(\boldsymbol{W}\boldsymbol{x}_t)\boldsymbol{x}_t^T\Big], \tag{5.242}$$

where

$$\boldsymbol{h}(\boldsymbol{z}) \equiv \left[h_1(z_1), \ldots, h_M(z_M)\right]^T, \;\; h_m(z) = d/dz\big[\ln g_m(z)\big]. \tag{5.243}$$

We can drop the expectation operation in (5.242) using the stochastic approximation, and then use equation (1.23) for the derivative of a determinant, and multiply the likelihood gradient on the right by $\boldsymbol{W}^T\boldsymbol{W}$ (for stability reasons), to obtain the final form of the iterative gradient ascent equation for $t = 1, \ldots, N$,

$$\boldsymbol{W}^{[t+1]} = \boldsymbol{W}^{[t]} + \mu \left[\boldsymbol{I} + \boldsymbol{h}(\boldsymbol{W}^{[t]}\boldsymbol{x}_t)\ (\boldsymbol{W}^{[t]}\boldsymbol{x}_t)^T\right]\boldsymbol{W}^{[t]}. \tag{5.244}$$

The only remaining issue is the form of the functions h_m that depend on unknown (possibly unknowable) density functions g_m. It turns out that so long as the density functions are either sub-Gaussian, which decay faster than a Gaussian, e.g., a uniform density that drops to zero in a finite interval, or super-Gaussian that decay more slowly than a Gaussian, e.g., the Laplace density (5.215), then we can use the following approximations [64]

$$g_m(z) = \tanh(z) - z \text{ **"sub"**}, \quad g_m(z) = -2\ \tanh(z) \text{ **"super"**}, \tag{5.245}$$

where a sub-Gaussian density decays faster than a Gaussian, e.g., a uniform density that drops to zero in a finite interval, and a super-Gaussian density decays more slowly than a Gaussian, e.g., the Laplace density (5.215).

Figure 5.16 shows three mixed speech signals on the left; on the right are shown all three original independent speech signals (black) and the recovered maximum likelihood independent components using the algorithm in (5.244) with $\mu = 0.001$.

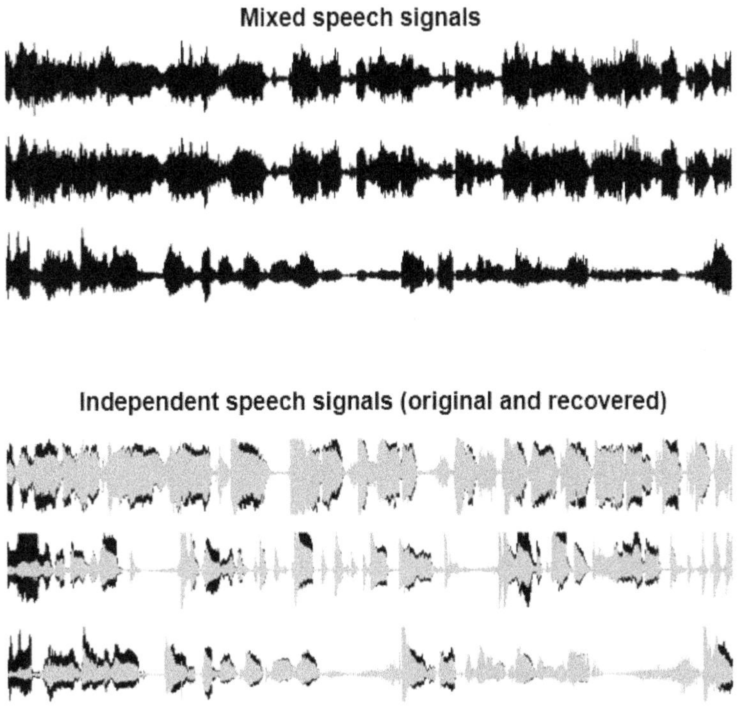

Figure 5.16: Maximum likelihood ICA example of three mixed speech signals.

CHAPTER 6

WSS Random Processes

6.1 Auto-Correlation and the Power Spectral Density

The *auto-correlation* function of a WSS process is

$$R_{xx}(t, t-\tau) = \mathrm{E}\left[x_t x_{t-\tau}^*\right] \equiv R_{xx}(\tau), \quad R_{xx}(\tau) = R_{xx}^*(-\tau). \tag{6.1}$$

Using the Cauchy-Schwartz inequality (1.9)

$$\left|\mathrm{E}\left[x_t x_{t-\tau}^*\right]\right|^2 \le \mathrm{E}\left[|x_t|^2\right] \mathrm{E}\left[|x_{t-\tau}|^2\right], \tag{6.2}$$

we have $R_{xx}(\tau) \le R_{xx}(0)$. If $R_{xx}(T) = R_{xx}(0)$ for some $T \neq 0$, then $R_{xx}(\tau)$ is periodic with period T. To see this, we use the Cauchy-Schwartz inequality (1.9):

$$\begin{aligned}\left(R_{xx}(T+\tau) - R_{xx}(\tau)\right)^2 &\le \mathrm{E}\left[|x_t|^2\right]\mathrm{E}\left[|x_{t-\tau} - x_{T+t-\tau}|^2\right] \\ &= 2R_{xx}(0)\left(R_{xx}(0) - R_{xx}(T)\right) = 0.\end{aligned} \tag{6.3}$$

In fact, it can be shown that the process x_t is periodic, i.e., $x_t = x_{t+T}$, with probability 1. As an example, consider the periodic SSS process $x_t = A\cos(\omega t + \phi)$ with period $T = 2\pi/\omega$, where ϕ is uniformly distributed in $[-\pi, \pi]$. The auto-correlation function is $R_{xx}(\tau) = A^2\cos(\omega\tau)/2$, which is periodic with period $T = 2\pi/\omega$.

Two processes are jointly WSS if each is WSS and their cross-correlation function depends only on the difference in times, i.e.,

$$R_{xy}(t, t-\tau) = \mathrm{E}[x_t y_{t-\tau}^*] \equiv R_{xy}(\tau), \quad R_{xy}(\tau) = R_{yx}^*(-\tau). \tag{6.4}$$

Although the Fourier transforms of infinite energy random signals do not exist (they may exist as generalized functions but not as functions in $L_2(\mathbb{R})$), the auto-covariance and auto-correlation functions of such signals often have finite Fourier transforms, and these frequency spectra have important roles in the analysis of linear systems with random inputs. At the heart of the frequency spectral representation of WSS random signals is the Wiener-Khinchin theorem [1].

Consider the auto-correlation function $R_{xx}(\tau)$ of a WSS random process x_t and assume that it is absolutely integrable, $R_{xx}(\tau) \in L_1(\mathbb{R})$; this is a weaker condition than square integrability; nevertheless, the Fourier transform of this auto-correlation function exists. The integrability condition on the auto-correlation function excludes WSS random processes with non-zero mean. If the random process of interest

[1] The original work of Wiener in 1930 related to non-random signals; in 1934 Khinchin extended Wiener's result to stationary random processes.

has non-zero mean, we use the auto-covariance function instead of the auto-correlation or simply subtract the mean and work with the mean-subtracted process that is now zero-mean. Henceforth, we will assume zero-mean WSS random processes unless otherwise stated. Now consider the random process $X_T(\omega)$ defined by

$$X_T(\omega) \equiv \int_{-T/2}^{T/2} x(t)\, e^{-i\omega t}\, dt, \tag{6.5}$$

and the corresponding power spectral density (PSD) function $|X_T(\omega)|^2/T$. It is then natural to define the PSD of the original process x_t by

$$S_{xx}(\omega) = \lim_{T\to\infty} \mathrm{E}\left[\frac{|X_T(\omega)|^2}{T}\right] = \lim_{T\to\infty} \frac{1}{T} \int_{-T/2}^{T/2}\int_{-T/2}^{T/2} \mathrm{E}\left[x_t {x^*}_{t'}\right] e^{-i\omega(t-t')}\, dtdt'. \tag{6.6}$$

Changing integration variables in the double integral to $\tau = t - t'$ and $\eta = t'$ we have

$$\mathrm{E}\left[\frac{1}{T}|X_T(\omega)|^2\right] = \int_{-T}^{T} \left(1 - \frac{|\tau|}{T}\right) R_{xx}(\tau)\, e^{-i\omega\tau}\, d\tau. \tag{6.7}$$

The apparently simple-minded method of taking the limit $T \to \infty$ is actually mathematically justified here by the Lebesgue dominated-convergence theorem [30] in view of the obvious inequality

$$\left|\left(1 - \frac{|\tau|}{T}\right) R_{xx}(\tau)\, e^{-i\omega\tau}\right| \le |R_{xx}(\tau)| \tag{6.8}$$

and the fact that $R_{xx}(\tau) \in L_1(\mathbb{R})$. Thus, we arrive at the Wiener-Khinchin theorem [67, 68].

Theorem 25. *The power spectral density (PSD) of a zero-mean WSS random process in continuous time is a real and non-negative function given by the Fourier transform of its auto-correlation function*

$$S_{xx}(\omega) = \int_{-\infty}^{\infty} R_{xx}(\tau)\, e^{-i\omega\tau}\, d\tau. \tag{6.9}$$

Using the inverse Fourier transform of the PSD (6.9) yields

$$R_{xx}(t - t') = \frac{1}{2\pi} \int_{-\infty}^{\infty} S_{xx}(\omega) e^{+i\omega(t-t')}\, d\omega. \tag{6.10}$$

For any complex valued function $a(t)$ with finite L_2 norm $\|a(t)\|_2 < \infty$, we find the condition

$$\int_{-\infty}^{\infty}\int_{-\infty}^{\infty} a^*(t) R_{xx}(t - t') a(t')\, dtdt' = \frac{1}{2\pi} \int_{-\infty}^{\infty} |A(\omega)|^2\, S_{xx}(\omega)\, d\omega \ge 0. \tag{6.11}$$

Thus, a valid auto-correlation function, guaranteeing that its Fourier transform is a real and non-negative function, must satisfy the positivity condition (6.11).

The cross power spectral density (CPSD) for joint zero-mean WSS random processes is a complex quantity and defined analogously to (6.7) and (6.9)

$$S_{xy}(\omega) = \lim_{T\to\infty} \mathrm{E}\left[\frac{X_T(\omega)Y_T^*(\omega)}{T}\right] = \int_{-\infty}^{\infty} R_{xy}(\tau)e^{-i\omega\tau}\, d\tau. \tag{6.12}$$

A frequency dependent correlation coefficient known as *magnitude squared coherence* (MSC) is defined by

$$\gamma_{xy}^2(\omega) = \gamma_{yx}^2(\omega) \equiv \frac{\left|S_{xy}(\omega)\right|^2}{S_{xx}(\omega)S_{yy}(\omega)}, \quad 0 \le \gamma_{xy}^2(\omega) \le 1. \tag{6.13}$$

The integral of the PSD over a band is equal to the power of the process in that band (energy in the band per unit time). A special case is zero-mean WGN ν_t whose auto-correlation function is $\sigma_\nu^2\delta(\tau)$ [see (5.52)] and whose PSD has the same value σ_ν^2 at all frequencies; thus, its power is the variance of the process. A WSS white process whose band is limited to the interval $[-B, +B]$ with power σ^2 has auto-correlation function $R_{xx}(\tau) = \sigma^2 \sin(2\pi B\tau)/\pi B\tau$, and its sample values at times $t \pm n/2B$ are uncorrelated. Taking the limit $B \to \infty$, we obtain the white noise process $\nu(t)$ with auto-correlation function $\sigma_\nu^2\delta(\tau)$; if this process is Gaussian, then it is known as *white Gaussian noise* (WGN) [see the definition of WGN as the derivative of the Wiener process and its Karhunen-Loéve expansion (5.95)].

Similar definitions follow for discrete time zero-mean WSS random processes; the auto-correlation at lag $k \in \mathbb{Z}$ is defined by

$$R_{xx}[k] = \mathrm{E}\left[x_{n+k}x_n^*\right] = \mathrm{E}\left[x_n x_{n-k}^*\right], \tag{6.14}$$

and its Fourier transform is the process PSD

$$S_{xx}\left(e^{i\omega}\right) = \sum_{k=-\infty}^{\infty} R_{xx}[k]\, e^{-ik\omega}, \quad S_{xy}\left(e^{i\omega}\right) = \sum_{k=-\infty}^{\infty} R_{xy}[k]\, e^{-ik\omega}. \tag{6.15}$$

The auto-correlation function of zero-mean WGN ν_n in discrete is $\sigma_\nu^2\delta_{0k}$, which produces uniform power σ_ν^2 at all frequencies. The Wiener-Khinchin theorem now takes the following form.

Theorem 26. *An absolutely summable sequence $R_{xx}[k] \in l_1(\mathbb{Z})$ is non-negative definite if, and only if, its Fourier transform is a non-negative function. If $R_{xx}[k]$ is the auto-correlation sequence of a discrete time zero-mean WSS random process x_n, then its Fourier transform is the power spectral density (PSD) of the process:*

$$S_{xx}\left(e^{i\omega}\right) = \sum_{l=-\infty}^{\infty} R_{xx}[l]e^{-ik\omega}. \tag{6.16}$$

The non-negative definite condition on $R_{xx}[k]$ means that for an arbitrary complex sequence a_k with finite L_2 norm, we have

$$\sum_{k=0}^{N-1}\sum_{l=0}^{N-1} a_l^* R_{xx}[l-k]a_k \ge 0. \tag{6.17}$$

Now we define the sequence

$$S_N\left(e^{i\omega}\right) = \frac{1}{N}\sum_{k=0}^{N-1}\sum_{l=0}^{N-1} e^{-il\omega}\, R_{xx}[l-k]\, e^{ik\omega}, \tag{6.18}$$

which using (6.17) must be non-negative and can be rewritten as

$$S_N\left(e^{i\omega}\right) = \frac{1}{N}\sum_{l=-(N-1)}^{N-1}\left(N-|l|\right)R_{xx}[l]\,e^{-il\omega} = \sum_{l=-(N-1)}^{N-1}\left(1-\frac{|l|}{N}\right)R_{xx}[l]\,e^{-il\omega}. \tag{6.19}$$

Thus,

$$\left|S_{xx}\left(e^{i\omega}\right) - S_N\left(e^{i\omega}\right)\right| \le \left|\sum_{|k|\ge N} R_{xx}[k]e^{-ik\omega}\right| + \left|\sum_{k=-(N-1)}^{N-1}\frac{|k|}{N}R_{xx}[k]e^{-ik\omega}\right|. \tag{6.20}$$

Both terms on the right hand side of (6.20) tend to 0 as $N \to \infty$ (proof for the first term is obvious from the fact that the sequence $R_{xx}[k]$ is absolutely summable; proof for the second term is far more complicated!). Now all $S_N(\omega)$ are non-negative functions, so their limit is also non-negative. This proves the "only if" part of theorem 26; the "if" part can be proven using the discrete time Fourier transform version of (6.11).

More generally, we define the Laplace transforms of auto-correlation and cross-correlation functions of two continuous time zero-mean WSS random processes by

$$S_{xx}(s) = \int_{-\infty}^{\infty} R_{xx}(\tau)\,e^{-s\tau}d\tau, \quad S_{xy}(s) = \int_{-\infty}^{\infty} R_{xy}(\tau)\,e^{-s\tau}d\tau, \tag{6.21}$$

and the **Z** transforms of auto-correlation and cross-correlation functions of two discrete time zero-mean WSS random processes by

$$S_{xx}(z) = \sum_{k=-\infty}^{\infty} R_{xx}[k]\,z^{-k}, \quad S_{xy}(z) = \sum_{k=-\infty}^{\infty} R_{xy}[k]\,z^{-k}. \tag{6.22}$$

In analogy with vector random variables, we define two vector random processes $\boldsymbol{x}_t$ and $\boldsymbol{y}_t$ to be jointly WSS when their auto-correlation and cross-correlation matrices are $\boldsymbol{R}_{xx}(\tau) = \mathrm{E}[\boldsymbol{x}_t\boldsymbol{x}_{t-\tau}^+]$, $\boldsymbol{R}_{yy}(\tau) = \mathrm{E}[\boldsymbol{y}_t\boldsymbol{y}_{t-\tau}^+]$, $\boldsymbol{R}_{xy}(\tau) = \mathrm{E}[\boldsymbol{x}_t\boldsymbol{y}_{t-\tau}^+]$, with $\boldsymbol{R}_{yx}(\tau) = \boldsymbol{R}_{xy}^+(-\tau)$. Similarly for discrete time random vectors $\boldsymbol{x}_n$ and $\boldsymbol{y}_n$ that are jointly WSS, we have $\boldsymbol{R}_{xx}[k] = \mathrm{E}[\boldsymbol{x}_n\boldsymbol{x}_{n-k}^+]$, $\boldsymbol{R}_{yy}[k] = \mathrm{E}[\boldsymbol{y}_n\boldsymbol{y}_{n-k}^+]$, $\boldsymbol{R}_{xy}[k] = \mathrm{E}[\boldsymbol{x}_n\boldsymbol{y}_{n-k}^+]$, and $\boldsymbol{R}_{yx}[k] = \boldsymbol{R}_{xy}^+[-k]$.

If the random vectors consist of time samples of individual processes, e.g., $\boldsymbol{x} = [x_n, \ldots, x_{n+M-1}]^T$ and $\boldsymbol{y} = [y_n, \ldots, y_{n+M-1}]^T$, then the $M \times M$ cross-correlation matrix of the two processes is $\boldsymbol{R}_{xy} = \mathrm{E}[\boldsymbol{x}\boldsymbol{y}^+] = \boldsymbol{R}_{yx}^+$ while the $M \times M$ auto-correlation matrices are $\boldsymbol{R}_{xx} = \mathrm{E}[\boldsymbol{x}\boldsymbol{x}^+]$ and $\boldsymbol{R}_{yy} = \mathrm{E}[\boldsymbol{y}\boldsymbol{y}^+]$. The auto-correlation matrix $\boldsymbol{R}_{xx}$ of a random vector representing samples of a zero-mean WSS process has a number of important properties summarized in the following theorem (cf. theorem 15).

Theorem 27. *Let $\boldsymbol{x} = [x_{n-(M-1)}, \ldots, x_n]^T$ (none of our results below depend on the arbitrary index n becauses of the WSS property) represent samples of a zero-mean WSS random process and define the $M \times M$ auto-correlation matrix $\boldsymbol{R}_{xx} = E[\boldsymbol{x}\boldsymbol{x}^+]$. Then*

- *$\boldsymbol{R}_{xx}$ is Hermitian and Toeplitz, i.e., $\boldsymbol{R}_{xx}^+ = \boldsymbol{R}_{xx}$ and $\left[\boldsymbol{R}_{xx}\right]_{jk} = R[j-k]$ where $R[l] = E[x_n x_{n-l}^*]$ and $R[l] = R^*[-l]$.*

- *$\boldsymbol{R}_{xx}$ is non-negative definite, i.e., $\boldsymbol{a}^+\boldsymbol{R}_{xx}\boldsymbol{a} \ge 0$ for any complex vector $\boldsymbol{a}$.*

- *$\boldsymbol{R}_{xx}$ has real and non-negative eigenvalues, and eigenvectors corresponding to distinct eigenvalues are orthogonal. If an eigenvalue is degenerate, i.e., it belongs to several distinct eigenvectors,* *the Gram-Schimdt process can be used to produce a set of orthonormal eigenvectors from the degenerate set. Thus, without loss of generality, we have the decomposition* $\boldsymbol{R}_{xx} = \boldsymbol{U}\boldsymbol{\Lambda}\boldsymbol{U}^+$ *where* $\boldsymbol{U} = [\boldsymbol{u}_1, \ldots, \boldsymbol{u}_M]$ *is the unitary matrix of orthonormal eigenvectors and* $\boldsymbol{\Lambda}$ *is the diagonal matrix of eigenvalues. The decomposition can also be written as* $\boldsymbol{R}_{xx} = \sum_{m=1}^{M} \lambda_m \boldsymbol{u}_m \boldsymbol{u}_m^+$ *with the inverse matrix (that exists if there are no zero eigenvalues)* $\boldsymbol{R}_{xx}^{-1} = \sum_{m=1}^{M} \lambda_m^{-1} \boldsymbol{u}_m \boldsymbol{u}_m^+$.

- *The eigenvalues of the auto-correlation matrix are bounded below and above by the minimum and maximum values of the process spectrum, i.e., if* $S(e^{i\omega}) = \sum_{l=-\infty}^{+\infty} R[l]e^{-il\omega}$, *then* $\boldsymbol{min}(S) \leq \lambda_m \leq \boldsymbol{max}(S)$, $1 \leq m \leq M$.
- *The auto-correlation matrix also has a Cholesky decomposition* $\boldsymbol{R}_{xx} = \boldsymbol{L}\boldsymbol{L}^+$ *where* $\boldsymbol{L}$ *is a lower triangular matrix, as described in section 5.5. This decomposition in the context of zero-mean WSS processes has applications to causal filtering.*

In analogy to the auto-covariance matrix of a random vector, equation (5.41) shows that the auto-correlation matrix of a zero-mean WSS process can only have a zero eigenvalue if the elements of $\boldsymbol{x}$ are linearly dependent and so the auto-correlation matrix is positive definite if they are linearly independent. To find the condition for linear dependence (and following the same method that led to (5.41)), we write the eigenvalue equation as

$$\Lambda = \mathrm{E}\left[\boldsymbol{y}\boldsymbol{y}^+\right], \quad \boldsymbol{y} = \boldsymbol{U}^+\boldsymbol{x}, \quad \boldsymbol{x} \equiv \left[x_{n-(M-1)}, \ldots, x_n\right]^T.$$

Let us assume we have one zero eigenvalue at diagonal location (k, k) on the left hand side; then using an argument identical to the discussion after equation (5.41) we obtain the equation $\boldsymbol{u}_k^+\boldsymbol{x}_n = 0$, where $\boldsymbol{u}_k$ is the k-th column of $\boldsymbol{U}$. Thus, the elements of $\boldsymbol{x}_n$ are linearly dependent

$$u_0^* x_{n-(M-1)} + \ldots + u_{M-1}^* x_n = 0.$$

Multiplying both sides with x_{n-l}^* and taking expectation values, we find

$$u_0^* R[l-(M-1)] + \ldots + u_{M-1}^* R[l] = 0. \tag{6.23}$$

Multiplying both sides of (6.23) with $e^{-il\omega}$ and summing over l gives the Fourier transform relation

$$S(e^{i\omega})\left(u_{M-1}^* + \ldots + u_0^* e^{-i(M-1)\omega}\right) = 0. \tag{6.24}$$

The trigonometric polynomial inside parentheses in (6.24) has at most M zeros in $[-\pi, +\pi]$ denoted by ω_m and so the solution to (6.24) must be a sum of delta distributions at those zeros; thus, its inverse Fourier transform (the auto-correlation function) must be a sum of complex exponentials at the same zeros

$$S(e^{i\omega}) = \sum_{m=1}^{M} \mathcal{P}_m\, \delta(\omega - \omega_m) \quad \Leftrightarrow \quad R[l] = \sum_{m=1}^{M} \mathcal{P}_m\, e^{il\omega_m}. \tag{6.25}$$

As we shall see in section 6.2, the spectrum and its associated correlation function in (6.25) correspond to a signal that is a sum of M complex exponentials with random and uncorrelated amplitudes A_m and individual powers $\mathcal{P}_m$. Thus, referring to the second statement of theorem 27 we have the following result.

Theorem 28. *The auto-correlation function has positive (non-zero) eigenvalues when the underlying WSS random signal x_n is not a sum of complex exponentials with uncorrelated amplitudes; if it is then there will be a zero eigenvalue and the auto-correlation function is non-negative definite.*

The bounds on the eigenvalues of the auto-correlation matrix stated in theorem 27 can be found as follows. Let $\boldsymbol{v}$ denote a normalized eigenvector of $\boldsymbol{R}_{xx}$ with eigenvalue λ, i.e., $\boldsymbol{R}_{xx}\boldsymbol{v} = \lambda\boldsymbol{v}$ and $\boldsymbol{v}^{+}\boldsymbol{R}_{xx}\boldsymbol{v} = \lambda\left\|\boldsymbol{v}\right\|^2 = \lambda$. Now consider $y_n = \boldsymbol{v}^{+}\boldsymbol{x} = v_0^* x_{n-(M-1)} + \ldots + v_{M-1}^* x_n$ for which we have

$$\mathrm{E}\big[|y_n|^2\big] = R_{yy}[0] = \boldsymbol{v}^{+}\mathrm{E}\big[\boldsymbol{x}\boldsymbol{x}^{+}\big]\boldsymbol{v} = \boldsymbol{v}^{+}\boldsymbol{R}_{xx}\boldsymbol{v} = \lambda. \tag{6.26}$$

In addition, we have

$$\mathrm{E}\big[y_n y_{n-l}^*\big] = R_{yy}[l] = \sum_{m=0}^{M-1}\sum_{k=0}^{M-1} v_m^* v_k \mathrm{E}\big[x_{n+m-(M-1)} x_{n-l+k-(M-1)}^*\big] = \sum_{m=0}^{M-1}\sum_{k=0}^{M-1} v_m^* v_k R[l+m-k],$$

where $R[l+m-k]$ denotes the auto-correlation sequence for $\boldsymbol{x}$. Multiplying with $e^{-il\omega}$ and summing over l we have

$$S_{yy}\left(e^{i\omega}\right) = \sum_{m=0}^{M-1}\sum_{k=0}^{M-1} v_m^* v_k \sum_{l=-\infty}^{+\infty} e^{-il\omega} R[l+m-k] = \left|V\left(e^{i\omega}\right)\right|^2 S\left(e^{i\omega}\right), \tag{6.27}$$

where S denotes the PSD of $\boldsymbol{x}$ and V denotes the spectrum of $\boldsymbol{v}$. In deriving the last equality we rewrote the exponent in the last sum as $l + (m-k) - (m-k)$, changed summation index from l to $q = l + m - k$, and moved the remaining exponential outside that sum and combined with $v_m^* v_k$ to produce the spectrum of $\boldsymbol{v}$. Taking the inverse Fourier transform of (6.27) and setting the auto-correlation lag to zero gives an alternative expression for the left hand side of (6.26)

$$R_{yy}[0] = (2\pi)^{-1}\int_{-\pi}^{+\pi}\left|V\left(e^{i\omega}\right)\right|^2 S_{xx}\left(e^{i\omega}\right)d\omega. \tag{6.28}$$

Clearly the bounds on (6.28) are found by replacing S_{xx} in the integrand by its maximum and minimum and noting that both bounds are then multiplied by $\left\|\boldsymbol{v}\right\|^2 = 1$. Hence, the penultimate statement in theorem 27 is proven. The eigenvalues and eigenvectors of the auto-correlation matrix of samples of a zero-mean WSS random process have numerous applications and we will discuss some in the rest of this chapter; applications to adaptive beamforming appear in chapter 11.

There are two main approaches to the problem of estimating the PSD of a WSS random process x_n from a finite number of observations $x_0, \ldots, x_{N-1}$. The *non-parametric* or classical approach is based on the Wiener-Khinchin theorem 26: the PSD is estimated by using the Fourier transform of an estimate of the auto-covariance function from the finite record length data, with no assumptions on the functional form of the PSD (the same approach can, of course, be used to estimate the CPSD). The *parametric* or model-based approach, on the other hand, makes specific assumptions about the underlying WSS random process, which imply a specific form for the PSD depending on a number of parameters, which, once estimated from the finite record length data, will fully determine the PSD. Classical methods often require relatively longer data records to achieve good frequency resolution, while parametric methods often achieve higher resolutions with shorter data record lengths, sometimes at the expense of certain artifacts. Both classical and model-based spectral estimation methods will be discussed in chapter 8.

6.2 Complex Sinusoids in Zero-Mean White Noise

One of the most important models in signal processing theory is the sum of multiple complex sinusoids and zero-mean WSS random noise. The discrete time signal model of a sum of K complex sinusoids is

$$s_n = \sum_{k=1}^{K} A_k\, e^{in\omega_k}, \quad n \in \mathbb{Z}, \tag{6.29}$$

where ω_k are K distinct frequencies and A_k are uncorrelated zero-mean complex random variables. When there are K real sinusoids we simply use the identities

$$2\cos(n\omega_k) \equiv e^{in\omega_k} + e^{-in\omega_k}, \quad 2i\sin(n\omega_k) \equiv e^{in\omega_k} - e^{-in\omega_k}$$

to write them as a sum of $2K$ complex sinusoids. Let M denote the size of the signal auto-correlation matrix, where $M > K$, and define the individual signal vectors $\boldsymbol{s}_k$,

$$\boldsymbol{s}_k \equiv [1, e^{i\omega_k}, e^{2i\omega_k}, \ldots, e^{(M-1)i\omega_k}]^T, \quad 1 \le k \le K, \tag{6.30}$$

to form the $M \times K$ signal matrix

$$\boldsymbol{S} = \left[\boldsymbol{s}_1, \boldsymbol{s}_2, \ldots, \boldsymbol{s}_K\right]. \tag{6.31}$$

The signal matrix (6.31) has full column rank K since the K frequencies are distinct. We define the total signal vector $\boldsymbol{s}$ [using the time samples in equation (6.29)]:

$$\boldsymbol{s} = \left[s_0, s_1, \ldots, s_{M-1}\right]^T, \tag{6.32}$$

which is a zero-mean random vector on account of its dependence on the random amplitudes. Its $M \times M$ auto-correlation matrix is

$$\boldsymbol{R}_{ss} = \mathrm{E}\left[\boldsymbol{s}\boldsymbol{s}^+\right] = \sum_{k=1}^{K}\sum_{j=1}^{K} \mathrm{E}\left[A_j A_k^*\right] \boldsymbol{s}_j \boldsymbol{s}_k^+, \tag{6.33}$$

which, since the complex amplitudes are uncorrelated, is

$$\boldsymbol{R}_{ss} = \sum_{k=1}^{K} \mathcal{P}_k \boldsymbol{s}_k \boldsymbol{s}_k^+ = \boldsymbol{SPS}^+, \tag{6.34}$$

where $\boldsymbol{P}$ is a diagonal $K \times K$ matrix of the signal powers $\mathcal{P}_k = \mathrm{E}\left[|A_k|^2\right]$ and $\boldsymbol{S}$ is the $M \times K$ signal matrix

$$\boldsymbol{P} = \begin{bmatrix} \mathcal{P}_1 & 0 & \cdots & 0 \\ 0 & \ddots & \ddots & \vdots \\ \vdots & \ddots & \ddots & 0 \\ 0 & \cdots & 0 & \mathcal{P}_K \end{bmatrix}, \quad \boldsymbol{S} = \begin{bmatrix} 1 & \cdots & 1 \\ e^{i\omega_1} & \cdots & e^{i\omega_K} \\ \vdots & \vdots & \vdots \\ e^{i(M-1)\omega_1} & \cdots & e^{i(M-1)\omega_K} \end{bmatrix}. \tag{6.35}$$

The signal matrix $\boldsymbol{S}$ and its Hermitian conjugate $\boldsymbol{S}^+$ have column rank K; the signal power matrix also has full column rank K (it is a $K \times K$ diagonal matrix with non-zero diagonal elements), so the product of all three, i.e., the $M \times M$ signal auto-correlation matrix $\boldsymbol{R}_{ss}$ has rank K. Therefore, only K of all M real eigenvalues of $\boldsymbol{R}_{ss}$, denoted by $\lambda_k^{(s)}$, are non-zero and we arrange them in descending order:

$$\boldsymbol{R}_{ss}\boldsymbol{v}_k = \lambda_k^{(s)}\boldsymbol{v}_k, \quad 1 \le k \le K, \quad \lambda_K^{(s)} \le \ldots \le \lambda_1^{(s)}, \quad \text{and} \quad \boldsymbol{R}_{ss}\boldsymbol{v}_k = \boldsymbol{0}, \quad K+1 \le k \le M. \tag{6.36}$$

The observation noise is assumed to be a zero-mean white sequence ν_n, i.e., if $\boldsymbol{\nu} = [\nu_0, \ldots, \nu_{M-1}]^T$ is the zero-mean $M \times 1$ noise vector, then its $M \times M$ auto-correlation matrix is

$$\boldsymbol{R}_{\nu\nu} = E\left[\boldsymbol{\nu}\boldsymbol{\nu}^+\right] = \sigma_\nu^2 \boldsymbol{I}. \tag{6.37}$$

The observed $M \times 1$ signal plus noise data $\boldsymbol{x}$ is a complex vector given by $\boldsymbol{x} = \boldsymbol{s} + \boldsymbol{\nu}$; it is zero-mean and its $M \times M$ auto-correlation matrix is

$$\boldsymbol{R}_{xx} = E\left[\boldsymbol{x}\boldsymbol{x}^+\right] = \boldsymbol{R}_{ss} + \boldsymbol{R}_{\nu\nu} = \boldsymbol{SPS}^+ + \sigma_\nu^2 \boldsymbol{I}, \tag{6.38}$$

where we use the assumption that the random signal amplitudes and the noise process are independently distributed and therefore uncorrelated [2].

An explicit formula for the inverse of the auto-correlation matrix (6.38) can be found by the ansatz

$$\boldsymbol{R}_{xx}^{-1} \equiv \frac{1}{\sigma_\nu^2}\,(\boldsymbol{SQS}^+ + \boldsymbol{I}), \tag{6.39}$$

for a $K \times K$ matrix $\boldsymbol{Q}$ to be determined by the requirement

$$\boldsymbol{I} = \boldsymbol{R}_{xx}^{-1}\boldsymbol{R}_x = \boldsymbol{I} + \frac{1}{\sigma_\nu^2}\,\boldsymbol{SPS}^+ + \boldsymbol{SQS}^+ + \frac{1}{\sigma_\nu^2}\,\boldsymbol{SQS}^+\boldsymbol{SPS}^+. \tag{6.40}$$

Setting the sum of the last three terms to zero we find

$$\boldsymbol{S}(\sigma_\nu^2\boldsymbol{Q} + \boldsymbol{QS}^+\boldsymbol{SP})\boldsymbol{S}^+ = -\boldsymbol{SPS}^+ \;\Rightarrow\; \boldsymbol{Q} = -\left(\sigma_\nu^2\boldsymbol{P}^{-1} + \boldsymbol{S}^+\boldsymbol{S}\right)^{-1}. \tag{6.41}$$

Thus, the inverse of the auto-correlation matrix (6.39) is

$$\boldsymbol{R}_{xx}^{-1} \equiv \frac{1}{\sigma_\nu^2}\left[\boldsymbol{I} - \boldsymbol{S}\left(\sigma_\nu^2\boldsymbol{P}^{-1} + \boldsymbol{S}^+\boldsymbol{S}\right)^{-1}\boldsymbol{S}^+\right]. \tag{6.42}$$

Important signal processing algorithms such as MUSIC (section 6.3) address the estimation of the following unknown quantities from a finite record length observation vector $\boldsymbol{x}$: the number of distinct frequencies K, the frequencies ω_k, their associated powers $\mathcal{P}_k$, and the white noise variance σ_ν^2.

6.3 The MUSIC Algorithm

The MUSIC (MUltiple SIgnal Classification) algorithm [69] uses the structure of the eigenvectors of the auto-correlation matrix of $\boldsymbol{x}$ (as described in section 6.2) to provide high-resolution estimates of the unknown frequencies and reliable values for the other unknowns when the SNR is high.

The number of frequencies K can be determined by an examination of the eigenvalues $\lambda_k^{(s)}$: the eigenvalues monotonically decrease and reach a flat level at the white noise variance σ_ν^2. For example, we consider a model consisting of three complex sinusoids with normalized frequencies $[\omega_1, \omega_2, \omega_3] = 2\pi[0.1, 0.15, 0.2]$ and amplitudes $[A_1, A_2, A_3] = [4, 5, 6]$,

$$x_n = 4e^{0.2i\pi n} + 5e^{0.3i\pi n} + 6e^{0.4i\pi n} + \nu_n, \;\; \nu_n \sim \mathcal{N}(0, \sigma_\nu^2). \tag{6.43}$$

[2] In practice, the observed signal plus noise auto-correlation values have to be estimated from an observed realization, when the dimension of the observed data N should be $\gg M$. Our discussion here relates to the structure of the exact auto-correlation matrix and so does not involve a finite data record length N.

The first column of the 7×7 signal auto-correlation matrix (6.34) is the signal auto-correlation vector for lags $0, \ldots, 6$, given by 77, $38.7635+63.868i$, $-31.9058+60.1536i$, $-57.8453+1.78205i$, $-22.0451-39.5281i$, $20-25i$, $18.4058+10.1388i$. Table 6.1 shows the first 7 eigenvalues of the auto-correlation matrix (6.38) for six values of σ_ν^2; as expected, the fourth through seventh eigenvalues (the last column) are equal to the noise variance in each case.

σ_ν^2	$\lambda_1^{(s)}$	$\lambda_2^{(s)}$	$\lambda_3^{(s)}$	$\lambda_4^{(s)}$-$\lambda_7^{(s)}$
0	434.4	101.1	3.5	0
4	438.4	105.1	7.5	4
9	443.4	110.1	12.5	9
16	450.4	117.1	19.5	16
25	459.4	126.1	28.5	25
36	470.4	137.1	39.5	36

Table 6.1: The eigenvalues of three complex sinusoids in white noise.

The eigenvectors $\boldsymbol{v}_1, \ldots, \boldsymbol{v}_K$ corresponding to the largest K eigenvalues span a subspace $\mathbb{V}^{(s)}$ known as the signal subspace, while the eigenvectors $\boldsymbol{v}_{K+1}, \ldots, \boldsymbol{v}_M$ span the noise subspace $\mathbb{V}^{(n)}$. The signal and noise subspaces are, of course, orthogonal to each other, i.e., $\boldsymbol{v}_k{}^+\boldsymbol{v}_m = 0$ for $1 \le k \le K$ and $K+1 \le m \le M$; this follows from the fact that eigenvectors corresponding to distinct eigenvalues of a Hermitian matrix are necessarily orthogonal. Using the eigenvalue equation $\boldsymbol{R}_{xx}\boldsymbol{v}_m = \sigma_\nu^2\boldsymbol{v}_m$, $K+1 \le m \le M$, we have

$$\left(\boldsymbol{R}_{xx} - \sigma_\nu^2\boldsymbol{I}\right)\boldsymbol{v}_m = \boldsymbol{R}_{ss}\boldsymbol{v}_m = \boldsymbol{S}\boldsymbol{P}\boldsymbol{S}^+\boldsymbol{v}_m = \boldsymbol{0},\ K+1 \le m \le M. \tag{6.44}$$

The matrix $\boldsymbol{S}$ has full column rank K and an equation of the form $\boldsymbol{S}\boldsymbol{v} = \boldsymbol{0}$ implies that $\boldsymbol{v} = \boldsymbol{0}$, since $\boldsymbol{S}\boldsymbol{v}$ is a linear combination of the K linearly independent columns of $\boldsymbol{S}$. Thus, we must have

$$\boldsymbol{P}\boldsymbol{S}^+\boldsymbol{v}_m = \boldsymbol{0},\ K+1 \le m \le M. \tag{6.45}$$

Since the $K \times K$ power matrix $\boldsymbol{P}$ is positive definite and diagonal, we may multiply both sides by its inverse to obtain

$$\boldsymbol{S}^+\boldsymbol{v}_m = \boldsymbol{0},\ K+1 \le m \le M. \tag{6.46}$$

The Hermitian conjugate of this equation shows that the columns of $\boldsymbol{S}$ are orthogonal to the noise eigenvectors, i.e.,

$$\boldsymbol{v}_m^+\boldsymbol{s}_k = 0,\ \ K+1 \le m \le M, \text{ and } 1 \le k \le K. \tag{6.47}$$

Therefore, all K linearly independent signal vectors $\boldsymbol{s}_1, \ldots, \boldsymbol{s}_K$ are in the signal subspace $\mathbb{V}^{(s)}$ and span it; thus, the noise subspace must be spanned by the noise eigenvectors:

$$\mathbb{V}^{(n)} = \textbf{span}\left\{\boldsymbol{v}_{K+1}, \ldots, \boldsymbol{v}_M\right\}, \quad \text{and} \quad \mathbb{V}^{(s)} = \textbf{span}\left\{\boldsymbol{s}_1, \ldots, \boldsymbol{s}_K\right\}. \tag{6.48}$$

The orthogonality of signal and noise subspaces allow for the construction of a *pseudo-spectrum* that will peak at every one of the K distinct frequencies; the conventional formula is:

$$\hat{S}_{\text{MUSIC}}\left(e^{i\omega}\right) = \left(\sum_{m=K+1}^{M} \left|\boldsymbol{e}_\omega^+ \cdot \boldsymbol{v}_m\right|^2\right)^{-1}, \quad \boldsymbol{e}_\omega \equiv \left[1, e^{i\omega}, e^{2i\omega}, \ldots, e^{(M-1)i\omega}\right]^T. \tag{6.49}$$

As $\omega \to \omega_k$, $\boldsymbol{e}_\omega \to \boldsymbol{s}_k$, $1 \le k \le K$, the sum in parentheses tends to 0, and the pseudo-spectrum peaks, although the peak values are not related to the actual power associated with each frequency; hence, (6.49) is not a true spectrum. Figure 6.1 shows the MUSIC pseudo-spectrum for the model with three sinusoids (see table 6.1) with $\sigma_\nu^2 = 9$. Since the MUSIC algorithm uses the noise eigenvectors to construct a pseudo-spectrum, it is known as a "noise subspace method."

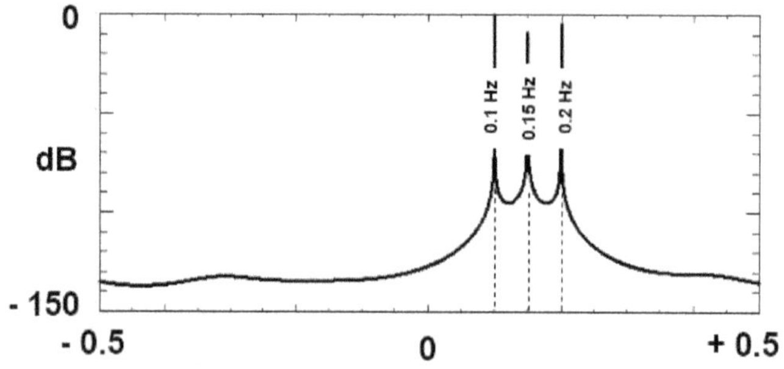

Figure 6.1: MUSIC pseudo-spectrum for three complex sinusoids in white noise.

To find the power $\mathcal{P}_k$ associated with each frequency and the white noise variance, we solve the $K+1$ linear equations

$$R_{xx}[l] = \mathrm{E}[x_n x_{n-l}^*] = \sum_{k=1}^{K} \mathcal{P}_k e^{il\omega_k} + \sigma_\nu^2 \delta_{0l}, \quad 0 \le l \le K, \tag{6.50}$$

which are equivalent to the $(K+1) \times (K+1)$ matrix equation

$$\begin{bmatrix} 1 & 1 & \cdots & 1 \\ 0 & e^{i\omega_1} & \cdots & e^{i\omega_K} \\ \vdots & \vdots & \vdots & \vdots \\ 0 & e^{iK\omega_1} & \cdots & e^{iK\omega_K} \end{bmatrix} \begin{bmatrix} \sigma_\nu^2 \\ \mathcal{P}_1 \\ \vdots \\ \mathcal{P}_K \end{bmatrix} = \begin{bmatrix} R_{xx}[0] \\ R_{xx}[1] \\ \vdots \\ R_{xx}[K] \end{bmatrix}. \tag{6.51}$$

For the example of three complex sinusoids (equation (6.43) with $\sigma_\nu^2 = 9$) (6.51) becomes

$$\begin{bmatrix} 1 & 1 & 1 & 1 \\ 0 & 0.809 + 0.588i & 0.588 + 0.809i & 0.309 + 0.951i \\ 0 & 0.309 + 0.951i & -0.309 + 0.951i & -0.809 + 0.588i \\ 0 & -0.309 + 0.951i & -0.951 + 0.309i & -0.809 - 0.588i \end{bmatrix} \begin{bmatrix} \sigma_\nu^2 \\ \mathcal{P}_1 \\ \mathcal{P}_2 \\ \mathcal{P}_3 \end{bmatrix} = \begin{bmatrix} 86 \\ 38.764 + 63.868i \\ -31.9058 + 60.154i \\ -57.8453 + 1.78205i \end{bmatrix},$$

whose solution vector gives $\sigma_\nu^2 = 9$, $\mathcal{P}_1 = 16$, $\mathcal{P}_2 = 25$, $\mathcal{P}_3 = 36$.

A closely related algorithm to estimate the frequencies is Root MUSIC; we will discuss it in section 11.9 and in the context of estimating direction of arrival of propagating plane waves using a uniform linear array.

6.4 Pisarenko Harmonic Decomposition (PHD)

This algorithm [70] relies on prior knowledge of the number of complex sinusoids in white noise. If K is known, we set $M = K+1$ and note that the noise subspace is spanned by only one vector $\boldsymbol{v}$, while the signal subspace is spanned by the vectors $\boldsymbol{v}_1, \ldots, \boldsymbol{v}_K$, or by the signal vectors $\boldsymbol{s}_1, \ldots, \boldsymbol{s}_K$, whose orthogonality relations with the noise basis vector $\boldsymbol{v}$ are

$$\boldsymbol{s}_k^+ \boldsymbol{v} = 0, \quad 1 \le k \le K. \tag{6.52}$$

Equation (6.52) for each k represents a polynomial equation in $1/z_k = e^{-i\omega_k}$

$$v_1 + v_2 z_k^{-1} + v_3 z_k^{-2} + \ldots + v_K z_K^{-K} = 0, \;\; \boldsymbol{v} = [v_1, \ldots, v_K]^T. \tag{6.53}$$

The K solutions that lie on the unit circle correspond to the frequencies of interest ω_k. The powers associated with the frequencies and the noise variance are found from (6.51).

As an example, consider the observation of three complex sinusoids in white noise:

$$x_n = \sum_{k=1}^{3} A_k e^{in\omega_k} + \nu_n, \;\; n \in \mathbb{Z}, \;\; \nu_n \sim \mathcal{N}(0, 0.4), \tag{6.54}$$

with normalized frequencies $0.4, 0.3, 0.2$ Hz, and powers $|A_1|^2 = 3, |A_2|^2 = 2, |A_3|^2 = 1$, respectively. The estimated auto-correlation values are $R_{xx}[0] = 6.4$, $R_{xx}[1] = -2.7361 + 4.6165\, i$, $R_{xx}[2] = -1.5 - 3.441\, i$, and $R_{xx}[3] = 1.7361 + 1.0898\, i$, and we use these to build the Hermitian and Toeplitz 4×4 auto-correlation matrix whose eigenvalues are 19.3124, 5.2817, 0.6058, and 0.4. The smallest eigenvalue yields the noise variance $\sigma_\nu^2 = 0.4$; the noise eigenvector $\boldsymbol{v} = [-0.2041 + 0.1483\, i, 0.2041 + 0.6282\, i, 0.6606, 0.078 - 0.24\, i]^T$ gives the polynomial

$$p(1/z) = (-0.2041 + 0.1483\, i) + (0.2041 + 0.6282\, i)z^{-1} + (0.6606)z^{-2} + (0.078 - 0.24\, i)z^{-3}.$$

The roots of this polynomial are $z_1 = e^{0.8i\pi}$, $z_2 = e^{0.6i\pi}$, $z_3 = e^{0.4i\pi}$, which correspond to the normalized frequencies $0.4, 0.3, 0.2$ Hz. Since the noise variance has been estimated (the smallest eigenvalue), we solve for the signal powers using the 3×3 sub-matrix of equation (6.51) (excluding the first row and the last column) and find $3, 2, 1$.

6.5 The ESPRIT Algorithm

The *Estimation of SIgnal Parameters by Rotational Invariance Techniques* (ESPRIT) [71] uses the same signal model of complex sinusoids in white noise described in section 6.2 and is based on the observation that for a discrete time complex signal of the form $s_n = \exp(i\omega n)$, we have $s_{n+1} = \exp(i\omega)s_n$, i.e., the signal at time step $n+1$ is obtained from the signal at the previous step n by multiplication with the complex exponential $\exp(i\omega)$, which corresponds to an anti-clockwise rotation in the complex z-plane by an angle ω.

Let us define the observed data vector and a time shifted version

$$\boldsymbol{x} \equiv [x_n, \ldots, x_{n+N-1}]^T, \;\; \boldsymbol{y} \equiv [x_{n+1}, \ldots, x_{n+N}]^T,$$

corresponding to the signals $\boldsymbol{s}_k$ and $\exp(i\omega_k)\boldsymbol{s}_k$, respectively. We now have the following correlation matrices:

$$\boldsymbol{R}_{xx} = \boldsymbol{SPS}^+ + \sigma_\nu^2 \boldsymbol{I}, \quad \boldsymbol{R}_{xy} = \boldsymbol{SPT}^+\boldsymbol{S}^+ + \boldsymbol{R}_\nu, \tag{6.55}$$

where $\boldsymbol{R}_\nu = \mathrm{E}[\boldsymbol{\nu}_n \boldsymbol{\nu}_{n+1}^+]$ is an $N \times N$ matrix whose only non-zero elements are all equal to σ_ν^2 on the line parallel to and just below the main diagonal, and $\boldsymbol{T}$ is a diagonal $K \times K$ matrix whose diagonal elements are

$$[\boldsymbol{T}]_{kk} = e^{i\omega_k}, \;\; 1 \le k \le K. \tag{6.56}$$

The method involves the generalized eigenvalue problem for a *matrix pencil* $\boldsymbol{A} - \lambda \boldsymbol{B}$ defined for a pair of square $N \times N$ matrices $\boldsymbol{A}$ and $\boldsymbol{B}$. The pencil eigenvalues λ and eigenvectors $\boldsymbol{v}$ satisfy the following equations [3]

$$Av = \lambda Bv, \text{ and } \left|A - \lambda B\right| = 0. \tag{6.57}$$

Using (6.55) consider the matrix pencil $A - \lambda B$ for

$$A \equiv R_{xx} - \sigma_\nu^2 I = SPS^+, \;\; B \equiv R_{xy} - R_\nu = SPT^+S^+, \tag{6.58}$$

whose generalized eigenvector v satisfies the equation

$$SPS^+v = \lambda SPT^+S^+v \;\; \Rightarrow \;\; SP(I - \lambda T^+)S^+v = 0, \tag{6.59}$$

with the generalized eigenvalues found from the determinantal equation

$$\left|SP(I - \lambda T^+)S^+\right| = 0. \tag{6.60}$$

Zero is one of the generalized eigenvalues, since the matrix SPS^+ is $N \times N$ but has column rank $K < N$, so $|SPS^+| = 0$. The generalized eigenvectors corresponding to the eigenvalue 0 satisfy the equation $SPS^+v = 0$ which, as we have seen before, is equivalent to $S^+v = 0$; thus, eigenvectors corresponding to eigenvalue 0 are in the noise subspace of the eigenspace of R_{xx}. To find the non-zero generalized eigenvalues, we note that the eigenvector equation (6.59) when multiplied on the left by $T(S^+SP)^{-1}S^+$ gives

$$TS^+v = \lambda S^+v, \tag{6.61}$$

where we used the relation $TT^+ \equiv I$. Thus, the generalized non-zero eigenvalues are simply the elements of the diagonal $K \times K$ matrix T and the eigenvectors are of the form $u = S^+v$, i.e.,

$$\lambda_k = e^{i\omega_k}, \;\; 1 \le k \le K, \;\; Tu_k = \lambda_k u_k. \tag{6.62}$$

Since T is diagonal, the eigenvector u_k is the kth column of the $K \times K$ identity matrix. But $u_k = S^+v_k$ where v_k is in the signal subspace and is therefore expressible as a linear combination of the columns of the $N \times K$ signal matrix S, i.e., $v_k = Sc_k$ for a vector set of coefficients c_k. Thus,

$$S^+v_k = S^+Sc_k, \tag{6.63}$$

whose solution is

$$c_k = \left(S^+S\right)^{-1} S^+v_k. \tag{6.64}$$

Using this in the equation $v_k = Sc_k$ we finally obtain

$$v_k = S\left(S^+S\right)^{-1} S^+v_k = S\left(S^+S\right)^{-1} u_k. \tag{6.65}$$

Since the vectors u_k are columns of the $K \times K$ identity matrix, the right hand side of the above equation represents the K columns of the matrix $S(S^+S)^{-1}$. Thus, the generalized eigenvectors v_k corresponding to the non-zero eigenvalues λ_k are linear combinations of the columns of the signal matrix S and therefore belong to the signal subspace; hence ESPRIT is known as a "signal subspace method."

As an example, consider the three complex sinusoids with normalized frequencies 0.2, 0.3, and 0.4 Hz in white noise with variance 0.4 described in section 6.4. The auto-correlation values are $R_{xx}[0] = 6.4$, $R_{xx}[1] = -2.7361 + 4.6165\,i$, $R_{xx}[2] = -1.5 - 3.441\,i$, and $R_{xx}[3] = 1.7361 + 1.0898\,i$, in addition to

[3] When B is non-singular, the eigenvalues are simply the eigenvalues of the matrix $B^{-1}A$. If, on the other hand, B is ill-conditioned, other methods must be used to find the pencil eigenvalues.

$R_{xx}[4] = -1.5 + 0.8123\, i$. The noise variance is found as the smallest eigenvalue of the auto-correlation matrix, $\sigma^2 = 0.4$, as expected. The matrix $\boldsymbol{R}_{xy}$ is

$$\boldsymbol{R}_{xy} = \begin{bmatrix} R_{-1} & R_{-2} & R_{-3} & R_{-4} \\ R_0 & R_{-1} & R_{-2} & R_{-3} \\ R_1 & R_0 & R_{-1} & R_{-2} \\ R_2 & R_1 & R_0 & R_{-1} \end{bmatrix}, \quad R_{k} = \boldsymbol{R}_{xx}[k], \;\; k = 0, 1, 2, 3,$$

and $R_{-k} = R_k^*$. Thus, we have

$$\boldsymbol{A} = \boldsymbol{R}_{xx} - \sigma^2 \boldsymbol{I} = \begin{bmatrix} R_0 - 0.4 & R_{-1} & R_{-2} & R_{-3} \\ R_1 & R_0 - 0.4 & R_{-1} & R_{-2} \\ R_2 & R_1 & R_0 - 0.4 & R_{-1} \\ R_3 & R_2 & R_{-1} & R_0 - 0.4 \end{bmatrix},$$

and

$$\boldsymbol{B} = \boldsymbol{R}_{xy} - \boldsymbol{R}_{\nu} = \begin{bmatrix} R_{-1} & R_{-2} & R_{-3} & R_{-4} \\ R_0 - 0.4 & R_{-1} & R_{-2} & R_{-3} \\ R_1 & R_0 - 0.4 & R_{-1} & R_{-2} \\ R_2 & R_1 & R_0 - 0.4 & R_{-1} \end{bmatrix}.$$

The generalized non-zero eigenvalues are found from the equation $(\boldsymbol{A} - \lambda \boldsymbol{B})\boldsymbol{v} = \boldsymbol{0}$ to be $e^{0.4 i\pi}$, $e^{0.6 i\pi}$, and $e^{0.8 i\pi}$, as expected from (6.62), corresponding to (normalized) frequencies 0.2, 0.3, and 0.4 Hz, respectively.

6.6 The Auto-Correlation Matrix for Time Reversed Signal Vectors

The definition of the auto-correlation sequence for a zero-mean WSS random process given in equation (6.14) is the one used to define the auto-correlation matrix in theorem 27. Given the forward time indexed vector $\boldsymbol{x} = \left[x_{n-(M-1)}, \ldots, x_n\right]^T$, we have the $M \times M$ Toeplitz auto-correlation matrix

$$\left[\boldsymbol{R}_{xx}\right]_{jk} = R[j-k] = \mathrm{E}\left[x_n x^*_{n-j+k}\right] = \mathrm{E}\left[x_{n-k} x^*_{n-j}\right] = \mathrm{E}\left[x_{n+j} x^*_{n+k}\right], \tag{6.66}$$

and

$$\boldsymbol{R}_{xx} = \mathrm{E}\left[\boldsymbol{x}\boldsymbol{x}^+\right] = \begin{bmatrix} R_0 & R_{-1} & \cdots \\ R_{+1} & R_0 & \\ \vdots & & \ddots \end{bmatrix}, \quad R_{-l} = R^*_{+l}, \quad \boldsymbol{x} \equiv \left[x_{n-(M-1)}, \ldots, x_n\right]^T. \tag{6.67}$$

This expression depends on the definition of the $M \times 1$ random process vector $\boldsymbol{x}$ whose elements are indexed forward in time. Another possible definition (often used in linear prediction or other filtered WSS random process applications) is to use the backward time indexed (time reversed) vector $\boldsymbol{x}_n$, which is related to the forward time indexed vector by $\boldsymbol{x}_n = \boldsymbol{J}\boldsymbol{x}$ where $\boldsymbol{J}$ is the $M \times M$ real and symmetric counter identity matrix

$$\boldsymbol{J} = \begin{bmatrix} 0 & \cdots & 0 & 1 \\ \vdots & .\cdot\,^{\cdot} & 1 & 0 \\ 0 & .\cdot\,^{\cdot} & .\cdot\,^{\cdot} & \vdots \\ 1 & 0 & \cdots & 0 \end{bmatrix}, \quad \boldsymbol{J}^2 \equiv \boldsymbol{I}. \tag{6.68}$$

Clearly $\boldsymbol{x} = \boldsymbol{J}\boldsymbol{x}_n$ and since $\boldsymbol{J}$ is symmetric we also have

$$\boldsymbol{x}^T \boldsymbol{J} = \left[x_n, \ldots, x_{n-(M-1)}\right] \equiv \boldsymbol{x}_n^T. \tag{6.69}$$

Using the time reversed vector $\boldsymbol{x}_n$ we define the $M \times M$ Hermitian and Toeplitz matrix

$$\tilde{\boldsymbol{R}}_{xx} = \mathrm{E}\left[\boldsymbol{x}_n \boldsymbol{x}_n^+\right] = \begin{bmatrix} R_0 & R_{+1} & \cdots \\ R_{-1} & R_0 & \\ \vdots & & \ddots \end{bmatrix}, \quad \tilde{R}_{-l} = \tilde{R}_{+l}^*, \quad \boldsymbol{x}_n \equiv \left[x_n, \ldots, x_{n-(M-1)}\right]^T = \boldsymbol{J}\boldsymbol{x}. \tag{6.70}$$

Now we have the following relations between the two definitions (6.67) and (6.70) (dropping the "xx" subscript for now):

$$\boldsymbol{R} = \boldsymbol{J}\,\tilde{\boldsymbol{R}}\,\boldsymbol{J} = \tilde{\boldsymbol{R}}^* = \tilde{\boldsymbol{R}}^T, \quad \tilde{\boldsymbol{R}} = \boldsymbol{J}\,\boldsymbol{R}\,\boldsymbol{J} = \boldsymbol{R}^* = \boldsymbol{R}^T. \tag{6.71}$$

In addition, we have

$$\boldsymbol{J}\boldsymbol{R} = \boldsymbol{R}^*\boldsymbol{J}, \;\; \boldsymbol{J}\boldsymbol{R}\boldsymbol{J} = \boldsymbol{R}^*, \;\; \left[\boldsymbol{R}^*\right]^{-1}\boldsymbol{J} = \boldsymbol{J}\left[\boldsymbol{R}\right]^{-1}, \;\; \left[\boldsymbol{R}\right]^{-1} = \boldsymbol{J}\left[\boldsymbol{R}^*\right]^{-1}\boldsymbol{J}, \tag{6.72}$$

and

$$\mathrm{E}\left[\boldsymbol{x}_n \boldsymbol{x}_n^+\right] = \boldsymbol{R}^*, \text{ and } \mathrm{E}\left[\boldsymbol{x}_n^* \boldsymbol{x}_n^T\right] = \boldsymbol{R}. \tag{6.73}$$

If x_n is real, then its auto-correlation matrix is real and symmetric, and both definitions coincide.

Throughout this text we use (6.67) to define the auto-correlation matrix $\boldsymbol{R}$; when using the backward time indexed vector $\boldsymbol{x}_n$ we use (6.73) and write all expressions in terms of $\boldsymbol{R}$ (instead of $\tilde{\boldsymbol{R}}$). For instance, calculating the variance of the filtered output $y_n = \boldsymbol{h}^T \boldsymbol{x}_n$ we find

$$\mathrm{E}\left[\left|y_n\right|^2\right] = \boldsymbol{h}^+\,\mathrm{E}\left[\boldsymbol{x}_n^* \boldsymbol{x}_n^T\right]\,\boldsymbol{h} = \boldsymbol{h}^+ \boldsymbol{R}_{xx} \boldsymbol{h},$$

where we used the identity $\boldsymbol{h}^T \boldsymbol{x}_n \equiv \boldsymbol{x}_n^T \boldsymbol{h}$.

CHAPTER 7

Linear Systems and Stochastic Inputs

7.1 Filtered Random Processes

When the input to a linear time-invariant system, defined by a system function h_t, is a WSS random process, then the output is also a WSS random process (see figure 3.1). We will derive relationships between correlation functions and associated power spectra of the WSS random input and output. Taking expectation values, we find the mean and auto-correlation function of the output, and the cross-correlation function between the input and the output:

$$\mu_y \equiv \mathrm{E}[y_t] = \int_{-\infty}^{\infty} h_\tau \mathrm{E}\left[x_{t-\tau}\right] d\tau = \mu_x \int_{-\infty}^{\infty} h_\tau \, d\tau, \tag{7.1a}$$

$$\begin{aligned} R_{yy}(\tau) = \mathrm{E}[y_t y_{t-\tau}^*] &= \mathrm{E}\left[\int_{-\infty}^{\infty} h_u x_{t-u} du \int_{-\infty}^{\infty} h_v^* x_{t-\tau-v}^* \, dv\right] \\ &= \int_{-\infty}^{\infty}\int_{-\infty}^{\infty} h_u h_v^* \mathrm{E}\left[x_{t-u} x_{t-\tau-v}^*\right] dudv \\ &= \int_{-\infty}^{\infty}\int_{-\infty}^{\infty} h_u h_v^* R_{xx}(\tau + v - u) \, dudv, \end{aligned} \tag{7.1b}$$

$$R_{xy}(\tau) = R_{yx}^*(-\tau) = \mathrm{E}\left[x_t y_{t-\tau}^*\right] = \int_{-\infty}^{\infty} h_v^* R_{xx}(\tau + v) \, dv. \tag{7.1c}$$

Taking Laplace transforms of (7.1a) – (7.1c), we find

$$S_{xy}(s) = H^*(-s^*) S_{xx}(s), \; S_{yx}(s) = H(s) S_{xx}(s), \; S_{yy}(s) = H^*(-s^*) H(s) S_{xx}(s), \tag{7.2}$$

while the corresponding Fourier transforms are found by setting $s = i\omega$, $\omega \in \mathbb{R}$. Analogous results for discrete time WSS random signals are

$$\mu_y = \mathrm{E}[y_n] = \mu_x \sum_{k=-\infty}^{\infty} h_k, \tag{7.3a}$$

$$R_{xy}[k] = R_{yx}^*[-k] = \mathrm{E}\big[x_n y_{n-k}^*\big] = \sum_{m=-\infty}^{\infty} h_m^* R_{xx}[k+m], \tag{7.3b}$$

$$R_{yx}[k] = \mathrm{E}\big[y_n x_{n-k}^*\big] = \sum_{m=-\infty}^{\infty} h_m R_{xx}[k-m], \tag{7.3c}$$

$$R_{yy}[k] = \mathrm{E}\big[y_n y_{n-k}^*\big] = \sum_{l=-\infty}^{\infty}\sum_{m=-\infty}^{\infty} h_l h_m^* R_{xx}[k+m-l], \tag{7.3d}$$

whose **Z** transforms give

$$S_{xy}(z) = H^*(1/z^*)S_{xx}(z),\ S_{yx}(z) = H(z)S_{xx}(z),\ S_{yy}(z) = H(z)H^*(1/z^*)S_{xx}(z), \tag{7.4}$$

and the discrete time Fourier transforms are found by setting $z = e^{i\omega}$, $\omega \in [-\pi, \pi]$.

As a simple example of the above relations let us calculate the average noise power for an RC circuit. The thermal noise in a resistor R can be modeled as a WGN voltage source whose PSD is $S_{\nu\nu}(\omega) = k_{_B} RT/\pi$ where $k_{_B}$ is Boltzmann's constant and T is temperature (in degrees Kelvin), for all frequencies ω. Figure 7.1 shows the RC circuit where the WGN voltage is in series with the resistor. The circuit transfer function is $H(\omega) = 1/(1 + i\omega RC)$ and the output PSD, in terms of input PSD, is (last equation in (7.2) with $s = i\omega$)

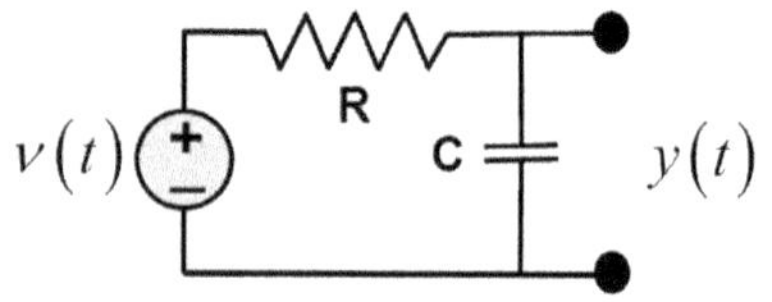

Figure 7.1: An RC circuit.

$$S_{yy}(\omega) = \big|H(\omega)\big|^2\, S_{\nu\nu}(\omega) = \frac{k_{_B}\, TR}{\pi\big(1+\omega^2R^2C^2\big)}.$$

The average power is

$$\mathrm{E}\big[|y|^2\big] = R_{yy}(0) = \int_{-\infty}^{+\infty} S_{yy}(\omega)\, d\omega = \frac{k_{_B} T}{C},$$

which is independent of the resistance R.

In sections 7.2 and 7.3 we show two important examples of filtered random processes and determination of optimal filters based on maximizing an output SNR appropriately defined for non-random known signals and for random signals, respectively.

7.2 Detection of a Known Non-Random Signal in WSS Noise

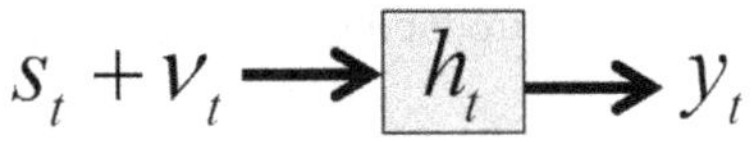

Figure 7.2: The matched filter.

In section 4.4 we learned to design a least squares filter to match a given input data to a desired signal (see figure 4.1). In this section we consider an input data consisting of a known non-random signal s_t plus an additive zero-mean WSS random noise ν_t whose auto-correlation function is known. We use the information contained in the auto-correlation function of the noise to design an optimal filter h_t, as shown in figure 7.2, to maximize the ratio of the filtered signal amplitude at some time T to the standard deviation of the filtered noise; this filter is known as the *matched filter*.

The filtered signal and noise are given by convolution integrals

$$\tilde{s}_t = \int_{-\infty}^{\infty} h_\tau s_{t-\tau} d\tau, \quad \tilde{\nu}_t = \int_{-\infty}^{\infty} h_\tau \nu_{t-\tau} d\tau, \tag{7.5}$$

while the noise power $P_{\tilde{\nu}}$ is defined by

$$P_{\tilde{\nu}} \equiv \mathrm{E}[\tilde{\nu}_t \tilde{\nu}_t^*] = \int_{-\infty}^{\infty}\int_{-\infty}^{\infty} h_\sigma^* R_{\nu\nu}(\sigma - \tau) h_\tau d\sigma d\tau. \tag{7.6}$$

The statistical basis for the optimality criterion is found in the following test of two hypotheses when the observation z_t is used to determine the existence of a known non-random signal s_t in the presence of random noise ν_t:

- $\mathbf{H}_0$: the known signal is absent and $y_t = \nu_t$.
- $\mathbf{H}_1$: the known signal is present and $y_t = s_t + \nu_t$.

We define the deflection signal to noise $\mathbf{SNR}_t$ relative to the two hypotheses by

$$\mathbf{SNR}_t = \left| E\left[y_t \,|\mathbf{H}_1\right] - E\left[y_t \,|\mathbf{H}_0\right] \right|^2 \Big/ \mathrm{E}\left[y_t y_t^* \,|\mathbf{H}_0\right]. \tag{7.7}$$

For instance, if the signal s_t is constant and the noise is zero-mean and white with variance σ_ν^2, then we have $\mathbf{SNR}_t = s^2/\sigma_\nu^2$ which is a constant independent of t.

In the more general case of a known non-constant and non-random signal immersed in non-white noise with known auto-correlation function, we seek a linear filter to maximize $\mathbf{SNR}_T$ for some arbitrary time T at the output of the filter in figure 7.2:

$$\mathbf{SNR}_T = \frac{\left| E\left[y_T \,|\mathbf{H}_1\right] - E\left[y_T \,|\mathbf{H}_0\right] \right|^2}{\mathrm{E}\left[y_T y_T^* \,|\mathbf{H}_0\right]} = \frac{\left| \int_{-\infty}^{\infty} h_\tau s_{T-\tau} d\tau \right|^2}{\int_{-\infty}^{\infty}\int_{-\infty}^{\infty} h_\sigma^* R_{\nu\nu}(\sigma - \beta) h_\beta d\sigma d\beta}. \tag{7.8}$$

Thus, the problem of maximizing $\mathbf{SNR}_T$ is equivalent to the problem of minimizing the denominator, i.e., the filtered noise power $P_{\tilde{\nu}}$, subject to the constraint of keeping the numerator fixed. The numerator constraint can be further simplified by keeping the filtered signal value at time T (or equivalently its complex conjugate) fixed. Introducing a Lagrange multiplier μ, the optimal filter minimizes the quantity

$$\int_{-\infty}^{\infty}\int_{-\infty}^{\infty} h_{\tau}^{*} R_{\nu\nu}(\tau-\sigma)\, h_{\sigma}\, d\tau d\sigma + \mu \int_{-\infty}^{\infty} h_{\tau}^{*} s_{T-\tau}^{*}\, d\tau. \tag{7.9}$$

Differentiating with respect to h_{κ}^{*} and setting the result equal to zero we find

$$\int_{-\infty}^{\infty} R_{\nu\nu}(\kappa-\sigma)\, h_{\sigma}\, d\sigma = -\mu s_{T-\kappa}^{*}, \tag{7.10}$$

whose solution is the matched filter. The left hand side of (7.10) is a convolution and so taking Fourier transforms yields

$$H_{\mathrm{mf}}(\omega) = -\mu\, e^{-i\omega T}\, S^{*}(\omega)\,/S_{\nu\nu}(\omega), \tag{7.11}$$

where $S(\omega)$ is the Fourier transform of the signal s_t and $S_{\nu\nu}(\omega)$ is the PSD of the noise ν_t. The overall gain factor $-\mu$ does not affect the SNR and can be discarded. If ν is white, the matched filter is simply the time-reversed version of the signal and filtering with the matched filter is equivalent to correlating with the signal.

The derivation of the discrete FIR matched filter $\boldsymbol{h}_{\mathrm{mf}}$ with M elements is similar except that the time T is now replaced by an index K and the quantity to be minimized is

$$\boldsymbol{h}^{+}\boldsymbol{R}_{\nu\nu}\boldsymbol{h} + \mu\boldsymbol{h}^{+}\boldsymbol{s}_{K}^{*}, \tag{7.12}$$

where $\boldsymbol{s}_{K} = [s_K, s_{K-1}, \ldots, s_{K-M+1}]^{T}$. The matched filter is now the solution to the matrix equation

$$\boldsymbol{R}_{\nu\nu}\boldsymbol{h} = -\mu\boldsymbol{s}_{K}^{*}, \tag{7.13}$$

where the noise correlation matrix $\boldsymbol{R}_{\nu\nu}$ is Hermitian and Toeplitz. The matched filter solution is

$$\boldsymbol{h}_{\mathrm{mf}} = -\mu\, \boldsymbol{R}_{\nu\nu}^{-1}\, \boldsymbol{s}_{K}^{*}, \tag{7.14}$$

and the overall gain factor can be dropped as before. As an example consider the data shown in figure 7.3: the left panel shows a 21-point known and non-random signal $\boldsymbol{s}$ whose duration is 0.05 seconds; the right panel shows a 2-second segment of data consisting of a 20-second realization of a zero-mean WSS random process with the signal symmetrically placed at its center at time 0, with SNR $= 0$ dB (the SNR is here defined as the square of the maximum signal magnitude divided by the variance of the random process). The process auto-correlation sequence is known and its 21 values for lags $0, \ldots, 20$ are $[12.2253, 9.25460, 2.60413, -3.56194, -6.68476, -6.64582, -4.51285, -1.41886, 1.62293, 3.65777, 4.06029, 2.89068, 0.866031, -1.07231, -2.22858, -2.36529, -1.65332, -0.504690, 0.606139, 1.29965, 1.40434]$.

Figure 7.4 shows filtered outputs: the matched filter output is the solid line and the SNR at signal center is 16 dB, while the dotted line shows the output of a filter that assumes the random process is white (i.e., the filter is the time-reversed version of the signal) and the SNR at signal center is now 10 dB. Thus, the knowledge of the correlation values for the random noise produces a 6 dB increase in SNR for the correct matched filter.

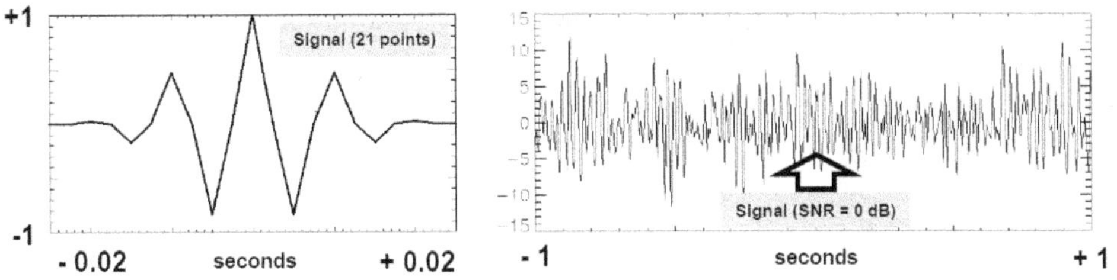

Figure 7.3: Signal (left) and noise plus signal (right) with SNR = 0 dB.

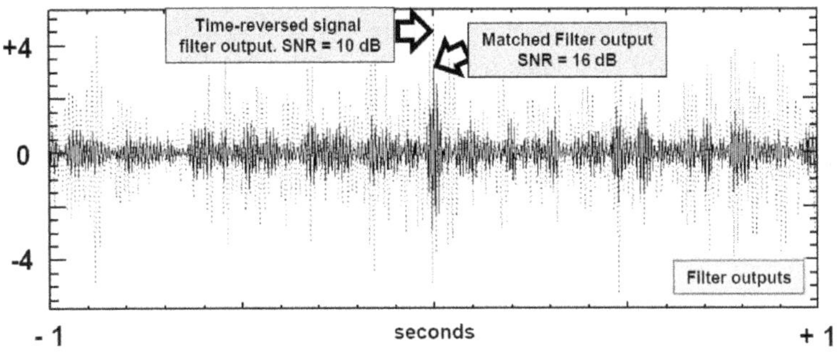

Figure 7.4: Matched filter (solid black) and time-reversed signal filter (dotted black) outputs.

7.3 Detection of a WSS Random Signal in WSS Random Noise

In section 7.2 we assumed a known non-random signal with finite energy in zero-mean WSS random noise with known auto-correlation function. Here we wish to find an optimal filter to maximize the ratio of signal power to noise power when the signal is a zero-mean WSS random process, in the presence of additive zero-mean WSS random noise. We consider the discrete time case with signal vector $\boldsymbol{s}$ whose auto-correlation matrix is $\boldsymbol{R}_{ss} = \mathrm{E}[\boldsymbol{s}\boldsymbol{s}^+]$. The auto-correlation matrix of the additive zero-mean WSS random noise $\boldsymbol{\nu}$ is denoted by $\boldsymbol{R}_{\nu\nu} = \mathrm{E}[\boldsymbol{\nu}\boldsymbol{\nu}^+]$. We consider the following two cases:

- Noise auto-correlation matrix $\boldsymbol{R}_{\nu\nu}$ is known.
- Signal auto-correlation matrix $\boldsymbol{R}_{ss}$ is known.

We seek an FIR filter $\boldsymbol{h} = [h_0, \ldots, h_{M-1}]^T$ that will maximize the SNR, which for the sum of a random signal and random noise is the ratio of signal power to noise power. The filtering process is the same as the previous section and is depicted in figure 7.2. Defining the time reversed signal vector $\boldsymbol{s}_n = [s_n, s_{n-1}, \ldots, s_{n-(M-1)}]^T$ and the noise vector $\boldsymbol{\nu}_n = [\nu_n, \nu_{n-1}, \ldots, \nu_{n-(M-1)}]^T$, the filtered signal and noise are $\boldsymbol{h}^T\boldsymbol{s}_n$ and $\boldsymbol{h}^T\boldsymbol{\nu}_n$, respectively. The corresponding powers are

$$\mathcal{P}_s = \mathrm{E}\left[\left|\boldsymbol{h}^T\boldsymbol{s}_n\right|^2\right] = \boldsymbol{h}^+\boldsymbol{R}_{ss}\boldsymbol{h}, \quad \mathcal{P}_\nu = \mathrm{E}\left[\left|\boldsymbol{h}^T\boldsymbol{\nu}_n\right|^2\right] = \boldsymbol{h}^+\boldsymbol{R}_{\nu\nu}\boldsymbol{h}, \tag{7.15}$$

where we have used the results of section 6.6. The SNR is **SNR** $= \mathcal{P}_s/\mathcal{P}_\nu$.

Consider first the case when the noise auto-correlation matrix is known. Then the problem is that of maximizing the numerator with a fixed denominator. The solution is found by choosing $\boldsymbol{h}$ to be the eigenvector of the signal auto-correlation matrix with the largest eigenvalue $\lambda_{\max}^{(s)}$, for which $\boldsymbol{R}_{ss}\boldsymbol{h}_{\text{opt}} = \lambda_{\max}^{(s)}\boldsymbol{h}_{\text{opt}}$ and $\mathbf{SNR}_{\max} = \lambda_{\max}^{(s)}/\mathcal{P}_{\nu}$.

If, on the other hand, the signal auto-correlation matrix is known, then the maximization of SNR is equivalent to the minimization of the denominator keeping the numerator fixed. This problem is solved by choosing the eigenvector of the noise auto-correlation matrix with the smallest eigenvalue $\lambda_{\min}^{(\nu)}$, for which $\boldsymbol{R}_{\nu\nu}\boldsymbol{h}_{\text{opt}} = \lambda_{\min}^{(\nu)}\boldsymbol{h}_{\text{opt}}$ and $\mathbf{SNR}_{\max} = \mathcal{P}_s/\lambda_{\min}^{(\nu)}$.

7.4 Canonical Factorization

In section 3.2 we discussed the minimum phase and maximum phase multiplicative factors of a pole-zero system function. A power spectral function, $S_{xx}(s)$ in continuous time or $S_{xx}(z)$ in discrete time, when evaluated on the imaginary axis for continuous time or the unit circle for discrete time, is a real and non-negative function of frequency, namely the PSD. Consider the zero-mean WSS random process x_n and its power spectral function $S_{xx}(z)$. Canonical factors of $S_{xx}(z)$ play important roles in much of signal processing [72, 73, 74].

Theorem 29. *If the power spectral function $S_{xx}(z)$ and its logarithm $\ln S_{xx}(z)$ are analytic in the annular ROC defined by $a < |z| < 1/a$, $0 < a < 1$, which includes the unit circle, and the Paley-Wiener condition [74]*

$$\int_{-\pi}^{+\pi} \left|\ln S(e^{i\omega})\right| \, d\omega < \infty \tag{7.16}$$

holds, then $S_{xx}(z)$ can be factored

$$S_{xx}(z) = S_{xx}^{(+)}(z) S_{xx}^{(-)}(z), \tag{7.17}$$

where $S_{xx}^{(+)}(z)$ is minimum phase and $S_{xx}^{(-)}(z)$ is maximum phase. That is, if the inverse **Z** *transform of $S_{xx}^{(+)}(z)$ is denoted by $R_{xx}^{(+)}[k]$, then the latter and its inverse are causal and stable discrete time functions. Furthermore, defining the* cepstral coefficients [1] *$c[n]$ to be the inverse Fourier transform of $\ln S(e^{i\omega})$*

$$\ln S(e^{i\omega}) = \sum_{n=-\infty}^{\infty} c[n] e^{-in\omega}, \quad c[n] = \frac{1}{2\pi} \int_{-\pi}^{+\pi} e^{+in\omega} \ln S(e^{i\omega}) d\omega, \tag{7.18}$$

and defining the geometric mean σ_g^2 of the power spectrum by

$$\ln \sigma_g^2 \equiv \frac{1}{2\pi} \int_{-\pi}^{+\pi} \ln S(e^{i\omega}) \, d\omega, \tag{7.19}$$

then

$$S_{xx}^{(+)}(z) = \sigma_g \exp\left(\sum_{n=1}^{\infty} c[n] z^{-n}\right), \quad |z| > a, \tag{7.20}$$

[1] In equation (3.76b) defining the real cepstrum we used the magnitude of the Fourier transform but noted that it is sometimes defined in terms of the squared magnitude of the Fourier transform. The difference between the two definitions is, of course, a factor of 2.

and $S_{xx}^{(-)}(z) = S_{xx}^{(+)*}(1/z^*)$. *Thus,*

$$S_{xx}^{(-)}(z) = \sigma_g \exp\Big(\sum_{n=1}^{\infty} c^*[n]z^n\Big), \quad |z| < 1/a. \tag{7.21}$$

To prove the results of theorem 29 we use the analyticity of $\ln S_{xx}(z)$ to write the Laurent series

$$\ln S_{xx}(z) = \sum_{n=-\infty}^{\infty} c[n]\, z^{-n}, \quad a < |z| < 1/a, \tag{7.22}$$

which can be written as

$$\begin{aligned} S_{xx}(z) &= \exp\Big(\sum_{n=-\infty}^{1} c[n]z^{-n} + c[0] + \sum_{n=1}^{\infty} c[n]z^{-n}\Big) \\ &= \sigma_g^2 \exp\Big(\sum_{n=1}^{\infty} c[n]z^{-n}\Big) \exp\Big(\sum_{n=1}^{\infty} c^*[n]z^n\Big), \end{aligned} \tag{7.23}$$

where we have used the relation $c[n] = c^*[-n]$, which follows from the fact that the PSD $S(e^{i\omega})$ is real and non-negative. The discrete time minimum phase and maximum phase filters can be obtained from the Fourier transforms of the cepstral coefficients $c^{(\pm)}[n]$

$$c^{(\pm)}[n] \equiv c[0]/2 + c[n]\theta[\pm n - 1], \quad n \in \mathbb{Z}, \tag{7.24}$$

through the relation

$$h^{(\pm)}[n] = \mathcal{F}^{-1}\Big[e^{\mathcal{F}\{c^{(\pm)}[n]\}}\Big]. \tag{7.25}$$

Equations (7.22) and (7.23) have an important consequence, namely, that if a system function $G(z)$ is minimum phase, then the inverse **Z** transform $\mathcal{Z}^{-1}\{\ln G(z)\}$ is causal; the minimum phase condition means that the exponential on the right hand side of (7.23) with the positive powers z^n will be absent and we find

$$G(z) \text{ is minimum phase} \quad \Leftrightarrow \quad \ln G(z) = c[0] + \sum_{n=1}^{\infty} c[n]z^{-n}. \tag{7.26}$$

A trivial, but useful, observation is that the cepstral coefficients for $\ln G(z)$ and those for $\ln 1/G(z)$ are related by a multiplicative minus sign, namely,

$$\ln G(z) \Leftrightarrow c[n], \quad \ln 1/G(z) \Leftrightarrow \; - c[n]. \tag{7.27}$$

The power spectral function $S_{xx}(s)$ of a zero-mean WSS random process x_t in continuous time has an analogous factorization in terms of minimum phase and maximum phase functions

$$S_{xx}(s) = S_{xx}^{(+)}(s) S_{xx}^{(-)}(s), \quad S_{xx}^{(-)}(s) = S_{xx}^{(+)*}(-s^*), \tag{7.28}$$

and we will use this result to derive the causal Wiener filter in section 7.5. The Paley-Wiener condition in this case (the analogue of (7.16)) is [75]

$$\int_{-\infty}^{+\infty} \frac{|\ln S(i\omega)|}{1+\omega^2}\, d\omega < \infty\,, \tag{7.29}$$

and the spectral function $S_{xx}(s)$ must have a region of convergence in the complex s-plane that includes the imaginary axis $s = i\omega$. For example, $S(i\omega) = \exp(-\omega^2)$ does not satisfy the Paley-Wiener condition, but $S(i\omega) = (1+\omega^2)^{-1/2}$ does.

Spectral factorization of general power spectral functions that are not pole-zero is a difficult, if not impossible, task. When $S_{xx}(s)$ or $S_{xx}(z)$ are legitimate power spectral functions (i.e., they are real and positive functions with the right symmetries when evaluated on the imaginary axis of the complex s-plane, or on the unit circle of the complex z-plane, respectively) that are pole-zero functions with known zeros and poles in the complex s-plane, or z-plane, then we merely group the terms with zeros and poles in the left half complex s-plane (for continuous time functions), or inside the unit circle in the complex z-plane (for discrete time functions), to build the minimum phase factors. The maximum phase factors will automatically include the corresponding zeros and poles in the right half complex s-plane, or outside the unit circle of the complex z-plane. For instance, the function

$$S_{xx}(z) = \frac{(1 - z_0/z)(1 - z_0^* z)}{(1 - p_0/z)(1 - p_0^* z)}, \quad |z_0| < 1, \ |p_0| < 1, \tag{7.30}$$

is a legitimate power spectral function since it is real and non-negative when evaluated on the unit circle $z = \exp(i\omega)$, and the spectral factors are

$$S_{xx}^{(+)}(z) = \frac{(1 - z_0/z)}{(1 - p_0/z)}, \quad S_{xx}^{(-)}(z) = \frac{(1 - z_0^* z)}{(1 - p_0^* z)} = S_{xx}^{(+)*}(1/z^*). \tag{7.31}$$

For signals in discrete time all legitimate power spectral functions in the complex z-plane when evaluated on the unit circle $z = \exp(i\omega)$ will be functions of $\cos(\omega)$. To find the spectral factors we replace $\cos(\omega)$ by $(z + z^{-1})/2$ and proceed to find the factors with zeros and poles inside the unit circle and group them together to produce the minimum phase spectral factor.

7.5 The Continuous-Time Causal Wiener Filter

The real-time linear prediction of a zero-mean WSS random process y_t using a correlated zero-mean WSS random process x_t can be described by a causal Wiener filter h_τ (which vanishes for $\tau < 0$):

$$\hat{y}_{t+T} = \int_0^\infty h_\tau x_{t-\tau} \, d\tau = \int_{-\infty}^t h_{t-\tau} x_\tau \, d\tau. \tag{7.32}$$

The pure prediction problem is for $T > 0$, the real-time estimation problem corresponds to $T = 0$, and the real-time interpolation (or smoothing) problem is defined by $T < 0$. The optimality criterion is that of minimum MSE (MMSE) $\|e_{t+T}\|_2^2 = \|y_{t+T} - \hat{y}_{t+T}\|_2^2$ subject to the filter causality requirement $h_\tau = 0$ when $\tau < 0$. The MMSE solution satisfies the orthogonality principle [see section 1.11] using the appropriate inner product:

$$\langle e_{t+T} x_\tau^* \rangle = \mathrm{E}\left[e_{t+T} x_\tau^*\right] = 0, \quad \tau \le t. \tag{7.33}$$

The solution is given by the integral equation

$$\int_{-\infty}^t R_{xx}(\tau - \sigma) \, h_{t-\tau} d\tau = R_{yx}(t + T - \sigma), \ \sigma \le t, \tag{7.34}$$

which after changing integration variables becomes the celebrated *Wiener-Hopf equation* [76] [2]:

$$\int_0^\infty R_{xx}(t-\tau)\,h_\tau d\tau = R_{yx}(t+T), \;\; t \geq 0. \tag{7.35}$$

The Wiener-Hopf equation is not a convolution and so cannot be solved by Fourier transformation. The key to solving this equation is the observation that if $R_{xx}(\tau) = \delta(\tau)$ then the solution is $h(t) = R_{yx}(t+T)\theta(t)$, where $\theta(t)$ is the step function. Thus, we first *whiten* the x_t process using a linear filter $g_w(\tau)$ as shown in 7.5. Using (7.2) and the requirement for ν_t to be a unit variance white process we have

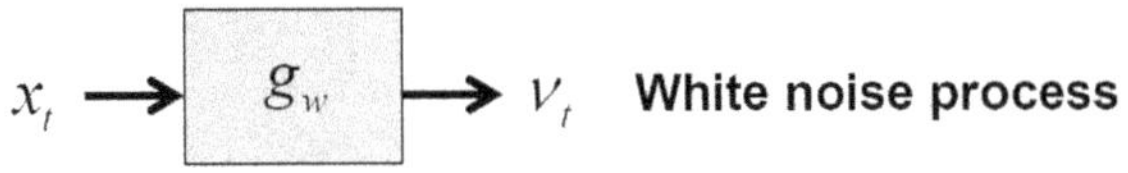

Figure 7.5: Whitening x_t using a linear filter g_w.

$$S_{\nu\nu}(s) = G_w^*(-s^*)\,G_w(s)\,S_{xx}(s) = 1. \tag{7.36}$$

Using the canonical factors (7.28),

$$S_{xx}(s) = S_{xx}^{(+)}(s)S_{xx}^{(-)}(s) = S_{xx}^{(+)}(s)S_{xx}^{(+)*}(-s^*), \tag{7.37}$$

and so the Laplace transform of the whitening (decorrelation) filter is

$$G_w(s) = 1/S_{xx}^{(+)}(s). \tag{7.38}$$

Since the spectral factor $S_{xx}^{(+)}(s)$ is a causal function whose inverse is also causal, the relationship between x_t and ν_t is causally invertible: ν_t is the causally equivalent white noise process for x_t; it is known as the *innovations representation* of x_t. This equivalence is shown in figure 7.6: x_t is obtainable from all values $\nu_{t'}$ for $t' < t$, and vice versa, and the only distinction between the two processes is that while x_t is a correlated process, ν_t is a white process with unit variance.

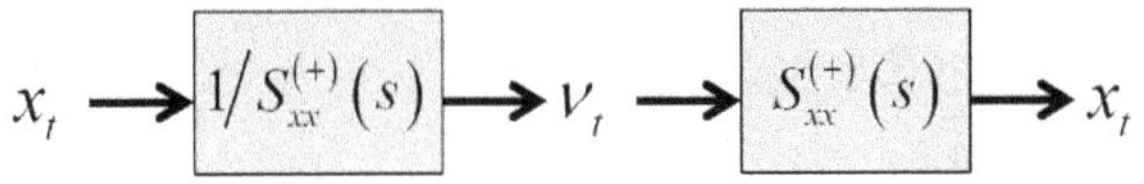

Figure 7.6: Process x_t and its equivalent white noise process ν_t.

Passing the white process ν_t through the causal filter $R_{y\nu}(t+T)\theta(t)$, the output will be $\hat{y}_{t+T}$. Thus, the optimal solution to the Wiener-Hopf equation is given by cascading the two filters: first the whitening filter $\mathcal{L}^{-1}\left\{1/S_{xx}^{(+)}(s)\right\}$, and next the filter $R_{y\nu}(t+T)\,\theta(t)$. The final result is easily expressed in the Laplace transform domain by multiplying the Laplace transforms of the two cascading filters. Taking the Laplace transform of $R_{y\nu}(t+T) = \mathrm{E}[y_{t+T+\tau}\nu_\tau^*]$, we find $\exp(sT)S_{y\nu}(-s) = G_w(-s)\exp(sT)S_{yx}(-s)$. But

[2] Wiener's original work on the derivation of this integral equation and its causal solution was the subject of a 1942 MIT OSRD SECRET report titled *The Extrapolation, Interpolation and Smoothing of Stationary Time Series with Engineering Applications* with a yellow jacket. It was distributed among radar and gunnery engineers who "couldn't make heads nor tails of it", and so the report became known as the *Yellow Peril*. It was declassified in 1946 and gained wide recognition by 1950 [77].

what we need is the Laplace transform of the causal part of the time function $R_{y\nu}(t+T)$, namely, $R_{y\nu}(t+T)\theta(t)$ for which we introduced the subscript "c" notation at the end of section 3.2,

$$\int_{-\infty}^{\infty} R_{yx}(t+T)\,\theta(t)\,e^{-st}dt = \left\{\frac{e^{sT}S_{yx}(s)}{S_{xx}^{(+)*}(-s^*)}\right\}_c = \left\{\frac{e^{sT}S_{yx}(s)}{S_{xx}^{(-)}(s)}\right\}_c. \tag{7.39}$$

The right hand side of (7.39) must be multiplied with the whitening filter to produce the optimal solution to the Wiener-Hopf equation:

$$H_{\text{opt}}(s) = \frac{1}{S_{xx}^{(+)}(s)}\left\{\frac{e^{sT}S_{yx}(s)}{S_{xx}^{(-)}(s)}\right\}_c \equiv \frac{1}{S_{xx}^{(+)}(s)}\mathcal{L}\left\{\theta(t)\mathcal{L}^{-1}\left\{\frac{e^{sT}S_{yx}(s)}{S_{xx}^{(-)}(s)}\right\}\right\}. \tag{7.40}$$

The MSE is found as follows

$$\begin{aligned}\mathrm{E}\left[\left|e_{t+T}\right|^2\right]_{\min} &= \mathrm{E}\left[e_{t+T}(y_{t+T}-\hat{y}_{t+T})^*\right] = R_{yy}[0] - \int_0^{\infty} h_\tau R_{xy}(-\tau-T)\,d\tau \\ &= R_{yy}[0] - \int_0^{\infty}\int_0^{\infty} h_\sigma^* R_{xx}(\sigma-\tau)h_\tau\,d\sigma d\tau,\end{aligned} \tag{7.41}$$

where we have used the orthogonality principle $\langle e_{t+T}, \hat{y}_{t+T}\rangle = \mathrm{E}\left[e_{t+T}\hat{y}_{t+T}^*\right] = 0$, equation (7.1c), and the Wiener-Hopf equation (7.35).

As an example, consider the Ornstein-Uhlenbeck process u_t describing the velocity of a particle undergoing Brownian motion in a fluid. The particle position is described by the Wiener process w_t whose equation of motion (5.46) we write in terms of the particle velocity $u_t \equiv \dot{w}_t$

$$\dot{u}_t + u_t/\tau_0 = \nu_t, \quad t \geq 0, \quad \nu_t \sim \mathcal{N}\left(0, \sigma_\nu^2\right). \tag{7.42}$$

Thus, the particle velocity u_t is a zero-mean Gaussian random process whose auto-correlation function can be found from the following set of differential equations

$$\frac{\partial R_{\nu u}(t,t')}{\partial t'} + \frac{1}{\tau_0}R_{\nu u}(t,t') = \sigma_\nu^2\,\delta(t-t'), \quad t \geq 0, \quad R_{\nu u}(0,t') = 0, \tag{7.43}$$

$$\frac{\partial R_{uu}(t,t')}{\partial t} + \frac{1}{\tau_0}R_{uu}(t,t') = R_{\nu u}(t,t'), \quad t' \geq 0, \quad R_{\nu u}(t,0) = 0. \tag{7.44}$$

The solutions to (7.43) is $R_{\nu u}(t,t') = \sigma_\nu^2\, e^{-(t'-t)/\tau_0}\,\theta(t'-t)$, which when substituted into (7.44) gives the auto-correlation function

$$R_{uu}(t,t') = \frac{\sigma_\nu^2\tau_0}{2}\left(e^{-|t-t'|/\tau_0} - e^{-(t+t')/\tau_0}\right), \quad t,t' \geq 0. \tag{7.45}$$

Writing $t' = t+\tau$ in (7.45) and taking the limit $t \to \infty$ (for the effects of zero initial conditions to vanish) gives the auto-correlation function $R_{uu}(\tau) = \sigma_\nu^2\tau_0/2\,\exp\left(-|\tau|/\tau_0\right)$ for the WSS Ornstein-Uhlenbeck process. Together with Gaussianity we conclude that the WSS Ornstein-Uhlenbeck process is SSS; it is, in fact, the only example of a Markov random process that is Gaussian and SSS (Doob's theorem [78]).

Now consider the pure prediction problem for u_t using it as its own reference, i.e., we wish to determine the optimal filter in the equation:

$$\hat{u}(t+T) = \int_{-\infty}^{t} h(\tau)\, u(t-\tau)\, d\tau. \tag{7.46}$$

The spectral factors of $S_{uu}(s)$, namely, $S_{uu}^{(+)}(s)$ and $S_{uu}^{(-)}(s)$, are found from

$$\frac{\sigma_\nu^2 \tau_0}{2} \int_{-\infty}^{\infty} e^{-s\tau} e^{-|\tau|/\tau_0}\, d\tau = \frac{\sigma_\nu}{1/\tau_0 + s} \times \frac{\sigma_\nu}{1/\tau_0 - s}, \tag{7.47}$$

and the minimum phase factor is found by taking the term with the LHP pole $s = -1/\tau_0$, while the maximum phase factor is $S_{uu}^{(-)}(s) = S_{uu}^{(+)*}(-s^*)$. Thus,

$$S_{uu}^{(+)}(s) = \sigma_\nu\, (1/\tau_0 + s)^{-1}, \quad S_{uu}^{(-)}(s) = \sigma_\nu\, (1/\tau_0 - s)^{-1}. \tag{7.48}$$

Setting $x_t = y_t = u_t$ in this problem we note that $S_{yx}(s) = S_{xx}(s) = S_{uu}(s)$, and so the optimal filter's Laplace transform is

$$H_{\text{opt}}(s) = \frac{1/\tau_0 + s}{\left\{ \frac{e^{sT}}{1/\tau_0 + s} \right\}_c}. \tag{7.49}$$

Using the causal extraction results for the example at the end of section 3.2 we find the Laplace transform of the optimal filter and its inverse transform,

$$H_{\text{opt}}(s) = e^{-T/\tau_0} \quad \leftrightarrow \quad h_{\text{opt}}(t) = \mathcal{L}^{-1}\left[H_{\text{opt}}(s)\right] = e^{-T/\tau_0}\, \delta(t), \tag{7.50}$$

which gives the prediction equation $\hat{u}(t+T) = e^{-T/\tau_0}\, u(t)$. Thus, the optimal future predictions of the Ornstein-Uhlenbeck process depend only on the present value and not on the past values of the process. The MSE is found from equations (7.41) and (7.50) to be $1 - \exp(-2T/\tau_0)$, which is 0 for $T = 0$, is < 1 for $T > 0$ and $\to 1$ as $T \to \infty$.

7.6 The Discrete-Time Causal Wiener Filter

The discrete time infinite impulse response causal Wiener filter is found using a similar procedure to that of the continuous time case of section 7.5. That is, we minimize the MSE function

$$\mathrm{E}\left[\left|e_{n+k}\right|^2\right] = \mathrm{E}\left[\left|y_{n+k} - \hat{y}_{n+k}\right|^2\right], \quad \hat{y}_{n+k} = \sum_{m=0}^{\infty} h_m x_{n-m}, \quad n \in \mathbb{Z}, \tag{7.51}$$

to obtain the discrete time equivalent of the Wiener-Hopf equation (7.35):

$$\sum_{m=0}^{\infty} h_m R_{xx}[n-m] = R_{yx}[n+k], \quad k \geq 0. \tag{7.52}$$

The optimal causal solution is found from equation (7.40) if we replace s with z, and the exponential factor e^{sT} by z^k to obtain

$$H_{\text{opt}}(z) = \frac{1}{S_{xx}^{(+)}(z)} \left\{ \frac{z^k S_{yx}(z)}{S_{xx}^{(-)}(z)} \right\}_c \equiv \frac{1}{S_{xx}^{(+)}(z)} \mathcal{Z}\left\{ \theta_n \mathcal{Z}^{-1}\left\{ \frac{z^k S_{yx}(z)}{S_{xx}^{(-)}(z)} \right\} \right\}, \tag{7.53}$$

where θ_n is the discrete time unit step function: 1 for $n \geq 0$ and 0 for $n < 0$. The MSE is

$$\mathrm{E}\left[\left|e_{n+k}\right|^2\right]_{\min} = \mathrm{E}\left[e_{n+k}(y_{n+k} - \hat{y}_{n+k})^*\right] = \mathrm{E}\left[e_{n+k}y^*_{n+k}\right] = R_{ey}[0]$$

$$= R_{yy}[0] - \sum_{l=0}^{\infty}\sum_{m=0}^{\infty} h^*_l R_{xx}[l-m]h_m, \tag{7.54}$$

where we used the orthogonality principle

$$\left\langle e_{n+k}, \hat{y}^*_{n+k}\right\rangle = \mathrm{E}\left[e_{n+k}\hat{y}^*_{n+k}\right] = 0,$$

the relation $R^*_{xx}[m-l] = R_{xx}[l-m]$, and the Wiener-Hopf equation (7.52). Note that the right hand side of (7.54) implicitly depends on k through the optimal filter solution of equation (7.52).

Equation (7.54) shows that the MMSE is equal to the correlation function $R_{ey}[0]$ which can be expressed as an inverse **Z** transform contour integral as described in equation (3.22). The cross-correlation for lag l is

$$R_{ey}[l] = \mathrm{E}\left[e_{n+k}y^*_{n+k-l}\right] = R_{yy}[l] - \sum_{m=0}^{\infty} h_m R_{xy}[l-m-k]. \tag{7.55}$$

Since the optimal Wiener filter h_m vanishes for $m < 0$, this equation is equivalent to the **Z** transform relation

$$S_{ey}(z) = S_{yy}(z) - z^{-k}H_{\text{opt}}(z)S_{xy}(z). \tag{7.56}$$

Thus, $R_{ey}[0]$ is found by using the inverse **Z** transform (3.22), i.e.,

$$R_{ey}[0] = \frac{1}{2i\pi}\oint_C \frac{S_{ey}(z)}{z}\,dz = \frac{1}{2i\pi}\oint_C \frac{S_{yy}(z) - z^{-k}H_{\text{opt}}(z)S_{xy}(z)}{z}\,dz. \tag{7.57}$$

As an example consider the optimal real-time estimation of the discrete time process y_n from $x_n = y_n + \nu_n$ with $\nu_n \sim \mathscr{N}(0, \sigma^2_\nu)$ uncorrelated with y_n, i.e., the $k = 0$ case in equation (7.53). Using the relations (7.4) we have

$$S_{yx}(z) = S_{yy}(z), \quad S_{xx}(z) = S_{yy}(z) + \sigma^2_\nu. \tag{7.58}$$

Let us assume that y_n is a stable first order Markov process given by

$$y_n = -a\,y_{n-1} + \varepsilon_n, \quad |a| < 1, \tag{7.59}$$

where ε_n is a white Gaussian process with variance σ^2_ε. Thus, we have the spectral functions

$$S_{yy}(z) = \frac{\sigma^2_\varepsilon}{(1 + a/z)(1 + a^*z)}, \quad S_{xx}(z) = S_{yy}(z) + \sigma^2_\nu \equiv \frac{\alpha(1 + b/z)(1 + b^*z)}{(1 + a/z)(1 + a^*z)}, \tag{7.60}$$

where α and b satisfy the following two equations

$$\alpha(1 + |b|^2) = \sigma^2_\nu(1 + |a|^2) + \sigma^2_\varepsilon, \quad \alpha b = \sigma^2_\nu a. \tag{7.61}$$

Equations (7.61) admit solutions for which $\alpha > 0$, $|a| < 1$ and $|b| < 1$, and that can be used to calculate the spectral factors of $S_{xx}(z)$ as required by equation (7.53), namely,

$$S^{(-)}_{xx}(z) = \frac{\sqrt{\alpha}\,(1 + b^*z)}{1 + a^*z}, \quad \frac{S_{yx}(z)}{S^{(-)}_{xx}(z)} = \frac{\sigma^2_\varepsilon}{\sqrt{\alpha}\,(1 + a/z)(1 + b^*z)}. \tag{7.62}$$

To find the additive causal part of the second equation in (7.62) (as prescribed by equation (7.53)), we use the inverse **Z** transform equation (3.22) to obtain $h[n]$ and then sum this sequence over indices $n \geq 0$. Thus,

$$h\left[n\right] = \frac{1}{2i\pi} \oint_C \frac{\sigma_\varepsilon^2}{\sqrt{\alpha}\,(1 + a/z)(1 + b^* z)} z^{n-1}\, dz = \frac{\sigma_\varepsilon^2 (-a)^n}{\sqrt{\alpha}\,(1 - ab^*)}, \tag{7.63}$$

which gives the following additive causal term:

$$\left\{\frac{S_{yx}\left(z\right)}{S_{xx}^{(-)}(z)}\right\}_c = \sum_{n=0}^{\infty} h\left[n\right] z^{-n} = \frac{\sigma_\varepsilon^2}{\sqrt{\alpha}\,(1 - ab^*)(1 + a/z)}. \tag{7.64}$$

Finally, setting $k = 0$ in equation (7.53), we find the optimal causal Wiener filter:

$$H_W(z) = \frac{\sigma_\varepsilon^2}{\alpha\,(1 - ab^*)(1 + b/z)} = \frac{\beta}{1 + b/z}, \quad \beta \equiv \frac{\sigma_\varepsilon^2}{\alpha\,(1 - ab^*)}. \tag{7.65}$$

Figure 7.7 illustrates the optimal Wiener filter for noisy observations of this first order Gaussian Markov process y_n. Using equation (7.57), the MSE of the example is found to be

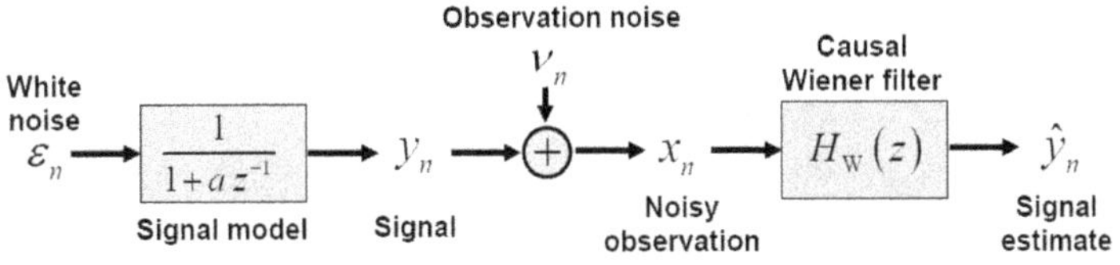

Figure 7.7: Wiener estimation of a first order Gaussian Markov process.

$$R_{ey}[0] = \frac{\sigma_\varepsilon^2 (1 - \beta - b/a)}{(1 - |a|^2)(1 - b/a)} - \frac{\sigma_\varepsilon^2 \beta}{(1 - a^*/b)(1 - ab)} = \frac{\sigma_\varepsilon^2 (b/a - 1)}{(1 - a^*/b)(1 - ab)}, \tag{7.66}$$

where we have used the result $\beta = 1 - b/a$, which follows from equations (7.61): the first is $\sigma_\nu^2 - \alpha\,|b|^2 - \alpha + \sigma_\nu^2\,|a|^2 + \sigma_\varepsilon^2 = 0$, which on using the second set of equations gives $(1 - ab^*)(\sigma_\nu^2 - \alpha) + \sigma_\varepsilon^2 = 0$ or equivalently $a\sigma_\nu^2(1 - ab^*) = a\alpha(1 - ab^*) - a\sigma_\varepsilon^2$, and again $\alpha b(1 - ab^*) = \alpha a(1 - ab^*) - a\sigma_\varepsilon^2$; dividing both sides by $\alpha(1 - ab^*)$ and using the definition of β leads to the result. The MSE (7.66) should be compared to σ_ν^2, which is the MSE in the absence of the Wiener filter when the noisy observation x_n is used as the best estimate of y_n. Figure 7.8 shows the estimation error improvement for the ratio $R_{ey}[0]/\sigma_\nu^2$ as a function of σ_ν^2.

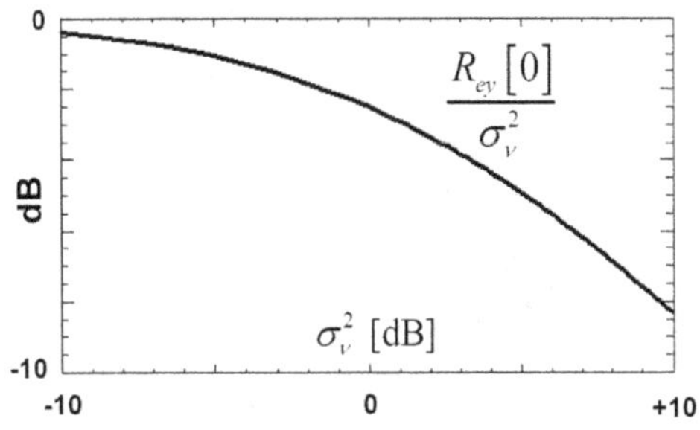

Figure 7.8: Estimation error improvement—equation (7.66).

7.7 The Causal Wiener Filter and the Kalman Filter

The signal estimate provided by the Wiener filter example illustrated in figure 7.7 is given by the equation

$$\hat{y}_n = -b\hat{y}_{n-1} + \beta x_n, \tag{7.67}$$

for appropriately defined values of b and β given by (7.61) and (7.65). The signal was assumed to be a first order Gaussian Markov process described by equation (7.59). The optimal Wiener filter is applied to the noisy observation $x_n = y_n + \nu_n$, as shown in figure 7.7. To illustrate the close connection between the Wiener filter and the Kalman filter using this specific example, we first introduce some notation:

- $\hat{y}_{n|n}$ is the signal estimate at time step n given the observations up to and including n.
- $\hat{y}_{n|n-1}$ is the signal prediction at time step n given the observations up to and including $n-1$.
- $\hat{x}_{n|n-1}$ is the observation prediction at time step n given the observation and signal estimate at $n-1$.

The idea of the Kalman filter is to use the *state transition equation* provided by the signal model $y_n = -ay_{n-1} + \varepsilon_n$ and the signal estimate $\hat{y}_{n-1|n-1}$ to predict the signal estimate $\hat{y}_{n|n-1}$, and then to use this prediction in the observation equation to predict the observation $\hat{x}_{n|n-1}$, namely,

$$\hat{y}_{n|n-1} = -a\hat{y}_{n-1|n-1} \quad \rightarrow \quad \hat{x}_{n|n-1} = \hat{y}_{n|n-1}, \tag{7.68}$$

and finally use the residual error $x_n - \hat{x}_{n|n-1}$ to obtain the estimate

$$\hat{y}_{n|n} = \hat{y}_{n|n-1} + K\left(x_n - \hat{x}_{n|n-1}\right), \tag{7.69}$$

where K is the Kalman gain. If the predictions are perfect then the observation residual error sequence is white. In order to relate the Wiener estimate to the Kalman prediction, we rewrite equation (7.67) in the equivalent form

$$\hat{y}_n = -a\hat{y}_{n-1} + \beta\big(x_n - \frac{a-b}{\beta}\,\hat{y}_{n-1}\big), \tag{7.70}$$

which upon using $\beta = 1 - b/a$ [see the discussion after (7.66)] becomes

$$\hat{y}_n = -a\hat{y}_{n-1} + \beta\big(x_n + a\hat{y}_{n-1}\big). \tag{7.71}$$

Finally, identifying $\hat{y}_n \leftrightarrow \hat{y}_{n|n}$ and using (7.68), the Wiener estimate is transformed into the Kalman form (7.69) provided we also identify β as the optimal value of the Kalman gain K; any value of $K \neq \beta$ will provide a possible prediction for the signal but the optimal value is $K = \beta$.

7.8 The Non-Causal Wiener Filter and the Coherence Function

The non-causal Wiener filter h_t to estimate the zero-mean WSS random process y_t from a correlated zero-mean WSS random process x_t is defined by the convolution integral

$$\hat{y}_t = \int_{-\infty}^{\infty} h_\tau x_{t-\tau}\, d\tau, \tag{7.72}$$

and has applications to off-line data analysis, or image processing, where causality, $h(\tau) = 0$ for $\tau < 0$, is not a requirement. Minimizing the L_2 norm of the error

$$\|e_t\|_2^2 \equiv \mathrm{E}\left[|e_t|^2\right] = \|y_t - \hat{y}_t\|_2^2, \tag{7.73}$$

or using the orthogonality principle

$$\langle e_t, x_{t-\tau} \rangle = \mathrm{E}\left[e_t, x^*_{t-\tau}\right] = 0, \quad -\infty < \tau < \infty, \tag{7.74}$$

we find the integral equation

$$\int_{-\infty}^{\infty} R_{xx}(t-\tau)\, h_\tau \, d\tau = R_{yx}(t), \tag{7.75}$$

whose solution is found by Fourier transformation,

$$H(\omega) = S_{yx}(\omega)/S_{xx}(\omega). \tag{7.76}$$

As an example, consider the problem of estimating a random process y_t from noisy observations $x_t = y_t + \nu_t$ where y_t and ν_t are uncorrelated. Now $S_{yx} = S_{yy}$ and $S_{xx} = S_{yy} + S_{\nu\nu}$, since $S_{y\nu} = 0$. The optimal Wiener filter in the frequency domain is

$$H(\omega) = \frac{S_{yy}(\omega)}{S_{yy}(\omega) + S_{\nu\nu}(\omega)}. \tag{7.77}$$

When the signal power is much stronger than the noise power, $S_{yy}(\omega) \gg S_{\nu\nu}(\omega)$, the filter response is nearly 1. On the other hand, if $S_{yy}(\omega) \ll S_{\nu\nu}(\omega)$ then the filter response is nearly 0. Thus, the optimal Wiener filter passes the signal and stops the noise depending SNR strength. Let us calculate the optimal Wiener filter (7.77) in the time domain when $S_{yy}(\omega) = 1/(1+\omega^2)$ and $S_{\nu\nu}(\omega) = 1/(\sigma^2 + \omega^2)$. The inverse Fourier transform of (7.77) gives

$$h(t) = \frac{\sigma^2}{1+\sigma^2}\,\delta(t) + \frac{1-\sigma^2}{\sqrt{2}(1+\sigma^2)^{3/2}}\, e^{-\sqrt{\frac{1+\sigma^2}{2}}\,|t|}.$$

The single channel solution in equation (7.76) can be generalized to N reference channels by introducing multiple filters whose Fourier transforms satisfy the equations

$$\sum_{k=1}^{N} S_{kn}(\omega)\, H_k(\omega) = S_{yn}(\omega), \quad 1 \le n \le N, \tag{7.78}$$

where indices k and n refer to x_k and x_n, respectively. Using the relation $S_{kn}(\omega) = S^*_{nk}(\omega)$, equation (7.78) can be written in matrix form

$$\begin{bmatrix} S^*_{11}(\omega) & \cdots & S^*_{1N}(\omega) \\ \vdots & \ddots & \vdots \\ S^*_{N1}(\omega) & \cdots & S^*_{NN}(\omega) \end{bmatrix} \begin{bmatrix} H_1(\omega) \\ \vdots \\ H_N(\omega) \end{bmatrix} = \begin{bmatrix} S_{y1}(\omega) \\ \vdots \\ S_{yN}(\omega) \end{bmatrix}. \tag{7.79}$$

The auto-correlation function of the error process e_t is $R_{ee}(\tau) = \mathrm{E}\left[(y_t - \hat{y}_t)(y^*_t - \hat{y}^*_t)\right]$, which upon Fourier transformation gives the error PSD:

$$\begin{aligned} S_{ee}(\omega) &= S_{yy}(\omega) + S_{xx}(\omega)\,|H(\omega)|^2 - H(\omega)S^*_{yx}(\omega) - H^*(\omega)S_{yx}(\omega) \\ &= S_{yy}(\omega)\left|H(\omega) - \frac{S_{yx}(\omega)}{S_{xx}(\omega)}\right|^2 + S_{yy}(\omega)(1 - \gamma^2_{yx}(\omega)), \end{aligned} \tag{7.80}$$

where $\gamma^2_{yx}(\omega) = \gamma^2_{xy}(\omega)$ is the MSC (6.13) characterizing the degree of linear correlation between y and x as a function of frequency. When the filter transfer function is the optimal solution (7.77), the error PSD attains its minimum value

$$S_{ee}^{\min}(\omega) = S_{yy}(\omega)(1 - \gamma^2_{yx}). \tag{7.81}$$

The *complex coherence function*, whose squared magnitude is the MSC, is defined by [79]

$$C_{yx}(\omega) \equiv \frac{S_{yx}(\omega)}{\sqrt{S_{xx}(\omega)S_{xx}(\omega)}}. \tag{7.82}$$

Using the Cauchy-Schwarz inequality (1.9) we have

$$0 \leq \left|C_{yx}(\omega)\right|^2 \equiv \gamma^2_{yx} = \gamma^2_{xy} \leq 1. \tag{7.83}$$

The single channel MSC can be extended to multiple reference channels $x_1, \ldots, x_N$ by defining the multiple MSC $\gamma^2_{y\boldsymbol{x}}$ through a generalization of (7.81)

$$S_{ee}^{\min}(\omega) \equiv S_{yy}(\omega)(1 - \gamma^2_{y\boldsymbol{x}}(\omega)). \tag{7.84}$$

Consider an example with $N = 2$ where H_1 and H_2 (suppressing the dependence of all functions on frequency ω) are solutions to the 2×2 matrix equation (7.79),

$$H_1 = \frac{S_{y1}/S_{11} - \gamma^2_{12} S_{y2}/S_{12}}{1 - \gamma^2_{12}}, \quad H_2 = \frac{S_{y2}/S_{22} - \gamma^2_{12} S_{y1}/S_{21}}{1 - \gamma^2_{12}}, \tag{7.85}$$

which together with the equation $S_{ee} = S_{yy} + |H_1|^2 S_{11} + |H_2|^2 S_{22} - 2\,\boldsymbol{Re}\left(H_1^* S_{y1}\right) - 2\,\boldsymbol{Re}\left(H_2^* S_{y2}\right) + 2\,\boldsymbol{Re}\left(H_1 S_{12} H_2^*\right)$ can be used to give a complicated expression for the multiple MSC for $N \geq 2$:

$$\gamma^2_{y\boldsymbol{x}} = 1 - \left(S_{yy} A_{yy}\right)^{-1}, \quad \boldsymbol{x} \equiv [x_1, \ldots, x_N]^T, \tag{7.86}$$

where A_{yy} is the top left-most element of the $(N+1) \times (N+1)$ matrix that is the inverse of the cross-correlation matrix of $y, x_1, \ldots, x_N$ (the elements of the first row of the latter matrix are $S_{yy}, S_{y1}, \ldots, S_{yN}$).

A more tractable formula amenable to simple computation can be obtained using the concept of *partial coherence* [79]. Suppose stationary processes u and v are both correlated with a third stationary process z, in addition to having their own correlation. Removing the correlations between u and z (using a Wiener filter on z and subtracting from u) results in a residual process e_u whose spectrum is denoted by $S_{uu;z}$. A similar procedure between v and z results in e_v with spectrum $S_{vv;z}$. The standard MSC between the two residuals, using the squared magnitude of the formula (7.82), is denoted by $\gamma^2_{uv;z}$:

$$\gamma^2_{uv;z} \equiv \frac{|S_{uv;z}|^2}{S_{uu;z} S_{vv;z}} = \frac{S_{uv;z} S_{vu;z}}{S_{uu;z} S_{vv;z}}. \tag{7.87}$$

For correlated processes y, x, and $\boldsymbol{z} = \{z_1, \ldots, z_N\}$, we can show

$$\gamma^2_{yx} = 1 - \frac{S_{yy;x}}{S_{yy}}, \quad \gamma^2_{y\boldsymbol{z}} = 1 - \frac{S_{yy;\boldsymbol{z}}}{S_{yy}}, \quad \gamma^2_{yx;\boldsymbol{z}} = 1 - \frac{S_{yy;x\boldsymbol{z}}}{S_{yy;\boldsymbol{z}}}, \tag{7.88}$$

and that

$$\gamma^2_{yz_1 \ldots z_N} = \gamma^2_{yz_N;z_1 \ldots z_{N-1}}\left(1 - \gamma^2_{yz_1 \ldots z_{N-1}}\right) + \gamma^2_{yz_1 \ldots z_{N-1}}. \tag{7.89}$$

Thus, to calculate the multiple coherence $\gamma^2_{y1 \ldots N}$ (indices $1, \ldots, N$ refer to $z_1 \ldots z_{N-1}$) for the primary channel y and N reference channels $z_1 \ldots z_{N-1}$, we start with γ^2_{y1} (i.e., the MSC for y and x_1), to find $\gamma^2_{y12} = \gamma^2_{y2;1}(1 - \gamma^2_{y1}) + \gamma^2_{y1}$, and then $\gamma^2_{y123} = \gamma^2_{y3;12}(1 - \gamma^2_{y12}) + \gamma^2_{y12}$, and so on, with partial coherence functions computed using (7.88).

7.9 Generalized Cross-Correlation and Time-Delay Estimation

An important application of CPSD and MSC is to the problem of time delay estimation between two geographically separated sensors whose data are modeled as

$$y(t) = s(t) + \nu_1(t), \text{ and } x(t) = \alpha\, s(t-T) + \nu_2(t), \tag{7.90}$$

for some unknown real parameter $\alpha > 0$. The noise processes $\nu_1(t)$ and $\nu_2(t)$ are real, jointly stationary and correlated, while both are uncorrelated with the stationary random signal $s(t)$. Assuming that signal and noises have Gaussian distributions, an MLE of the time delay T is found as that value for which the function $\mathcal{R}(\tau)$ has a peak [80]:

$$\mathcal{R}(\tau) = \int_{-\infty}^{\infty} \frac{\gamma_{yx}^2(\omega)}{1-\gamma_{yx}^2(\omega)}\, e^{i\phi(\omega)+i\omega\tau}\, d\omega, \tag{7.91}$$

where $\gamma_{yx}^2(\omega)$ is the MSC (7.83) and $\phi(\omega)$ is the phase of the CPSD, i.e.,

$$S_{yx}(\omega) = \left|S_{yx}(\omega)\right| e^{i\phi(\omega)}. \tag{7.92}$$

Equation (7.91) is one example of the generalized cross-correlation (GCC) function

$$\mathcal{R}(\tau) = \int_{-\infty}^{\infty} W(\omega) S_{yx}(\omega)\, e^{i\omega\tau}\, d\omega, \tag{7.93}$$

where $S_{yx}(\omega)$ is the CPSD function of the two processes y and x and $W(\omega)$ is a frequency domain window function. In practice, estimates of the CPSD and the MSC (e.g., using Welch's method desribed in equation (8.37)) are used to calculate $\mathcal{R}(\tau)$. The maximum likelihood weight inferred from equation (7.91), and other popular weighting functions are shown in table 7.1 [80].

Method	$W(\omega)$
SCC (standard cross-correlation)	1
ROTH	$1/S_{yy}(\omega)$
Wiener	$\gamma_{yx}^2(\omega)$
SCOT (smoothed coherence transform)	$1/\sqrt{S_{yy}(\omega)S_{xx}(\omega)}$
PHAT (phase transform)	$1/\left\|S_{yx}(\omega)\right\|$
ML (maximum likelihood)	$\gamma_{yx}^2(\omega)\left[\left(1-\gamma_{yx}^2(\omega)\right)\left\|S_{yx}(\omega)\right\|\right]^{-1}$

Table 7.1: Weights for generalized coherence functions.

When estimating time delay of a bandlimited random signal in white Gaussian noise, then the appropriate window function is the maximum likelihood weight. The smoothed coherence transform (SCOT) weights are often used for time-delay estimation of broadband signals with tonal contamination, an example of which will be shown in section 8.8.

7.10 Random Fields

When random signals are functions of both time and space, as in chapter 11 when we study linear spatial array data, we refer to them as random fields; $v(t,\boldsymbol{r})$ is a random field when there is an associated probability density function $f_v(v)$. In addition, at two sets of coordinates $t,\boldsymbol{r}$ and $t',\boldsymbol{r}'$ the random field will take on values $v = v(t,\boldsymbol{r})$ and $v' = v(t',\boldsymbol{r}')$ with an associated joint density function $f_{v,v'}(v,v')$, and so on. Thus, the mean and correlation functions are defined by

$$\mathrm{E}\big[v(t,\boldsymbol{r})\big] = \int v f_{v'}(v')\, dv',$$
$$R_{vv'}(t,t';\boldsymbol{r},\boldsymbol{r}') = \mathrm{E}\big[v(t,\boldsymbol{r})v^*(t',\boldsymbol{r}')\big] = \int vv'^* f_{v,v'}(v,v')\, dvdv'. \tag{7.94}$$

We generalize the concept of WSS random functions of one variable to random fields that depend on four space-time variables by the requirement that the mean and the correlation function depend only on the difference between the space and time coordinates, respectively, i.e.,

$$R_{vv'}(t,t';\boldsymbol{r},\boldsymbol{r}') = \mathrm{E}\big[v(t,\boldsymbol{r})v^*(t-\tau,\boldsymbol{r}-\boldsymbol{\chi})\big] = R_{vv'}(\tau;\boldsymbol{\chi}). \tag{7.95}$$

The power spectrum is defined by the four-dimensional Fourier transform of the correlation function (theorem 25),

$$S_{vv'}(\omega,\boldsymbol{k}) = \int R_{vv'}(\tau;\boldsymbol{\chi})e^{-i\omega\tau+i\boldsymbol{k}\cdot\boldsymbol{\chi}}\, d\tau d^3\boldsymbol{\chi}, \tag{7.96}$$

with the inverse relation

$$R_{vv'}(\tau;\boldsymbol{\chi}) = (2\pi)^{-4}\int S_{vv'}(\omega,\boldsymbol{k})e^{+i\omega\tau-i\boldsymbol{k}\cdot\boldsymbol{\chi}}\, d\omega d^3\boldsymbol{k}. \tag{7.97}$$

Analogously to equation (7.2) we have the following (Fourier) relationship between the power spectra of random fields into and out of a linear spatial and time-invariant system defined by system function $H(\omega,\boldsymbol{k})$,

$$S_{\text{out}}(\omega,\boldsymbol{k}) = \big|H(\omega,\boldsymbol{k})\big|^2 S_{\text{in}}(\omega,\boldsymbol{k}). \tag{7.98}$$

An example of a random field that is often used in array processing is provided by isotropic noise ν, which consists of a multitude of random waves in all directions and with equal probability, and whose PSD is

$$S_{\nu\nu}(\omega,\boldsymbol{k}) = P_\nu(\omega)\,\delta(k-\omega/c),\quad k \equiv |\boldsymbol{k}|, \tag{7.99}$$

where $P_\nu(\omega)$ describes the distribution of power as a function of temporal frequency ω, and the delta distribution imposes the plane wave dispersion relation. To find the correlation function we use equation (7.97),

$$R_{\nu\nu}(\tau;\chi) = \frac{1}{(2\pi)^4}\int_{\omega=-\infty}^{\infty}\int_{k=0}^{\infty}\int_{\theta=0}^{\pi}\int_{\phi=0}^{2\pi} P_\nu(\omega)\delta(k-\omega/c)e^{+i\omega\tau-ik\chi\cos\theta}\times$$
$$k^2\sin\theta\, d\phi\, d\theta\, dk\, d\omega, \tag{7.100}$$

where we have used spherical polar coordinates in $\boldsymbol{k}$ space with $\boldsymbol{\chi}$ along the z-axis so that $\boldsymbol{k} \cdot \boldsymbol{\chi} = k\chi \cos\theta$. Performing the ϕ integral first followed by the θ integral we find

$$
\begin{aligned}
R_{\nu\nu}(\tau;\chi) &= \frac{1}{4\pi^3 \chi} \int_{\omega=-\infty}^{\infty} \int_{k=0}^{\infty} P_\nu(\omega)\delta(k - \omega/c)e^{+i\omega\tau} k \sin(k\chi)\, dk\, d\omega \\
&= \frac{1}{4\pi^3 c\chi} \int_{\omega=-\infty}^{\infty} \omega \sin(2\pi\chi/\lambda) P_\nu(\omega) e^{+i\omega\tau}\, d\omega,
\end{aligned} \tag{7.101}
$$

where we used $\omega/c = 2\pi/\lambda$. Now the integrand vanishes whenever $\chi = n\lambda/2$, $n = \pm 1, \pm 2, \ldots$. Thus, half-wavelength sampling of isotropic noise field, as in a linear half-wavelength sensor array, ensures that the noise samples on sensors are uncorrelated. If isotropic noise is also Gaussian, then noise samples on linear sensor arrays that are half wavelength apart are independent.

CHAPTER 8

Power Spectral Density Estimation and Signal Models

8.1 Introduction

Estimates of correlation and covariance functions can be deduced from a single realization of a stationary process whose statistical properties do not change with time by replacing the expectation operation with the time-averaging operation if we assume that the stationary process is ergodic. Of particular importance to the analysis of WSS random signals is the concept of ergodicity in the mean and auto-correlation, which is described in this chapter. We discuss how estimates of the cross- and auto-spectral density functions (CPSD and PSD) can be found by using estimated values for cross- and auto-correlation sequences; these techniques are often referred to as non-parametric or classical methods of spectral estimation and generally require long data records to produce estimates with low variances and good resolution.

In addition, we will discuss the minimum variance distortionless spectral estimate (MVDSE), motivated by a filtering interpretation of the MLE of the amplitude of a complex exponential in WGN. We introduce the maximum entropy principle to indefinitely extend a finite set of known auto-correlation sequence values for a zero-mean random WSS signal, and show that the solution is an AR model of some order P, denoted by AR(P), whose parameters include a $P \times 1$ vector $\boldsymbol{a}$ that satisfies the Yule-Walker equation, and a positive number σ^2. PSD estimation methods based on modeling the underlying process are referred to as parametric or model-based methods. We will discuss the Levinson-Durbin algorithm to solve the Yule-Walker equation iteratively; this algorithm is also used for linear prediction of signals.

8.2 Ergodicity

Ergodicity is a general requirement applicable to all possible statistics of a random process. If the statistical properties of the process can be deduced from any single realization of that process, then the process is ergodic. Ergodicity implies stationarity but the converse is not true. For instance, the process $\cos(\omega t + \phi)$, where ω and ϕ have a uniform joint probability density $1/0.6\pi^2$ in $[0.2\pi, 0.5\pi] \times [-\pi, +\pi]$, is stationary but not ergodic. Most stationary processes encountered in real life, however, are ergodic. Since we mainly work with first and second order statistics of WSS random processes (mean, covariance and correlation), we need only assume that a process is *ergodic* in mean and covariance.

The mean $\mathrm{E}[x_t] = \mu_x$ of a WSS random process x_t is an ensemble average computed (in theory) using the distribution of x_t. The time-averaged quantity

$$\hat{\mu}_x = \frac{1}{T} \int_{-T/2}^{+T/2} x_t \, dt \tag{8.1}$$

is not necessarily equal to the mean; it serves as an estimate of the mean, and it can be shown that $\hat{\mu}_x \to \mu_x$ as $T \to \infty$ provided that the variance of the estimate approaches 0 as $T \to \infty$, or equivalently

$$\frac{1}{T} \int_{-T/2}^{+T/2} \left(1 - \frac{|\tau|}{T}\right) C_{xx}(\tau) \, d\tau \to 0 \text{ as } T \to \infty, \tag{8.2}$$

where C_{xx} is the true process auto-covariance function. If (8.2) holds, the process is said to be ergodic in the mean. Similarly, the auto-correlation function can be estimated by

$$\hat{R}_{xx}(\tau) = \frac{1}{T} \int_{-T/2}^{+T/2} x_t x_{t-\tau}^* dt, \tag{8.3}$$

which can be shown to approach the true process auto-correlation $R_{xx}(\tau)$ as $T \to \infty$ provided that

$$\frac{1}{T} \int_{-T/2}^{+T/2} \left(1 - \frac{|\tau|}{T}\right) C_{xx}^{(2)}(\tau) \, d\tau \to 0 \text{ as } T \to \infty, \tag{8.4}$$

where $C_{xx}^{(2)}(\tau)$ is the auto-covariance of the random process $x_t x_{t-\tau}^*$. If (8.4) holds, the process is said to be *auto-correlation ergodic*. Analogous results hold for the cross-correlation function, namely,

$$\hat{R}_{xy}(\tau) = \frac{1}{T} \int_{-T/2}^{+T/2} x_t y_{t-\tau}^* dt, \tag{8.5}$$

which can be shown to approach the true cross-correlation $R_{xy}(\tau)$ as $T \to \infty$ provided that

$$\frac{1}{T} \int_{-T/2}^{+T/2} \left(1 - \frac{|\tau|}{T}\right) C_{xy}^{(2)}(\tau) \, d\tau \to 0 \text{ as } T \to \infty, \tag{8.6}$$

where $C_{xy}^{(2)}(\tau)$ is the cross-covariance of the random process $x_t y_{t-\tau}^*$.

The process $\cos(\omega t + \phi)$ introduced in the beginning of this section is ergodic in the mean but not in covariance. Applying equation (8.1), we find the value 0 as $T \to \infty$, and this is the correct mean. Equation (8.3) for the auto-correlation gives the value $\cos(\omega_0 \tau)/2$ for a given realization with frequency ω_0, whereas the true correlation is $\big(\sin(0.5\pi\tau) - \sin(0.2\pi\tau)\big)/0.6\pi\tau$.

8.3 Sample Estimates of Mean and Correlation Functions

When dealing with sampled data x_n, $0 \le n \le N-1$, we use the sample mean as an unbiased estimate of the true mean, i.e.,

$$\hat{\mu}_x = \frac{1}{N}\sum_{n=0}^{N-1} x_n, \quad \mathrm{E}\big[\hat{\mu}_x\big] = \mu_x. \tag{8.7}$$

There are, however, two different estimates we can use for the correlation function of two random processes: one unbiased and the other biased (indicated by superscripts (u) and (b), respectively),

$$\hat{R}^{(u)}_{yx}[l \ge 0] = \frac{1}{N-l}\sum_{n=l}^{N-1} y_n x^*_{n-l}, \quad \hat{R}^{(b)}_{yx}[l \ge 0] = \frac{1}{N}\sum_{n=l}^{N-1} y_n x^*_{n-l}, \tag{8.8}$$

with the values for negative lags found from $R_{yx}[-l] = R^*_{xy}[l]$. Taking expectation values, we find (for both positive and negative lags)

$$\mathrm{E}\big[\hat{R}^{(u)}_{yx}[l]\big] = R_{yx}[l], \quad \mathrm{E}\big[\hat{R}^{(b)}_{yx}[l]\big] = \left(1 - \frac{|l|}{N}\right) R_{yx}[l]. \tag{8.9}$$

The biased auto-correlation estimate $\hat{R}^{(b)}_{xx}$ (computed after the removal of the mean estimate) is preferred because it is guaranteed to produce a positive spectrum; theorem 30 states a more general result [81].

Theorem 30. *The Toeplitz sample auto-correlation matrix*

$$\hat{\boldsymbol{R}}_{xx} \equiv \begin{bmatrix} \hat{R}_0 & \hat{R}_{-1} & \cdots & \hat{R}_{-(N-1)} \\ \hat{R}_{+1} & \hat{R}_0 & \ddots & \vdots \\ \vdots & \ddots & \ddots & \hat{R}_{-1} \\ \hat{R}_{+(N-1)} & \cdots & \hat{R}_{+1} & \hat{R}_0 \end{bmatrix} \tag{8.10}$$

whose elements are the biased sample auto-correlation sequence elements $\hat{R}_l \equiv R^{(b)}_{xx}[l]$, $0 \le l \le N-1$, $\hat{R}_{-l} = \hat{R}^*_l$, is positive definite.

The theorem follows from the identity $\hat{\boldsymbol{R}}_{xx} \equiv N^{-1}\boldsymbol{Q}\boldsymbol{Q}^+$, where

$$\boldsymbol{Q} = \begin{bmatrix} 0 & 0 & \cdots & 0 & x_0 & x_1 & \cdots & x_{N-1} \\ 0 & \cdots & 0 & x_0 & x_1 & \cdots & x_{N-1} & 0 \\ \vdots & \vdots & \vdots & \vdots & \vdots & \vdots & \vdots & \vdots \\ 0 & x_0 & x_1 & \cdots & x_{N-1} & 0 & \cdots & 0 \end{bmatrix} \tag{8.11}$$

is an $N \times 2N$ matrix. Thus, given a complex vector $\boldsymbol{c} \ne \boldsymbol{0}$ of dimension $N \times 1$,

$$\boldsymbol{c}^+\hat{\boldsymbol{R}}_{xx}\boldsymbol{c} = \frac{1}{N}\boldsymbol{c}^+\boldsymbol{Q}\boldsymbol{Q}^+\boldsymbol{c} = \left|\boldsymbol{Q}^+\boldsymbol{c}\right|^2 > 0, \tag{8.12}$$

where the form of the matrix $\boldsymbol{Q}^+$ ensures that $\boldsymbol{Q}^+\boldsymbol{c} = \boldsymbol{0}$ implies that $\boldsymbol{c} = \boldsymbol{0}$ assuming $x_{N-1} \ne 0$.

In general, we estimate the correlation functions using a maximum lag $L \ll N$, in which case there is very little difference between the biased and unbiased estimates. A rule of thumb is to use the biased estimate and a maximum lag L that is less than $10\% - 20\%$ of N and $N > 50$. Henceforth, we will drop the superscripts "u" and "b" and simply refer to the "correlation estimate" indicating the biased quantity.

Based on the Wiener-Khinchin theorem (see theorem 25) [1], an estimate of the true CPSD $S_{yx}(e^{i\omega})$ for two WSS random processes x, y of dimension $N \times 1$ and an estimate of true PSD of x, can be obtained using the DFTs of the cross-correlation estimates of the two processes,

$$\hat{S}_{yx}(e^{i\omega}) = \sum_{l=-L}^{L} \hat{R}_{yx}[l]\; e^{-il\omega}, \quad \hat{S}_{xx}(e^{i\omega}) = \sum_{l=-L}^{L} \hat{R}_{xx}[l]\; e^{-il\omega}. \tag{8.13}$$

$\hat{S}_{xx}$ is known as the *correlogram* when $L < N-1$, or the *periodogram* when $L = N-1$. The periodogram and various modifications of it are the classical techniques of estimating power spectra (see section 8.5).

A different approach to estimating the PSD of a WSS random process using the maximum likelihood (or minimum variance) estimator of the amplitude of a complex sinusoid in WGN, discussed in section 5.16, leads to an interpretation of the PSD values as outputs of narrow-band filters centered at each desired frequency. Together with a constraint that the minimum variance filters should pass a complex sinusoid without distortion, the resulting spectrum is known as the minimum variance distortionless spectral estimate (MVDSE) and is discussed in section 8.9. Although the filters are determined using the minimum variance distortionless property, the final formula for the spectrum does not involve the filters themselves and only uses the input data auto-correlation matrix that can be estimated from the data itself as described above.

Finally, in section 8.10 we introduce the AR model of order P, AR(P), which provides a formula for its PSD in terms of the autoregressive model parameters $\{P, \boldsymbol{a}, \sigma_\nu^2\}$. The AR model and ways to estimate its parameters are the subject of several spectral estimation techniques collectively known as maximum entropy method spectral estimates, or MEM spectra, which will be discussed in section 8.12; they tend to have high-resolutions (but with artifacts such as splitting of lines) and are useful for small observation length N. The theoretical basis for high resolution property of MEM spectra is the fact that the model includes all necessary information to compute its auto-correlation sequence indefinitely as a function of lag.

8.4 The Periodogram

Given a zero-mean WSS random process x_n, $n \in \mathbb{Z}$, we define its PSD through the Wiener-Khinchin theorem (theorem 25) by

$$S_{xx}\left(e^{i\omega}\right) \equiv \sum_{l=-\infty}^{\infty} R_{xx}[l]\; e^{-il\omega}, \tag{8.14}$$

where $R_{xx}[l] = \mathrm{E}[x_n x_{n-l}^*]$ is the (true) auto-correlation function of the random process x_n. Defining the sequence $x_N[n]$ to be that part of x_n limited to the finite range $0 \le n \le N-1$ and using the biased estimate introduced in section 8.2, we arrive at the *periodogram* estimate of the PSD [82]:

$$\hat{S}_{\mathrm{P}}\left(e^{i\omega}\right) \equiv \sum_{l=-N+1}^{N-1} \hat{R}_{xx}[l]\; e^{-il\omega}. \tag{8.15}$$

[1]Since the Wiener-Khinchin theorem demands the absolute integrability (or summability for discrete time) of the correlation functions, the data must have its mean removed by subtracting the associated estimated means. In practice this is equivalent to removing the DC component of the spectrum at zero frequency.

The negative lags are, of course, given by $R_{xx}[-l] = R^*_{xx}[l]$. The sequence $x_N[n]$ with finite number of elements is obtained by applying a rectangular window function $w_R[n]$ of unit height to the infinite sequence x_n,

$$x_N[n] = w_R[n]\, x_n, \quad w_R[n] = 1, \text{ for } 0 \le n \le N-1, \text{ and 0 otherwise.} \tag{8.16}$$

The auto-correlation estimate is then given by the following convolution equation

$$\hat{R}_{xx}[l] = \frac{1}{N} \sum_{n=-\infty}^{\infty} x_N[n]\, x^*_N[n-l], \tag{8.17}$$

whose Fourier transform gives the periodogram

$$\hat{S}_{\mathrm{P}}\left(e^{i\omega}\right) = \frac{1}{N}\left|X_N\left(e^{i\omega}\right)\right|^2, \quad X_N\left(e^{i\omega}\right) \equiv \sum_{n=0}^{N-1} x[n]\, e^{-in\omega}. \tag{8.18}$$

The identity [1]

$$\sum_{n=0}^{N-1} x_N[n]\, e^{-in\omega_k} \equiv e^{-i(N-1)\omega_k} \sum_{n=0}^{N-1} x_N[N-1-n]\; e^{in\omega_k} \tag{8.19}$$

shows that the periodogram at frequency values $\omega_k = 2\pi k/N$ is the squared magnitude of the convolution of x_N with an FIR filter $h_k[n] = w_R[n]\exp(in\omega_k)/\sqrt{N}$, where $w_R[n]$ is the rectangular window limited to the interval $[0, N-1]$; this is illustrated in figure 8.1. The Fourier transform of the filter $h_k[n]$ is

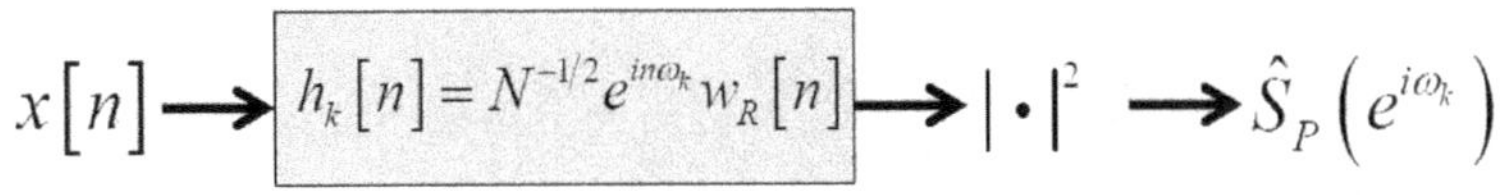

Figure 8.1: Estimate of the periodogram obtained by a linear band-pass filter.

$$H_k\left(e^{i\omega}\right) = \frac{1}{\sqrt{N}} \sum_{n=0}^{N-1} h_k[n]\, e^{-in\omega} = \frac{1}{\sqrt{N}} e^{-i(N-1)(\omega-\omega_k)/2}\, \frac{\sin\left(N\frac{\omega-\omega_k}{2}\right)}{\sin\left(\frac{\omega-\omega_k}{2}\right)}, \tag{8.20}$$

and its magnitude is shown in figure 8.2 for $N = 128$ and $\omega_k = 0.1$. Thus, the periodogram estimate at ω_k is equal to the square of the magnitude of the output of a band-pass filter as shown in figure 8.1, with the filter centered at ω_k having a width in the frequency domain of approximately $2\pi/N$.

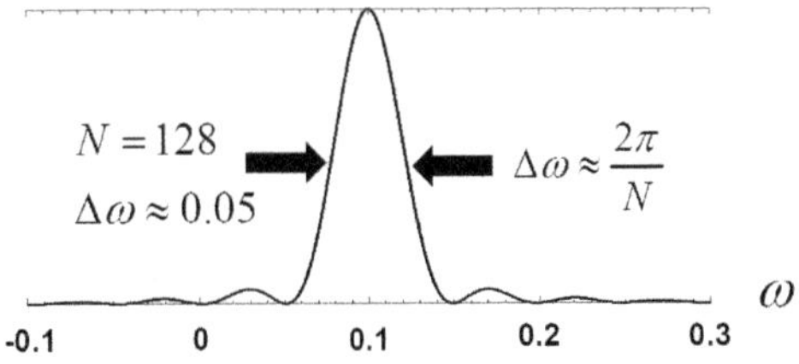

Figure 8.2: Spectrum of the filter $h_k[n] = w_R[n]e^{in\omega_k}/N$.

The interpretation of the periodogram estimate of the PSD of a random signal at frequency values ω_k as the output of a number of band-pass filters, each centered at the corresponding frequency ω_k, will be used later to derive the minimum variance distortionless spectral estimate (MVDSE).

Equations (8.17) and (8.18) can be generalized to two correlated WSS random processes with finite number of elements to give

$$\hat{R}_{yx}[l] = \frac{1}{N} \sum_{n=-\infty}^{\infty} y_{N}[n] x_{N}^{*}[n-l], \quad \hat{S}_{yx}(e^{i\omega}) = \frac{1}{N} Y_N(e^{i\omega}) X_N^{*}(e^{i\omega}). \tag{8.21}$$

8.5 Statistical Properties of the Periodogram

In order to study the mean and covariance of the periodogram, we begin with the biased auto-correlation estimate whose expectation value is

$$\mathrm{E}\big[\hat{R}_{xx}[l]\big] = \left(1 - \frac{|l|}{N}\right) R_{xx}[l] \equiv w_B\,[l]\, R_{xx}\,[l]\,, \tag{8.22}$$

defining the Bartlett (or the triangle) window function $w_B[l]$. The expectation value of the periodogram is, therefore, the convolution of the Fourier transform of the (true) auto-correlation function and the Fourier transform of the Bartlet (triangle) window function. Thus,

$$\mathrm{E}\left[\hat{S}_{\mathrm{P}}\left(e^{i\omega}\right)\right] = \frac{1}{2\pi N} \int_{-\pi}^{\pi} \left| \frac{\sin\left(N(\sigma-\omega)/2\right)}{\sin\left((\sigma-\omega)/2\right)} \right|^2 S_{xx}\left(e^{i\sigma}\right) d\sigma. \tag{8.23}$$

Using the relation

$$\lim_{N\to\infty} \frac{1}{N} \left| \frac{\sin\left(N\omega/2\right)}{\sin\left(\omega/2\right)} \right|^2 = 2\pi \sum_{n=-\infty}^{\infty} \delta\left(\omega - 2n\pi\right), \tag{8.24}$$

we find that the periodogram is asymptotically unbiased [1]

$$\sum_{N\to\infty} \mathrm{E}\big[\hat{S}_{\mathrm{P}}\left(e^{i\omega}\right)\big] \to S_{xx}\left(e^{i\omega}\right) \quad \text{as } N\to\infty. \tag{8.25}$$

Equation (8.24) can be proven by using the discrete time Fourier transform of the triangle function

$$f_{N}[n] = 1 - |n|/N, \quad -N \le n \le N, \text{ and zero otherwise} \quad \leftrightarrow \quad F_{N}(\omega) = \frac{1}{N} \left| \frac{\sin(N\omega/2)}{\sin(\omega/2)} \right|^2, \tag{8.26}$$

when by the Fourier convolution theorem for discrete time sequences we have

$$\int_{-\pi}^{\pi} F_{N}\left(\omega\right) G\left(\omega\right) e^{ik\omega} d\omega = 2\pi \sum_{m=-\infty}^{\infty} \left(1 - \frac{|k-m|}{N}\right) g\,[m], \tag{8.27}$$

for an arbitrary function $G(\omega)$ that is the Fourier transform of some discrete time sequence $g[m] \in l_2(\mathbb{Z})$. Setting $k=0$ we find

$$\int_{-\pi}^{\pi} F_{N}\left(\omega\right) G\left(\omega\right) d\omega \to 2\pi \sum_{m=-\infty}^{\infty} g\,[m] = 2\pi G\left(0\right) \quad \text{as } N\to\infty.$$

Thus, $F_N(\omega) \to 2\pi\delta(\omega)$ as $N \to \infty$ for $\omega \in [-\pi, +\pi]$. Equation (8.24) then follows from the fact that $F_N(\omega)$ is 2π-periodic.

Next, we calculate the periodogram auto-correlation at two distinct frequencies ω_1 and ω_2

$$\mathrm{E}\left[\hat{S}_{\mathrm{P}}\left(e^{i\omega_1}\right)\hat{S}_{\mathrm{P}}\left(e^{i\omega_2}\right)\right] = \frac{1}{N^2}\sum_{k,l,m,n=0}^{N-1} \mathrm{E}\left[x_k x_l^* x_m x_n^*\right] e^{-i(k-l)\omega_1} e^{-i(m-n)\omega_2}. \tag{8.28}$$

Here we assume that the process is jointly Gaussian and use the result [83]

$$\mathrm{E}\left[x_k x_l^* x_m x_n^*\right] = \mathrm{E}\left[x_k x_l^*\right]\mathrm{E}\left[x_m x_n^*\right] + \mathrm{E}\left[x_k x_n^*\right]\mathrm{E}\left[x_m x_l^*\right], \tag{8.29}$$

in addition to a further assumption that x_n is white, i.e., $R_{xx}[l] = \sigma^2\delta[l]$, to find

$$\mathrm{E}\left[\hat{S}_{\mathrm{P}}\left(e^{i\omega_1}\right)\hat{S}_{\mathrm{P}}\left(e^{i\omega_2}\right)\right] = \sigma^4\left(1 + \left|\frac{\sin\left(N\frac{\omega_1+\omega_2}{2}\right)}{N\sin\left(\frac{\omega_1+\omega_2}{2}\right)}\right|^2 + \left|\frac{\sin\left(N\frac{\omega_1-\omega_2}{2}\right)}{N\sin\left(\frac{\omega_1-\omega_2}{2}\right)}\right|^2\right),$$

giving the auto-covariance at the two distinct frequencies [1]:

$$\mathbf{Cov}\left(\hat{S}_{\mathrm{P}}\left(e^{i\omega_1}\right)\hat{S}_{\mathrm{P}}\left(e^{i\omega_2}\right)\right) = \sigma^4\left|\frac{\sin\left(N\frac{\omega_1+\omega_2}{2}\right)}{N\sin\left(\frac{\omega_1+\omega_2}{2}\right)}\right|^2 + \sigma^4\left|\frac{\sin\left(N\frac{\omega_1-\omega_2}{2}\right)}{N\sin\left(\frac{\omega_1-\omega_2}{2}\right)}\right|^2. \tag{8.30}$$

Thus, the variance of the periodogram at a single frequency ω is

$$\mathbf{Var}\left(\hat{S}_{\mathrm{P}}\left(e^{i\omega}\right)\right) = \sigma^4\left(1 + \left|\frac{\sin(N\omega)}{N\sin(\omega)}\right|^2\right), \tag{8.31}$$

which does not vanish as $N \to \infty$; i.e., the periodogram is not a *consistent estimate* of the true PSD. Note that for frequency values of a discrete Fourier transform, we have $\omega_1 = 2\pi k_1/N$ and $\omega_2 = 2\pi k_2/N$, and the covariance vanishes when $k_1 \neq k_2$, i.e., the periodogram values separated by integer multiples of $2\pi/N$ are uncorrelated. As $N \to \infty$, those uncorrelated frequency values get closer together while the variance approaches a non-zero constant; hence, the fluctuations in the periodogram increase with increasing N.

For colored Gaussian processes, it can be shown that the variance of the periodogram is approximately proportional to the square of the true spectrum so the periodogram is generally not a consistent estimate and can fluctuate substantially about the true spectral value [84]. These results tend to hold for a broad range of random processes and so it is crucial to find methods to reduce the variance of the spectral estimate; for the periodogram, this means a modification to the periodogram itself, as discussed in section 8.6.

8.6 Reducing the Periodogram Variance

Variance reduction methods applied to the periodogram adversely affect the frequency resolution of the spectral estimate. We therefore digress to introduce the resolution of a spectral estimate, which is a measure of its ability to resolve two closely spaced spectral lines. For instance, consider the signal

$$x_n = A_1\sin(n\omega_1 + \phi_1) + A_2\sin(n\omega_2 + \phi_2) + \nu_n, \quad 0 \le n \le N-1, \tag{8.32}$$

where ν_n is WGN with variance σ_ν^2 and phases $\phi_{1,2}$ that are uniformly distributed in $[-\pi, \pi]$. The mean of the periodogram estimate is

$$\begin{aligned}\mathrm{E}\big[\hat{S}_{\mathrm{P}}\left(e^{i\omega}\right)\big] &= \frac{1}{2\pi}\int_{-\pi}^{\pi} S_{xx}\left(e^{i\sigma - i\omega}\right) W_B\left(e^{i\sigma}\right)\, d\sigma \\ &= \sigma_\nu^2 + 0.25\, A_1^2\Big(W_B\big(e^{i\omega - i\omega_1}\big) + W_B\big(e^{i\omega + i\omega_1}\big)\Big) \\ &+ 0.25\, A_2^2\Big(W_B\big(e^{i\omega - i\omega_2}\big) + W_B\big(e^{i\omega + i\omega_1}\big)\Big),\end{aligned} \tag{8.33}$$

where $W_B(e^{i\omega})$ is the Fourier transform of the triangle function (the Bartlett window). Equation (8.33) shows that the lines at $\omega_{1,2}$ are now represented by the peaks of the function W_B shifted to those frequencies. Figure 8.3 shows examples of these shifted peaks for $N = 64$: on the left, we have $\omega_1 = 1$ and $\omega_2 = 1.05$, while on the right, we have $\omega_1 = 1$ and $\omega_2 = 1.09$, with the vertical axis in dB scale. Thus, two frequencies are resolvable when their spectral tails cross below 3 dB from the peak value: the frequencies on the left of figure 8.3 are unresolvable, while those on the right are. For the Bartlett window, the resolution $\Delta\omega$ is approximately $(0.89)2\pi/N$, which equals 0.087 in this example, confirming that the two frequencies separated by 0.09 are in fact resolvable.

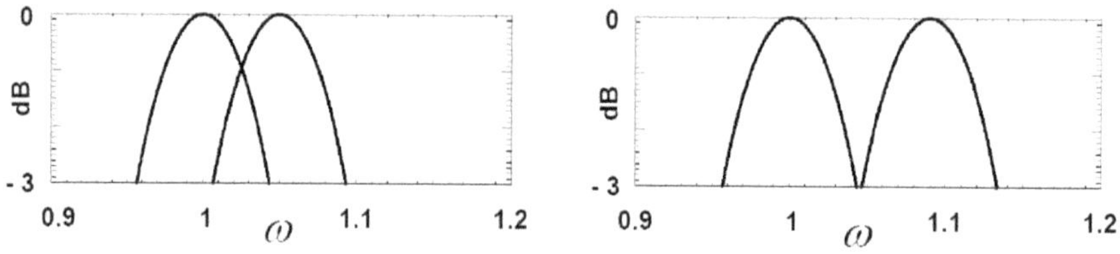

Figure 8.3: Frequency resolution of the periodogram.

There are essentially two methods of reducing the variance of the periodogram. The first method relies on smoothing the periodogram by either averaging it over small bands of frequencies (a moving average) or by windowing the associated auto-correlation sequence, which is equivalent to convolving the periodogram with the Fourier transform of the window function; this amounts to a weighted average in the frequency domain,

$$\hat{S}_{\mathrm{sP}}\left(e^{i\omega_l}\right) = \frac{1}{2K+1}\sum_{k=-K}^{K} \hat{S}_P\left[e^{i\omega_{l-k}}\right]. \tag{8.34}$$

Since the periodogram samples are nearly uncorrelated, this reduces the variance of the periodogram by a factor of $2K+1$. However, the averaging operation worsens the resolution by the same factor.

Blackman and Tukey [85], on the other hand, preferred to multiply the auto-correlation sequence by a *lag window* $w[l]$ before taking a Fourier transform,

$$\hat{S}_{\mathrm{BT}}\left(e^{i\omega}\right) = \sum_{l=-L+1}^{L-1} \hat{R}_{xx}\left[l\right] w\left[l\right]\, e^{-il\omega}. \tag{8.35}$$

When $L \ll N$, the Blackman-Tukey smoothed correlogram (8.35) has the following statistical measures:

$$\mathrm{E}\left[\hat{S}_{\mathrm{BT}}(e^{i\omega})\right] \approx \frac{1}{2\pi}\int_{-\pi}^{\pi} S_{xx}\left(e^{i\sigma}\right) W\left(e^{i\sigma-i\omega}\right)\, d\sigma,$$

$$\mathbf{Var}\left[\hat{S}_{\mathrm{BT}}(e^{i\omega})\right] \approx \frac{1}{N}\, S^2_{xx}\left(e^{i\omega}\right) \sum_{l=-L+1}^{L-1} w^2\,[l], \tag{8.36}$$

where $W(e^{i\omega})$ is the Fourier transform of the lag window [84]. A small bias requires a narrow main-lobe of the window in the frequency domain, which translates to a large value for L. On the other hand, a small variance requires a small value for the window energy, which requires a small value for L. A reasonable compromise in practice is achieved by choosing $L \sim N/5$. Note also that the Blackman-Tukey method worsens the spectral resolution of the periodogram from $1/N$ to approximately $1/L$ with the final resolution depending on the choice of the window.

The second (and more popular) method of variance reduction, known as Welch's method [86], is to divide the record length N of the signal into J overlapping sub-records of length L, apply a fixed window function to each sub-record before computing the associated periodograms, and finally average all of them. Denoting by $x_j[l]$, $0 \le l \le L-1$, the $L \times 1$ data in the j-th section and using an $L \times 1$ data window w with average energy u_w, the Welch spectral estimate is

$$\hat{S}_{\mathrm{W}}\left(e^{i\omega}\right) = \frac{1}{JLu_w}\sum_{j=1}^{J}\left|\sum_{l=0}^{L-1} w\,[l]\, x_j\,[l]\, e^{-il\omega}\right|^2, \quad u \equiv L^{-1}\sum_{l=0}^{L-1} w^2\,[l]. \tag{8.37}$$

The Bartlett method is a special case of Welch's method when there is no overlap between sections and a rectangular data window is used: the associated spectral estimate is then denoted by $\hat{S}_{\mathrm{B}}$. The Bartlett estimate has the same expectation value as the periodogram and therefore is asymptotically unbiased. Its variance, however, is reduced by a factor of J (the number of non-overlapped sections) at the expense of frequency resolution worsened by the same factor. Using a non-zero overlap increases the correlation between sub-records and adversely affects variance reduction; in practice, sub-records are overlapped by 50%.

A figure of merit that combines both resolution and variability is

$$\gamma \equiv v\,\Delta\omega, \quad v = \mathbf{Var}\left[\hat{S}_{xx}\left(e^{i\omega}\right)\right] \Big/ \left(\mathbf{E}\left[\hat{\mathbf{S}}_{\mathbf{xx}}\left(\mathbf{e}^{i\omega}\right)\right]\right)^2, \tag{8.38}$$

where v is the coefficient of variation and $\Delta\omega$ is the frequency resolution. Table 8.1 summarizes the values of variability v, frequency resolution $\Delta\omega$, and figure of merit γ for all the classical spectral estimation methods discussed here. Thus, all classical methods of variance reduction result in roughly the same figure of merit $\gamma \approx 2\pi/N$, and they are all fundamentally limited by the contradictory requirements of lower variance and finer resolution for fixed record length; modern spectral estimation techniques try to overcome this limitation.

Both the Blackman-Tukey and the Welch methods can be generalized to the case of estimating the CPSD between two WSS random processes with finite number of elements. Thus, in analogy to equations (8.35) and (8.37), we have the following expressions for the CPSD estimate:

$$\sum_{l=-L+1}^{L-1} \hat{R}_{yx}\,[l]\, w\,[l]\; e^{-il\omega}, \quad \text{or} \quad \frac{1}{JLu_w}\sum_{j=1}^{J} Y_j X_j^*, \tag{8.39}$$

where Y_j and X_j are the Fourier transforms of the corresponding windowed data sections of length L.

Method	v	$\Delta\omega$	γ
Periodogram	1	$(0.89)2\pi/N$	$(0.89)2\pi/N$
Bartlett	$1/J$	$(0.89)2\pi J/N$	$(0.89)2\pi/N$
Welch (50%)	$9/(8J)$	$(1.28)2\pi J/N$	$(0.72)2\pi/N$
Blackman-Tukey	$2M/(3N)$	$(0.64)2\pi/M$	$(0.43)2\pi/N$

Table 8.1: Figures of merit for classical spectral estimates [1].

8.7 The Multitaper Method

Consider an $N \times 1$ sample of a WSS random process x_n, $0 \le n \le N-1$. Figure 8.2 implies that the periodogram is a smoothed estimate of the true spectrum, where the smoothing kernel is the squared magnitude of the filter Fourier transform; thus, large side-lobes cause significant leakage from the main-lobe. On the other hand, low side-lobes come at the expense of a wide main-lobe. The issues of bias and variance can be addressed optimally by using multiple data window functions; this is referred to as the multitaper method [87].

In multitaper spectrum estimation, we use several different windows and average the resulting spectra, which, if uncorrelated, will reduce the variance of the estimate by a factor equal to the number of windows (tapers). If $w_n^{(k)}$, $1 \le k \le K$, denotes K tapers used in obtaining K spectral estimates $\hat{S}^{(k)}(\omega)$, then the spectral estimates are approximately uncorrelated when [88]

$$\sum_{n=0}^{N-1} w_n^{(l)} w_n^{(k)} = \delta_{lk}, \quad 1 \le, l, k \le K, \tag{8.40}$$

and the multitaper spectral estimate is

$$\hat{S}_{\text{mt}}(\omega) = \frac{1}{K}\sum_{k=1}^{K} \hat{S}^{(k)}(\omega) = \frac{1}{K}\sum_{k=1}^{K} \left| \sum_{n=0}^{N-1} w_n^{(k)} x_n \, e^{-in\omega} \right|^2 . \tag{8.41}$$

Since the variance is reduced by a factor K, we need to find specific window functions to minimize the bias of the spectral estimate. Denoting the (normalized) linear bandwidth parameter Δf, the minimization problem becomes [89]

$$\underset{\boldsymbol{w}^{(1)},\ldots,\boldsymbol{w}^{(K)}}{\arg\max} \sum_{k=1}^{K} \boldsymbol{w}^{(k)\,T} \boldsymbol{D}\, \boldsymbol{w}^{(k)}, \quad \left[\boldsymbol{D}\right]_{lk} = \frac{\sin\left[\pi \Delta f \,(l-k)\right]}{\sin\left[\pi (l-k)\right]} . \tag{8.42}$$

The solutions to (8.42) are the K orthogonal eigenvectors corresponding to the K largest eigenvalues of the $N \times N$ real symmetric matrix $\boldsymbol{D}$, but these are the first K DPSSs of equation (1.55). The largest $K \approx 2N\,\Delta f$ eigenvalues are close to 1, while the rest are nearly zero (see figure 1.4). Δf determines the resolution of the estimated spectrum; so to resolve two peaks that are Δf_0 apart, we must take $\Delta f < \Delta f_0$. Smaller bandwidths result in smaller K favoring the bias of the estimate over its variance. A common choice [87] is $K = 2N\,\Delta f - 1$.

An alternative way to arrive at the DPSS is to use the spectral representation theorem, also known as Cramér's representation [90].

Theorem 31. *For any zero-mean WSS random process x_n, there exists a process $Z(\omega)$, $\omega \in [-\pi, +\pi]$, such that*

$$x_n = \int_{-\pi}^{+\pi} e^{+in\omega}\, dZ(\omega), \tag{8.43}$$

where $Z(\omega)$ has orthogonal increments in the sense that for any two non-intersecting intervals $[\omega_1, \omega_2]$ and $[\omega_3, \omega_4]$, $E\big[(Z(\omega_2) - Z(\omega_1))(Z(\omega_4) - Z(\omega_3))\big] = 0$.

Cramér's representation is a generalization of the Fourier series

$$x_n = \sum_{k=0}^{N} \left(a_k \cos(n\omega_k) + b_k \sin(n\omega_k)\right), \quad \omega_k = 2\pi k/N, \quad n = 0, \ldots, N-1, \tag{8.44}$$

to WSS random processes (we assume real processes for now) by taking the limit $N \to \infty$ when the coefficients a_k, b_k will tend to 0. An alternative representation is then available by using differentials $a_k \to dA(\omega_k)$ and $b_k \to dB(\omega_k)$, where $A(\omega)$ and $B(\omega)$ are step functions with discontinuities at ω_k. Taking the limit $N \to \infty$ and replacing the sum in (8.44) with an integral we obtain

$$x_n = \int_0^{\pi} \Big(\cos(n\omega)\, dA(\omega) + \sin(n\omega)\, dB(\omega)\Big), \tag{8.45}$$

where we now consider n to have infinite range $n \in \mathbb{Z}$, and the right hand side of (8.45) is interpreted as a Fourier-Stieltjes integral [30]. If x_n is a zero-mean random process, so are $A(\omega)$ and $B(\omega)$ over the interval $[0, \pi]$. We further assume that the two processes are mutually uncorrelated with uncorrelated (orthogonal) increments and that the variance of the increments define the spectral distribution and density functions. Thus,

$$\mathrm{E}\big[dA(\omega)\, dB(\omega')\big] = 0, \quad \omega, \omega' \in [0, \pi], \tag{8.46a}$$

$$\mathrm{E}\big[dA(\omega)\, dA(\omega')\big] = 0, \quad \mathrm{E}\big[dB(\omega)\, dB(\omega')\big] = 0, \ \omega \neq \omega', \tag{8.46b}$$

$$\mathbf{var}\big[dA(\omega)\big] = \mathbf{var}\big[dB(\omega)\big] = 2\, dP(\omega) = 2\, S(\omega)\, d\omega, \tag{8.46c}$$

where $P(\omega)$ is the spectral distribution function and its derivative $S(\omega)$ is the spectral density function.

We may write (8.45) in terms of complex exponentials by defining a complex random processes

$$dZ(\omega) = dA(\omega) - i\, dB(\omega) \tag{8.47}$$

and extending the domain of definition of A and B to the interval $[-\pi, +\pi]$ with $A(\omega) = A(-\omega)$ and $B(\omega) = -B(-\omega)$ (the symmetry and anti-symmetry conditions ensure that x is a real process). Then $dZ^*(\omega) = dZ(-\omega)$, and the relations (8.46) become

$$\mathrm{E}\big[dZ(\omega)\, dZ^*(\omega')\big] = 0, \ \omega \neq \omega', \ \text{ and } \mathrm{E}\big[dZ(\omega)\, dZ^*(\omega)\big] = S(\omega)\, d\omega. \tag{8.48}$$

In terms of dZ, equation (8.45) becomes the Cramér's spectral representation (8.43) for a real process. The complex process $dZ(\omega)$ can be written less formally as $z(\omega)\, d\omega$ in terms of a zero-mean complex process $z(\omega)$ whose variance is the signal spectrum,

$$\mathrm{E}\Big[\big|z(\omega)\big|^2\Big] = S(\omega). \tag{8.49}$$

Using $dZ(\omega) = z(\omega)\, d\omega$ and normalized linear frequency $\nu = \omega/2\pi$ in (8.43) and substituting the Cramér's spectral representation of x_n in its Fourier transform $X(f)$, we find

$$X(f) = \sum_{n=0}^{N-1} e^{-2i\pi nf} x_n = \int_{-1/2}^{+1/2} e^{-i(N-1)\pi(f-\nu)} \frac{\sin[N\pi(f-\nu)]}{\sin[\pi(f-\nu)]}\, z(\nu)\, d\nu. \tag{8.50}$$

The exponential term multiplying the integral can be removed by shifting the data x_n by $(N-1)/2$ in (8.43). Using the time shifted data, we arrive at a Fredholm integral equation of the first kind with a real symmetric kernel [12] for the random process $z(\nu)$:

$$X(f) = \int_{-1/2}^{+1/2} \frac{\sin[N\pi(f-\nu)]}{\sin[\pi(f-\nu)]}\, z(\nu)\, d\nu. \tag{8.51}$$

The integral equation (8.51) is similar to a real symmetric matrix equation, except that we are now dealing with an infinite dimensional matrix whose rows f and columns ν are continuously labeled. We also note that the kernel approaches the Dirac delta distribution $\delta(f-\nu)$ as $N \to \infty$. Thomson [87] solved the integral equation (8.51) in a "local interval" about a given frequency f, namely, $[f - \Delta f, f + \Delta f]$. Since the kernel is already centered at $\nu = f$, equation (8.51) written for the local interval becomes

$$X(f) = \int_{-\Delta f}^{+\Delta f} \frac{\sin[N\pi(f-\nu)]}{\sin[\pi(f-\nu)]}\, z(\nu + f)\, d\nu. \tag{8.52}$$

A series solution based on the eigenfunctions of the integral operator in (8.52) is

$$z(\nu) = \sum_k \frac{1}{\lambda_k} \Big[\int_{-\Delta f}^{+\Delta f} X(f+\nu)\psi^{(k)}(f)df \Big]\, \psi^{(k)}(\nu). \tag{8.53}$$

Using the original integral equation (8.51), which shows the projection from z onto X, together with the integral equation defining the discrete prolate spheroidal functions, we may write the k-th coefficient in (8.53) as

$$z^{(k)}(\nu) \equiv \int_{-1/2}^{+1/2} X(f+\nu)\psi^{(k)}(f)df. \tag{8.54}$$

The right hand side of (8.54) can be written as the DFT of the product of time domain functions; changing ν to f, we have

$$z^{(k)}(f) = \sum_{n=0}^{N-1} x_n \psi^{(k)}[n]\, e^{-2i\pi f(n-(N-1)/2)}. \tag{8.55}$$

Equation (8.55) is the DFT of the original data (the phase shift can be absorbed in the discrete prolate spheroidal sequence) windowed by the k-th DPSS. The multitaper spectral estimate is then found by averaging the squared magnitudes of the right hand side of (8.55).

A 500-point realization of a noisy signal containing two real sinusoids at 0.11 and 0.22 Hz (normalized frequencies) is shown at the top of figure 8.4. Two spectral estimates are shown at the bottom of the same figure: a multitaper spectral estimate with $c/\pi = 2.5$ and $K = 2c/\pi - 1 = 4$ largest eigenvalue DPSS, and a periodogram estimate using Welch's method with 100-point sections, 50% overlap, and a Kaiser-Bessel window.

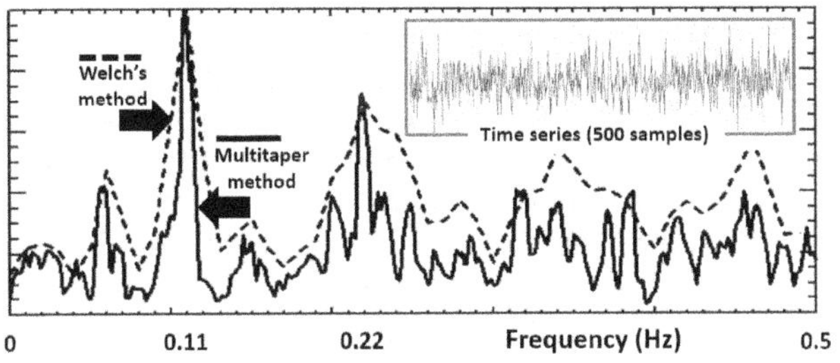

Figure 8.4: Comparison between multitaper and Welch's methods.

8.8 Example Applications of Classical Spectral Estimation

Consider three widely separated sensors (for instance, seismometers, hydrophones, or magnetometers), one of which is situated close to a signal emitter of interest; we will call this sensor the primary channel. The other two, called the reference channels, are placed sufficiently far away that they record only background noise and no signal. Alternatively, we imagine the signal emitter to be a slowly moving object passing close to the first sensor (the primary channel) but very far from the other two sensors (the reference channels). An example of a pure signal $s(t)$ and its spectrum are shown in figure 8.5; the signal consists of three sinusoids at 0.01, 0.025, and 0.05 Hz (data is assumed to be sampled at 1 Hz)[2]. While the signal of interest

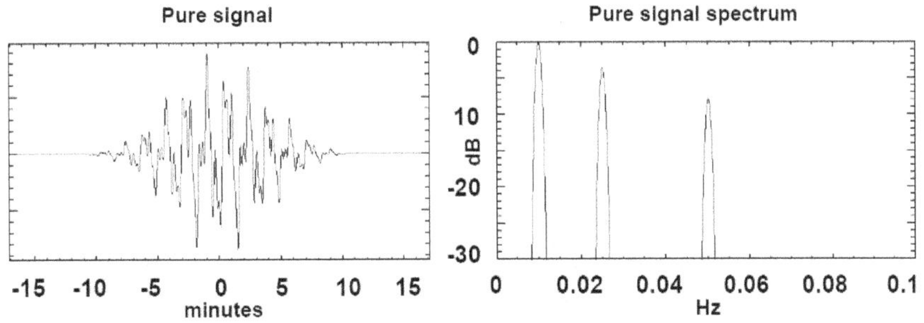

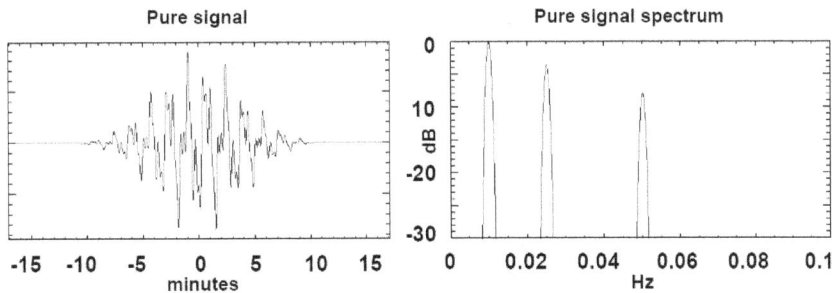

Figure 8.5: Pure signal and its spectrum.

is of finite duration and is assumed to be only on the primary channel, the background noise is recorded on all three sensors. The noise is not identical on all sensors, but its recorded versions retain statistical correlations that can be used to enhance the signal by canceling those correlations. In addition to the background noise, each sensor is subject to some receiver noise which, in practice, limits the recording of signal and background noise below a threshold referred to as the system noise level. This residual noise on each receiver is of thermal origin and is completely uncorrelated among all the receivers and cannot be removed. The background noise, on the other hand, has a physical origin in the environment, for instance, the Earth's magnetic field (when the sensors are magnetometers), or the low frequency surface waves known as microseisms (when the sensors are seismometers).

A trivial situation occurs when all sensors record exactly the same background data, in which case only one reference channel is needed. The primary channel data $s(t) + b(t) + n_1(t)$, where $b(t)$ is a background

[2] Although we use the continuous time notation, the actual data here is a sampled version. The spectrum of the signal is computed using a finite segment of the recorded data and shows some width for each sinusoidal component.

time series independent of location and $n_1(t)$ is the receiver noise, while the data on the reference channel is $b(t)+n_2(t)$. Cancellation of the background is achieved by simply subtracting the two data sets to obtain the best possible signal estimate as $s(t) + n_1(t) - n_2(t)$: the background masking the signal of interest disappears, and the signal should be quite discernible in as much as the residual noise is sufficiently low. Unfortunately, background data are usually inhomogeneous (differ from place to place); even though the data on different channels might share the same origin, the propagation channels cause differences on spatially separated sensors. In many circumstances, however, data recorded on widely separated channels retain some coherence (statistical correlations as a function of frequency as discussed in section 7.8), which allow us to remove a part of the background (as indicated by the coherence) from the primary channel, revealing the signal of interest, using the non-causal Wiener filter formulation of section 7.8 and equation (7.79).

For instance, consider the signal shown in figure 8.5. The primary channel recording, including background noise, and its spectrum are shown in figure 8.6; the background noise on this channel has completely buried the signal of interest. The two reference channels' time series are shown in figure 8.7. The expected

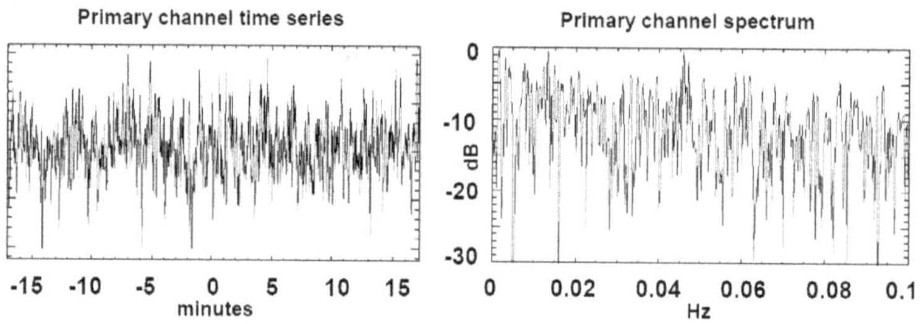

Figure 8.6: Data on the first sensor (signal plus background) and its spectrum.

amount of background noise cancellation is, of course, directly related to the coherence of the background data among all channels. In this example, data segments are 34 minutes long and sampled at 1.5625 Hz.

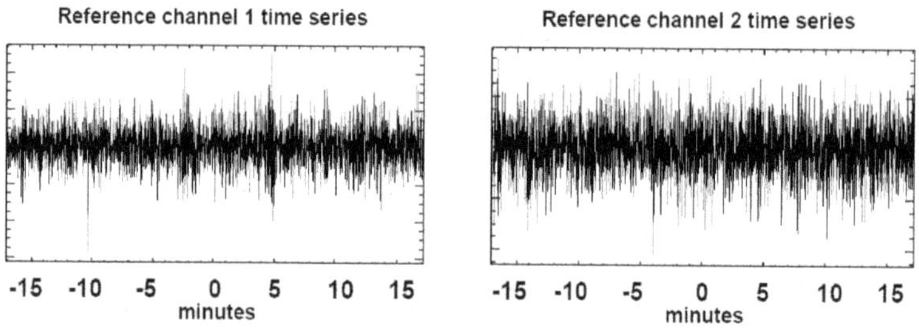

Figure 8.7: Data on the two reference channels.

Two Wiener filters $\boldsymbol{h}^{(1)}$ and $\boldsymbol{h}^{(2)}$ with 151 samples (approximately 1.6-minute duration) were derived from the matrix equation (7.79) whose entries were computed using Welch's method of auto- and cross-spectral estimation with a Kaiser window and sections with 151 samples, overlapped by 50%. The two filters were applied to the two reference channels shown in figure 8.7, and the sum was subtracted from the primary channel of figure 8.6 to obtain an *error sequence* e_n, where

$$e_n = y_n - \hat{y}_n,\ \hat{y}_n = \sum_{m=-75}^{75} h_m^{(1)} x_{n-m}^{(1)} + \sum_{m=-75}^{75} h_m^{(2)} x_{n-m}^{(2)}. \tag{8.56}$$

Figure 8.8 shows substantial cancellation of the correlated noise, unmasking the signal in the time domain and revealing the three signal frequencies well above the noise level in the frequency domain.

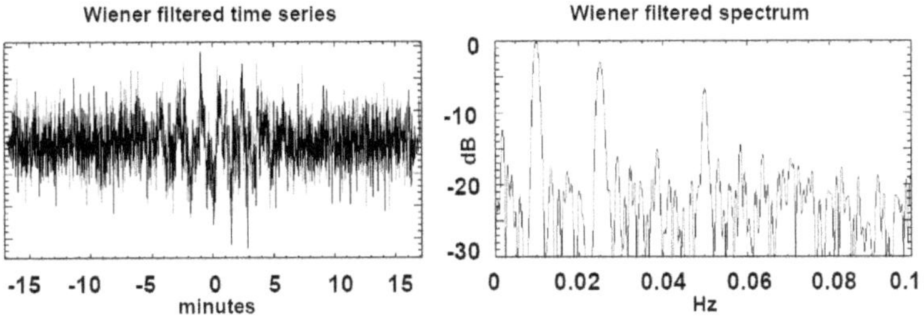

Figure 8.8: Wiener-filtered output and its spectrum.

Another application is to the matched filter solution of section 7.2, namely equation (7.11). Once the noise PSD $S_{\nu\nu}(\omega)$ is estimated from data, the matched filter is found by inverse Fourier transformation. Using the same data set in section 7.2, the noise spectral estimate was calculated using 201-point sections at 50% overlap with a Kaiser-Bessel window, as shown on the left of figure 8.9 (maximum is set at 0 dB). Once again, two filtered outputs were calculated: one with the matched filter and the other with the time-reversed signal used as an approximate matched filter (i.e., assuming a white noise PSD). The distribution function of the difference between the output SNRs in dB for 1000 realizations of the background noise process is shown on the right of figure 8.9.

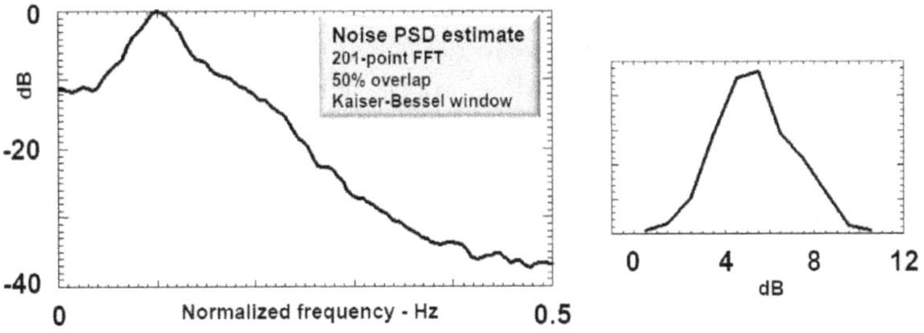

Figure 8.9: Application of Welch's method to the matched filter.

Finally, we apply the classical estimation techniques to the problem of time delay estimation discussed in section 7.9. Consider 5 seconds of a broadband ($[0, 40]$ Hz) Gaussian random signal with variance $\sigma_s^2 = 1$ sampled at 200 Hz and contaminated by two tones at 7 and 11 Hz with equal amplitudes of 3.5, in addition to WGN on both channels with equal variances of 1. The SCC (standard cross-correlator—see table 7.1) is clearly dominated by the contaminating tones. The SCOT weighted generalized cross-correlation suppresses the tonal contamination and correctly estimates the time delay at 0.1 seconds, as illustrated in figure 8.10.

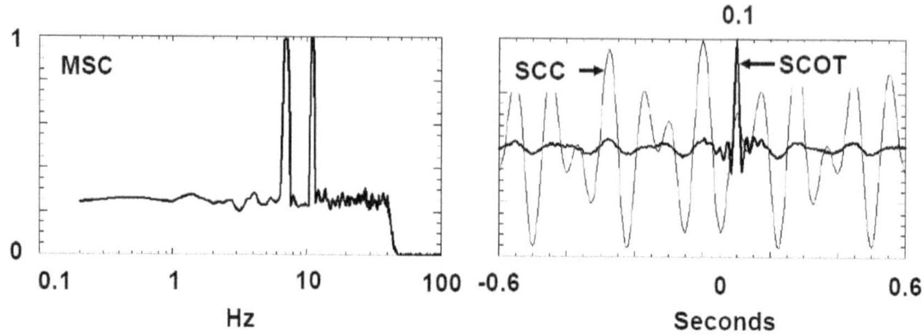

Figure 8.10: Time delay estimation: Magnitude Square Coherence (left) and generalized cross-correlator using SCOT weights.

8.9   Minimum Variance Distortionless Spectral Estimator

As observed in section 5.16, the ML (or the equivalent linear MVU) estimate of the amplitude of a complex sinusoid at a known frequency ω in Gaussian noise has a filtering interpretation so that if the input data vector is a pure sinusoid at the same frequency ω, then it passes through the filter undistorted; we called this the distortionless property of the linear minimum variance estimator. In addition, the minimum variance property of the filter corresponds to the minimization of the noise power at the output of the filter; the output power is then the best estimate of the signal power at the frequency ω. This interpretation of the MLE of the amplitude of a pure sinusoid in Gaussian noise is the basis for the minimum variance distortionless spectral estimate (MVDSE) [91]. The MVDSE is often used in spatial processing of arrays to find direction of arrival of a signal of interest; it has been known as the ML spectral estimate in exploration geophysics since the work of Lacoss [92].

Figure 8.1 showed the periodogram as the output of a linear bank of known filters $h_k[n]$, each centered at frequency ω_k at which the spectral estimate is sought. Thus, we view the filter as an unknown narrow-band FIR sequence of unknown dimension M centered at the frequency of interest ω_k, as shown in figure 8.2. Using the Fourier transform relation between the PSD and the auto-correlation function (6.9) and assuming the signal to be a zero-mean WSS random process with variance σ^2, we have

$$\sigma^2 = R_{xx}[0] = \frac{1}{2\pi} \int_{-\pi}^{\pi} S_{xx}\left(e^{i\omega}\right) d\omega. \tag{8.57}$$

Thus, the signal variance σ^2 is equal to the total signal power over the signal's entire band. Similarly, the variance of the (zero-mean WSS random) band-passed signal σ_k^2 (when properly normalized by the filter bandwidth) is an estimate of the signal power at the desired frequency ω_k, since it is equal to the signal power integrated over the pass band $\Delta\omega$ of the filter, as shown in figure 8.11.

$$\boxed{\sigma_x^2 = \frac{1}{2\pi} \int_{-\pi}^{\pi} S_{xx}(\omega)\, d\omega} \quad \mathbf{X} \longrightarrow \boxed{\mathbf{h}_k} \longrightarrow \mathbf{x}^{(k)} \boxed{\sigma_k^2 = \frac{1}{2\pi} \int_{\omega_k - \Delta\omega/2}^{\omega_k + \Delta\omega/2} S_{xx}(\omega)\, d\omega}$$

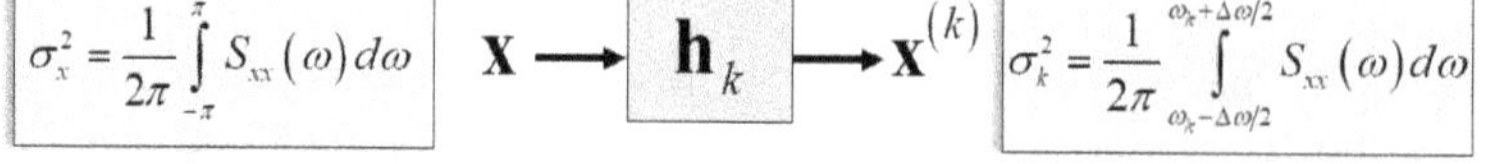

Figure 8.11: Spectral estimate at ω_k as the variance of the output of a linear band-pass filter centered at ω_k.

Let $\boldsymbol{x}$ denote the $M \times 1$ data sequence $[x_{n-(M-1)}, \ldots, x_n]^T$ and define the time-reversed data sequence

$\boldsymbol{x}_n \equiv [x_n, \ldots, x_{n-(M-1)}]^T = \boldsymbol{J}\boldsymbol{x}$. Then the convolution output $x_n^{(k)}$ of the filter $h_k[n] = [h_0, \ldots, h_{M-1}]^T$ for input $\boldsymbol{x}$ is $x_n^{(k)} = \boldsymbol{h}_k^T \boldsymbol{x}_n$ [3]. The filtered output variance can be written in terms of the filter and the input auto-correlation matrix

$$\mathrm{E}\left[\left|x_n^{(k)}\right|^2\right] \equiv \sigma_k^2 = \mathrm{E}\left[\boldsymbol{h}_k^T \boldsymbol{x}_n \boldsymbol{h}^+ \boldsymbol{x}_n^*\right] = \boldsymbol{h}_k^+ \,\mathrm{E}\left[\boldsymbol{x}_n^* \boldsymbol{x}_n^T\right] \boldsymbol{h}_k = \boldsymbol{h}_k^+ \boldsymbol{R}^{(M)} \boldsymbol{h}_k, \tag{8.58}$$

where we have used (6.73) (that is, the auto-correlation matrix is calculated using the forward time indexed vector $\boldsymbol{x}$ and not $\boldsymbol{x}_n$) and the identity $\boldsymbol{h}_k^T \boldsymbol{x}_n \equiv \boldsymbol{x}_n^T \boldsymbol{h}_k$. The constraint of no distortion means that the filter output for an input of a complex sinusoid at frequency ω_k must equal a pure phase, i.e.,

$$\boldsymbol{h}_k^T \boldsymbol{J} \boldsymbol{e}_k = e^{i\phi}, \quad \boldsymbol{e}_k \equiv \left[1, e^{i\omega_k}, \ldots, e^{i(M-1)\omega_k}\right]^T, \tag{8.59}$$

where we have used the counter identity matrix $\boldsymbol{J}$ to perform the time-reversal operation on $\boldsymbol{e}_k$ required by convolution. Defining

$$\boldsymbol{d}_k \equiv \boldsymbol{J}\boldsymbol{e}_k^* = [e^{-i(M-1)\omega_k}, e^{-i(M-2)\omega_k}, \ldots, e^{-i\omega_k}, 1]^T, \quad \Rightarrow \boldsymbol{d}_k = e^{-i(M-1)\omega_k}\, \boldsymbol{e}_k, \tag{8.60}$$

we write the distortionless property as

$$\boldsymbol{h}_k^T \boldsymbol{d}_k^* = \boldsymbol{d}_k^+ \boldsymbol{h}_k = e^{i\phi} \quad \Rightarrow \quad \boldsymbol{h}_k^+ \boldsymbol{d}_k = e^{-i\phi}. \tag{8.61}$$

The optimal filter is found by introducing a Lagrange multiplier λ for the constraint and minimizing the quantity

$$\boldsymbol{h}_k^+ \boldsymbol{R}^{(M)} \boldsymbol{h}_k + \lambda \left(e^{-i\phi} - \boldsymbol{h}_k^+ \boldsymbol{d}_k\right) \tag{8.62}$$

with respect to $\boldsymbol{h}_k$ (or its Hermitian conjugate $\boldsymbol{h}_k^+$) without a constraint. Differentiating with respect to $\boldsymbol{h}_k^+$ and setting the result equal to zero, we find

$$\boldsymbol{R}^{(M)} \boldsymbol{h}_k - \lambda \boldsymbol{d}_k = 0, \quad \Rightarrow \quad \boldsymbol{h}_k = \lambda [\boldsymbol{R}^{(M)}]^{-1} \boldsymbol{d}_k. \tag{8.63}$$

Multiplying both sides of the above on the left by $\boldsymbol{d}_k^+$ and using the equivalent form of the constraint in (8.61) we find the Lagrange multiplier

$$\lambda = \frac{e^{i\phi}}{\boldsymbol{d}_k^+ [\boldsymbol{R}^{(M)}]^{-1} \boldsymbol{d}_k} = \frac{e^{i\phi}}{\boldsymbol{e}_k^+ [\boldsymbol{R}^{(M)}]^{-1} \boldsymbol{e}_k}, \tag{8.64}$$

where the last equality follows from (8.60) and (6.72) which were used to write

$$\boldsymbol{d}_k^+ [\boldsymbol{R}^{(M)}]^{-1} \boldsymbol{d}_k = \boldsymbol{e}_k^T \boldsymbol{J} [\boldsymbol{R}^{(M)}]^{-1} \boldsymbol{J} \boldsymbol{e}_k^* = \boldsymbol{e}_k^T [\boldsymbol{R}^{(M)*}]^{-1} \boldsymbol{e}_k^* = \boldsymbol{e}_k^+ [\boldsymbol{R}^{(M)}]^{-1} \boldsymbol{e}_k.$$

Substituting for λ in the equation for the filter, we find the optimal minimum variance filter

$$\boldsymbol{h}_k = e^{i\phi} e^{-i(M-1)\omega_k} \frac{[\boldsymbol{R}^{(M)}]^{-1} \boldsymbol{e}_k}{\boldsymbol{e}_k^+ [\boldsymbol{R}^{(M)}]^{-1} \boldsymbol{e}_k} \tag{8.65}$$

and the filtered output variance

$$\sigma_k^2 = \frac{1}{\boldsymbol{e}_k^+ [\boldsymbol{R}^{(M)}]^{-1} \boldsymbol{e}_k}. \tag{8.66}$$

---

[3] The formulation of the filtering operation in this section differs from the one in section 5.16 where (replacing N with M) it is defined as $\boldsymbol{h}^+ \boldsymbol{x}$ with $\boldsymbol{h} = [h_{M-1}, \ldots, h_0]^T$ and $\boldsymbol{x} = [x_0, \ldots, x_{M-1}]^T$.

To interpret this result as a true PSD we must normalize it by the filter bandwidth; setting $\Delta\omega = 2\pi/M$ we obtain *the minimum variance distortionless spectral estimate* (MVDSE) at frequency ω_k

$$\hat{S}_{\text{MVD}}\left(e^{i\omega_k}\right) = \frac{M}{\boldsymbol{e}_k^+[\boldsymbol{R}^{(M)}]^{-1}\boldsymbol{e}_k}, \quad \boldsymbol{e}_k \equiv [1, e^{i\omega_k}, e^{2i\omega_k}, \ldots, e^{i(M-1)\omega_k}]^T. \tag{8.67}$$

A simple example is provided by white noise ν_n with variance σ_ν^2 and $\boldsymbol{R} = \sigma_\nu^2 \boldsymbol{I}$. Using $\boldsymbol{e}_k^+\boldsymbol{e}_k = M$ we find $\sigma_k^2 = \sigma_\nu^2/M$, i.e., the spectral estimate is constant and independent of frequency. Total power in bandwidth $\Delta\omega$ is then $\sigma_\nu^2\Delta\omega$.

As another example, consider a real zero-mean WSS random process and assume that its auto-correlation sequence for 29 lags $(0-28)$ are known (we will discuss later in section 8.12 how the data was generated, but for now we only know the given 29 auto-correlation sequence values, as shown in figure 8.12). The MVDSE for this process using the indicated lags and a set of 1001 equally spaced frequencies in the range $[-0.5, +0.5]$ Hz is shown on the left of figure 8.12. We note that when using the first three lags $0, 1, 2$, only one peak is visible at 0.125 Hz, and a hint of two frequency peaks appears when we use 7 lags $0-6$; the two frequencies become distinctly visible when we use 9 lags $0-8$. The two frequencies' separation continues to improve when more lags are included; the right hand side of figure 8.12 shows the estimate when all 29 lags are used.

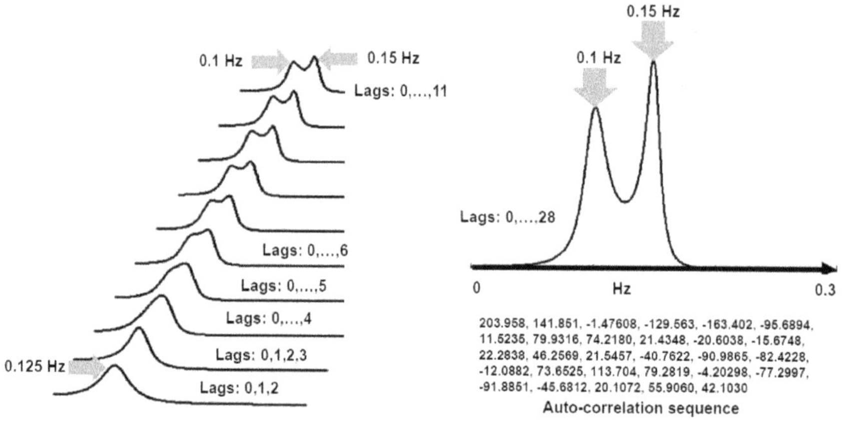

Figure 8.12: Minimum Variance Spectral Estimates for a model whose auto-correlation sequence for the first 29 lags $(0-28)$ is known.

This example illustrates an important principle: frequency resolution improves by increasing the maximum lag for known true auto-correlation sequence values. In practice, however, the true auto-correlation sequence values are unknown for any lag. In addition, we often have short data segments to work with and so estimates of the true auto-correlation sequence values with reasonable statistical significance can be obtained for far fewer lags than the data record length. In such circumstances, we resort to parametric data models such as the AR(P) model, which have proven useful in obtaining high spectral resolution for short data segments.

8.10 Autoregressive Moving Average (ARMA) Signal Models

The second half of figure 7.6 shows that a discrete time zero-mean WSS random signal x_n, $n \in \mathbb{Z}$, is the output of a minimum phase filter, namely, $S_{xx}^{(+)}(z)$, driven by zero-mean white noise process ν_n with unit variance; this can be used to introduce a parametric signal model by generating x_n using an appropriate minimum phase filter $H(z)$ driven by zero-mean white noise. The simplest method is to use a minimum phase filter whose **Z** transform is a pole-zero function

$$H(z) = \prod_{m=1}^{Q} \left(1 - z_m z^{-1}\right) \Big/ \prod_{k=1}^{P} \left(1 - p_k z^{-1}\right), \quad Q < P, \tag{8.68}$$

and whose zeros and poles are inside the unit circle of the complex z-plane, i.e., $|z_l| < 1$ and $|p_k| < 1$; the filter is then driven by a zero-mean white noise sequence ν_n with variance σ_ν^2. The filter $H(z)$ can be written as a ratio of polynomials parametrized by the vectors $\boldsymbol{b} \equiv [1, b_1, \ldots, b_Q]^T$ and $\boldsymbol{a} \equiv [a_1, \ldots, a_P]^T$, shown in figure 8.13, known as an *autoregressive moving average* (ARMA) model, denoted by ARMA(P,Q), and described by the following time domain equation:

$$x_n = -\sum_{k=1}^{P} a_k x_{n-k} + \sum_{m=0}^{Q} b_m \nu_{n-m}, \quad b_0 \equiv 1, \tag{8.69}$$

with an equivalent representation in terms of a causal filter sequence h_m with $h_0 = 1$ and $h_m = 0$ for $m < 0$, which is the inverse **Z** transform of $H(z)$,

$$x_n = \sum_{m=0}^{\infty} h_m \nu_{n-m}, \quad H(z) = \sum_{m=0}^{\infty} h_m z^{-m}. \tag{8.70}$$

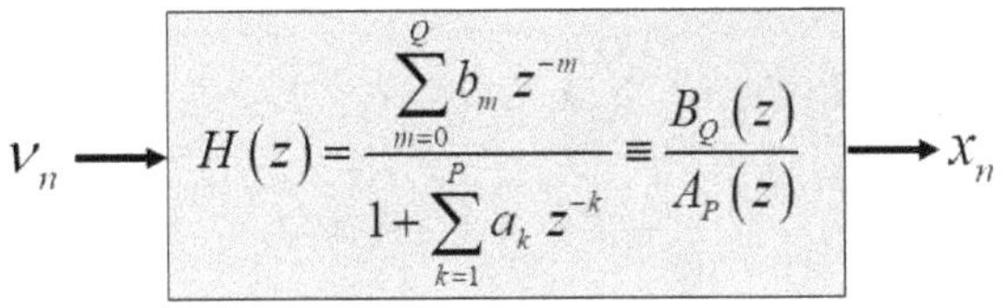

Figure 8.13: ARMA(P, Q) model of x_n as the output of a minimum phase pole-zero filter function $H(z)$, $b_0 = 1$, driven by white noise with variance σ_ν^2.

An ARMA(P, Q) model is fully specified by the values of the parameters P, Q, σ_ν^2, and the coefficient vectors $\boldsymbol{a}$ and $\boldsymbol{b}$. The problem of estimating these parameters from an $N \times 1$ observation vector $\boldsymbol{x}$ is central to much of modern signal processing. An ARMA(P, Q) model's PSD is an exact expression:

$$S_{\text{ARMA}}(e^{i\omega}) = \sigma_\nu^2 \left| \frac{1 + \sum_{m=1}^{Q} b_m e^{-im\omega}}{1 + \sum_{k=1}^{P} a_k e^{-ik\omega}} \right|^2. \tag{8.71}$$

To estimate the ARMA(P, Q) parameter vectors $\boldsymbol{a}$ and $\boldsymbol{b}$, we use the relationship between the process auto-correlation sequence R_l and the vectors $\boldsymbol{a}$ and $\boldsymbol{b}$. Using equations (7.4) and (8.68), the spectrum $S(z)$ of the ARMA(P, Q) process is

$$S(z) = \sum_{l=-\infty}^{\infty} R_l z^{-l} = \sigma_\nu^2 \, H(z)H^*(1/z^*) = \sigma_\nu^2 \frac{B_Q(z)B_Q^*(1/z^*)}{A_P(z)A_P^*(1/z^*)}, \tag{8.72}$$

where the region of convergence is an annular region in the complex plane. Using the definition of the auto-correlation sequence $R_l = \mathrm{E}[x_n x_{n-l}^*]$, the causal filter equation (8.70) for x_n, and $\mathrm{E}[\nu_k \nu_l^*] = \sigma_\nu^2 \delta_{kl}$, we find

$$R_l = \sigma_\nu^2 \sum_{m=0}^{\infty} h_{m+l} h_m^*. \tag{8.73}$$

Next we multiply both sides of (8.69) by x_{n-l}^*, $l \geq 0$, and take expectation values to obtain

$$R_l = -\sum_{k=1}^{P} a_k R_{l-k} + \sum_{m=0}^{Q} b_m \eta_{l-m}, \quad \eta_l \equiv \mathrm{E}[\nu_k x_{k-l}^*], \quad l \geq 0. \tag{8.74}$$

Taking the complex conjugate of the causal filter relation (8.70), changing $n \to n-l$, multiplying both sides with ν_n and taking expectation values gives

$$\eta_l = \sum_{m=0}^{\infty} h_m^* \mathrm{E}\left[\nu_n \nu_{n-l-m}^*\right] = \sigma_\nu^2 \, h_{-l}^*, \quad l \geq 0, \;\; h_0 = 1, \;\; \text{and} \;\; \eta_l = 0 \;\text{for}\; l > 0. \tag{8.75}$$

which when used in (8.74) leads to the final relation between the process auto-correlation sequence R_l and the ARMA parameter vectors $\boldsymbol{a}$ and $\boldsymbol{b}$

$$R_l = \begin{cases} -\sum\limits_{k=1}^{P} a_k R_{l-k} + \sigma_\nu^2 \sum\limits_{k=l}^{Q} b_k h_{l-k}^* & 0 \leq l \leq Q \text{ with } b_0 = h_0 = 1, \\ -\sum\limits_{k=1}^{P} a_k R_{l-k} & l > Q, \end{cases} \tag{8.76}$$

with $R_{-l} = R_l^*$ for $l > 0$. Now suppose that the auto-correlation sequence values are known for $l > Q$. Choosing $l = Q+1, \ldots, Q+P$ eliminates the second sum on the right hand side of the first equation in (8.76) because of the filter causality condition and gives us P equations for the unknown $P \times 1$ parameter vector $\boldsymbol{a}$:

$$\begin{bmatrix} R_Q & \cdots & R_{Q-P+1} \\ \vdots & \ddots & \vdots \\ R_{Q+P-1} & \cdots & R_Q \end{bmatrix} \begin{bmatrix} a_1 \\ \vdots \\ a_P \end{bmatrix} = - \begin{bmatrix} R_{Q+1} \\ \vdots \\ R_{Q+P} \end{bmatrix}. \tag{8.77}$$

Thus, $\boldsymbol{a}$ can be found from the Toeplitz matrix equation (8.77); when substituted back into (8.76), however, the resulting equations for $\boldsymbol{b}$ are non-linear because the filter sequence h_m depends on $\boldsymbol{b}$. This is one reason AR(P) or ARMA(P, 0) models are favored over ARMA(P, Q); a compelling theoretical reason for choosing AR models follows from theorem 33 in section 8.11.

8.11 Autoregressive Signal Models

A zero-mean WSS random process x_n is said to be *deterministic* if it can be predicted from its past values with zero prediction error; otherwise, the process is called *regular*. Herman Wold (a mathematical economist) in 1938 showed that any zero-mean WSS regular process x_n can be written as the sum of an MA(∞) process and a deterministic process that can be represented as a (possibly infinite) linear combination of the past values of x_n; this result is known as the *Wold decomposition theorem* [4] and is stated in theorem 32 without proof [94].

Theorem 32. *Any regular zero mean stationary discrete time process x_n, $n \in \mathbb{Z}$, has a unique representation in terms of a causal MA process and a deterministic process, i.e.,*

$$x_n = \sum_{k=0}^{+\infty} b_k u_{n-k} + d_k, \quad b_0 = 1, \tag{8.78}$$

where u_k is a white process and d_k is a deterministic process that can be written as a (possibly infinite) linear combination of all past values of x_n.

If the deterministic part is removed (detrending) or the process is purely *indeterministic* (i.e., the deterministic part d_k is zero), then the minimum phase filter associated with the MA(∞) representation has a convergent, causal, and minimum phase inverse; the inverse filter representation of the indeterministic process is then an AR(∞) process described in theorem 33, providing a compelling reason to use AR models [95].

Theorem 33. *If a stationary discrete time process x_n, $n \in \mathbb{Z}$, with zero mean and absolutely summable auto-correlation sequence $R_{xx}[l]$ (i.e., an $l_1(\mathbb{Z})$ sequence) has a strictly positive and absolutely continuous PSD $S_{xx}(e^{i\omega})$,*

$$\sum_{l=-\infty}^{+\infty} \left| R_{xx}[l] \right| < \infty, \qquad S_{xx}(e^{i\omega}) > 0, \quad -\pi \leq \omega \leq \pi, \tag{8.79}$$

then it can be represented as an infinitely long AR process,

$$x_n = -\sum_{k=1}^{\infty} a_k x_{n-k} + \nu_n, \quad n \in \mathbb{Z}, \tag{8.80}$$

where the AR coefficients are absolutely summable (i.e., an $l_1(\mathbb{Z})$ sequence),

$$\sum_{k=1}^{\infty} |a_k| < \infty, \tag{8.81}$$

and ν_n is a zero-mean white process with finite variance σ_ν^2.

AR(P) models (also referred to as *all-pole*), as shown in figure 8.14, are often used to model data that are known to have spectral peaks partly because MA(Q) models require many more coefficients to produce the same peaks. AR(P) models are said to produce high-resolution spectral estimates because of

[4] Theorem 32 states the causal and minimum phase form of the decomposition for which the decomposition is unique. The non-causal form of the theorem does not have a unique representation and includes maximum phase and mixed phase representations [93].

the sharpness of spectral peaks and general separation of closely spaced peaks that is usually difficult or impossible to achieve using traditional methods of spectral estimation due to generally short observation times. However, estimated AR(P) models are known to produce artifacts in the form of split spectral lines [96].

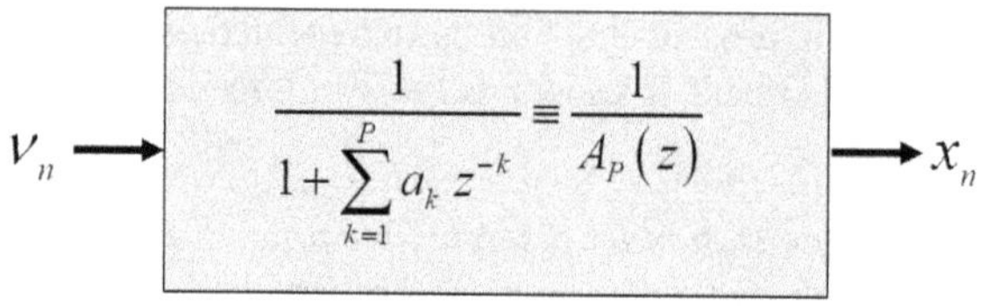

Figure 8.14: AR(P): Autoregressive (all-pole) model of order P.

Useful methods of AR parameter estimation produce coefficient vectors $\boldsymbol{a}$ whose associated poles are inside the unit circle of the complex z-plane, i.e., the associated filters are minimum phase. Once the parameters $\{P, \boldsymbol{a}, \sigma_\nu^2\}$ for an AR(P) model have been estimated, the model PSD estimate is given by the formula (equation (8.71) with $Q = 0$):

$$\hat{S}_{\text{AR}}(e^{i\omega}) = \sigma_\nu^2 \left| 1 + \sum_{k=1}^{P} a_k e^{-ik\omega} \right|^{-2}, \tag{8.82}$$

also known as the MEM (maximum entropy method) spectral estimate.

There are two other ways to arrive at the AR(P) model: the maximum entropy and the spectral flatness principles, discussed in sections 8.12 and 8.13, respectively. The basis for high spectral resolution of an AR(P) process is the formula which allows indefinitely extending its $P+1$ auto-correlation lags $R_0, \ldots, R_P$; using (8.76) with $Q = 0$, we have

$$R_l = -\sum_{k=1}^{P} a_k R_{l-k} + \sigma_\nu^2 \delta_{0l}, \quad l \geq 0 \quad \text{and} \quad R_{-l} = R_l^*, \tag{8.83}$$

Indeed, the maximum entropy solution to the problem of extending a set of known auto-correlation sequence values is the AR(P) process, as shown in section 8.12. Thus, the spectrum of a zero-mean WSS random AR(P) process with known parameters $\{P, \sigma_\nu^2, \boldsymbol{a}\}$ achieves the highest possible frequency resolution by effectively including the auto-correlation sequence values for all lags.

8.12 Maximum Entropy and the AR(P) Process

In this section, we will show that the solution to the problem of extending a known set of auto-correlation values for lags $0, \ldots, P$, when combined with the requirement of *maximum entropy*, is the AR(P) process. First, we show that the entropy rate of a zero-mean WSS complex Gaussian sequence (the proof is the same for a real Gaussian process) is related to the logarithm of its power spectrum.

Consider the zero-mean $(P+1) \times 1$ Gaussian random vector $\boldsymbol{x}$ whose distribution function is [see (5.65)]

$$f_{\boldsymbol{x}}(\boldsymbol{x}) = \frac{1}{\pi^{P+1} |\boldsymbol{R}|} e^{-\boldsymbol{x}^+ \boldsymbol{R}^{-1} \boldsymbol{x}}, \tag{8.84}$$

where $\boldsymbol{R}$ is the $(P+1)\times(P+1)$ Toeplitz auto-correlation matrix associated with auto-correlation sequence values $R_0, \ldots, R_P$ of the process and $|\boldsymbol{R}|$ is its determinant. The entropy is [see equation (5.213) in section 5.18]

$$H_{\boldsymbol{x}} \equiv -\int f_{\boldsymbol{x}}(\boldsymbol{x}) \ln f_{\boldsymbol{x}}(\boldsymbol{x})\, d\boldsymbol{x} = \ln|\boldsymbol{R}|, \tag{8.85}$$

where we have neglected a constant term. To avoid the divergence of the expression (8.85) for $P \to \infty$ we define the entropy rate:

$$h_{\boldsymbol{x}} \equiv \lim_{P\to\infty} \frac{H_{\boldsymbol{x}}}{P+1} = \lim_{P\to\infty} \frac{1}{P+1} \ln|\boldsymbol{R}|. \tag{8.86}$$

Since the correlation matrix $\boldsymbol{R}$ is Hermitian and positive definite, its eigenvalues $\lambda_1, \ldots, \lambda_{P+1}$ are real and positive, and $|\boldsymbol{R}| = \lambda_1 \ldots \lambda_{P+1}$; hence,

$$h_{\boldsymbol{x}} = \lim_{P\to\infty} \frac{1}{P+1} \sum_{k=1}^{P+1} \ln \lambda_k. \tag{8.87}$$

The entropy rate is related to the PSD of the process by Szegö's theorem [97].

Theorem 34. *If $\boldsymbol{x}$ is a zero-mean WSS random process whose PSD is $S(e^{i\omega})$ and whose $(P+1)\times(P+1)$ Hermitian and Toeplitz auto-correlation matrix has (real) eigenvalues $\lambda_1, \ldots, \lambda_{P+1}$, and if $f(u)$ is an arbitrary but continuous function of u, then*

$$\lim_{P\to\infty} \frac{1}{P+1} \sum_{k=1}^{k=P+1} f(\lambda_k) = \frac{1}{2\pi} \int_{-\pi}^{\pi} f\left(S(e^{i\omega})\right) d\omega. \tag{8.88}$$

Using the natural logarithm function (the natural logarithm for all spectral quantities refers to the *principal branch*) for f in theorem 34, equation (8.87) becomes

$$h_{\boldsymbol{x}} = \frac{1}{2\pi} \int_{-\pi}^{\pi} \ln\left|S\left(e^{i\omega}\right)\right| d\omega. \tag{8.89}$$

Thus, the problem becomes that of indefinitely extending a given set of auto-correlation sequence values $R_0, \ldots, R_P$ so that the entire set is positive definite [58]. The maximum entropy argument is that the extrapolation of the given auto-correlation sequence values should be performed in such a way as to keep the spectrum of the associated time series as flat as possible (i.e., the associated time series be as "white" as possible); such a sequence would be the most random among all possible solutions to the extrapolation problem—hence the maximum entropy principle. Thus, we maximize the entropy rate subject to the $P+1$ constraints provided by the $P+1$ given auto-correlation values:

$$\frac{1}{2\pi} \int_{-\pi}^{\pi} S\left(e^{i\omega}\right) e^{il\omega}\, d\omega = R_l, \quad 0 \le l \le P. \tag{8.90}$$

Differentiating the entropy rate with respect to the unspecified auto-correlation values R_l, $|l| > P$ and setting the result to zero [5], we have

$$\frac{\partial h_{\boldsymbol{x}}}{\partial R_l} = \frac{1}{2\pi} \int_{-\pi}^{+\pi} \frac{1}{S\left(e^{i\omega}\right)} e^{+il\omega} d\omega = 0, \quad |l| > P. \tag{8.91}$$

[5] Although the zero derivatives are not always sufficient for maximization, they are in this case [98, 99].

Equation (8.91) is an inverse Fourier transform showing that the discrete time sequence s_l whose DFT is $1/S(e^{i\omega})$ must vanish for $|l| > P$. Thus, we have the DFT relation

$$\frac{1}{S(e^{i\omega})} = \sum_{l=-P}^{+P} s_l e^{-il\omega}. \tag{8.92}$$

Since the left hand side of (8.92) is real and positive, we must have $s_{-l} = s_l^*$, and we may use the Fejér-Riesz factorization theorem [5].

Theorem 35. *Any real and non-negative trigonometric polynomial function*

$$W\left(e^{i\omega}\right) = \sum_{l=-P}^{P} w_l e^{-il\omega}, \quad w_{-l} = w_l^* \tag{8.93}$$

can be written in the form

$$W\left(e^{i\omega}\right) \equiv \left|B\left(e^{i\omega}\right)\right|^2, \tag{8.94}$$

where the polynomial

$$B\left(z\right) \equiv \sum_{k=0}^{P} b_k z^{-k} \tag{8.95}$$

is unique except for an arbitrary phase and has all of its roots inside or on the unit circle, i.e., the region $|z| \leq 1$.

To see this, consider

$$\left(W(1/z^*)\right)^* = \sum_{l=-P}^{+P} w_l^* z^l = \sum_{l=-P}^{P} w_{-l} z^l = \sum_{l=-P}^{P} w_l z^{-l} = W\left(z\right), \tag{8.96}$$

which shows that if $W(z_j) = 0$, then $W(1/z_j^*) = 0$. In addition, we can write

$$W\left(z\right) = z^{-P} w_{-P}\left(z^{2P} + \cdots + \frac{w_P}{w_{-P}}\right), \tag{8.97}$$

which shows that the product of all the roots is w_P/w_{-P} whose magnitude is 1 because $w_{-l} = w_l^*$. Hence, the roots must occur in pairs of the form $(z_j, 1/z_j^*)$, $j = 0, \ldots, P$, where we assume that the roots have been suitably ordered to have $\left|z_j\right| \leq 1$. Thus,

$$W\left(z\right) = w_{-P}\left(\prod_{k=0}^{P} z_k^*\right)^{-1} \prod_{j=0}^{P}\left[\left(1 - z_j/z\right)\left(z z_j^* - 1\right)\right]. \tag{8.98}$$

The theorem follows by choosing the roots that are on or inside the unit circle to construct $B(z)$, i.e.,

$$B\left(z\right) = \sqrt{w_{-(P)}}\left(\prod_{k=0}^{P} z_k^*\right)^{-1/2} \prod_{j=0}^{P}\left(1 - z_j/z\right), \tag{8.99}$$

to find

$$W(z) = B(z)\Big(B(1/z^*)\Big)^* \Rightarrow W\left(e^{i\omega}\right) \equiv \left|B\left(e^{i\omega}\right)\right|^2 = \left|\sum_{k=0}^{P} b_k e^{-ik\omega}\right|^2 . \tag{8.100}$$

Applying the result of theorem 35 to $W(e^{i\omega}) = 1/S(e^{i\omega})$ of equation (8.92) we find

$$S(e^{i\omega}) = \sigma^2 \left|1 + \sum_{k=1}^{P} a_k e^{-ik\omega}\right|^{-2} , \tag{8.101}$$

where we have defined the new coefficients $a_k = b_k/b_0$, $1 \le k \le P$, and $a_0 = 1$. Thus, we have shown that the AR(P) model spectrum is the maximum entropy rate solution to the problem of indefinitely extending a given set of $P+1$ auto-correlation sequence values (at lags $0, \ldots, P$) of a zero-mean WSS random process.

Theorem 36. *The maximum entropy rate zero-mean WSS random process x_n whose first $P+1$ auto-correlation coefficients have known values given by*

$$E\left[x_n x^*_{n-l}\right] = R_l, \quad 0 \le l \le P \tag{8.102}$$

is the AR(P) process, where

$$x_n = -\sum_{k=1}^{P} a^{(P)}_k x_{n-k} + \nu_n \tag{8.103}$$

and $\nu_n \sim \mathcal{N}(0, \sigma^2_\nu)$. The parameter vector $\boldsymbol{a}^{(P)}$ satisfies the Yule-Walker Toeplitz matrix equation $\boldsymbol{R}^{(P)}\boldsymbol{a}^{(P)} = -\boldsymbol{r}_P$,

$$\begin{bmatrix} R_0 & R_{-1} & \cdots & R_{-(P-1)} \\ R_1 & \ddots & \ddots & \vdots \\ \vdots & \ddots & \ddots & R_{-1} \\ R_{P-1} & \cdots & R_1 & R_0 \end{bmatrix} \begin{bmatrix} a_1 \\ a_2 \\ \vdots \\ a_P \end{bmatrix} = - \begin{bmatrix} R_1 \\ R_2 \\ \vdots \\ R_P \end{bmatrix}, \tag{8.104}$$

where $\boldsymbol{r}_P \equiv [R_1, \ldots, R_P]^T$, and σ^2_ν is found from the $l = 0$ term in (8.83)

$$\sigma^2_\nu = R_0 + \sum_{k=1}^{P} a^{(P)}_k R_{-k}, \quad R_{-k} = R^*_k. \tag{8.105}$$

The AR(P) model can be represented as the output of a minimum phase all-pole system function whose input ν_n is white noise with variance σ^2_ν, as shown in figure 8.14, and the output spectrum

$$S_{AR(P)}(e^{i\omega}) = \sigma^2_\nu \left|1 + \sum_{k=1}^{P} a^{(P)}_k e^{-ik\omega}\right|^{-2} \equiv \sigma^2_\nu \left|A_P(e^{i\omega})\right|^{-2},$$

$$A_P(e^{i\omega}) \equiv \sum_{k=0}^{P} a^{(P)}_k e^{-ik\omega}, \quad a^{(P)}_0 \equiv 1. \tag{8.106}$$

The superscript P in the vector of coefficients $a_k^{(P)}$ indicates the order of the process (the number of coefficients). The Yule-Walker equation (8.104) can be put in the following equivalent form:

$$\boldsymbol{R}^{(P+1)}\begin{bmatrix} 1 \\ \boldsymbol{a}^{(P)} \end{bmatrix} \equiv \begin{bmatrix} R_0 & R_{-1} & \cdots & R_{-P} \\ R_1 & \ddots & \ddots & \vdots \\ \vdots & \ddots & \ddots & R_{-1} \\ R_P & \cdots & R_1 & R_0 \end{bmatrix}\begin{bmatrix} 1 \\ a_1^{(P)} \\ \vdots \\ a_P^{(P)} \end{bmatrix} = \begin{bmatrix} \sigma_\nu^2 \\ 0 \\ \vdots \\ 0 \end{bmatrix}, \tag{8.107}$$

and the coefficients $[a_0^{(P)}, \boldsymbol{a}^{(P)}]^T$, with $a_0^{(P)} \equiv 1$, define the *prediction error filter* (PEF) of linear prediction theory (see chapter 9). The minimum phase property of the all-pole system function will be shown in section 9.7. The AR(0) model is defined as a white noise process, and its spectrum is the variance of the white noise sequence σ_ν^2. A fast recursive method for solving the Yule-Walker equation (8.104) for all orders $1, \ldots, P$ is the Levinson-Durbin algorithm, which will be discussed in section 8.15. For now, consider the solution to (8.107) using the eigenvector decomposition $\boldsymbol{U\Lambda U}^+$ for the auto-correlation matrix and the $(P+1) \times 1$ vector $\boldsymbol{v}_1 = [1, 0, \ldots, 0]^T$,

$$\begin{bmatrix} 1 \\ \boldsymbol{a}^{(P)} \end{bmatrix} = \sigma_\nu^2 \left[\boldsymbol{R}^{(P+1)}\right]^{-1} \boldsymbol{v}_1 = \sigma_\nu^2 \sum_{k=1}^{P+1} \lambda_k^{-1} \left(\boldsymbol{v}_1^T \boldsymbol{u}_k^*\right) \boldsymbol{u}_k. \tag{8.108}$$

Multiplying on the left with $\boldsymbol{v}_1^T$ gives

$$1 = \sigma_\nu^2 \sum_{k=1}^{P+1} \lambda_k^{-1} \left|\boldsymbol{v}_1^T \boldsymbol{u}_k\right|^2.$$

To prove equation (8.104) we write (8.83) for $1 \le l \le P$ arriving at the Yule-Walker matrix equation for the parameter vector $\boldsymbol{a}$. Multiplying the AR(P) process equation by ν_n^* and taking expectation values we find $\mathrm{E}\left[x_n \nu_n^*\right] = \sigma_\nu^2 = \mathrm{E}\left[\nu_n x_n^*\right]$. Finally, multiplying by x_n^* and taking expectation values, we find equation (8.105). Equation (8.83) is, of course, the basis for the high resolution of AR model spectra. Theorem 37 is a modified form of theorem 36.

Theorem 37. *Let y_n and $\hat{y}_n$ be two zero-mean WSS random processes with power spectral densities $S(e^{i\omega})$, $\hat{S}(e^{i\omega})$, and auto-correlation sequences R_l, $\hat{R}_l$, respectively. If $\hat{y}_n$ is an AR(P) process whose first $P+1$ auto-correlation values are the same as those of y_n, i.e., $\hat{R}_l = R_l$ for $0 \le l \le P$, then*

$$\int_{-\pi}^{\pi} \ln\left(\hat{S}(e^{i\omega})\right) d\omega \ge \int_{-\pi}^{\pi} \ln\left(S(e^{i\omega})\right) d\omega. \tag{8.109}$$

If, on the other hand, $\hat{y}_n$ is the optimal AR(P) approximation to y_n, i.e., denoting the optimal PEF parameters by $\{\sigma^2, \boldsymbol{a}^{(P)}\}$ with PEF polynomial $A_P(z)$ the approximating process $\hat{y}_n$ is generated by passing white noise with variance σ^2 through the all-pole filter $1/A_P(z)$, then $R[l] = \hat{R}[l]$ for $0 \le l \le P$ (this is known as the auto-correlation matching property).

8.13 Spectral Flatness and the AR(P) Process

In this section we will describe the maximum *spectral flatness* property of an AR(P) process. To motivate a measure of spectral flatness, consider the geometric and arithmetic means of a sequence of positive values

$x_k > 0, k = 1, \ldots, P$. The two means satisfy the following inequality:

$$\Big(\prod_{k=1}^{P} x_k\Big)^{1/P} \leq \frac{1}{P}\sum_{k=1}^{P} x_k,$$

where equality holds if, and only if, we have a "flat" sequence of values, i.e., when all x_k are equal. We define a measure of flatness for a sequence of positive values by the ratio of its geometric mean to its arithmetic mean, denoted by κ_x, with a range $(0, 1]$; the closer κ_x is to 1, the flatter the underlying positive sequence values. The geometric mean can be equivalently expressed as the exponential of the arithmetic mean of the natural logarithm of the (positive) sequence values,

$$\Big(\prod_{k=1}^{P} x_k\Big)^{1/P} \equiv \exp\Big(\frac{1}{P}\sum_{k=1}^{P} \ln(x_k)\Big).$$

It is straightforward to extend these results to the case of positive and continuous PSD values $S(e^{i\omega})$ of a zero-mean WSS random process (with corresponding auto-correlation sequence R_l) and to obtain the spectral flatness measure [100]:

$$\kappa_x^2 \equiv \frac{\exp\Big(\frac{1}{2\pi}\int_{-\pi}^{\pi} \ln\, S(e^{i\omega})\, d\omega\Big)}{\frac{1}{2\pi}\int_{-\pi}^{\pi} S(e^{i\omega})\, d\omega}, \tag{8.110}$$

with $0 < \kappa_x^2 \leq 1$ [30]; note that the denominator is the process power R_0 by Parseval's relation. Suppose x_n is an AR(P) process with known coefficients $a_k^{(P)}$. Then the process equation (8.103) has the **Z** transform

$$X(z) = N(z)/A_P(z), \quad A_P(z) \equiv 1 + \sum_{k=1}^{P} a_k^{(P)} z^{-k}, \tag{8.111}$$

where $N(z)$ represents a flat spectrum since the noise process ν_n is white (see figure 8.14). Now consider figure 8.15 (the inverse of figure 8.14), which is equivalent to

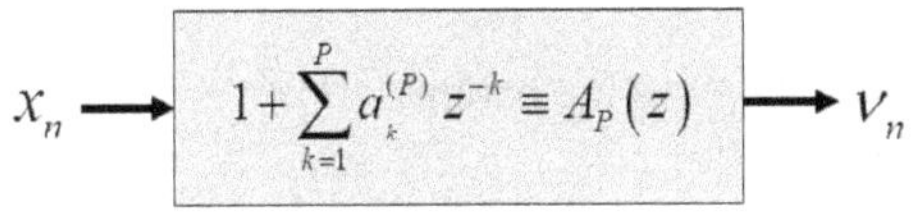

Figure 8.15: Spectral flatness property of AR(P) model (the inverse of figure 8.14).

$$N(z) = X(z)A_P(z), \quad A_P(z) \equiv 1 + \sum_{k=1}^{P} a_k^{(P)} z^{-k}. \tag{8.112}$$

Equation (8.112) describes the flat spectrum property of the noise process, which when used as input to the filter function $1/A_P(z)$ would produce the AR(P) process x_n. Thus, maximizing the output spectral flatness measure with respect to the coefficients of a (minimum phase) filter $\hat{A}_P(z) \equiv 1 + \sum_{k=1}^{P} \hat{a}_k^{(P)} z^{-k}$ with input $X(z)$ will lead to the Yule-Walker equation (8.104) of theorem 36 for the coefficients $\hat{\boldsymbol{a}}$. If the input sequence is an AR(P) process with coefficients $\boldsymbol{a}^{(P)}$, then $\hat{\boldsymbol{a}} = \boldsymbol{a}^{(P)}$. When the zero-mean WSS process is not necessarily AR we have the more general result of theorem 38.

Theorem 38. *If a zero-mean WSS random process with PSD $S(e^{i\omega})$ is filtered by $A_P(z) \equiv 1+\sum_{k=1}^{P} a_k^{(P)} z^{-k}$, and if we denote the mean square output power by σ^2, then*

$$\sigma^2 \geq \exp\left(\frac{1}{2\pi}\int_{-\pi}^{\pi} \ln\left(S(e^{i\omega})\right) d\omega\right), \tag{8.113}$$

with equality if, and only if, the process is AR(P).

In view of the bounds on the spectral flatness measure $0 < \kappa^2 \leq 1$, the output mean square power

$$\sigma^2 = \frac{1}{2\pi}\int_{-\pi}^{\pi} S(e^{i\omega})\left|A_P(e^{i\omega})\right|^2 d\omega,$$

satisfies the inequality

$$\sigma^2 \geq \exp\left(\frac{1}{2\pi}\int_{-\pi}^{\pi} \ln\left\{S(e^{i\omega})\left|A_P(e^{i\omega})\right|^2\right\} d\omega\right),$$

which using the result $\int_{-\pi}^{+\pi} \ln\left|A_P\left(e^{i\omega}\right)\right|^2 d\omega = 0$ (this is proven later in theorem 40) proves theorem 38. If the process is AR(P) then equality holds in (8.113) and so the process spectral flatness measure is $\kappa^2 = \sigma^2/R_0$.

8.14 AR(P) Process Examples

Consider the AR(2) model

$$x_n = -a_1^{(2)} x_{n-1} - a_2^{(2)} x_{n-2} + \nu_n. \tag{8.114}$$

Factorizing the polynomial $1 + a_1^{(2)} z^{-1} + a_2^{(2)} z^{-2}$, we find the poles

$$p_{1,2}^{(2)} = \frac{1}{2}\left(-a_1^{(2)} \pm \sqrt{[a_1^{(2)}]^2 - 4a_2^{(2)}}\right). \tag{8.115}$$

The Yule-Walker equation (8.104) then gives the parameters $a_{1,2}^{(2)}$

$$\begin{bmatrix} a_1^{(2)} \\ a_2^{(2)} \end{bmatrix} = -\frac{1}{R_0^2 - |R_1|^2}\begin{bmatrix} R_0 & -R_{-1} \\ -R_1 & R_0 \end{bmatrix}\begin{bmatrix} R_1 \\ R_2 \end{bmatrix}, \tag{8.116}$$

guaranteeing the minimum phase condition $\left|p_{1,2}^{(2)}\right| < 1$. For instance, consider the complex AR(2) model described by $a_1^{(2)} = -0.965318 - 1.57526\, i$ and $a_2^{(2)} = -0.409726 + 0.804133\, i$ corresponding to two poles inside the unit circle with equal magnitudes 0.95 at normalized frequencies 0.125 and 0.2 Hz, respectively, and complex white noise variance with $\sigma_\nu^2 = 1$. This process has auto-correlation sequence values $94.6124, 48.0061 + 78.3381\, i, -38.2965 + 75.1619\, i$ at lags $0, 1, 2$, respectively. The real part of 1000 samples (seconds) of a realization of this data and its exact AR spectrum (8.106) are shown in figure 8.16.

The example shown in figure 8.12 is a real AR(4) process with four poles: two poles have magnitude 0.96 at frequencies ± 0.1 Hz, and the other two have magnitude 0.98 at frequencies ± 0.15 Hz. The white noise input has unit variance. The AR parameter vector elements are $a_1^{(4)} = -2.70537$, $a_2^{(4)} = 3.67151$,

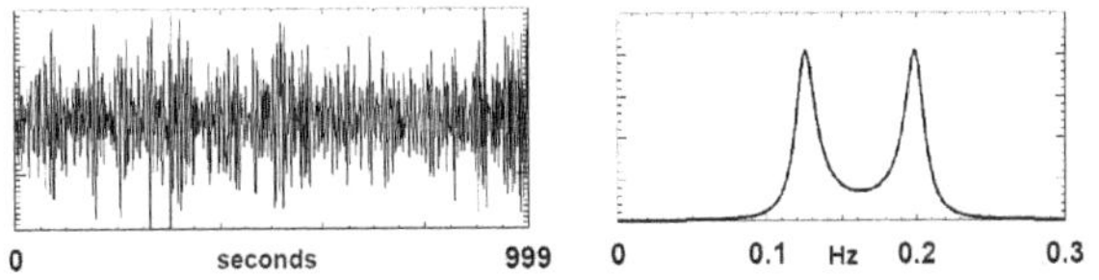

Figure 8.16: A complex AR(2) process and its exact AR spectrum.

$a_3^{(4)} = -2.55354$, and $a_4^{(4)} = 0.885105$. The exact AR spectrum shows sharp peaks at 0.1 and 0.15 Hz, similar to the right hand panel of figure 8.12.

The simplest way to estimate the auto-correlation sequence from data of finite duration is to use the sample estimates (see section 8.3)

$$\hat{R}_l = \frac{1}{\alpha} \sum_{n=l}^{N-1} x_n x_{n-l}^*, \quad l \geq 0, \quad \text{and} \quad \hat{R}_{-l} = \hat{R}_l^*, \tag{8.117}$$

where $\alpha = N-l$ for the unbiased estimate, and $\alpha = N$ for the biased estimate. Theorem 30 showed that the estimated Toeplitz auto-correlation matrix of a zero-mean WSS random process is positive definite when the biased estimate is used, and, as we shall see later, the biased estimate also ensures the minimum phase property of the prediction error filter (PEF) of the one-step ahead linear predictor. In the context of spectral estimation, however, one can use either the biased or the unbiased estimate of the auto-correlation sequence. Using the auto-correlation estimates (8.117) in the Yule-Walker equation for the AR(P) coefficient vector $\boldsymbol{a}$ is equivalent to using the estimated Yule-Walker equation; this is referred to as the auto-correlation method of linear prediction. Other methods of estimating the AR coefficient vector $\boldsymbol{a}$, namely, the covariance method and the Burg method, will be discussed in the context of linear prediction in chapter 9.

For now, consider the example of a real AR(4) process whose four poles have magnitude 0.99 at normalized frequencies ± 0.11 and ± 0.22 Hz. Figure 8.17 shows the AR(4) spectrum (solid black) using the exact coefficients, $a_1^{(4)} = -1.89663$, $a_2^{(4)} = 2.52623$, $a_3^{(4)} = -1.85889$, and $a_4^{(4)} = 0.960596$, as well as three periodograms and three Yule-Walker method estimates based on 20, 100, and 300 data samples, respectively.

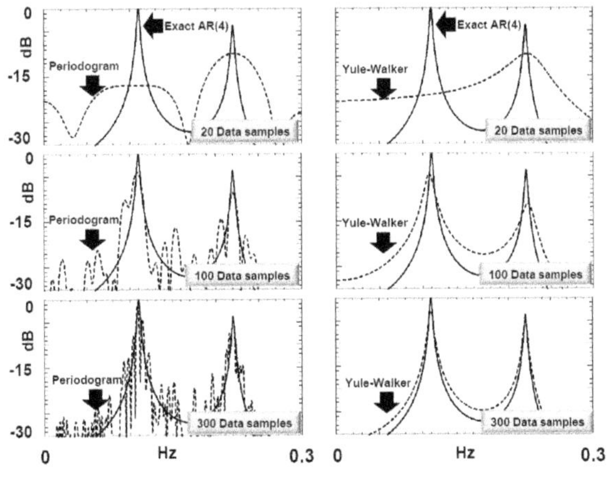

Figure 8.17: AR(4) example model spectra.

The auto-correlation method of estimating the AR parameter vector $\boldsymbol{a}$ is the same as the MLE of $\boldsymbol{a}$ for large data records; this will be discussed in section 8.22.

8.15 The Levinson-Durbin Algorithm

The *Levinson-Durbin* algorithm is a recursive $\mathcal{O}(P^2)$ method [as opposed to $\mathcal{O}(P^3)$ using the inverse of a $P \times P$ matrix] to solve the Yule-Walker matrix equation (8.104) for an AR(P) process. Consider the AR($P+1$) Yule-Walker equation and relate its solution to that of the AR(P) process that shares its $P+1$ auto-correlation values at lags $0, \dots, P$ with the AR($P+1$) process:

$$\boldsymbol{R}^{(P+1)}\boldsymbol{a}^{(P+1)} \equiv \begin{bmatrix} R_0 & R_{-1} & \cdots & R_{-P} \\ R_1 & \ddots & \ddots & \vdots \\ \vdots & \ddots & \ddots & R_{-1} \\ R_P & \cdots & R_1 & R_0 \end{bmatrix} \begin{bmatrix} a_1^{(P+1)} \\ a_2^{(P+1)} \\ \vdots \\ a_{P+1}^{(P+1)} \end{bmatrix} = - \begin{bmatrix} R_1 \\ R_2 \\ \vdots \\ R_{P+1} \end{bmatrix}, \tag{8.118}$$

where $R_l = \mathrm{E}[x_n x_{n-l}^*]$ and $R_{-l} = R_l^*, 0 \le l \le P$. The first P elements of the right hand side are the last P elements of the first column of the $(P+1) \times (P+1)$ Hermitian and Toeplitz auto-correlation matrix $\boldsymbol{R}^{(P+1)}$, and the last element of the right hand side is the next auto-correlation value at lag P.

Defining the vector $\boldsymbol{r}_P = [R_1, \dots, R_P]^T$, the first P elements of the last row of the auto-correlation matrix are the elements of $\boldsymbol{r}_P$ in reverse order; in addition, the first P elements of the last column are the elements of the complex conjugate vector $\boldsymbol{r}_P^*$ in reverse order. We use the counter identity matrix $\boldsymbol{J}^{(P)}$ of order P, defined in (6.68), to perform the element reversing operation.

The first $P+1$ auto-correlation values at lags $0, \dots, P$ constitute the AR(P) process whose solution is $\boldsymbol{a}^{(P)}$; our task is to express the solution $\boldsymbol{a}^{(P+1)}$ in terms of order P solution. In general, the two solutions have no common elements. The two processes, however, share $P+1$ auto-correlation values at lags $0, \dots, P$. Thus, the top left $P \times P$ submatrix of $\boldsymbol{R}^{(P+1)}$ is the matrix $\boldsymbol{R}^{(P)}$ associated with the AR(P) process, and so we partition the $(P+1) \times (P+1)$ auto-correlation matrix $\boldsymbol{R}^{(P+1)}$:

$$\boldsymbol{R}^{(P+1)} = \begin{bmatrix} \boldsymbol{R}^{(P)} & \boldsymbol{J}^{(P)}\boldsymbol{r}_P^* \\ \boldsymbol{r}_P^T\boldsymbol{J}^{(P)} & R_0 \end{bmatrix}. \tag{8.119}$$

We will use the solution to the AR(P) Yule-Walker equation, $\boldsymbol{a}^{(P)} = -[\boldsymbol{R}^{(P)}]^{-1}\boldsymbol{r}_P$, to obtain the solution vector $\boldsymbol{a}^{(P+1)}$ of the AR($P+1$) model. The partitioned matrix equation is

$$\begin{bmatrix} \boldsymbol{R}^{(P)} & \boldsymbol{J}^{(P)}\boldsymbol{r}_P^* \\ \boldsymbol{r}_P^T\boldsymbol{J}^{(P)} & R_0 \end{bmatrix} \begin{bmatrix} \tilde{\boldsymbol{a}} \\ \gamma_{P+1} \end{bmatrix} = - \begin{bmatrix} \boldsymbol{r}_P \\ R_{P+1} \end{bmatrix}, \tag{8.120}$$

where we have denoted the first P elements of $\boldsymbol{a}^{(P+1)}$ by $\tilde{\boldsymbol{a}}$, and its last element by γ_{P+1}, which is known as the *reflection coefficient* of order $P+1$, i.e.,

$$\tilde{\boldsymbol{a}} \equiv [a_1^{(P+1)}, \dots, a_P^{(P+1)}]^T, \quad \text{and} \quad \gamma_{P+1} \equiv a_{P+1}^{(P+1)}. \tag{8.121}$$

Equation (8.120) is equivalent to two separate equations

$$\boldsymbol{R}^{(P)}\tilde{\boldsymbol{a}} + \gamma_{P+1}\boldsymbol{J}^{(P)}\boldsymbol{r}_P^* = -\boldsymbol{r}_P, \quad \text{and} \quad \boldsymbol{r}_P^T\boldsymbol{J}^{(P)}\tilde{\boldsymbol{a}} + \gamma_{P+1}R_0 = -R_{P+1}. \tag{8.122}$$

The reflection coefficient can be written in terms of $\tilde{\boldsymbol{a}}$ using the second equation in (8.122):

$$\gamma_{P+1} = -\frac{R_{P+1}}{R_0} - \frac{\boldsymbol{r}_P^T\boldsymbol{J}^{(P)}\tilde{\boldsymbol{a}}}{R_0}. \tag{8.123}$$

Multiplying the first (matrix) equation in (8.122) on the left by $[\boldsymbol{R}^{(P)}]^{-1}$, using (6.72), and the results

$$[\boldsymbol{R}^{(P)}]^{-1}\boldsymbol{r}_P = -\boldsymbol{a}^{(P)}, \quad [\boldsymbol{R}^{(P)\,*}]^{-1}\boldsymbol{r}_{P\,*} = -\boldsymbol{a}^{(P)\,*}, \tag{8.124}$$

we obtain

$$\tilde{\boldsymbol{a}} = \boldsymbol{a}^{(P)} + \gamma_{P+1}\boldsymbol{J}^{(P)}\boldsymbol{a}^{(P)\,*}. \tag{8.125}$$

Multiplying the above equation on the left by $\boldsymbol{J}^{(P)}$ we find

$$\boldsymbol{J}^{(P)}\tilde{\boldsymbol{a}} = \boldsymbol{J}^{(P)}\boldsymbol{a}^{(P)} + \gamma_{P+1}\boldsymbol{a}^{(P)\,*}, \tag{8.126}$$

which when substituted into (8.123) gives

$$\gamma_{P+1} = -\frac{R_{P+1} + \boldsymbol{r}^{(P)\,T}\boldsymbol{J}^{(P)}\boldsymbol{a}^{(P)}}{R_0 + \boldsymbol{r}^{(P)\,T}\boldsymbol{a}^{(P)*}} = -\frac{R_{P+1} + \boldsymbol{r}^{(P)\,T}\boldsymbol{J}^{(P)}\boldsymbol{a}^{(P)}}{R_0 + \boldsymbol{a}^{(P)+}\boldsymbol{r}_P}. \tag{8.127}$$

Thus, given the solution $\boldsymbol{a}^{(P)}$ we first compute γ_{P+1} and then

$$\boldsymbol{a}^{(P+1)} = [\boldsymbol{a}^{(P)} + \gamma_{P+1}\boldsymbol{J}^{(P)}\boldsymbol{a}^{(P)\,*},\ \gamma_{P+1}]^T, \tag{8.128}$$

starting with $P = 1$ when $\boldsymbol{a}^{(1)}$ has one element, namely, $a_1^{(1)} = \gamma_1 = -R_1/R_0$. Finally, the white noise variance σ^2_{P+1} of the AR$(P+1)$ process is found using

$$\sigma^2_{P+1} = R_0 + \sum_{k=1}^{P+1} a_k^{(P+1)} R_{-k} = R_0 + \boldsymbol{r}_{P+1\,+}\,\boldsymbol{a}^{(P+1)}, \tag{8.129}$$

where we have used the relation $R_{-k} = R_k^*$. Since σ^2_{P+1} and R_0 are real (and positive), we conclude that $\boldsymbol{r}_{(P+1)\,+}\,\boldsymbol{a}^{(P+1)} = \boldsymbol{a}^{(P+1)\,+}\,\boldsymbol{r}_{(P+1)}$.

Using partitioned vectors

$$\boldsymbol{r}_{P+1} = [\boldsymbol{r}_P, R_{P+1}]^T, \quad \text{and} \quad \boldsymbol{a}^{(P+1)} = [\boldsymbol{a}^{(P)} + \gamma_{P+1}\boldsymbol{J}^{(P)}\boldsymbol{a}^{(P)\,*},\ \gamma_{P+1}]^T, \tag{8.130}$$

and defining $\rho_P \equiv \sigma^2_P$, we find

$$\begin{aligned}
\rho_{P+1} &= \sigma^2_{P+1} = R_0 + [\boldsymbol{r}_{P+}, R^*_{P+1}]\,[\boldsymbol{a}^{(P)} + \gamma_{P+1}\boldsymbol{J}^{(P)}\boldsymbol{a}^{(P)\,*}, \gamma_{P+1}]^T \\
&= R_0 + \boldsymbol{r}_{P+}\,\boldsymbol{a}^{(P)} + \gamma_{P+1}\left(R^*_{P+1} + \boldsymbol{r}_{P+}\,\boldsymbol{J}^{(P)}\boldsymbol{a}^{(P)\,*}\right) \\
&= \sigma^2_P + \gamma_{P+1}\left(-\gamma^*_{P+1}\sigma^2_P\right) = \rho_P\left(1 - |\gamma_{P+1}|^2\right).
\end{aligned} \tag{8.131}$$

The Levinson-Durbin recursion in component notation is

- $\gamma_1 = a_1^{(1)} = -R_1/R_0$
- $\rho_1 = \sigma_1^2 = R_0 + \gamma_1 R_{-1} = R_0 - |R_1|^2/R_0$

  ⋆ $\gamma_{P+1} \equiv a_{P+1}^{(P+1)} = -\rho_P^{-1}\left(R_{P+1} + \sum_{k=1}^{P} R_k a_{P-k+1}^{(P)}\right)$

  ⋆ $a_k^{(P+1)} = a_k^{(P)} + \gamma_{P+1} a_{P-k+1}^{(P)\,*}, \quad k = 1, \ldots, P$

  ⋆ $\rho_{P+1} = \sigma^2_{P+1} = \rho_P\left(1 - |\gamma_{P+1}|^2\right)$

8.16 The Relationship Between MVD and AR Spectra

Suppose that the $(P+1)\times(P+1)$ Hermitian and Toeplitz auto-correlation matrix $\boldsymbol{R}^{(P+1)}$ for a zero-mean WSS random process x_n is either known or has been estimated from available data; thus, the (j,k)-element $[\boldsymbol{R}^{(P+1)}]_{jk}$ denoted by R_{j-k}, $1\le j,k\le P+1$, is either known from $\mathrm{E}[x_{n-k}x^*_{n-j}]$ or is the latter's estimate,

$$\boldsymbol{R}^{(P+1)} = \begin{bmatrix} R_0 & \cdots & R_{-P} \\ \vdots & \ddots & \vdots \\ R_P & \cdots & R_0 \end{bmatrix}, \quad R_{-k} = R_k^*,\ 0\le k\le P. \tag{8.132}$$

$\boldsymbol{R}^{(P+1)}$ can then be used to calculate the normalized MVDSE [see (8.67)]

$$S_{\mathrm{MVD}}\left(e^{i\omega}\right) = \frac{P+1}{\boldsymbol{e}_\omega^+\,[\boldsymbol{R}^{(P+1)}]^{-1}\,\boldsymbol{e}_\omega}, \quad \boldsymbol{e}_\omega \equiv [1, e^{i\omega}, e^{2i\omega}, \ldots, e^{Pi\omega}]^T. \tag{8.133}$$

We can, on the other hand, use the correlation matrices $\boldsymbol{R}^{(k)}$ to calculate AR spectral estimates $S_{\mathrm{AR}(k)}$ for all orders $0\le k\le P$, where the AR(0) spectral estimate is defined to equal the variance of the process σ_x^2. The MVD spectral estimate will, in general, show lower resolution than the higher order AR spectral estimates. Our aim in this section is to provide a theoretical explanation for the lower resolution of the MVD spectrum through a relationship between the two spectra.

Consider the case $P=1$. The AR(0) model is defined by the process variance $R_0=\sigma_x^2\equiv\rho_0$. The AR(1) model parameter $a_1^{(1)}$ is the solution to the Yule-Walker equation $R_0\,a=-R_1$, i.e., $a_1^{(1)}=-R_1/R_0$, while the variance of the AR(1) model noise ρ_1 is given by (8.105), which using $R_{-1}=R_1^*$ becomes

$$\rho_1\equiv\sigma_1^2 = R_0 + a_1^{(1)}R_{-1} = R_0\left(1-\frac{|R_1|^2}{R_0^2}\right) = \rho_0\left(1-\left|a_1^{(1)}\right|^2\right). \tag{8.134}$$

The auto-correlation matrix $\boldsymbol{R}^{(2)}$ and its inverse are

$$\boldsymbol{R}^{(2)} \equiv \begin{bmatrix} R_0 & R_{-1} \\ R_1 & R_0 \end{bmatrix}, \quad [\boldsymbol{R}^{(2)}]^{-1} = \frac{1}{R_0^2-|R_1|^2}\begin{bmatrix} R_0 & -R_{-1} \\ -R_1 & R_0 \end{bmatrix}. \tag{8.135}$$

Now, $R_0=\rho_1+|R_1|^2/R_0$, which upon multiplication by R_0 gives $R_0^2-|R_1|^2=R_0\,\rho_1$. Thus, the inverse auto-correlation matrix is

$$\begin{aligned}
[\boldsymbol{R}^{(2)}]^{-1} &= \frac{1}{R_0\,\rho_1}\begin{bmatrix} \rho_1+|R_1|^2/R_0 & -R_{-1} \\ -R_1 & R_0 \end{bmatrix} \\
&= \begin{bmatrix} 1 & -R_{-1}/R_0 \\ 0 & 1 \end{bmatrix}\begin{bmatrix} 1/\rho_0 & 0 \\ 0 & 1/\rho_1 \end{bmatrix}\begin{bmatrix} 1 & 0 \\ -R_1/R_0 & 1 \end{bmatrix}. \\
&= \begin{bmatrix} 1 & a_1^{(1)*} \\ 0 & 1 \end{bmatrix}\begin{bmatrix} 1/\rho_0 & 0 \\ 0 & 1/\rho_1 \end{bmatrix}\begin{bmatrix} 1 & 0 \\ a_1^{(1)} & 1 \end{bmatrix}.
\end{aligned} \tag{8.136}$$

In order to calculate the MVDSE we use $\boldsymbol{e}_\omega=[1,e^{i\omega}]^T$ and

$$\begin{bmatrix} 1 & 0 \\ a_1^{(1)} & 1 \end{bmatrix}\begin{bmatrix} 1 \\ e^{i\omega} \end{bmatrix} = \begin{bmatrix} 1 \\ e^{i\omega}A_1(e^{i\omega}) \end{bmatrix}, \quad A_1(e^{i\omega})\equiv 1+a_1^{(1)}e^{-i\omega}, \tag{8.137}$$

to find

$$\frac{2}{S_{\mathrm{MVD}}} = \boldsymbol{e}_\omega^+[\boldsymbol{R}^{(2)}]^{-1}\boldsymbol{e}_\omega = \frac{1}{\rho_0}+\frac{1}{\rho_1}\left|A_1(e^{i\omega})\right|^2 = \frac{1}{S_{\mathrm{AR}(0)}}+\frac{1}{S_{\mathrm{AR}(1)}}. \tag{8.138}$$

Thus, the normalized MVDSE for $P = 1$ is the harmonic mean of the AR(0) and AR(1) spectra [101, 102]; theorem 39 generalizes this result to $P > 1$.

Theorem 39. *If $\boldsymbol{R}^{(P+1)}$ denotes the auto-correlation matrix of a zero-mean WSS random process x_n, then the corresponding normalized MVDSE S_{MVD} is the harmonic mean of all the process AR(k) spectra for $0 \le k \le P$,*

$$\frac{P+1}{S_{MVD}(e^{i\omega})} = \sum_{k=0}^{P} \frac{1}{S_{AR(k)}(e^{i\omega})}, \tag{8.139}$$

where S_{MVD} is the normalized MVDSE and $S_{AR(k)}$ are the various AR spectra,

$$S_{MVD}\left(e^{i\omega}\right) \equiv (P+1)/(\boldsymbol{e}_\omega^+ [\boldsymbol{R}^{(P+1)}]^{-1} \boldsymbol{e}_\omega), \;\; \boldsymbol{e}_\omega \equiv [1, e^{i\omega}, e^{2i\omega}, \ldots, e^{Pi\omega}]^T,$$

$$S_{AR(0)}(e^{i\omega}) \equiv \sigma_x^2 \equiv \rho_0, \;\; S_{AR(k)}(e^{i\omega}) \equiv \rho_k \left| 1 + \sum_{m=1}^{k} a_m^{(k)} e^{-im\omega} \right|^{-2}, \tag{8.140}$$

and the $k \times 1$ coefficient vectors $\boldsymbol{a}^{(k)}$, $1 \le k \le P$, are solutions of the Yule-Walker equations:

$$\boldsymbol{R}^{(k)}\boldsymbol{a}^{(k)} = -\boldsymbol{r}_k \Leftrightarrow \begin{bmatrix} R_0 & \cdots & R_{-(k-1)} \\ \vdots & \ddots & \vdots \\ R_{k-1} & \cdots & R_0 \end{bmatrix} \begin{bmatrix} a_1^{(k)} \\ \vdots \\ a_k^{(k)} \end{bmatrix} = - \begin{bmatrix} R_1 \\ \vdots \\ R_k \end{bmatrix}. \tag{8.141}$$

In addition,

$$\rho_k = R_0 + \boldsymbol{a}^{(k)+}\boldsymbol{r}_k = R_0 + \boldsymbol{r}_k^+ \boldsymbol{a}^{(k)} = R_0 - \boldsymbol{r}_k^+ [\boldsymbol{R}^{(k)}]^{-1} \boldsymbol{r}_k, \;\; 1 \le k \le P. \tag{8.142}$$

The proof of theorem 39 is essentially a generalization of the example preceding it. Writing $[\boldsymbol{R}^{(P+1)}]^{-1} = \boldsymbol{L}_P^+ \boldsymbol{\rho}_P^{-1} \boldsymbol{L}_P$ where

$$\boldsymbol{L}_P \equiv \begin{bmatrix} 1 & 0 & \cdots & \cdots & 0 \\ a_1^{(1)} & 1 & 0 & \cdots & 0 \\ a_2^{(2)} & a_1^{(2)} & 1 & \ddots & \vdots \\ \vdots & \vdots & \vdots & \ddots & 0 \\ a_P^{(P)} & a_{P-1}^{(P)} & \cdots & a_1^{(P)} & 1 \end{bmatrix}, \;\; \boldsymbol{\rho}_P \equiv \begin{bmatrix} \rho_0 & 0 & \cdots & \cdots & 0 \\ 0 & \rho_1 & 0 & \cdots & 0 \\ 0 & 0 & \rho_2 & \ddots & \vdots \\ \vdots & \vdots & \vdots & \ddots & 0 \\ 0 & 0 & \cdots & 0 & \rho_P \end{bmatrix},$$

$$\boldsymbol{L}_P \boldsymbol{e}_\omega = \begin{bmatrix} 1 \\ e^{i\omega} A_1(e^{i\omega}) \\ e^{2i\omega} A_2(e^{i\omega}) \\ \vdots \\ e^{Pi\omega} A_P(e^{i\omega}) \end{bmatrix}, \;\; A_k(e^{i\omega}) \equiv 1 + \sum_{m=1}^{k} a_m^{(k)} e^{-im\omega},$$

$$\boldsymbol{e}_\omega^+ [\boldsymbol{R}^{(P+1)}]^{-1} \boldsymbol{e}_\omega = \boldsymbol{e}_\omega^+ \boldsymbol{L}_P^+ \boldsymbol{\rho}_P^{-1} \boldsymbol{L}_P \boldsymbol{e}_\omega = (\boldsymbol{L}_P \boldsymbol{e}_\omega)^+ (\boldsymbol{L}_P \boldsymbol{e}_\omega) \boldsymbol{\rho}_P^{-1} = 1/\rho_0 + \sum_{k=1}^{P} \left| A_k\left(e^{i\omega}\right)\right|^2 / \rho_k.$$

The left hand side is $(P+1)/S_{\text{MVD}}\left(e^{i\omega}\right)$, while the right hand side is the sum of the inverses of the AR(k) spectra for $0 \le k \le P$, i.e., the harmonic mean of the associated AR spectra. Equation (8.142) is proven in chapter 9, theorem 41.

Thus, an MEM spectrum always has higher resolution than an MVDSE for the same model order since the MVD estimate is equivalent to a system of parallel resistors that average the lowest to the highest possible resolutions of the corresponding MEM estimates. Historically, in spectral estimation problems with sharp peaks the MVD spectral estimate has found less use than the MEM spectral estimate, even though the peaks may not have been actually generated by the underlying physical processes! However, if the data is known to consist of pure sinusoids in WGN, then the MVD (or ML) spectral estimate is the optimal estimator.

8.17 Autoregressive Model of a Zero-Mean WSS Random Signal

If a zero-mean WSS random process x_n is known to be AR(P), then the Yule-Walker equation (8.104) and the associated equation (8.105) for the white noise variance define the AR(P) process so long as the first $P+1$ auto-correlation values $R_0, \ldots R_P$ are known. If, on the other hand, x_n is not known to be AR(P) but its $P+1$ auto-correlation values are known (or have been estimated from the observed process values), then x_n may be modeled as an AR(P) process whose parameters are calculated using (8.104) and (8.105). Here we will show that the Yule-Walker equation (8.104) results from minimizing the model MSE, i.e., given the AR(P) model equation

$$\hat{x}_n = -\sum_{k=1}^{P} a_k x_{n-k} + e_n, \tag{8.143}$$

with error sequence e_n, the MSE minimization problem

$$\arg\min_{\boldsymbol{a}} \mathrm{E}\left[\left|x_n - \hat{x}_n\right|^2\right] = \mathrm{E}\left[\left|x_n + \sum_{k=1}^{P} a_k x_{n-k}\right|^2\right], \tag{8.144}$$

is solved by the parameter vector $\boldsymbol{a}$ that satisfies the Yule-Waller equation (8.104).

The modeling equation (8.143) can be written as

$$\hat{x}_n = -\boldsymbol{a}^T \boldsymbol{x}_{n-1} + e_n, \quad \boldsymbol{a} = [a_1, \ldots, a_P]^T, \quad \boldsymbol{x}_{n-1} \equiv [x_{n-1}, \ldots, x_{n-P}]^T, \tag{8.145}$$

where $\boldsymbol{x}_{n-1}$ is the time-reversed data vector, and the MSE is:

$$\begin{aligned}\mathrm{E}\left[\left|e_n\right|^2\right] &= R_0 + \boldsymbol{a}^T \mathrm{E}\left[\boldsymbol{x}_{n-1} x_n^*\right] + \boldsymbol{a}^+ \mathrm{E}\left[\boldsymbol{x}_{n-1}^* x_n\right] + \mathrm{E}\left[\boldsymbol{a}^+ \boldsymbol{x}_{n-1}^* \boldsymbol{a}^T \boldsymbol{x}_{n-1}\right] \\ &= R_0 + \boldsymbol{a}^T \mathrm{E}\left[\boldsymbol{x}_{n-1} x_n^*\right] + \boldsymbol{a}^+ \mathrm{E}\left[x_n \boldsymbol{x}_{n-1}^*\right] + \boldsymbol{a}^+ \mathrm{E}\left[\boldsymbol{x}_{n-1}^* \boldsymbol{x}_{n-1}^T\right] \boldsymbol{a},\end{aligned} \tag{8.146}$$

where we used the identity $\boldsymbol{a}^T \boldsymbol{x}_{n-1} \equiv \boldsymbol{x}_{n-1}^T \boldsymbol{a}$. Using equation (6.73) (applicable to the time reversed vector $\boldsymbol{x}_{n-1}$) we find

$$\mathrm{E}\left[\boldsymbol{x}_{n-1}^* \boldsymbol{x}_{n-1}^T\right] = \left(\mathrm{E}\left[\boldsymbol{x}_{n-1} \boldsymbol{x}_{n-1}^+\right]\right)^* = \boldsymbol{R}^{(P)}, \tag{8.147}$$

which when substituted in (8.146) gives

$$\mathrm{E}\left[\left|e_n\right|^2\right] = R_0 + \boldsymbol{a}^+ \left[R_1, \ldots, R_P\right]^T + \left[R_1, \ldots, R_P\right]^* \boldsymbol{a} + \boldsymbol{a}^+ \boldsymbol{R}^{(P)} \boldsymbol{a}. \tag{8.148}$$

Setting the derivative of (8.148) with respect to $\boldsymbol{a}^+$ equal to zero leads to the Yule-Walker equation (8.104) for the optimal coefficient vector $\boldsymbol{a}$, and the minimum MSE is then given by (8.105). Thus, the only inputs to this procedure are the AR model order P and the $P+1$ auto-correlation sequence values

$R_0, \ldots, R_P$. As noted before, the Yule-Walker matrix equation (8.104) has an equivalent form (8.107) in terms of the $(P+1)$-element parameter vector $\boldsymbol{a}' \equiv [a_0, a_1, \ldots, a_P]$ with $a_0 \equiv 1$. It is sometimes useful to write the MSE minimization problem in terms of $\boldsymbol{a}'$ with a constraint on its first element,

$$\arg\min_{\boldsymbol{a}'} \left(\mathrm{E}\left[|e_n|^2 \right] = R_0 + \boldsymbol{a}'^{+} \boldsymbol{R}^{(P+1)} \boldsymbol{a}' \right) \text{ subject to } a_0 = 1. \tag{8.149}$$

The constraint equation can be written in the form $\boldsymbol{\alpha}^{+}\boldsymbol{a}' = \boldsymbol{a}'^{+}\boldsymbol{\alpha} = 1$ where $\boldsymbol{\alpha} \equiv [1, 0, \ldots, 0]^T$. Introducing a Lagrange multiplier λ problem (8.149) is equivalent to the unconstrained problem

$$\arg\min_{\boldsymbol{a}'} \left[R_0 + \boldsymbol{a}'^{+} \boldsymbol{R}^{(P+1)} \boldsymbol{a}' + \lambda (1 - \boldsymbol{a}'^{+} \boldsymbol{\alpha}) \right]. \tag{8.150}$$

Differentiating with respect to $\boldsymbol{a}'^{+}$ and setting the result equal to zero gives

$$\boldsymbol{R}^{(P+1)} \boldsymbol{a}' = \lambda \boldsymbol{\alpha} \Rightarrow \boldsymbol{a}' = \lambda \left[\boldsymbol{R}^{(P+1)} \right]^{-1} \boldsymbol{\alpha}. \tag{8.151}$$

Multiplying on the left by $\boldsymbol{\alpha}^{+}$ and using the constraint, we find the Lagrange multiplier

$$\lambda = 1 / \boldsymbol{\alpha}^{+} [\boldsymbol{R}^{(P+1)}]^{-1} \boldsymbol{\alpha}. \tag{8.152}$$

In some important signal models, e.g., complex sinusoids in white noise, the exact form of the inverse of the auto-correlation matrix is known [see equations (6.39), (6.41), and (6.42)], in which case equations (8.151) and (8.152) can be used to solve for the AR(P) model parameters.

8.18 Autoregressive Model of a Complex Sinusoid in White Noise

Consider a single complex sinusoid in white noise,

$$x_n = A_0 e^{in\omega_0} + \nu_n, \quad n \in \mathbb{Z}, \tag{8.153}$$

where $\mathrm{E}[\nu_m \nu_n^*] = \delta_{mn} \sigma_\nu^2$ and A_0 is a zero-mean uncorrelated complex random variable (i.e., the variance of the real part is equal to the variance of the imaginary part and the real and imaginary parts are uncorrelated) with variance $\mathcal{P}_0$. As seen in section 6.2, the auto-correlation sequence for x_n is

$$R_l = \mathrm{E}[x_n x_{n-l}^*] = \mathcal{P}_0 \, e^{il\omega_0} + \sigma_\nu^2 \, \delta_{0l}, \quad l \in \mathbb{Z}, \quad R_{-l} = R_l^*, \tag{8.154}$$

and the PSD is

$$S_{xx} = \sum_{l=-\infty}^{\infty} e^{-il\omega} R_l = \sigma_\nu^2 + 2\pi \mathcal{P}_0 \, \delta(\omega - \omega_0), \tag{8.155}$$

i.e., a delta distribution at the frequency ω_0 on a constant background level of σ_ν^2 with SNR $= \mathcal{P}_0/\sigma_\nu^2$. The auto-correlation sequence of x_n is a complex sinusoid with the same frequency ω_0 and never decays, yet it is possible to model the data as an AR process of order 1 for high SNR. The AR(1) model for this data is found by solving for the optimal coefficient a_1 using the $P = 1$ Yule-Walker equation (8.104) and the relation $R_{-1} = R_1^*$,

$$a_1 = -\frac{R_1}{R_0} = -\frac{\mathcal{P}_0}{\mathcal{P}_0 + \sigma_\nu^2} e^{i\omega_0} = -\frac{\text{SNR}}{\text{SNR} + 1} e^{i\omega_0}. \tag{8.156}$$

The AR(1) model spectrum is

$$S_{\mathrm{AR}(1)}(e^{i\omega}) = \frac{\sigma_e^2}{|1 + a_1 e^{-i\omega}|^2} = \frac{\sigma_e^2}{\left|1 - \frac{\mathrm{SNR}}{\mathrm{SNR}+1} e^{-i(\omega-\omega_0)}\right|^2}, \tag{8.157}$$

where σ_e^2 is found from (8.105):

$$\sigma_e^2 = R_0 + a_1 R_{-1} = R_0 - \frac{|R_1|^2}{R_0} = \sigma_\nu^2\left(1 + \mathrm{SNR} - \frac{\mathrm{SNR}^2}{1 + \mathrm{SNR}}\right). \tag{8.158}$$

Evaluating the AR(1) spectrum (8.157) at ω_0, we find

$$S_{\mathrm{AR}(1)}(e^{i\omega_0}) = \sigma_\nu^2(1 + \mathrm{SNR})(1 + 2\,\mathrm{SNR}). \tag{8.159}$$

Figure 8.18 shows the AR(1) spectra, equation (8.157), for a single complex sinusoid of frequency 0.1 Hz in white noise with variance 1 for four SNR values 0, 5, 10, and 15 dB; the spectral peak becomes sharper as SNR increases in agreement with equation (8.159).

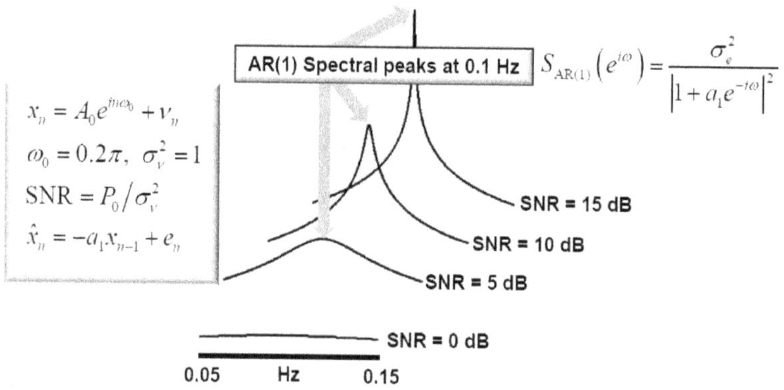

Figure 8.18: AR(1) model PSD for one complex sinusoid in white noise.

8.19 Autoregressive Model of Multiple Complex Sinusoids in White Noise

When the signal is the sum of P complex sinusoids with uncorrelated complex random amplitudes in white noise,

$$x_n = s_n + \nu_n = \sum_{k=1}^{P} A_k\, e^{in\omega_k} + \nu_n, \quad n \in \mathbb{Z}, \quad \nu_n \sim \mathcal{CN}(0, \sigma_\nu^2), \tag{8.160}$$

its $P \times P$ auto-correlation matrix $\boldsymbol{R}^{(P)}$ and the auto-correlation matrix inverse are given by equations (6.38), (6.39), (6.41), and (6.42); these equations, when extended to the $(P+1) \times (P+1)$ auto-correlation matrix $\boldsymbol{R}^{(P+1)}$, can be used to solve for the AR vector parameter $\boldsymbol{a}'$ in equations (8.151):

$$\boldsymbol{a}' = \frac{\lambda}{\sigma_\nu^2}\left(\boldsymbol{I} - \boldsymbol{S}\left(\sigma_\nu^2 \boldsymbol{P}^{-1} + \boldsymbol{S}^{+}\boldsymbol{S}\right)^{-1}\boldsymbol{S}^{+}\right)\boldsymbol{\alpha}. \tag{8.161}$$

Note that using the definition (6.35) for the $(P+1) \times P$ signal matrix $\boldsymbol{S}$, we have $\boldsymbol{S}^{+}\boldsymbol{\alpha} = [1, 1, \ldots, 1]^T$, and so

$$\boldsymbol{a}' = \frac{\lambda}{\sigma_\nu^2}\left(\boldsymbol{\alpha} - \boldsymbol{S}\left(\sigma_\nu^2 \boldsymbol{P}^{-1} + \boldsymbol{S}^{+}\boldsymbol{S}\right)^{-1}\boldsymbol{1}\right), \quad \boldsymbol{1} \equiv [1, 1, \ldots, 1]^T. \tag{8.162}$$

The Lagrange multiplier λ can be determined using (8.152) or from the requirement that the first element of $\boldsymbol{a}'$ is 1. In either case, the AR(P) model spectrum

$$S_{\text{AR}(P)}(e^{i\omega}) = \frac{\sigma_e^2}{|A'(e^{i\omega})|^2}, \quad A'(e^{i\omega}) \equiv \sum_{k=0}^{P} a'_k e^{-ik\omega} \tag{8.163}$$

depends on the ratio σ_ν^2/λ^2, which can be neglected when studying the spectral peak locations.

Figure 8.19 shows the result of fitting a real AR(6) model to the sum of six complex exponentials at ± 0.1, ± 0.15, and ± 0.2 Hz (corresponding to three real sinusoids) with equal amplitudes in white noise with unit variance for five SNR values, namely 0, 5, 10, 15, and 20 dB. The auto-correlation values for lags $0-6$ and the corresponding parameter vector $\boldsymbol{a}$ for each of the five spectra are also displayed (note that $\boldsymbol{a}' = [1, \boldsymbol{a}]^T$). Clearly, the sinusoid resolutions increase with increasing SNR delineating the spectral peaks at 0.1, 0.15, and 0.2 Hz.

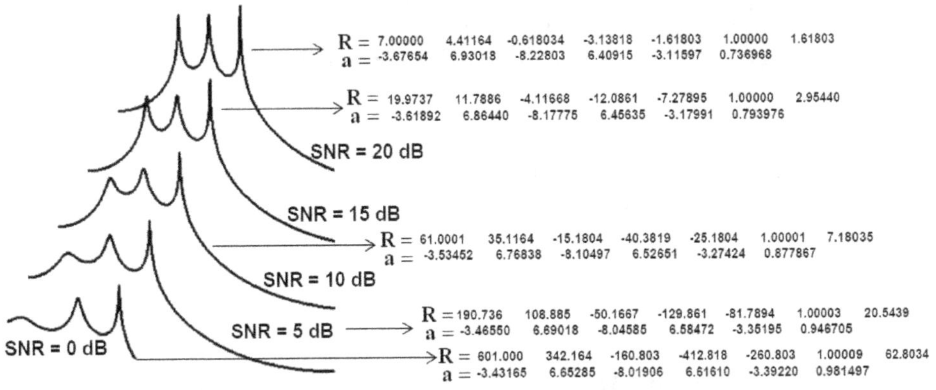

Figure 8.19: AR(6) model spectra for three real sinusoids in white noise.

8.20 Resolution of AR Models

In order to study the resolution properties of AR(P) models we use the following approximate form [96]:

$$\Delta f_{\text{AR}(P)} = \frac{1.03 f_s}{P\left((1+P)\,10^{\text{SNR}/10}\right)^{0.31}}, \tag{8.164}$$

where f_s is the sampling frequency in Hz, SNR is in dB, P is the order of the AR process used to model the data, and resolution is measured in Hz; the approximation is valid for SNR $> 10(1 - \text{Log}_{10} P)$ and is independent of data record length. Note that resolution increases with SNR but it also increases with increasing AR model order P. The resolution of the periodogram is inversely proportional to the observed record length N and is approximated by $\Delta f_{\text{Pgm}} = 0.86/N$. For instance, consider two complex sinusoids with equal amplitudes in white noise. The correct AR model order for two complex sinusoids is 2, and we have

$$\frac{\Delta f_{\mathrm{AR(2)}}}{\Delta f_{\mathrm{Pgm}}} = \frac{20.1\, N}{10^{0.031 \times \mathrm{SNR}}}, \tag{8.165}$$

which is valid for SNR > 7 dB. Figure 8.20 shows the ratio of the two resolutions for SNR values between 10 and 200 dB and record lengths between 10 and 1000 samples. The AR(2) spectral estimate is better than the periodogram where the ratio is less than 1; the bar on the right hand side refers only to the area where AR resolution is better than that of the periodogram, and darker values (close to 0) are better than lighter values (close to 1).

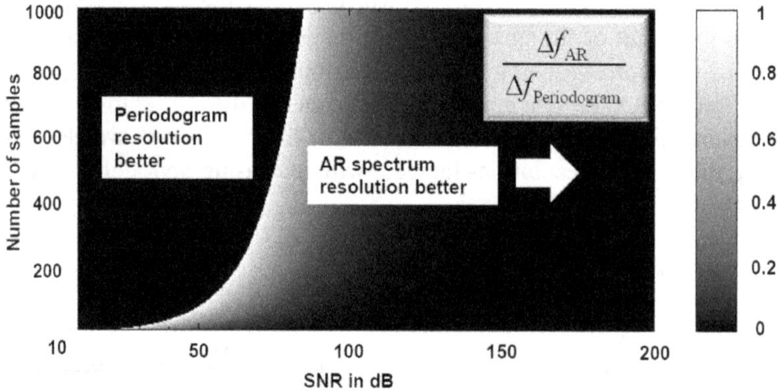

Figure 8.20: The ratio of the AR(2) and periodogram resolutions for SNR values between 10 and 200 dB, and record lengths between 10 and 1000 samples (the bar on the right refers only to the area where AR resolution is better).

8.21 AR Model Parameter Estimation

Whether an AR model agrees sufficiently with the data is an important first question that can be investigated using the error sequence (or the model residuals) that theoretically should represent white noise. Thus, any departure from a flat spectrum in the residual process should be of concern. A graph of the residuals (scaled by the estimated standard deviation) should show no indication of deviation from zero mean or unit variance, and the data values should mostly lie between ± 1.96 [see (5.184)]. The sample auto-correlation function of the residuals, however, must be estimated and shown to have the right properties to represent a spike at zero lag. If the residual samples are independent and identically distributed (iid) with finite variance, then the normalized sample auto-correlation values (i.e., all values divided by the estimated value at zero lag) based on N samples of the residuals should be approximately iid and $\sim \mathcal{N}(0, 1/N)$ [81], and most values should fall within $\pm 1.96/\sqrt{N}$.

There are two main approaches to the problem of estimating AR parameters for application to time series modeling and spectral estimation. The first approach is to estimate the auto-correlation matrix which, when used in the Yule-Walker equation (8.104), leads to estimated values for the parameter vector $\boldsymbol{a}^{(P)}$. The second approach is to estimate the reflection coefficients (8.123) and use the Levinson-Durbin algorithm to produce estimated values for the parameter vector $\boldsymbol{a}^{(P)}$; this approach is based on linear prediction (see chapter 9). In this section, we will discuss the estimation of auto-correlation sequence values from observed data with finite number of points.

In section 4.4, we studied two methods of solving the least squares filtering problem: the auto-correlation and the covariance methods. We use an appropriately modified formula for the matrix $\boldsymbol{A}_{\text{cov}}^{+}\boldsymbol{A}_{\text{cov}}$, where $\boldsymbol{A}_{\text{cov}}$ is the data matrix for the data vector $\boldsymbol{x} = [x_0, \ldots, x_{N-1}]^T$, to define the *covariance method* of estimating the auto-correlation matrix,

$$\left[\hat{\boldsymbol{R}}_{\text{cov}}\right]_{km} = \frac{1}{N-P}\sum_{j=P}^{N-1} x_{j-k}^{*}\, x_{j-m}. \tag{8.166}$$

The factor before the summation in (8.166) ensures that the right hand side is an unbiased estimate of the true auto-correlation matrix; $\hat{\boldsymbol{R}}$ is Hermitian, and positive semi-definite (it is singular when the data is a noise-less combination of complex sinusoids) but not Toeplitz. The estimated matrix (8.166) is then used in equation (8.104) to solve for the AR parameter vector $\hat{\boldsymbol{a}}$:

$$\sum_{m=1}^{P}\left[\hat{\boldsymbol{R}}_{\text{cov}}\right]_{km}\hat{a}_m = -\left[\hat{\boldsymbol{R}}_{\text{cov}}\right]_{k0}, \quad k = 1, \ldots, P, \tag{8.167}$$

with the estimated white noise variance [see (8.105)] [100]

$$\hat{\sigma}^2 = \left[\hat{\boldsymbol{R}}_{\text{cov}}\right]_{00} + \sum_{k=1}^{P}\left[\hat{\boldsymbol{R}}_{\text{cov}}\right]_{0k}\hat{a}_k. \tag{8.168}$$

Equation (8.167) with the specific form (8.166) is the result of the following minimization problem:

$$\arg\min_{\boldsymbol{a}} \frac{1}{N-P}\sum_{n=P}^{N-1}\left|x_n + \sum_{k=1}^{P} a_k x_{n-k}\right|^2, \tag{8.169}$$

and it is solved using Cholesky decomposition, but the associated poles are not guaranteed to lie inside the unit circle [103], i.e., the estimate (8.166) with the specific summation range $[P, N-1]$ does not guarantee a minimum phase (and hence stable) AR model. To obtain a minimum phase model, we modify the summation range in (8.169) to $[-\infty, +\infty]$, assume that the data outside the range $[0, N-1]$ is zero, and divide by N instead of $N - P$; this results in the biased auto-correlation method estimate of the true auto-correlation matrix [100]:

$$[\hat{\boldsymbol{R}}_{\text{ac}}]_{km} = \frac{1}{N}\sum_{j=0}^{N-1-(k-m)} x_{j+k-m}\, x_j^{*}, \tag{8.170}$$

and the associated Hermitian Toeplitz matrix equation:

$$\sum_{m=1}^{P}\hat{R}[k-m]\,\hat{a}_m = -\hat{R}[k], \quad k = 1, \ldots, P, \quad \hat{R}[k-m] \equiv [\hat{\boldsymbol{R}}_{\text{ac}}]_{km}, \tag{8.171}$$

with the estimated white noise variance

$$\hat{\sigma}^2 = \hat{R}[0] + \sum_{k=1}^{P}\hat{R}^{*}[k]\,\hat{a}_k, \quad \text{and } \hat{R}^{*}[k] = \hat{R}[-k]. \tag{8.172}$$

Although the auto-correlation method is guaranteed to produce poles that lie inside the unit circle, it generally produces AR models with poorer resolution than other methods [103].

The statistical justification for the above two methods of solving for AR parameters is that the associated estimators of the true auto-correlation matrix are approximate MLEs: the auto-correlation method matrix is an approximate maximum likelihood estimator for long data records, while the covariance method matrix is the maximum likelihood outcome of a conditional likelihood function. We will discuss these in sections 8.22 and 8.23.

A third method of estimating the auto-correlation matrix and solving for the AR parameters is known as the *modified covariance method* [103] and is defined by the minimization problem

$$\arg\min_{\boldsymbol{a}} \frac{1}{2(N-P)}\left[\sum_{n=P}^{N-1}\left|x_n + \sum_{k=1}^{P} a_k x_{n-k}\right|^2 + \sum_{n=0}^{N-1-P}\left|x_n + \sum_{k=1}^{P} a_k^* x_{n+k}\right|^2\right]. \tag{8.173}$$

The second term in (8.173) represents a *backward prediction error*, as opposed to the first term, which is the *forward prediction error*; both types of prediction were originally introduced in the context of linear predictive coding of speech (LPC) and will be discussed in more detail in chapter 9. The matrix equation is identical in form to equation (8.167) for the covariance method, but the estimated matrix is now given by [104]

$$\left[\hat{\boldsymbol{R}}_{\text{mcov}}\right]_{km} = \frac{1}{2(N-P)}\left[\sum_{j=P}^{N-1} x_{j-k}^* x_{j-m} + \sum_{j=0}^{N-1-P} x_{j+k} x_{j+m}^*\right], \tag{8.174}$$

which when used in equation (8.168) (instead of the covariance method matrix) gives the estimated white noise variance. This method offers higher resolution spectral estimates and is particularly useful for complex sinusoids in white noise with no line splitting (an artifact of some high-resolution spectral estimates such as Burg's method of section 9.8) [105], and its estimated reflection coefficients always have magnitudes less than 1, but its associated poles are not guaranteed to lie inside the unit circle. The latter, although important in some applications, is not relevant to the problem of spectral estimation.

8.22 Maximum Likelihood AR Parameter Estimation: the Auto-Correlation Method

Practical modeling of a time series as an AR(P) process requires the estimates of the auto-correlation sequence. As discussed in section 8.3, the biased estimate is often preferred because it leads to minimum phase prediction filters. Here we will show that the AR parameter vector $\boldsymbol{a}^{(P)}$, the solution to the matrix Yule-Walker equation (8.104) with the biased estimate of the auto-correlation sequence, is also the MLE of the same parameter vector for large data records.

When $\boldsymbol{x}$ is an $N \times 1$ realization of a zero-mean WSS random process with auto-correlation matrix $\boldsymbol{R}_{xx}$ that is Hermitian and Toeplitz, the log-likelihood function for large record length N has the following asymptotic form [106]:

$$\ln f_{\boldsymbol{x}}(\boldsymbol{x};\boldsymbol{\theta}) \approx -\frac{N}{2}\ln 2\pi - \frac{N}{4\pi}\int_{-\pi}^{+\pi}\left[\ln S_{xx}\left(e^{i\omega}\right) + \frac{\hat{S}_{\text{pgm}}\left(e^{i\omega}\right)}{S_{xx}\left(e^{i\omega}\right)}\right] d\omega, \tag{8.175}$$

where $\hat{S}_{\text{pgm}}(e^{i\omega})$ is the periodogram of the data (we choose the "pgm" subscript here to avoid confusion with the AR order parameter P):

$$\hat{S}_{\text{pgm}}(e^{i\omega}) = \frac{1}{N}\left|\sum_{n=0}^{N-1} x_n e^{-in\omega}\right|^2, \tag{8.176}$$

and $S_{xx}(e^{i\omega})$ is the true PSD of the data [6]. For an AR(P) process, the vector $[1, \boldsymbol{a}^{(P)}]^T$ defines the *prediction error filter* (PEF) [also defined after equation (8.107)]. The PEF must be a minimum phase sequence, i.e., its **Z** transform:

$$A_P(z) = \sum_{k=0}^{P} a_k^{(P)} z^{-k}, \quad a_0 \equiv 1, \tag{8.177}$$

must have all its roots inside the unit circle (see section 9.7 for more details), and its inverse $1/A_P(z)$ must be causal and stable.

Theorem 40. *The minimum phase PEF* (8.177) *defined by coefficients* $[1, \boldsymbol{a}^{(P)}]^T$ *satisfies the equation*

$$\int_{-\pi}^{+\pi} \ln\left|A_P(e^{i\omega})\right|^2 d\omega = 0, \tag{8.178}$$

where the integrand refers to the principal branch of the logarithm.

The minimum phase property implies that

$$A_P(e^{i\omega}) = \prod_{k=1}^{P}\left(1 - z_k e^{-i\omega}\right), \quad |z_k| < 1,$$

and so

$$\ln\left|A_P(e^{i\omega})\right|^2 = \sum_{k=1}^{P} \ln(1 - z_k e^{-i\omega}) + \sum_{k=1}^{P} \ln(1 - z_k^* e^{+i\omega}).$$

Since the zeros are inside the unit circle, $|z_k| < 1$, we have

$$\ln(1 - z_k e^{-i\omega}) + \ln(1 - z_k^* e^{+i\omega}) = -\sum_{n=1}^{\infty} n^{-1} z_k^n e^{-in\omega} - \sum_{n=1}^{\infty} n^{-1} z_k^{*\,n} e^{+in\omega},$$

and

$$\int_{-\pi}^{+\pi} \ln\left|A_P(e^{i\omega})\right|^2 d\omega = -\sum_{k=1}^{P}\int_{-\pi}^{+\pi}\sum_{n=1}^{\infty} n^{-1}\left(z_k^n e^{-in\omega} + z_k^{*\,n}\, e^{+in\omega}\right) d\omega.$$

Both series converge uniformly inside the unit circle, and so we can interchange the integral and the sum operations; each integral vanishes, and we obtain equation (8.178).

Thus, for an AR(P) process whose parameter vector $\boldsymbol{a}^{(P)}$ is to be estimated, using the AR(P) spectrum (theorem 36),

$$S_{\text{AR}(P)}(e^{i\omega}) = \frac{\sigma_\nu^2}{\left|A_P(e^{i\omega})\right|^2}, \quad A_P(e^{i\omega}) \equiv \sum_{k=0}^{P} a_k^{(P)} e^{-ik\omega}, \quad a_0^{(P)} \equiv 1, \tag{8.179}$$

[6]The approximation (8.175) refers to a real random process. If $\boldsymbol{x}$ is a complex process then the formula is modified slightly by $N \to 2N$ and removing the 2 in $\ln 2\pi$.

the approximation (8.175) to the log-likelihood function becomes

$$\ln f_{\boldsymbol{x}}\left(\boldsymbol{x};\boldsymbol{a}^{(P)},\sigma_\nu^2\right) = -\frac{N}{2}\ln 2\pi - \frac{N}{2}\ln\sigma_\nu^2 - \frac{N}{4\pi\sigma_\nu^2}\int_{-\pi}^{+\pi}\left|A_P\left(e^{i\omega}\right)\right|^2 \hat{S}_{\text{pgm}}\left(e^{i\omega}\right)d\omega\,.$$

Differentiating with respect to σ_ν^2 and setting the result to zero, we find the estimated noise variance:

$$\hat{\sigma}_\nu^2 = \int_{-\pi}^{+\pi}\left|A_P\left(e^{i\omega}\right)\right|^2 \hat{S}_{\text{pgm}}\left(e^{i\omega}\right)d\omega\,. \tag{8.180}$$

Differentiating with respect to $\boldsymbol{a}^{(P)}$ and setting the result to zero, we find the following equation for the MLE of the coefficient vector $\hat{\boldsymbol{a}}^{(P)}$,

$$\sum_{k=1}^{P}\left[\int_{-\pi}^{+\pi}\hat{S}_{\text{pgm}}\left(e^{i\omega}\right)e^{i\omega(l-k)}d\omega\right]\hat{a}_k^{(P)} = -\int_{-\pi}^{+\pi}\hat{S}_{\text{pgm}}\left(e^{i\omega}\right)e^{i\omega l}d\omega,\quad l=0,\ldots,N-1.$$

The integral on the right hand side is the inverse DFT of the periodogram: it vanishes for $|l|\geq N$ and

$$\int_{-\pi}^{+\pi}\hat{S}_{\text{pgm}}\left(e^{i\omega}\right)e^{i\omega l}d\omega = \frac{1}{N}\sum_{n=|l|}^{N-1}x_n x_{n-|l|}^*\,,\quad |l|\leq N-1. \tag{8.181}$$

The right hand side is the biased estimate of the auto-correlation sequence values of the AR(P) process for data $\boldsymbol{x}$ with N points. Thus, the estimated Yule-Walker method of solving for the AR parameter vector $\boldsymbol{a}^{(P)}$, when the biased auto-correlation estimates are used in equation (8.104), is equivalent to the MLE of the same vector for large data record length N. This method of estimating the auto-correlation sequence values to solve the Yule-Walker equation is termed the *auto-correlation method* of AR parameter estimation. In section 8.23, we show that the covariance method estimate (8.166) can be derived as an approximate MLE of the AR parameters and is valid for all N.

8.23 Maximum Likelihood AR Parameter Estimation: the Covariance Method

For the observed $N\times 1$ data vector $\boldsymbol{x}=[x_0,\ldots,x_{N-1}]$, define $\boldsymbol{x}_P\equiv[x_P,\ldots,x_{N-1}]$ and $\boldsymbol{x}_0\equiv[x_0,\ldots,x_{P-1}]$. Then the joint probability density function of the observed data is

$$f(\boldsymbol{x};\boldsymbol{a}^{(P)},\sigma_\nu^2) = f_C(\boldsymbol{x}_P\mid\boldsymbol{x}_0;\boldsymbol{a}^{(P)},\sigma_\nu^2)\,f_0(\boldsymbol{x}_0;\boldsymbol{a}^{(P)},\sigma_\nu^2). \tag{8.182}$$

Derivation of an MLE of the AR(P) parameters for $\boldsymbol{x}$ using the left hand side of (8.182) leads to non-linear equations. An approximate MLE is found by using the conditional density on the right hand side, namely, $f_C(\boldsymbol{x}_P\mid\boldsymbol{x}_0;\boldsymbol{a}^{(P)},\sigma_\nu^2)$, since for large data records, the probability density of the initial conditions $f_0(\boldsymbol{x}_0;\boldsymbol{a}^{(P)},\sigma_\nu^2)$ can be neglected so long as the poles are not too close to the unit circle [107].

Rewriting the AR(P) process equation, the noise process

$$\nu_n = x_n + \sum_{k=1}^{P}a_k^{(P)}x_{n-k}\,, \tag{8.183}$$

is white and so the logarithm of the conditional density is [see (5.65)]

$$\ln f_c(\boldsymbol{x}_P \mid \boldsymbol{x}_0; \boldsymbol{a}^{(P)}, \sigma_\nu^2) = -(N-P)\ln \pi - (N-P)\ln \sigma_\nu^2 + \frac{1}{\sigma_\nu^2}\sum_{n=P}^{N-1}\left| x_n + \sum_{k=1}^{P} a_k^{(P)} x_{n-k} \right|^2. \tag{8.184}$$

The conditional density is maximized by minimizing the exponent with respect to the AR parameters $\boldsymbol{a}^{(P)}, \sigma_\nu^2$. Differentiating with respect to $\boldsymbol{a}^{(P)}$ and setting the result equal to zero gives the estimated Yule-Walker equation

$$\begin{bmatrix} \hat{R}_{11} & \hat{R}_{12} & \cdots & \hat{R}_{1P} \\ \hat{R}_{21} & \hat{R}_{22} & \cdots & \hat{R}_{2P} \\ \vdots & \vdots & \ddots & \vdots \\ \hat{R}_{P1} & \hat{R}_{P2} & \cdots & \hat{R}_{PP} \end{bmatrix} \begin{bmatrix} \hat{a}_1^{(P)} \\ \hat{a}_2^{(P)} \\ \vdots \\ \hat{a}_P^{(P)} \end{bmatrix} = - \begin{bmatrix} \hat{R}_{10} \\ \hat{R}_{20} \\ \vdots \\ \hat{R}_{P0} \end{bmatrix}, \tag{8.185}$$

where the auto-correlation matrix estimate $\hat{\boldsymbol{R}}$ is exactly as given in (8.166):

$$\left[\hat{\boldsymbol{R}}\right]_{km} \equiv \hat{R}_{km} = \frac{1}{N-P}\sum_{j=P}^{N-1} x_{j-k}^* x_{j-m}, \tag{8.186}$$

and it is Hermitian but not Toeplitz. Furthermore, for an arbitrary $P \times 1$ vector $\boldsymbol{d}$ we have

$$\boldsymbol{d}^+ \hat{\boldsymbol{R}} \boldsymbol{d} = \frac{1}{N-P}\sum_{n=P}^{N-1}\left|\sum_{k=1}^{P} d_k x_{n-k}\right|^2, \tag{8.187}$$

which is positive when $\boldsymbol{x}$ is not a noise-free sum of complex exponentials.

Differentiating the logarithm of the density with respect to σ_ν^2 and setting the result equal to zero gives the estimated white noise variance:

$$\hat{\sigma}_\nu^2 = \hat{R}_{00} + \sum_{k=1}^{P} \hat{a}_k^{(P)} \hat{R}_{0k}. \tag{8.188}$$

The approximate MLE covariance method estimates (8.185), (8.186) and (8.188) of AR(P) parameters $(\hat{\boldsymbol{a}}, \hat{\sigma}_\nu^2)$ are unbiased:

$$\mathrm{E}\left[\hat{\boldsymbol{a}}\right] = \boldsymbol{a}, \;\; \mathrm{E}\left[\hat{\sigma}_\nu^2\right] = \sigma_\nu^2, \tag{8.189}$$

and the correlation estimates (8.186) can be found using the data matrix $\boldsymbol{A}_{\mathrm{cov}}$ of equation (4.33) in section 4.4.

Note that the AR auto-correlation method estimate of the auto-correlation matrix (assuming large number of data samples N), described by equation (8.181) in section 8.22, is equal to a modified form of the covariance method estimate (8.186), where we change the summation range appropriately and divide by N instead of $N-P$,

$$\hat{R}_{lm} = \frac{1}{N}\sum_{k=0}^{N-1-(l-m)} x_k x_{k-(l-m)}^*, \quad l \geq m, \text{ and } \hat{R}_{lm} = \hat{R}_{ml}^* \text{ for } l < m. \tag{8.190}$$

The estimated auto-correlation matrix (8.190) is now Toeplitz and the estimate can be found using the data matrix $\boldsymbol{A}_{\mathrm{ac}}$ of equation (4.36) in section 4.4.

The approximation to the conditional density (8.184) results in the Fisher information matrix [see section 5.13] for the MLE vector $\left[\hat{\boldsymbol{a}}, \hat{\sigma}_\nu^2\right]$ [107],

$$\mathbf{I} = \frac{N}{2\sigma_\nu^4}\begin{bmatrix} 2\sigma_\nu^2 \boldsymbol{R}_{xx} & \mathbf{0} \\ \mathbf{0} & 1 \end{bmatrix}, \quad N \gg P, \tag{8.191}$$

where $\boldsymbol{R}_{xx}$ is the true auto-correlation matrix of the data vector $\boldsymbol{x}$. The CRLB [see section 5.13] for the MLE of the parameter vector is given by the diagonal elements of the inverse of the Fisher information matrix (8.191).

8.24 Model Order Selection

Model selection methods can identify "useful" models [7] that ought to penalize for excess complexity, i.e., they should be the simplest models that describe the data well. However, they should also be generalizable, i.e., the models should fit unseen data from the same underlying process; in other words, useful models should have predictive accuracy. More complex models tend to overfit data [8], as illustrated in figure 8.21.

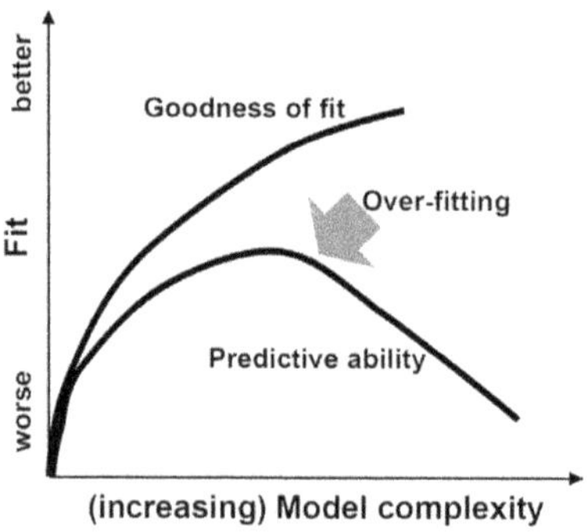

Figure 8.21: An overfit model.

As a general rule, higher order (larger P) AR models can better fit the data, but they may overfit the data and have no predictive ability. Consider the example of linear prediction (LP) (discussed in chapter 9) of a sound of the vowel "a" spoken by a male speaker, using 100 samples of data y_n at 8 kHz. Figure 8.22 shows the actual time series and two fits to the data for $P = 3$ and $P = 30$ (the Burg algorithm described later in section 9.8 was used to calculate the AR coefficient vector $\boldsymbol{a}$). The panel on the right displays the residual MSE (for unit mean signal energy) as a function of AR model order P from 3 to 30, showing a decrease as model order increases. Up to some point, the reduction might be attributed to the actual structure of the sound signal, but beyond that point, the predictor is simply taking advantage of the extra coefficients, and the decrease is due to overfitting. If the goal is to merely minimize the mean squared prediction error, then the answer is the largest P possible (30 in this case). With the multiplicity of available parameters, it is very possible to overfit to this particular data set, and lose all ability to fit another instance of the data, even by the same speaker, much less for a different speaker. Another important area in which model selection is important is the design and training of neural networks (described in chapter 12). A large neural network may overfit the training data with very small error yet fail to generalize.

[7] "All models are wrong, but some are useful." [108].

[8] "Overfitting is a sin precisely because it undermines the goal of predictive accuracy." [109].

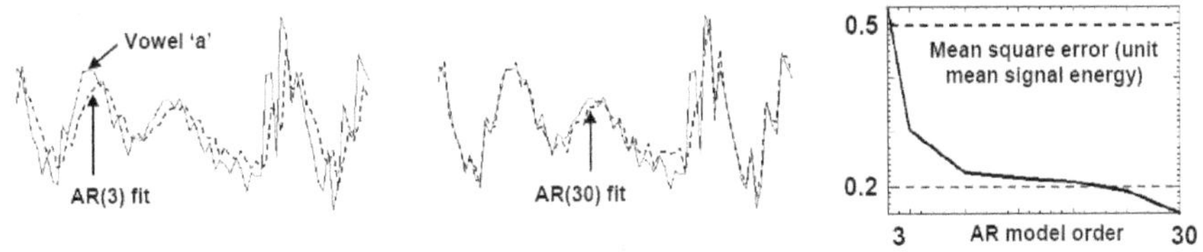

Figure 8.22: Sound of vowel "a" (male speaker): 12.5 ms sampled at 8 kHz, together with AR(3) and AR(30) model fits, and MSE (normalized by mean signal energy) as a function of increasing AR model order.

In all model fitting problems, there is a method of learning the model parameters and a method of determining how well the model fits. However, there is a larger problem of selecting among competing models.

8.25 Akaike Information Criterion

The Akaike Information Criterion (AIC) is a method which aims to select a correct model order without overfitting [110, 111, 112]. We will illustrate its use for regression and AR modeling problems.

In a regression problem, let observations $\boldsymbol{y} = [y_1, \ldots, y_N]^T$ be produced according to an operating model $\boldsymbol{y} = \boldsymbol{b} + \boldsymbol{\nu}$, where $\boldsymbol{b} = [b_1, \ldots, b_N]^T$ and $\boldsymbol{\nu} = [\nu_1, \ldots, \nu_N]^T$, $\nu_n \sim \mathcal{N}(0, \sigma_\nu^2)$. The likelihood function is

$$\begin{aligned} f(\boldsymbol{y}|\boldsymbol{b}) &= (2\pi\sigma_\nu^2)^{-N/2} \exp\Big[-\frac{1}{2\sigma_\nu^2}(\boldsymbol{y}-\boldsymbol{b})^+(\boldsymbol{y}-\boldsymbol{b})\Big] \\ &= (2\pi\sigma_\nu^2)^{-N/2} \exp\Big[-\frac{1}{2\sigma_\nu^2}\boldsymbol{\nu}^+\boldsymbol{\nu}\Big]. \end{aligned} \tag{8.192}$$

The data is to be fitted by an approximating linear model $\boldsymbol{y} = \boldsymbol{X}\boldsymbol{\theta} + \boldsymbol{u}$, where $\boldsymbol{u} = [u_1, \ldots, u_N]^T$, $u_n \sim \mathcal{N}(0, \sigma^2)$. $\boldsymbol{X}$ is an $N \times M$ matrix, and $\boldsymbol{\theta}$ is an $M \times 1$ parameter vector. For instance, suppose that we wish to fit the data using an AR(2) model. Then the parameter vector is $\boldsymbol{\theta} = -[a_1, a_2]^T$, the linear model is

$$\begin{bmatrix} y_1 \\ y_2 \\ \vdots \\ y_N \end{bmatrix} = \begin{bmatrix} y_0 & y_{-1} \\ y_1 & y_0 \\ \vdots & \vdots \\ y_{N-1} & y_{N-2} \end{bmatrix} \begin{bmatrix} -a_1 \\ -a_2 \end{bmatrix} + \begin{bmatrix} u_1 \\ u_2 \\ \vdots \\ u_N \end{bmatrix}, \tag{8.193}$$

and the likelihood function for the parameters is

$$g(\boldsymbol{y}|\boldsymbol{\theta}, \sigma^2) = (2\pi\sigma^2)^{-N/2} \exp\Big[-\frac{1}{2\sigma^2}(\boldsymbol{y}-\boldsymbol{X}\boldsymbol{\theta})^+(\boldsymbol{y}-\boldsymbol{X}\boldsymbol{\theta})\Big]. \tag{8.194}$$

The standard measure of discrepancy between the two distributions f and g of equations (8.192) and (8.194) is the Kullback-Leibler divergence (5.219) given by the difference between the entropy associated with f and the cross-entropy of f and g. The entropy of f is independent of all parameters, and so instead of using the Kullback-Leibler divergence, we use the negative of the cross entropy as our measure of discrepancy between the true model and the approximating linear model. Introducing an overall factor of 2 (for convenience), we minimize the quantity

$$\Delta(\sigma^2, \boldsymbol{\theta}) = -2\,\mathbf{E}_f\big[\ln g(\boldsymbol{y}|\boldsymbol{\theta}, \sigma^2)\big], \tag{8.195}$$

which upon replacing the unknown parameters $\boldsymbol{\theta}$ and σ^2 by their MLEs

$$\hat{\boldsymbol{\theta}} = \arg\min_{\boldsymbol{\theta}}\left[(\boldsymbol{y} - \boldsymbol{X\theta})^+(\boldsymbol{y} - \boldsymbol{X\theta})\right], \quad \hat{\sigma}^2 = \left[(\boldsymbol{y} - \boldsymbol{X}\hat{\boldsymbol{\theta}})^+(\boldsymbol{y} - \boldsymbol{X}\hat{\boldsymbol{\theta}})\right]/N, \tag{8.196}$$

becomes

$$\Delta(\hat{\sigma}^2, \hat{\boldsymbol{\theta}}) - N\ln 2\pi = N\ln\hat{\sigma}^2 + N\sigma_\nu^2/\hat{\sigma}^2 + (\boldsymbol{b} - \boldsymbol{X}\hat{\boldsymbol{\theta}})^+(\boldsymbol{b} - \boldsymbol{X}\hat{\boldsymbol{\theta}})/\hat{\sigma}^2. \tag{8.197}$$

When there are multiple competing models from which to choose, a reasonable choice is the model which minimizes $\Delta(\hat{\sigma}^2, \hat{\boldsymbol{\theta}})$. The *AIC* for regression and autoregression

$$\text{AIC}(M) = N\ln\hat{\sigma}^2 + N + 2(M+1) \tag{8.198}$$

was designed to provide an approximately unbiased estimate of $\mathrm{E}_f\left[\Delta(\hat{\sigma}^2, \hat{\boldsymbol{\theta}})\right]$. The first term is a measure of how well the model fits the data (goodness of fit), while the last term penalizes more complex models (models with more parameters). In practice, the AIC is computed for several models, i.e., several values of M, and the model with the lowest AIC is selected as the best fit to the data. When N is fairly small, the AIC(M) may be biased, leading to model order estimates that are too large. A criterion with less bias for small values of N is known as the *corrected AIC*, or AIC_C [111],

$$\text{AIC}_\text{C}(M) = N\ln\hat{\sigma}^2 + N\,\frac{1 + M/N}{1 - (M+2)/N} = \text{AIC} + \frac{2(M+1)(M+2)}{N - M - 2}, \tag{8.199}$$

which is valid for both regression and autoregression. The last term on the right hand side is an additional penalty which prohibits larger values of M. As N becomes larger, the extra correction term loses its significance accordingly. Figure 8.23 shows the AIC and the AIC_C as functions of model order for the speech data of figure 8.22; both criteria choose AR model order $P = 10$.

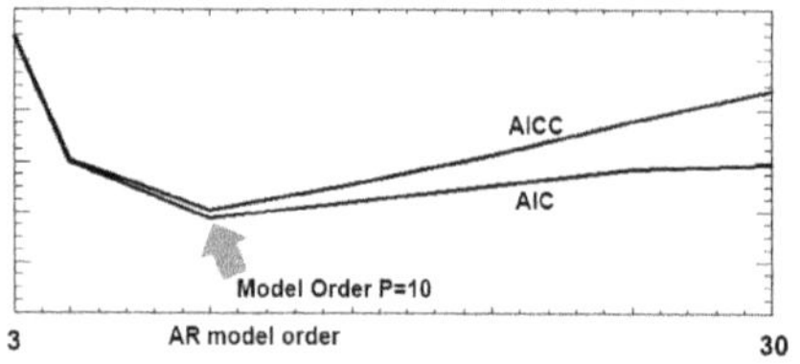

Figure 8.23: AIC and AIC_C for the speech data of figure 8.22.

The left hand side of figure 8.24 shows 11 data points to be fitted by a polynomial. The AIC and the AIC_C as functions of polynomial order from 1 to 6 are shown on the right hand side. The AIC suffers from bias because of the small number of data points, while the AIC_C clearly chooses the best polynomial fit as a cubic.

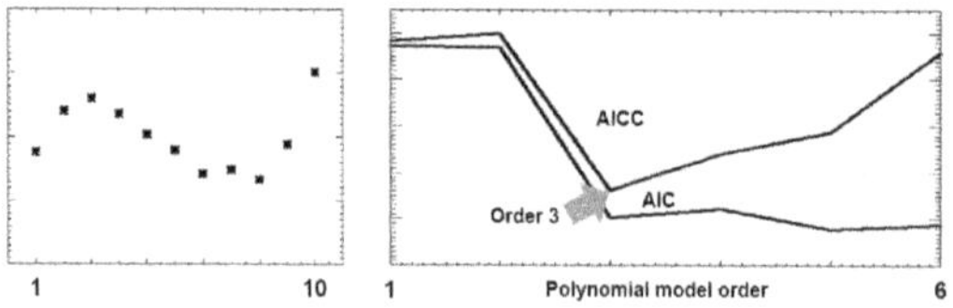

Figure 8.24: AIC and AIC_C for polynomial fits to 11 data points.

8.26 Bayesian Model Order Selection

In Bayesian model selection, each model k is evaluated based on its marginal likelihood, namely,

$$P(\boldsymbol{y}_{\text{obs}}|k) = \int f(\boldsymbol{y}_{\text{obs}}|\boldsymbol{\theta}, k)\pi(\boldsymbol{\theta}|k)\, d\boldsymbol{\theta}. \tag{8.200}$$

The choice between two models l and k is then based on the ratio of two marginal likelihoods, known as *Bayes factor* [113]:

$$\text{BF}_{lk} \equiv P(\boldsymbol{y}_{\text{obs}}|l)/P(\boldsymbol{y}_{\text{obs}}|k), \tag{8.201}$$

which, assuming equal model priors, gives

$$\text{BF}_{lk} = P(l|\boldsymbol{y}_{\text{obs}})/P(k|\boldsymbol{y}_{\text{obs}}). \tag{8.202}$$

For large N we find the Bayesian Information Criterion (BIC) [114]:

$$\text{BIC} = -2\ln P(\boldsymbol{y}_{\text{obs}}|k) = -2\ln f(\boldsymbol{y}_{\text{obs}}|\hat{\boldsymbol{\theta}}, k) + M\ln N. \tag{8.203}$$

8.27 Minimum Description Length

The *Minimum Description Length* (MDL) is a principle rather than a particular algorithm; it states that the best model is the one that provides the tightest compression in the form of the shortest description length of the data in bits: the best model is that which takes advantage of the structure (e.g., regularity or redundancy) in the data and requires fewer bits to describe the data [115, 116].

Regularities are features that can be represented well by a model. Let D represent a data set, such as a sequence of observations $\{y_1, y_2, \ldots, y_N\}$, and let $\mathcal{H}$ denote a set of hypotheses (that is, different models). Under MDL, regularity is viewed as the ability to compress the data. If a model represents the data perfectly, then it may perform good compression, because all that is needed to represent the data is a description of the model. A model with worse representation could require some description of model residuals. In any case, the ability to compress is determined by both the description of the model and the description of the data under the model. More complicated models require more description, and this might compress less.

Crude MDL [117] can be stated as follows: let $\mathcal{H}^{(1)}, \mathcal{H}^{(2)}, \ldots$ form a set of candidate models. For instance, $\mathcal{H}^{(k)}$ might be the set of polynomials of degree k. Each set contains hypotheses (such as an individual polynomial). Let $H \in \mathcal{H}^{(1)} \cup \mathcal{H}^{(2)} \cup \ldots$ be a particular hypothesis being examined to explain the data D. Associated with H are two quantities: the description length of the model itself $DL(H)$ (for example, the number of bits used to describe all the coefficients of the polynomial model) and the description length of the data when encoded (modeled) using the hypothesis (model) $DL(D|H)$. For each hypothesis under consideration, the sum $DL(H)+DL(D|H)$ is computed, and the best model (hypothesis) is the one that minimizes this quantity.

For parametric models, such as AR models, two major implementations of MDL exist. The first is the *Normalized Maximum Likelihood* (NML) distribution, originally formulated as the optimal universal distribution $\tilde{g}(\boldsymbol{x})$ for the *minimax* problem [118],

$$\tilde{g}(\boldsymbol{x}) = \arg\min_{g}\max_{h} \mathbf{E}_h\left[\ln \frac{f(\boldsymbol{x}|\hat{\boldsymbol{\theta}})}{g(\boldsymbol{x})}\right], \tag{8.204}$$

where h and g can be any distributions, and $\hat{\boldsymbol{\theta}}(\boldsymbol{x})$ is the MLE of the parameter vector. The solution for the observed data $\boldsymbol{y}$ is

$$\tilde{g}(\boldsymbol{y}) = -\ln \frac{f(\boldsymbol{y}|\hat{\boldsymbol{\theta}})}{\int f(\boldsymbol{y}'|\hat{\boldsymbol{\theta}}(\boldsymbol{y}'))\, d\boldsymbol{y}'}. \tag{8.205}$$

The NML criterion is then

$$\text{NML} = -\ln \tilde{g}(\boldsymbol{y}). \tag{8.206}$$

While the NML has many desirable properties, the denominator in (8.205) is difficult to evaluate. The NML solution actually resides outside the model class, yet it is the one distribution that is universally representative of the entire model class in the minimax sense of its definition in (8.204). For instance, let the true distribution be a binomial:

$$f(y|\theta) = \frac{N!}{y!(N-y)!}\theta^y(1-\theta)^{N-y}, \quad y = 0, 1, \ldots, N, \ n\text{and}\ 0 \le \theta \le 1.$$

Using the MLE $\hat{\theta} = y/N$, the NML distribution is

$$\tilde{g}(y|\theta) = \frac{f(y|\hat{\theta})}{S}, \quad S \equiv \sum_{n=0}^{N} \binom{N}{n} \left(\frac{n}{N}\right)^n \left(1 - \frac{n}{N}\right)^{N-n},$$

which is clearly not a binomial distribution; $S \approx 4.66$ for $N = 10$.

The second implementation of MDL, often used instead of NML, is the *Fisher Information Approximation* (FIA):

$$\text{FIA} = -\ln f(\boldsymbol{y}|\hat{\boldsymbol{\theta}}) + \frac{M}{2}\ln\frac{N}{2\pi} + \ln \int \sqrt{|\boldsymbol{I}_{\theta\theta}|}\, d\boldsymbol{\theta}, \tag{8.207}$$

where $\mathbf{I}_{\theta\theta}$ is the Fisher information matrix (5.164). The second and third terms in equation 8.207 represent penalties for model complexity: the first is a penalty due to the number of parameters (much the same as in AIC and BIC), while the next is a penalty due to the functional form of the model equation.

The sum of the last two terms of (8.207) has an interpretation in terms of *geometric complexity* in modern information geometry [119], which considers the family of probability distributions as a Riemannian manifold with a metric tensor given by the Fisher information matrix of the true distribution; similar distributions form clusters in the manifold, and the dissimilarity is measured by the metric. The Riemannian volume measure is then precisely of the form $\sqrt{|\mathbf{I}_{\theta\theta}|}\, d\boldsymbol{\theta}$. A simple model occupies a relatively small volume compared to a complex model. A model is considered "good" if it contains many distributions in a cluster near the true distribution. The log volume ratio between the good distributions and the whole volume gives exactly the last two terms of the FIA (8.207).

Assuming certain regularity conditions [120], it can be shown that the FIA asymptotically converges to $-\ln$ NML; so, although the optimality of NML is guaranteed for all N, the FIA is only asymptotically optimal. The FIA, however, is much easier to compute, since the integration is performed only over the parameter space.

The strongest appeal of MDL is the fact that it avoids the assumption that the "truth" is to be found within the considered set of models. In fact, it does not even rely on the existence of a "true" distribution generating the data! Instead, it is based on efficient coding of the observed data and the ability of the model to generalize, i.e., to have predictive accuracy.

CHAPTER 9

Discrete Time Wiener Filter and Linear Prediction

9.1 Introduction

In this chapter, we develop the theory of the FIR Wiener filter in discrete time. This naturally leads to the problems of one-step forward and backward prediction with applications to the processing and coding of speech, referred to as linear prediction (LP) or linear predictive coding (LPC). LP models have many applications including forecasting, speech and video coding, and speech recognition. A linear model to predict a signal from its past values can be built on knowledge of the signal's auto-correlation function. Most signals of interest such as speech, are only partially predictable and are often modeled as the output of a minimum phase filter driven by white noise; the input white noise models the unpredictable part of the signal while the filter models its predictable part.

9.2 The Discrete Time FIR Wiener Filter

Consider the problem of estimating a discrete time zero-mean WSS random process y_{n+k} from a correlated discrete time zero-mean WSS random process x_n using an FIR filter vector $\boldsymbol{h}$ with M elements,

$$\hat{y}_{n+k} = \sum_{m=0}^{M-1} h_m x_{n-m}, \quad \boldsymbol{h} = [h_0, \ldots, h_{M-1}]^T, \tag{9.1}$$

where $k > 0$ defines the pure prediction of y from x, $k = 0$ is the real-time estimation problem, and $k < 0$ is the smoothing problem (also known as "backward prediction"). Defining the time-reversed data vector

$$\boldsymbol{x}_n = \left[x_n, \ldots, x_{n-(M-1)}\right]^T, \tag{9.2}$$

the optimal filter is the solution to the mean square minimization problem,

$$\arg\min_{\boldsymbol{h}} \mathrm{E}\left[\left|e_{n+k}\right|^2\right] = \arg\min_{\boldsymbol{h}} \mathrm{E}\left[\left|y_{n+k} - \boldsymbol{h}^T\boldsymbol{x}_n\right|^2\right], \tag{9.3}$$

where

$$\mathrm{E}\left[\left|e_{n+k}\right|^2\right] = \left|y_{n+k}\right|^2 - \boldsymbol{h}^+\mathrm{E}\left[y_{n+k}\boldsymbol{x}_n^*\right] - \mathrm{E}\left[y_{n+k}^*\boldsymbol{x}_n^T\right]\boldsymbol{h} + \boldsymbol{h}^+\mathrm{E}\left[\boldsymbol{x}_n^*\boldsymbol{x}_n^T\right]\boldsymbol{h}. \tag{9.4}$$

Using (6.73) for the time reversed vector $\boldsymbol{x}_n$ and defining the cross-correlation vector

$$\boldsymbol{r}_{yx} \equiv \mathrm{E}\left[y_{n+k}\boldsymbol{x}_n^*\right] \equiv [r_k, \ldots, r_{k+M-1}]^T, \tag{9.5}$$

minimization with respect to $\boldsymbol{h}$ leads to an equation similar to the Yule-Walker equation (8.104),

$$\begin{bmatrix} R_0 & R_{-1} & \cdots & R_{-(M-1)} \\ R_1 & \ddots & \ddots & \vdots \\ \vdots & \ddots & \ddots & R_{-1} \\ R_{M-1} & \cdots & R_1 & R_0 \end{bmatrix} \begin{bmatrix} h_0 \\ h_1 \\ \vdots \\ h_{M-1} \end{bmatrix} = \begin{bmatrix} r_k \\ r_{k+1} \\ \vdots \\ r_{k+M-1} \end{bmatrix} \equiv \boldsymbol{r}_{yx}, \tag{9.6}$$

where we have denoted the (m,n) element of the Hermitian Toeplitz auto-correlation matrix $\boldsymbol{R}_{xx}$ by R_{m-n}. The orthogonality principle

$$\mathrm{E}\left[e^*_{n+k}\boldsymbol{x}_n\right] = \mathrm{E}\left[e_{n+k}\boldsymbol{x}^*_n\right] = 0 \tag{9.7}$$

would, of course, lead to the same matrix equation (9.6), whose solution is the optimal Wiener filter:

$$\boldsymbol{h}_{\text{opt}} = \boldsymbol{R}^{-1}_{xx}\boldsymbol{r}_{yx}. \tag{9.8}$$

The Levinson-Durbin algorithm of section 8.15 described a fast recursive method to solve a restricted Hermitian and Toeplitz matrix equation [the Yule-Walker equation (8.104)] whose right hand side is a specific portion of the first column of the auto-correlation matrix. If, on the other hand, the right hand side is an arbitrary but known non-zero vector $\boldsymbol{b}$, then a fast recursive solution of $\boldsymbol{Rh} = \boldsymbol{b}$, where $\boldsymbol{R}$ is a Hermitian Toeplitz matrix, is based on the solution of the restricted Yule-Walker equation. Denoting the filter with M elements by $\boldsymbol{h}^{(\text{M})}$, we partition (9.6) into

$$\begin{bmatrix} \boldsymbol{R}^{(\text{M-1})} & \boldsymbol{J}^{(\text{M-1})}\boldsymbol{r}^*_{M-1} \\ \boldsymbol{r}^T_{M-1}\boldsymbol{J}^{(\text{M-1})} & R_0 \end{bmatrix} \begin{bmatrix} \boldsymbol{q}_{M-1} \\ \alpha_M \end{bmatrix} = \begin{bmatrix} \boldsymbol{b}_{M-1} \\ b_M \end{bmatrix}, \tag{9.9}$$

where we have denoted the first $M-1$ elements of $\boldsymbol{h}^{(\text{M})}$ by $\boldsymbol{q}_{M-1}$ and its last element by α_M, i.e., $\boldsymbol{h}^{(\text{M})} \equiv [\boldsymbol{q}_{M-1}, \alpha_M]^T$, while the last $M-1$ elements of the first column of $\boldsymbol{R}$ form the vector $\boldsymbol{r}_{M-1}$,

$$\boldsymbol{r}_{M-1} \equiv [R_1, \ldots, R_{M-1}]^T. \tag{9.10}$$

Equation (9.9) is equivalent to two separate equations

$$\boldsymbol{R}^{(\text{M-1})}\boldsymbol{q}_{M-1} + \alpha_M\boldsymbol{J}^{(\text{M-1})}\boldsymbol{r}^*_{M-1} = \boldsymbol{b}_{M-1} \quad \text{and} \quad \boldsymbol{r}^T_{M-1}\boldsymbol{J}^{(\text{M-1})}\boldsymbol{q}_{M-1} + \alpha_M R_0 = b_M. \tag{9.11}$$

Multiplying the first equation on the left by $[\boldsymbol{R}^{(\text{M-1})}]^{-1}$, using the identity [see (6.73)]

$$[\boldsymbol{R}^{(\text{M-1})}]^{-1}\boldsymbol{J}^{(\text{M-1})} \equiv \boldsymbol{J}^{(\text{M-1})}[\boldsymbol{R}^{(\text{M-1})*}]^{-1},$$

and the solution to the Yule-Walker equation (8.104), namely, $\boldsymbol{a}^{(\text{M-1})} = -[\boldsymbol{R}^{(\text{M-1})}]^{-1}\boldsymbol{r}_{M-1}$, and its complex conjugate, we find

$$\boldsymbol{q}_{M-1} = \boldsymbol{h}^{(\text{M-1})} + \alpha_M\,\boldsymbol{J}^{(\text{M-1})}\boldsymbol{a}^{(\text{M-1})*}. \tag{9.12}$$

Substituting (9.12) into the second of (9.11) (equation for α_M), we find

$$\alpha_M = \frac{b_M - \boldsymbol{r}^T_{M-1}\boldsymbol{J}^{(\text{M-1})}\boldsymbol{h}^{(\text{M-1})}}{R_0 + \boldsymbol{r}^T_{M-1}\boldsymbol{a}^{(\text{M-1})*}}. \tag{9.13}$$

Thus, the recursion begins with $M = 1$, for which $h^{(1)}_1 = b_1/R_0$, and

- $\alpha_{M+1} = \frac{b_{M+1} - \boldsymbol{r}_M\boldsymbol{J}^{(\text{M})}\boldsymbol{h}^{(\text{M})}}{R_0 + \boldsymbol{r}^T_M\boldsymbol{a}^{(\text{M})*}}$
- $\boldsymbol{q}_M = \boldsymbol{h}^{(\text{M})} + \alpha_{M+1}\,\boldsymbol{J}^{(\text{M})}\boldsymbol{a}^{(\text{M})*}$
- $\boldsymbol{h}^{(\text{M+1})} = [\boldsymbol{q}_M,\ \alpha_{M+1}]^T$

9.3 The Forward Prediction Problem

A one-step ahead linear predictor for zero-mean WSS random data y_n using a finite number P of its past values $\boldsymbol{y}_{n-1} \equiv [y_{n-1}, \ldots, y_{n-P}]^T$ is given by

$$\hat{y}_n^{(f)} = -\sum_{k=1}^{P} a_k^{(P)} y_{n-k} = -\boldsymbol{a}^{(P)T} \boldsymbol{y}_{n-1}, \quad \boldsymbol{a}^{(P)} = [a_1^{(P)}, \ldots, a_P^{(P)}]^T, \tag{9.14}$$

and illustrated in figure 9.1.

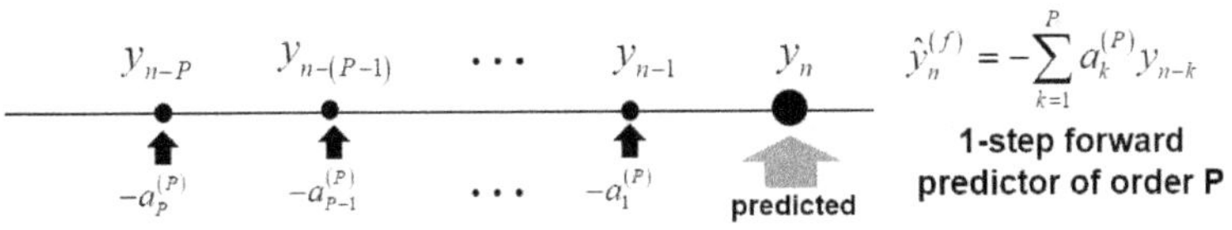

Figure 9.1: One-step ahead prediction.

In order to solve for the parameter vector $\boldsymbol{a}^{(P)}$ we assume that the process auto-correlation sequence is either known or can be estimated from the data. Defining the one-step forward prediction error sequence of order P

$$e_P^{(f)}[n] \equiv y_n - \hat{y}_n^{(f)} = y_n + \sum_{k=1}^{P} a_k^{(P)} y_{n-k}, \tag{9.15}$$

the one-step ahead prediction problem is equivalent to the prediction error filtering operation shown in figure 9.2 when the prediction error sequence at time step n is the output of a linear filter with y_n as input. The PEF in figure 9.2

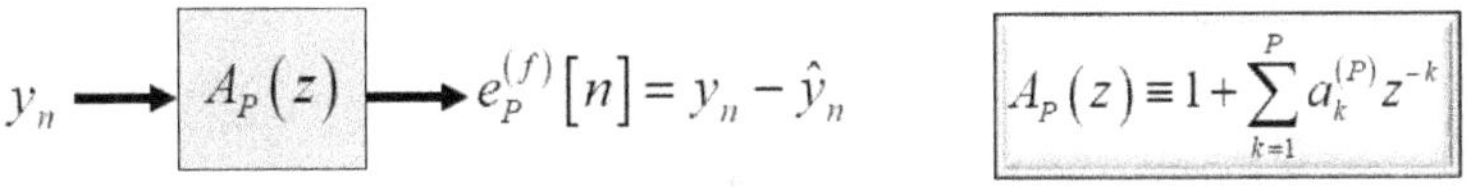

Figure 9.2: One-step ahead prediction filter.

$$A_P(z) = 1 + \sum_{k=1}^{P} a_k^{(P)} z^{-k}, \tag{9.16}$$

is an all-zero filter; hence, its inverse is an all-pole filter. All-pole filters are used in the processing of speech in what is collectively known as linear predictive coding or LPC (see section 9.9). The optimal parameter vector $\boldsymbol{a}^{(P)}$ is found by minimizing the mean square forward prediction error $\rho_P^{(f)} \equiv [\sigma_P^{(f)}]^2$,

$$\rho_P^{(f)} = \mathrm{E}\left[\left| e_P^{(f)}[n] \right|^2 \right] = R_0 + \boldsymbol{r}_P^{+} \boldsymbol{a}^{(P)} + \boldsymbol{a}^{(P)+} \boldsymbol{r}_P + \boldsymbol{a}^{(P)+} \boldsymbol{R}^{(P)} \boldsymbol{a}^{(P)}, \tag{9.17}$$

where

$$\boldsymbol{r}_P \equiv [R_1, \ldots, R_P]^T, \quad R_l = \mathrm{E}[y_n y_{n-l}^*], \quad 1 \le l \le P, \tag{9.18}$$

with respect to the parameter vector $\boldsymbol{a}^{(\mathrm{P})}$ (or its Hermitian conjugate). In deriving the right hand side of (9.18), we used the following relations for the time reversed vector $\boldsymbol{y}_n$,

$$\boldsymbol{a}^{(\mathrm{P})T}\boldsymbol{y}_{n-1} \equiv \boldsymbol{y}^T_{n-1}\boldsymbol{a}^{(\mathrm{P})}, \;\; \mathrm{E}[\boldsymbol{y}_{n-1}\boldsymbol{y}^{+}_{n-1}] = \boldsymbol{R}^{(\mathrm{P})*} \;\Leftrightarrow\; \mathrm{E}[\boldsymbol{y}^{*}_{n-1}\boldsymbol{y}^{T}_{n-1}] = \boldsymbol{R}^{(\mathrm{P})}. \tag{9.19}$$

Minimization of the forward prediction error results in the order P Yule-Walker equation (8.104) where the order P is yet another unknown parameter of the linear prediction model; the same equation can be obtained using the orthogonality principle:

$$\mathrm{E}\left[e^{(f)}_P y^*_{n-k}\right] = 0, \;\; 1 \le k \le P. \tag{9.20}$$

Equation (8.131) shows that the variance of the one-step forward linear predictor error decreases as the number of filter points increases so long as the PEF is minimum phase, i.e., $|\gamma_P| < 1$ (see section 9.7 for a proof of the last statement), i.e., $\rho^{(f)}_{P+1} < \rho^{(f)}_P$. Hence, as filter order increases, all that can be predicted will be predicted, and the error will become a white noise sequence that is no longer predictable from y_n; this is consistent with the Wold decomposition theorem 32 and theorem 33.

9.4 The Backward Prediction Problem

The backward prediction problem is useful in signal modeling. We use the term *backward prediction* of order P to refer to the one-step postdiction problem described by the coefficient vector $\boldsymbol{b}^{(\mathrm{P})}$ and the $P \times 1$ future values vector $\boldsymbol{y}_n \equiv [y_n, \ldots, y_{n-P+1}]^T$, as illustrated in figure 9.3.

$$\hat{y}^{(\mathrm{b})}_{n-P} = -\sum_{k=1}^{P} b^{(\mathrm{P})}_k y_{n-k+1} = -\boldsymbol{b}^{(\mathrm{P})T}\boldsymbol{y}_n, \;\; \boldsymbol{b}^{(\mathrm{P})} = [b^{(\mathrm{P})}_1, \ldots, b^{(\mathrm{P})}_P]^T. \tag{9.21}$$

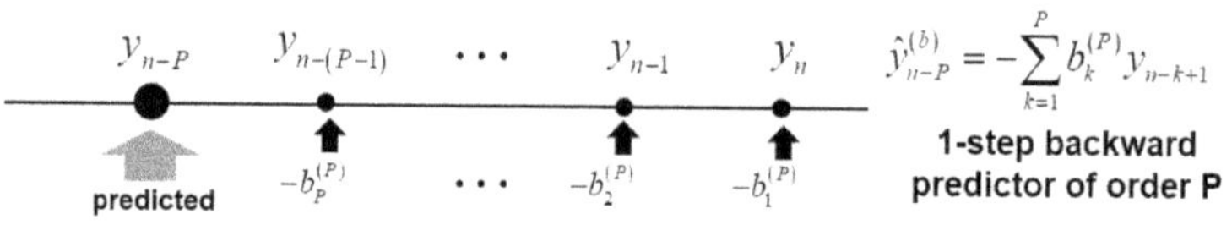

Figure 9.3: One-step backward prediction of order P.

For the one-step backward prediction error sequence of order P,

$$e^{(\mathrm{b})}_P[n] \equiv y_{n-P} - \hat{y}^{(\mathrm{b})}_{n-P} = y_{n-P} + \sum_{k=1}^{P} b^{(\mathrm{P})}_k y_{n-k+1}, \tag{9.22}$$

we have the equivalent backward prediction error filtering operation shown in figure 9.4 with the *backward prediction error filter*:

$$B_P(z) = z^{-P} + \sum_{k=1}^{P} b^{(\mathrm{P})}_k \, z^{-(k-1)}. \tag{9.23}$$

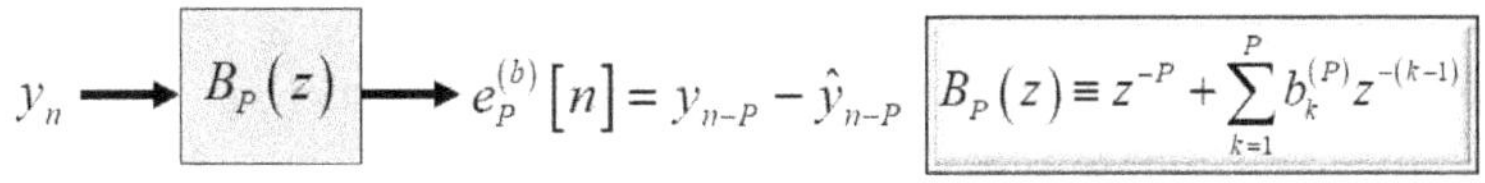

Figure 9.4: One-step backward prediction filter.

The optimal parameter vector is found by minimizing the mean square backward prediction error with respect to the parameter vector (or its Hermitian conjugate),

$$\rho_P^{(b)} = \mathrm{E}\left[\left|e_P^{(b)}[n]\right|^2\right] = R_0 + \boldsymbol{b}^{(P)+}\mathrm{E}[y_{n-P}\boldsymbol{y}_n^*] + \mathrm{E}[y_{n-P}^*\boldsymbol{y}_n^T]\boldsymbol{b}^{(P)} + \boldsymbol{b}^{(P)+}\boldsymbol{R}^{(P)}\boldsymbol{b}^{(P)}, \tag{9.24}$$

or equivalently,

$$\rho_P^{(b)} = R_0 + \boldsymbol{b}^{(P)+}\boldsymbol{J}^{(P)}\boldsymbol{r}_P^* + \boldsymbol{r}_P^T\boldsymbol{J}^{(P)}\boldsymbol{b}^{(P)} + \boldsymbol{b}^{(P)+}\boldsymbol{R}^{(P)}\boldsymbol{b}^{(P)}, \tag{9.25}$$

and solving the resulting Yule-Walker equation of order P, where the order P is yet another unknown parameter of the linear prediction model. The backward prediction Yule-Walker equation has a different right hand side from (8.104):

$$\begin{bmatrix} R_0 & R_{-1} & \cdots & R_{-(P-1)} \\ R_1 & \ddots & \ddots & \vdots \\ \vdots & \ddots & \ddots & R_{-1} \\ R_{P-1} & \cdots & R_1 & R_0 \end{bmatrix} \begin{bmatrix} b_1^{(P)} \\ b_2^{(P)} \\ \vdots \\ b_P^{(P)} \end{bmatrix} = - \begin{bmatrix} R_{-P} \\ R_{-P+1} \\ \vdots \\ R_{-1} \end{bmatrix}, \tag{9.26}$$

and is equivalent to the matrix equation

$$\boldsymbol{R}^{(P)}\boldsymbol{b}^{(P)} = -\boldsymbol{J}^{(P)}\boldsymbol{r}_P^*, \quad \boldsymbol{r}_P^* = [R_{-1}, \ldots, R_{-P}]^T. \tag{9.27}$$

Multiplying on the left by $\boldsymbol{J}^{(P)}$ and using the relations $\boldsymbol{J}^2 = \boldsymbol{I}$ and $\boldsymbol{J}\boldsymbol{R} = \boldsymbol{R}^*\boldsymbol{J}$ we find

$$\boldsymbol{R}^{(P)*}\boldsymbol{J}^{(P)}\boldsymbol{b}^{(P)} = -\boldsymbol{r}_P^*, \tag{9.28}$$

which after complex conjugation and comparison with the forward one-step predictor equation (8.104) leads to

$$\boldsymbol{J}^{(P)}\boldsymbol{b}^{(P)*} = \boldsymbol{a}^{(P)} \quad \Rightarrow \quad \boldsymbol{b}^{(P)} = \boldsymbol{J}^{(P)}\boldsymbol{a}^{(P)*}. \tag{9.29}$$

Thus, the optimal one-step backward linear predictor of order P is equal to the complex conjugate of the optimal one-step forward linear predictor of order P in reverse:

$$b_k^{(P)} = a_{P-k+1}^{(P)*}, \quad 1 \le k \le P. \tag{9.30}$$

The backward prediction equation written in terms of the forward predictor coefficient vector is

$$\hat{y}_{n-P}^{(b)} = -\sum_{k=1}^{P} a_k^{(P)*} y_{n-P+k}, \quad \text{or} \quad \hat{y}_n^{(b)} = -\sum_{k=1}^{P} a_k^{(P)*} y_{n+k}, \tag{9.31}$$

which when substituted in (9.22) gives

$$e_P^{(b)}[n] = y_{n-P} + \sum_{k=1}^{P} a_k^{(P)*} y_{n-P+k} \quad \Leftrightarrow \quad B_P(z) = z^{-P} A_P^*(1/z^*). \tag{9.32}$$

The mean square forward and backward prediction errors of the same order P are equal; this and expression (8.129) for the optimal value of the MSE follow from theorem 41.

Theorem 41. *The mean square forward and backward prediction errors $\rho_P^{(f)}$ and $\rho_P^{(b)}$ are equal and given by the following equivalent expressions:*

$$\rho_P^{(f)} = \rho_P^{(b)} = R_0 + \boldsymbol{a}^{(P)+}\boldsymbol{r}_P = R_0 + \boldsymbol{r}_P^+\boldsymbol{a}^{(P)} = R_0 - \boldsymbol{r}_P^+\left[\boldsymbol{R}^{(P)}\right]^{-1}\boldsymbol{r}_P \tag{9.33}$$

where $\boldsymbol{r}_P = [R_1, \ldots, R_P]^T$ and $\boldsymbol{a}^{(P)}$ is the optimal vector satisfying the Yule-Walker equation $\boldsymbol{R}^{(P)}\boldsymbol{a}^{(P)} = -\boldsymbol{r}_P$.

Using $\boldsymbol{b}^{(P)} = \boldsymbol{J}^{(P)}\boldsymbol{a}^{(P)*}$, $\boldsymbol{J}^{(P)2} = \boldsymbol{I}$, and $\boldsymbol{J}^{(P)}\boldsymbol{R}^{(P)}\boldsymbol{J}^{(P)} = \boldsymbol{R}^{(P)*}$, we have

$$\begin{aligned}\rho_P^{(b)} &= R_0 + \boldsymbol{a}^{(P)T}\boldsymbol{r}_P^* + \boldsymbol{r}_P^T\boldsymbol{a}^{(P)*} + \boldsymbol{a}^{(P)T}\boldsymbol{R}^{(P)*}\boldsymbol{a}^{(P)*} \\ &= R_0 + \boldsymbol{r}_P^+\boldsymbol{a}^{(P)} + \boldsymbol{a}^{(P)+}\boldsymbol{r}_P + \boldsymbol{a}^{(P)+}\boldsymbol{R}^{(P)}\boldsymbol{a}^{(P)} = \rho_P^{(f)},\end{aligned} \tag{9.34}$$

where we used the identities

$$\boldsymbol{a}^{(P)T}\boldsymbol{r}_P^* \equiv \boldsymbol{r}_P^+\boldsymbol{a}^{(P)}, \quad \boldsymbol{a}^{(P)T}\boldsymbol{R}^{(P)*}\boldsymbol{a}^{(P)*} \equiv \boldsymbol{a}^{(P)+}\boldsymbol{R}^{(P)}\boldsymbol{a}^{(P)}, \tag{9.35}$$

and the solution to the Yule-Walker equation, namely, $\boldsymbol{a}^{(P)} = -\left[\boldsymbol{R}^{(P)}\right]^{-1}\boldsymbol{r}_P$. The last equality in theorem 41 follows from

$$\boldsymbol{r}_P^+\boldsymbol{a}^{(P)} = \boldsymbol{a}^{(P)+}\boldsymbol{r}_P = -\boldsymbol{r}_P^+\left[\boldsymbol{R}^{(P)}\right]^{-1}\boldsymbol{r}_P. \tag{9.36}$$

9.5 Prediction Error Sequences and Partial Correlations

Consider the forward and backward predicted sequences $\hat{y}_n^{(f)}$ and $\hat{y}_{n-P-1}^{(b)}$

$$\hat{y}_n^{(f)} = -\sum_{k=1}^{P} a_k^{(P)} y_{n-k}, \quad \hat{y}_{n-P-1}^{(b)} = -\sum_{k=1}^{P} a_k^{(P)*} y_{n-P+k-1}. \tag{9.37}$$

Both predictions use the same set of data values $\{y_{n-P}, \ldots, y_{n-1}\}$ to predict the values at time steps n and $n-1$, respectively, as shown in figure 9.5. The (normalized) correlation coefficient between the two values

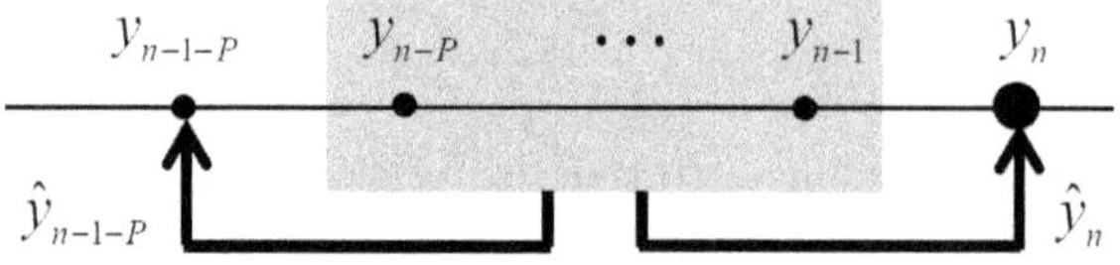

Figure 9.5: Forward prediction of y_n and backward prediction of y_{n-1-P}.

y_n and y_{n-1}, after the removal of each one's correlations with the common data values $\{y_{n-P}, \ldots, y_{n-1}\}$, is known as a *partial correlation* of order P; it is equal to the negative of the reflection coefficient $\gamma_{P+1} \equiv a_{P+1}^{(P+1)}$. Since the removal of the correlations produces the forward and backward prediction error sequences $e_P^{(f)}[n]$ and $e_P^{(b)}[n-1]$, we have the following result.

Theorem 42. *The reflection coefficient* $\gamma_{P+1} \equiv a^{(P+1)}_{P+1}$ *in (8.123) is given by the negative of the partial correlation of order P,*

$$\gamma_{P+1} \equiv a^{(P+1)}_{P+1} = -\frac{\langle e^{(f)}_P[n], e^{(b)}_P[n-1]\rangle}{\|e^{(f)}_P[n]\| \, \|e^{(b)}_P[n-1]\|}, \tag{9.38}$$

where

$$\langle e^{(f)}_P[n], e^{(b)}_P[n-1]\rangle \equiv E\Big[e^{(f)}_P[n] e^{(b)*}_P[n-1]\Big],$$
$$\|e^{(f)}_P[n]\|^2 \equiv E\Big[|e^{(f)}_P[n]|^2\Big], \quad \|e^{(b)}_P[n-1]\|^2 \equiv E\Big[|e^{(b)}_P[n-1]|^2\Big]. \tag{9.39}$$

Equation (9.38) can be proven by writing

$$e^{(b)}_P[n-1] = y_{n-P-1} + \boldsymbol{a}^{(P)+} \boldsymbol{J}^{(P)} \boldsymbol{y}_{n-1}, \quad e^{(f)}_P[n] = y_n + \boldsymbol{a}^{(P)T} \boldsymbol{y}_{n-1}, \tag{9.40}$$

where $\boldsymbol{y}_{n-1} \equiv [y_{n-1}, \ldots, y_{n-P}]^T$ is the time reversed data vector and $\boldsymbol{J}^{(P)}$ is the $P \times P$ counter identity matrix. Then

$$\begin{aligned} \mathrm{E}\Big[e^{(f)}_P[n] e^{(b)*}_P[n-1]\Big] &= R_{P+1} + 2\boldsymbol{a}^{(P)T} \boldsymbol{J}^{(P)} \boldsymbol{r}_P + \boldsymbol{a}^{(P)T} \mathrm{E}\big[\boldsymbol{y}_{n-1} \boldsymbol{y}^+_{n-1}\big] \boldsymbol{J}^{(P)} \boldsymbol{a}^{(P)} \\ &= R_{P+1} + 2\boldsymbol{a}^{(P)T} \boldsymbol{J}^{(P)} \boldsymbol{r}_P + \boldsymbol{a}^{(P)T} \boldsymbol{J}^{(P)} \boldsymbol{R}^{(P)} \boldsymbol{a}^{(P)} \\ &= R_{P+1} + \boldsymbol{a}^{(P)T} \boldsymbol{J}^{(P)} \boldsymbol{r}_P, \end{aligned} \tag{9.41}$$

where $\boldsymbol{r}_P = [R_1, \ldots, R_P]^T$, and we used the relation $\boldsymbol{a}^{(P)T} \boldsymbol{J}^{(P)} \boldsymbol{y}^*_{n-1} \equiv \boldsymbol{y}^+_{n-1} \boldsymbol{J}^{(P)} \boldsymbol{a}^{(P)}$, equations in section 6.6, and the Yule-Walker equation (8.104) and its solution $\boldsymbol{R}^{(P)} \boldsymbol{a}^{(P)} = -\boldsymbol{r}_P$. Now we use the definitions

$$\rho^{(f)}_P = \|e^{(f)}_P[n]\|^2, \quad \rho^{(b)}_P = \|e^{(b)}_P[n-1]\|^2, \tag{9.42}$$

and expression (8.127) for γ_{P+1} to arrive at (9.38). In addition, the Cauchy-Schwartz inequality applied to (9.38) gives (after changing $P+1$ to P)

$$|\gamma_P| \le 1, \quad P \ge 1. \tag{9.43}$$

Theorem 42, together with the recursive prediction error sequence relations to be derived next (section 9.6), forms the basis of the geometric mean algorithm for estimating AR(P) parameters to construct a high resolution AR(P) spectral estimate. The geometric mean and the Burg (or the harmonic mean) algorithms will be discussed in section 9.8.

9.6 Lattice Filters

The forward and backward predictors of order P for zero-mean WSS random signal y_n,

$$\hat{y}^{(f)}_n = -\sum_{k=1}^{P} a^{(P)}_k y_{n-k}, \quad \hat{y}^{(b)}_n = -\sum_{k=1}^{P} a^{(P)*}_k y_{n+k}, \tag{9.44}$$

have associated prediction error sequences:

$$e^{(f)}_P[n] \equiv y_n - \hat{y}^{(f)}_n, \quad e^{(b)}_P[n] \equiv y_{n-P} - \hat{y}^{(b)}_{n-P}. \tag{9.45}$$

Applying the Levinson-Durbin algorithm to the forward predictor of order P, we compute all prediction filter vectors $\boldsymbol{a}^{(k)}$ for all orders $1 \leq k \leq P$. The direct form filter realization (figure 3.5) for the prediction equations and the associated one-step forward and backward prediction error sequences are shown in figure 9.6, where $b_k^{(P)} = a_{P-k+1}^{(P)*}$ [equation (9.30)]. The main defect of the direct form filter realization is that it

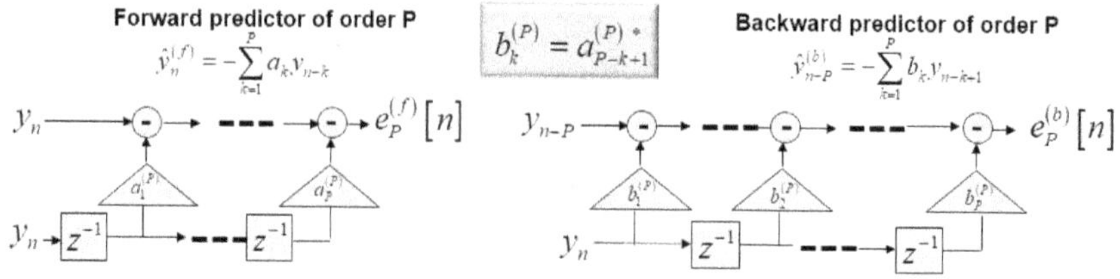

Figure 9.6: Direct form realization of prediction filters.

is not possible to increase the filter order P by simply appending new sub-structures to the existing filter coefficients: all filter coefficients have to be recomputed for the new order, i.e., to go from order P to order $P+1$, all filter coefficients in the triangles of figure 9.6 will have to be computed, since the new filter vector $\boldsymbol{a}^{(P+1)}$ has, in general, no common elements with the filter vector $\boldsymbol{a}^{(P)}$. The same, of course, holds for all the backward prediction filter vectors.

A realization of the prediction filters that can be extended by appending sub-structures without recalculation of all previous structures' coefficients is known as a *lattice filter*. To illustrate, we will describe the example of the passage from $P = 1$ to $P = 2$. The first order prediction filter $P = 1$ has one component $a_1^{(1)}$, while the second order $P = 2$ filter has two components $[a_1^{(2)}, a_2^{(2)}]^T$. The prediction errors are

$$\begin{aligned} e_1^{(f)}[n] &\equiv y_n + a_1^{(1)} y_{n-1}, \quad & e_1^{(b)}[n] &\equiv y_{n-1} + a_1^{(1)*} y_n, \\ e_2^{(f)}[n] &\equiv y_n + \sum_{k=1}^{2} a_k^{(2)} y_{n-k}, \quad & e_2^{(b)}[n] &\equiv y_{n-2} + \sum_{k=1}^{2} a_k^{(2)*} y_{n-2+k}. \end{aligned} \tag{9.46}$$

Using the reflection coefficient of order 1, namely, $\gamma_1 \equiv a_1^{(1)}$, the first set of prediction equations can be represented as a lattice filter, as shown in figure 9.7. Using the Levinson-Durbin recursion, we have

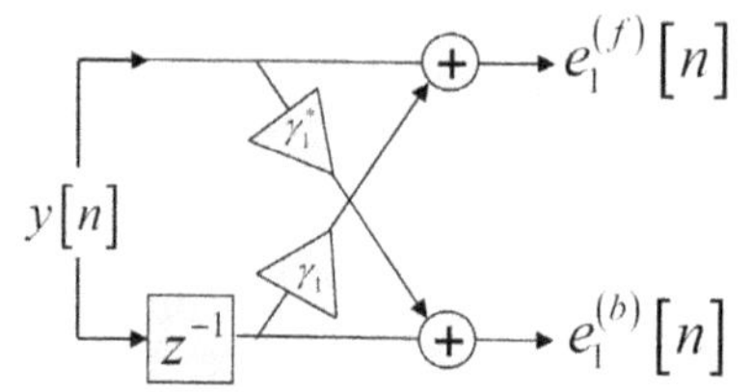

Figure 9.7: Lattice form realization of the order 1 forward and backward prediction filters.

$$a_1^{(2)} = \gamma_2 \gamma_1^* + \gamma_1, \quad a_2^{(2)} = \gamma_2, \tag{9.47}$$

which, when substituted into the equations for prediction errors of order 2, and using the equations for prediction errors of order 1, gives

$$\begin{aligned} e_2^{(f)}[n] &= y_n + (\gamma_2 \gamma_1^* + \gamma_1) y_{n-1} + \gamma_2 y_{n-2}, \\ e_2^{(b)}[n] &= y_{n-2} + (\gamma_2^* \gamma_1^* + \gamma_1) y_{n-1} + \gamma_2^* y_n, \end{aligned} \tag{9.48}$$

where the reflection coefficient of order 2 is $\gamma_2^* = a_2^{(2)}$. These two equations have the lattice filter realization shown in figure 9.8. Thus, the order 2 lattice form of the prediction filter is found by simply appending the same structure as that shown in figure 9.7 to the right hand side of it, with the exception that the reflection coefficient of order 2, namely, $\gamma_2 \equiv a_2^{(2)}$, must be used in the appended structure without having to calculate any new quantities. We now generalize the result to the passage from order $P-1$ to P. Using

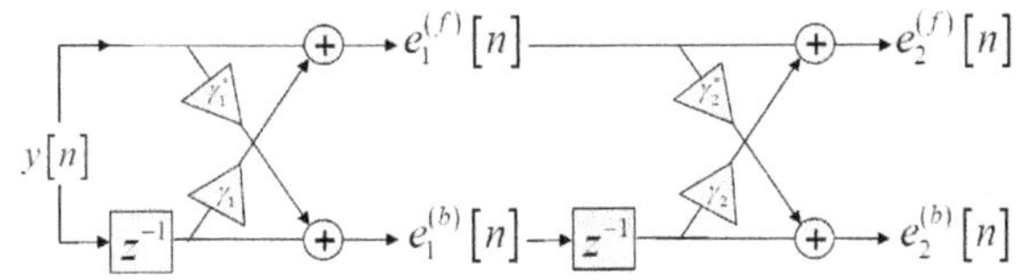

Figure 9.8: Lattice form realization of the orders 1, 2 forward and backward prediction filters.

the Levinson-Durbin recursion,

$$\gamma_P = a_P^{(P)} \text{ and } a_k^{(P)} = a_k^{(P-1)} + \gamma_P a_{P-k+1}^{(P-1)*}, \quad 1 \le k \le P-1, \tag{9.49}$$

the forward prediction error sequence of order P becomes

$$\begin{aligned} e_P^{(f)}[n] &= y_n + \sum_{k=1}^{P-1} a_k^{(P)} y_{n-k} + \gamma_P \, y_{n-P} \\ &= y_n + \sum_{k=1}^{P-1} a_k^{(P-1)} y_{n-k} + \gamma_P \sum_{k=1}^{P-1} a_{P-k+1}^{(P-1)*} y_{n-k} + \gamma_P y_{n-P} \\ &= e_{P-1}^{(f)}[n] + \gamma_P \left(y_{n-1-(P-1)} + \sum_{k=1}^{P-1} b_k^{(P-1)} y_{n-1-(k-1)} \right) \\ &= e_{P-1}^{(f)}[n] + \gamma_P \, e_{P-1}^{(b)}[n-1]. \end{aligned} \tag{9.50}$$

Taking **Z** transform on both sides, we have

$$E_P^{(f)}(z) = E_{P-1}^{(f)}(z) + \gamma_P z^{-1} E_{P-1}^{(b)}(z). \tag{9.51}$$

Similarly, for the backward prediction error sequence we find

$$e_P^{(b)}[n] = e_{P-1}^{(b)}[n-1] + \gamma_P^* e_{P-1}^{(f)}[n], \;\; E_P^{(b)}(z) = z^{-1} E_{P-1}^{(b)}(z) + \gamma_P^* E_{P-1}^{(f)}(z). \tag{9.52}$$

The lattice filter shown in figure 9.9 is a representation of the two **Z** transform relations. Note that the initial conditions when $P = 1$ are $e_0^{(f)}[n] = e_0^{(b)}[n] = y_n$.

Equations (9.51) and (9.52) can be cast in terms of the forward PEF $A_P(z)$ and the backward PEF $B_P(z)$ [equations (9.44) and (9.46)] using

$$E_P^{(f)}(z) = Y(z) A_P(z), \;\; E_P^{(b)}(z) = Y(z) B_P(z). \tag{9.53}$$

Thus,

$$A_P(z) = A_{P-1}(z) + \gamma_P z^{-1} B_{P-1}(z), \tag{9.54a}$$

$$B_P(z) = z^{-1} B_{P-1}(z) + \gamma_P^* A_{P-1}(z). \tag{9.54b}$$

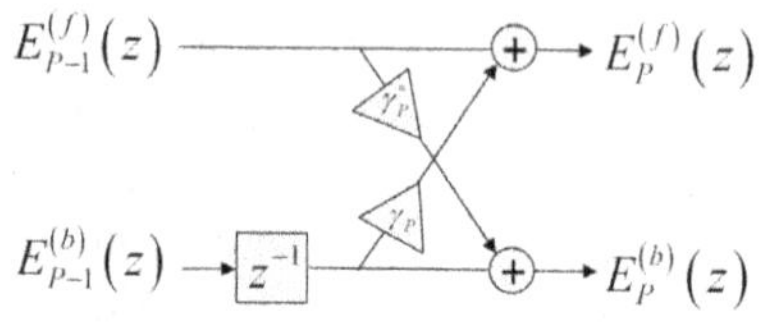

Figure 9.9: Lattice form realization of the one-step forward and backward consecutive order $P-1$ and P prediction filters.

9.7 The Minimum Phase Property of the Forward PEF

Theorem 36 showed that the maximum entropy rate requirement on a zero-mean WSS random process, with known values for its first $P+1$ auto-correlation sequence, is the AR(P) process whose $P \times 1$ parameter vector $\boldsymbol{a}$ satisfies the Yule-Walker equation (8.104), and that the AR(P) process is the output of a minimum phase all-pole filter whose input is a zero-mean Gaussian white noise process with variance σ_ν^2. We will now prove the minimum phase property of the PEF (all-pole filter) in the context of linear prediction using figure 9.10. Suppose that y_n is an AR(P) process defined by

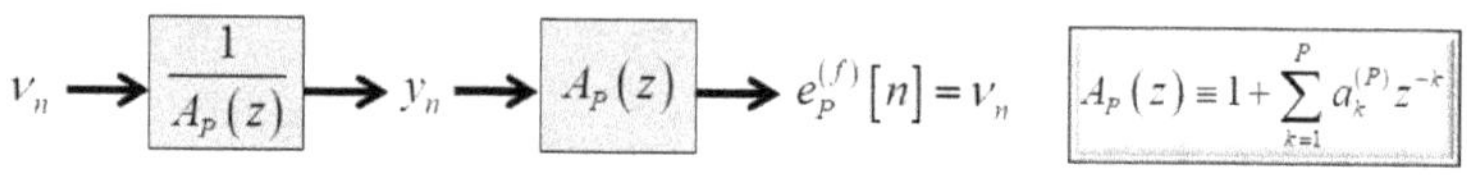

Figure 9.10: AR(P) process and its one-step forward order P predictive model.

$$y_n = -\sum_{k=1}^{P} a_k^{(P)} y_{n-k} + \nu_n, \quad \nu_n \sim \mathcal{N}(0, \sigma_\nu^2). \tag{9.55}$$

Then its parameter vector $\boldsymbol{a}^{(P)}$ satisfies the Yule-Walker equation $\boldsymbol{R}^{(P)}\boldsymbol{a}^{(P)} = -\boldsymbol{r}_P$ [see (8.104)]. Thus, y_n is the output of the all-pole filter $1/A_P(z)$, where $A_P(z) = 1 + \sum_{k=1}^{P} a_k^{(P)} z^{-k}$, driven by ν_n. Then the optimal order P one-step forward predictor of y_n is characterized by the PEF $A_P(z)$ and a prediction error sequence that is identical to ν_n. The proof of this result is obvious if we note that the parameter vectors in each case satisfy the same Yule-Walker equation and the prediction error sequence has the same variance as ν_n.

In order to show that $1/A_P(z)$ is minimum phase, we will show that its inverse, the all-zero filter $A_P(z)$, is minimum phase. Let us write $A_P(z)$ in its factored form in terms of its zeros z_k:

$$A_P(z) = 1 + \sum_{k=1}^{P} a_k^{(P)} z^{-k} = \prod_{k=1}^{P} \left(1 - z_k z^{-1}\right). \tag{9.56}$$

The optimal PEF coefficients satisfy the Yule-Walker equation (8.104), which is equivalent to the orthogonality relations

$$\mathrm{E}\left[e_P^{(f)}[n] y_{n-k}^*\right] = 0, \quad 1 \le k \le P. \tag{9.57}$$

Consider an arbitrary zero z_k and write

$$A_P(z) = (1 - z_k z^{-1})\, G_{P-1}(1/z), \tag{9.58}$$

where $G_{P-1}(1) = 1$. Now let $G_{P-1}(1/z)Y(z) = W(z)$, i.e.,

$$\begin{aligned} w_n &= y_n + \text{linear combination}\left\{y_{n-1}, \ldots, y_{n-(P-1)}\right\} \quad \Rightarrow \\ w_{n-1} &= y_{n-1} + \text{linear combination}\left\{y_{n-2}, \ldots, y_{n-P}\right\}. \end{aligned} \tag{9.59}$$

Therefore, $\mathrm{E}\left[e_P^{(f)}[n]w_{n-1}^*\right] = 0$, and since $e_P^{(f)}[n] = w_n - z_k w_{n-1}$, we have [121]

$$\mathrm{E}\left[(w_n - z_k w_{n-1})w_{n-1}^*\right] = 0 \;\Rightarrow\; R_{ww}[1] - R_{ww}[0]\, z_k = 0. \tag{9.60}$$

Thus,

$$\begin{aligned} \mathrm{E}\left[\left|e_P^{(f)}[n]\right|^2\right] &= \mathrm{E}\left[e_P^{(f)}[n]w_{n-1}^*\right] - z_k \mathrm{E}\left[e_P^{(f)}[n]w_{n-1}^*\right] \\ &= \mathrm{E}\left[e_P^{(f)}[n]w_{n-1}^*\right] = R_{ww}[0] - R_{ww}[-1] = \left(1 - \left|z_k\right|^2\right) R_{ww}[0]. \end{aligned} \tag{9.61}$$

If the process y_n is not fully predictable, then $\mathrm{E}\left[\left|e_P^{(f)}[n]\right|^2\right] > 0$ and so $|z_k| < 1$ and the chosen zero z_k is inside the unit circle. Therefore, all the zeros of the PEF are inside the unit circle, and the PEF is minimum phase. Figure 9.10 is, therefore, an expression of the following theorem.

Theorem 43. *A zero-mean WSS random AR(P) process y_n whose parameters $\left\{\boldsymbol{a}^{(P)}, \sigma_\nu^2\right\}$ satisfy the Yule-Walker equation* (8.104) *is the output of a minimum phase all-pole filter*

$$\frac{1}{A_P(z)} = \frac{1}{1 + \sum_{k=1}^{P} a_k^{(P)} z^{-k}}, \tag{9.62}$$

driven by a white noise sequence $\nu_n \sim \mathcal{N}(0, \sigma_\nu^2)$ with

$$\sigma_\nu^2 = R_0 + \sum_{k=1}^{P} a_k^{(P)} R_{-k}, \quad R_{-k} = R_k^*. \tag{9.63}$$

Furthermore, the order P one-step forward predictor of y_n has a prediction error sequence $e_P^{(f)}[n]$ which is the output of the minimum phase all-zero filter

$$A_P(z) = 1 + \sum_{k=1}^{P} a_k^{(P)} z^{-k}, \tag{9.64}$$

with input y_n, and $e_P^{(f)}[n] = \nu_n$.

Clearly the minimum phase property of the PEF holds for a one-step ahead predictor $\hat{y}_n$ of any order P for any zero-mean WSS random process y_n (whether AR or not) since the optimal predictor parameter vector $\boldsymbol{a}^{(P)}$ will satisfy the Yule-Walker equation (8.104) with the auto-correlation sequence of the process y_n. Thus, theorem 43 can be stated as follows.

Theorem 44. *The PEF*

$$A_P(z) = 1 + \sum_{k=1}^{P} a_k^{(P)} z^{-k}, \tag{9.65}$$

of an order P linear predictor of a zero-mean WSS random process whose coefficient vector $\boldsymbol{a}^{(P)}$ satisfies the Yule-Walker equation (8.104) *is minimum phase.*

Theorem 45. *The backward PEF of an order P linear predictor of a zero-mean WSS random process is maximum phase.*

Equation (9.32) shows that the zeros of the forward and the backward PEFs are inverse points with respect to the unit circle in the complex z-plane. The minimum phase property of the forward PEF $A_P(z)$ then implies that the backward PEF $B_P(z)$ is maximum phase, i.e., all its zeros are outside the unit circle.

A test for the minimum phase property of the PEF is based on a theorem [1] that describes the conditions under which a complex polynomial of degree P has zeros inside the unit circle. Consider the polynomial $g(z) = 1 + c_1^* z + \ldots + c_P^* z^P$ and an associated polynomial $\bar{g}(z) = z^P g^*(1/z*) = c_P + c_{P-1} z + \ldots + z^P$. Let us define the operator $\mathcal{T}$,

$$\mathcal{T}[g](z) \equiv g(z) - c_P^* \, \bar{g}(z), \tag{9.66}$$

so that the right hand side is a polynomial of degree at most $P-1$ in z. Clearly $\mathcal{T}[g](0) = 1 - |c_P|^2$ is real and if it is non-zero then we can continue to apply the operator $\mathcal{T}$ to $g(z)$ until we obtain 0 for $\mathcal{T}^K[g](0)$. Now we have the following theorem [122].

Theorem 46. *If, for some $l > 0$, $\mathcal{T}^l[g](0) < 0$, then $g(z)$ has at least one root inside the unit circle. If, instead, $\mathcal{T}^m[g](0) > 0$ for $1 \leq m < K$ and $\mathcal{T}^{K-1}[g]$ is a constant, then $g(z)$ has no root inside the unit circle.*

The polynomials $g(z)$ and $\bar{g}(z)$ defined above by the coefficients c_k are general polynomials whose coefficient vector $\boldsymbol{c}$ can be arbitrary and have no particular properties. To relate theorem 46 to the forward PEF of an order P linear predictor, we set $g(z) = z^P B_P(z) = A_P^*(1/z^*)$. Then $\bar{g}(z) = z^P A_P(z)$ and $\mathcal{T}[g](z) = g(z) - \gamma_P^* z^P g^*(1/z^*)$ where $\gamma_P \equiv a_P^{(P)}$. The Schur-Lehmer test for the minimum phase property of a given PEF is to first apply theorem 46 to $g(z)$ and then:

- if $g(z)$ has at least one zero inside the unit circle, then $B_P(z)$ is not maximum phase and, therefore, $A_P(z)$ is not minimum phase.

- if $g(z)$ is found not to have any zeros inside the unit circle, then $B_P(z)$ is maximum phase and $A_P(z)$ is minimum phase.

An equivalent form of theorem 46 for a given polynomial $A_P(z) = 1 + a_1^{(P)} z^{-1} + \ldots + a_P^{(P)} z^{-P}$ (where the coefficient vector $\boldsymbol{a}^{(P)}$ is arbitrary) is known as the Schur-Cohn test.

Theorem 47. *Let $A_P(z) = 1 + a_1^{(P)} z^{-1} + \ldots + a_P^{(P)} z^{-P}$, $A_0 = 1$, and define the recursion*

$$A_{k-1}(z) = \frac{A_k(z) - \gamma_k B_k(z)}{1 - |\gamma_k|^2}, \quad \gamma_k \equiv a_k^{(k)}, \quad P \leq k \leq 1. \tag{9.67}$$

Then all zeros of $A_P(z)$ are inside the unit circle if, and only if, $|\gamma_k| < 1$ for $k = P, P-1, \ldots, 1$.

The recursion (9.67) is motivated by the forward and backward PEF relations (9.54a) and (9.54b); note that no Yule-Walker equation is involved in the calculation of any of the functions $A_{k-1}(z)$ and the original coefficient vectors $\boldsymbol{a}^{(P)}$ are arbitrary. As an example, consider the polynomial $A_2(z) = 1 - 5/6\, z^{-1} + 1/6\, z^{-2}$ (whose zeros $1/2$ and $1/3$ are inside the unit circle). Proceeding with the Schur-Cohn test, $\gamma_2 = 1/6 < 1$, $B_2(z) = 1/6 - 5/6\, z^{-1} + z^{-2}$, $A_1(z) = 1 - 175/216\, z^{-1}$, $\gamma_1 = 175/216 < 1$, and so both zeros are inside the unit circle.

[1] The theorem was first proven by Schur in 1918 and modified by Lehmer [122].

Theorem 48. *The recursions* (9.54a) *and* (9.54b) *together with* $|\gamma_k| < 1$ *for* $k = P, P-1, \ldots, 1$ *imply that the PEF* $A_P(z)$ *is minimum phase, i.e., all its zeros are inside the unit circle.*

Using (9.32) we have $B_k(z)B_k^*(1/z^*) = A_k^*(z)A_k(1/z^*)$, which when evaluated on the unit circle $|z| = 1$ gives $|B_k(e^{i\omega})|^2 = |A_k(e^{i\omega})|^2$. In addition, when evaluated on the unit circle and using $|\gamma_k| < 1$ we have

$$|\gamma_k z^{-1} B_{k-1}(z)| < |B_{k-1}(z)| = |A_{k-1}(z)|, \quad |z| = 1.$$

According to Rouché's theorem [30], *two holomorphic (analytic) functions* $f(z)$ *and* $g(z)$ *have the same number of zeros in a closed region* Ω *of the complex plane bounded by a closed curve* C *if* $|f(z) - g(z)| < |f(z)|$ *for all* $z \in C$. A trivially modified form of this theorem follows by the replacements $f - g \to F_1$ and $g \to F_2$: if $F_1(z)$ and $F_2(z)$ are analytic in Ω and if $|F_1| < |F_2(z)|$ on the boundary C then $F_1(z) + F_2(z)$ has the same number of zeros within Ω as $F_2(z)$. Now consider the region $\Omega = \{z : |z| \geq 1\}$ (outside and on the unit circle) and the functions $F_1(z) \equiv A_{k-1}(z)$ and $F_2(z) \equiv \gamma_k z^{-1} B_{k-1}(z)$; both functions satisfy the conditions of Rouché's theorem and accordingly their sum $F_1(z) + F_2(z)$ has the same number of zeros outside the unit circle as $F_1(z)$. Equation (9.54a) shows that their sum is $A_k(z)$ and so $A_k(z)$ has the same number of zeros outside the unit circle as $A_{k-1}(z)$. Starting with $k = 1$ we have $A_0(z) = 1$ which has no zeros (anywhere) and so $A_1(z)$ will have no zeros in Ω so long as $\gamma_1 < 1$. Next, since $\gamma_2 < 1$ it follows that $A_2(z)$ will have no zeros in Ω, and so on. Thus, all zeros of $A_k(z)$, $1 \leq k \leq P$, are inside the unit circle since $\gamma_k < 1$.

In light of the results of this section it can be argued that acceptable AR (or linear prediction) parameter estimation methods are those which produce minimum phase all-pole filters. For instance, the biased auto-correlation estimate described in theorem 30 certainly does so; this is the auto-correlation method of parameter vector estimation discussed in section 8.22. Another important method of AR or LP parameter estimation proceeds via the reflection coefficients using the Levinson-Durbin recursions and lattice filter relations (9.54a) and (9.54b).

9.8 AR Parameter Estimation: the Burg Method

The modified covariance method described by (8.173) minimizes the sum of the forward and backward prediction errors with respect to the AR parameter vector. The Burg algorithm solves a constrained minimization problem based on the Levinson-Durbin recursion for the AR parameter vector so that the minimization is performed with respect to the reflection coefficient $\gamma_P = a_P^{(P)}$ instead of the entire parameter vector.

The simplest method to estimate the reflection coefficient γ_P is to estimate the numerator and the denominator of the expression for partial correlation (9.38). For observed data with N elements y_n, $0 \leq n \leq N-1$, the range of n on the right hand side of equations (9.15) and (9.32) for $e_P^{(f)}[n]$ and $e_P^{(b)}[n]$, is $[P, N-1]$, and we may use the estimates

$$\alpha \left\{ \sum_{n=P}^{N-1} e_{P-1}^{(f)}[n] e_{P-1}^{(b)*}[n-1], \; \sum_{n=P}^{N-1} \left| e_{P-1}^{(f)}[n] \right|^2, \; \sum_{n=P}^{N-1} \left| e_{P-1}^{(b)}[n-1] \right|^2 \right\}, \tag{9.68}$$

for any normalization α, since equation (9.38) for the reflection coefficient is independent of normalization. The magnitude of the estimated reflection coefficient is less than 1, by the Cauchy-Schwartz inequality, for all orders, which implies that the corresponding PEF is minimum phase. With the estimated reflection coefficient, we use the Levinson-Durbin algorithm to compute the AR model parameter vector and the associated PEF. The reflection coefficients method of AR parameter estimation begins by first calculating the estimates:

$$\hat{\rho}_0 = \frac{1}{N}\sum_{n=0}^{N-1} |y_n|^2, \quad \hat{e}^{(f)}_0[n] = \hat{e}^{(b)}_0[n] = y_n,$$

and then for $k = 1, \ldots, P$,

- $\hat{\gamma}_k = -\dfrac{\sum_{n=k}^{N-1} \hat{e}^{(f)}_{k-1}[n]\, \hat{e}^{(b)*}_{k-1}[n-1]}{\sqrt{\sum_{n=k}^{N-1} \left|\hat{e}^{(f)}_{k-1}[n]\right|^2}\sqrt{\sum_{n=k}^{N-1} \left|\hat{e}^{(b)}_{k-1}[n-1]\right|^2}}$
- $\hat{\rho}_k = \hat{\rho}_{k-1}\left(1 - |\hat{\gamma}_k|^2\right).$
- $\hat{a}^{(k)}_k = \hat{\gamma}_k, \; \hat{a}^{(k)}_l = \hat{a}^{(k-1)}_l + \hat{\gamma}_k\, \hat{a}^{(k-1)*}_{k-l}, \quad 1 \le l \le k-1$
- $\hat{e}^{(f)}_k[n] = \hat{e}^{(f)}_{k-1}[n] + \hat{\gamma}_k\, \hat{e}^{(b)}_{k-1}[n-1], \quad k \le n \le N-1$
- $\hat{e}^{(b)}_k[n] = \hat{e}^{(b)}_{k-1}[n-1] + \hat{\gamma}^*_1\, \hat{e}^{(f)}_{k-1}[n], \quad k \le n \le N-1$
- $k+1 \leftarrow k$

The reflection coefficient estimate $\hat{\gamma}_k$ is the geometric mean of the estimated forward and backward prediction errors of order $k-1$. The Burg algorithm [101, 96] uses an alternative estimate for the reflection coefficient derived from a least squares minimization of the average of the forward and backward prediction error powers of order k,

$$\arg\min_{\gamma_k} \left[\sum_{n=k+1}^{N-1} \left|e^{(f)}_k[n]\right|^2 + \left|e^{(b)}_k[n]\right|^2\right], \tag{9.69}$$

whose solution is an estimated reflection coefficient which is the harmonic mean of the estimated forward and backward prediction errors of order $k-1$,

$$\hat{\gamma}_k = -\frac{2\sum_{n=k}^{N-1} e^{(f)}_{k-1}[n] e^{(b)*}_{k-1}[n-1]}{\sum_{n=k}^{N-1} \left|e^{(f)}_{k-1}[n]\right|^2 + \sum_{n=k}^{N-1} \left|e^{(b)}_{k-1}[n-1]\right|^2}. \tag{9.70}$$

The Burg method has been known to show bias in the presence of pure sinusoids. The bias can be reduced by using a non-uniform weighting function when calculating the reflection coefficient harmonic mean estimate. A common practice [96] is to use a window derived from the data

$$w_{k-1}[n] = \sum_{l=n-k+1}^{n-1} |y_l|^2, \quad k \ge 2,$$

so that

- $\hat{\gamma}_k = -\dfrac{2\sum_{n=k}^{N-1} w_{k-1}[n] e^{(f)}_{k-1}[n] e^{(b)*}_{k-1}[n-1]}{\sum_{n=k}^{N-1} w_{k-1}[n]\left|e^{(f)}_{k-1}[n]\right|^2 + \sum_{n=k}^{N-1} w_{k-1}[n]\left|e^{(b)}_{k-1}[n-1]\right|^2}$

9.9 Linear Prediction and Speech Recognition

As mentioned in section 3.12, cepstral coefficients (and the mel cepstrum) have more or less replaced the direct use of linear prediction parameters in the processing of speech. Nevertheless, a basic understanding of *linear predictive coding* of speech is essential in advanced signal processing. We will briefly describe the source-filter model of voiced and unvoiced speech and LPC-based speech processing, including the derivation of the cepstral coefficients using LPC parameters.

The human speech system consists of the following parts and functions:

- the vocal cords: they act as an oscillator.
- the mouth, tongue and throat: they act as filters and also shape the sound.

Speech processing may be divided into three categories: speech synthesis, which converts text into natural sounding speech; speech recognition, through which a machine can understand speech; and speech coding and compression, whose goal is to represent speech in the smallest number of bits in order to allow for storage and transmission of vast amounts of speech data.

Linear prediction (LP or LPC) is based on a source-filter model [123] to approximate voiced and unvoiced sounds generated by humans, as shown in figure 9.11. Voiced sounds are produced by forcing periodic pulses of air through the vocal cords while unvoiced sounds are the result of forcing air through a constricted vocal tract producing turbulence. Thus, unvoiced speech is modeled as the output of an all-pole filter with a zero-mean white noise input, while voiced speech is modeled as the output of an all-pole filter with a pulse train input. In addition to the all-pole filter coefficients, LP or LPC parameters include the

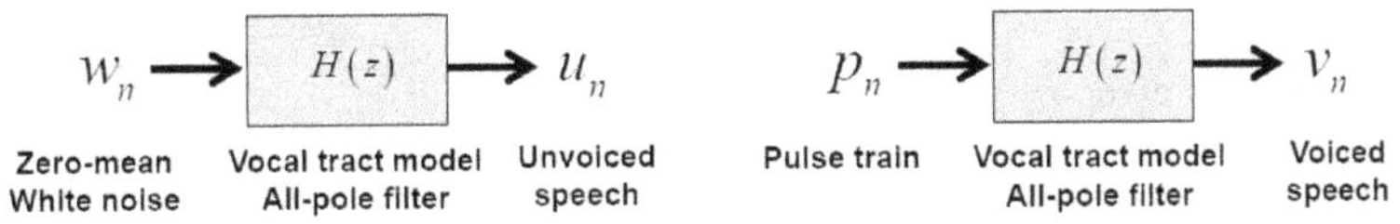

Figure 9.11: Linear prediction model for voiced and unvoiced speech.

variance of the zero-mean white noise for unvoiced speech and the pulse period, or its inverse, the pitch frequency, for voiced speech (see figure 3.12 for an example of pitch frequency determination using cepstral coefficients of a speech frame). The average pitch frequency for male speakers is about 125 Hz, while the average female pitch frequency is 250 Hz. Linear prediction is also related to an acoustic model of the vocal tract based on a number of connected tubes with different acoustic impedances. The physical length of the vocal tract V_L and the order of the predictor P are related by $P = 2V_L f_s/c$, where f_s is the sampling frequency, and c is the speed of sound in air (for an air temperature of $35°C$, $c = 350$ m/s). The mean lengths of vocal tracts for females and males are approximately 0.15 and 0.17 meters, respectively. For instance, at 8 kHz, the predictor order is $P = 10$.

Figure 9.12 shows a block diagram of a speech recognition system [123, 124] based on LPC parameter estimation, and cepstral coefficients c_k as the output. Cepstral coefficients (3.76) have been shown to provide robust performance in speech recognition applications. Other methods (generally with worse performance) use distribution of powers within words, or analysis of formants instead of cepstral coefficients (in formant analysis the first three positive frequencies for an AR model fit to each data frame are used as distinguishing features).

Speech signal s_n is initially filtered with a first order FIR filter, e.g., $1 - az^{-1}$, to flatten the signal spectrum and remove the DC component; the constant a is usually taken to be $15/16$. Next, different words

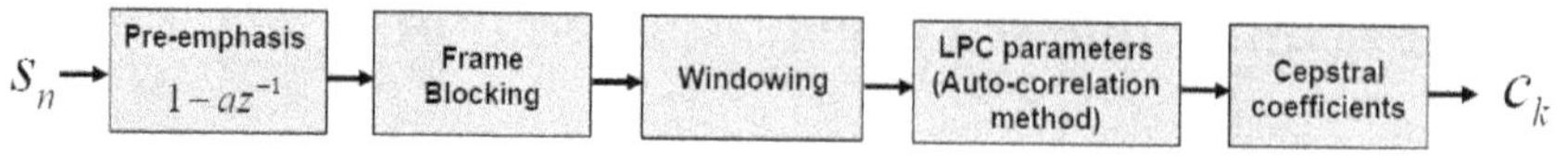

Figure 9.12: Speech recognition system based on LP cepstral coefficients.

in speech are temporally aligned using spectrograms, or dynamic time warping. Speech is then divided up into overlapping frames of some number of samples and each frame is windowed (e.g., Hamming window) to ensure signal continuity at the edges. For speech sampled at 8 kHz, each section can be 30 ms long with an overlap of 10 ms. Frame LPC parameters (AR model coefficients of order $P = 10$, say) are then computed using the auto-correlation method and are used to derive a number of cepstral coefficients (for example, 15) for that frame. The cepstral coefficients are directly calculated from the LP parameter vector $\boldsymbol{a}$, as shown in theorem 49.

Theorem 49. *Let $H(z)$ represent a minimum phase AR(P) model*

$$H(z) = \frac{\rho_0}{1 + \sum_{k=1}^{P} a_k z^{-k}}, \tag{9.71}$$

with associated (complex) cepstrum $C(z)$ which, according to equation (7.27), *corresponds to a causal set of coefficients*

$$C(z) = \ln H(z) \equiv \sum_{k=0}^{\infty} c_k z^{-k} = \ln \rho_0 - \ln\left(1 + \sum_{k=1}^{P} a_k z^{-k}\right). \tag{9.72}$$

The cepstral coefficients are given by the recursion

$$c_k = \begin{cases} -a_k - \sum_{l=1}^{k-1} \left(1 - \frac{l}{k}\right) c_{k-l}\, a_l, & 1 \le k \le P, \\ -\sum_{l=1}^{k-1} \left(1 - \frac{l}{k}\right) c_{k-l}\, a_l, & k > P\,. \end{cases} \tag{9.73}$$

Differentiating (9.71) with respect to z and multiplying by z, we have $zH'(z) = zC'(z)H(z)$ whose inverse **Z** transform is given by a causal sequence [since $H(z)$ is minimum phase, its inverse **Z** transform must be causal]:

$$h_n = \sum_{k=0}^{\infty} \left(1 - \frac{k}{n}\right) h_k c_{n-k} = h_0 c_n + \sum_{k=1}^{n-1} \left(1 - \frac{k}{n}\right) h_k c_{n-k}, \quad n \ge 1, \tag{9.74}$$

which gives the following recursion for the complex cepstral coefficients:

$$c_n = \frac{h_n}{h_0} - \sum_{k=1}^{n-1} \left(1 - \frac{k}{n}\right) \frac{h_k}{h_0}\, c_{n-k}, \quad n \ge 1. \tag{9.75}$$

If in equation (9.71) we had used $1/H(z)$ instead of $H(z)$, we would have obtained the same equation for the cepstral coefficients except that the filter coefficients h_n would then be given by

$$h_0 = 1, \quad h_k = a_k, \; 1 \le k \le P, \quad h_k = 0, \quad k > P, \tag{9.76}$$

and the cepstral coefficients would be the negative of those for $H(z)$, proving the result (9.73).

A cepstral distance between two spectra S_{xx} and S_{yy} in the frequency domain is defined by

$$\mathrm{D}^{c}_{xy} \equiv \frac{1}{2\pi}\int_{-\pi}^{+\pi}\left|\ln\left(S_{xx}(e^{i\omega})/S_{yy}(e^{i\omega})\right)\right|^2 d\omega, \tag{9.77}$$

which in the time domain is equivalent to

$$\mathrm{D}^{c}_{xy} = \sum_{k=-\infty}^{+\infty}\left|c_x[k] - c_y[k]\right|^2. \tag{9.78}$$

CHAPTER 10

Adaptive Filters

10.1 Introduction

Consider the discrete time single reference channel FIR Wiener filter with M elements, shown in figure 10.1 and zero-mean WSS complex random inputs. This corresponds to the real-time filtering problem $k = 0$ of section 9.2. The filter design criterion is the minimization of the output error sequence variance:

$$\|e_n\|_2^2 = \mathrm{E}\big[\,|e_n|^2\,\big], \;\; e_n \equiv y_n - \hat{y}_n. \tag{10.1}$$

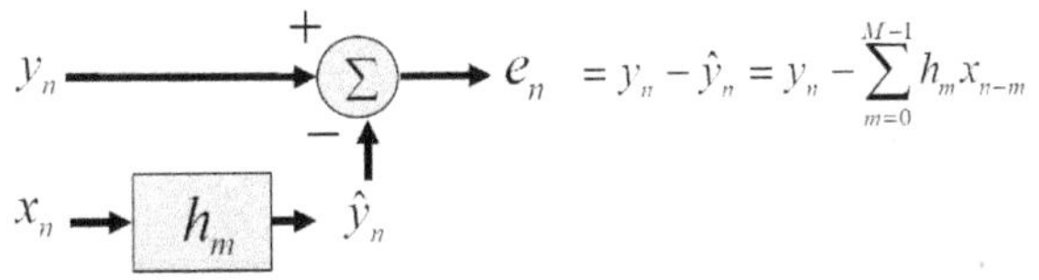

Figure 10.1: Single reference channel FIR causal Wiener filter.

Defining the filter vector $\boldsymbol{h}$ and the time-reversed data vector $\boldsymbol{x}_n$ [see (6.73)],

$$\boldsymbol{h} = [h_0, \ldots, h_{M-1}]^T, \;\; \boldsymbol{x}_n = [x_n, \ldots, x_{n-(M-1)}]^T, \;\; \boldsymbol{R}_{xx} = \mathrm{E}\big[\boldsymbol{x}_n^* \boldsymbol{x}_n^T\big], \tag{10.2}$$

we have $\hat{y}_n = \boldsymbol{h}^T \boldsymbol{x}_n$, and the optimal Wiener filter is the solution to

$$\boldsymbol{R}_{xx}\, \boldsymbol{h} = \boldsymbol{r}_{yx}, \quad \boldsymbol{r}_{yx} = [r_0, \ldots, r_{M-1}]^T \equiv \mathrm{E}\big[y_n \boldsymbol{x}_n^*\big]. \tag{10.3}$$

Here we address the problem of an adaptive implementation of this solution in order to extend the application of the optimal Wiener filter to data that have sufficiently slowly varying non-stationarities. The three most important issues in any adaptive algorithm are convergence speed, computational complexity, and numerical stability. We illustrate the basic principles of an adaptive implementation by considering a primary channel y_n with a single reference x_n that are real, zero-mean, and WSS processes, and using them to derive an optimal real single coefficient h. The estimation equation is $\hat{y}_n = h x_n$ and the variance of the error $\mathcal{E}_h$ now becomes a simple quadratic function of the real variable h,

$$\mathcal{E}_h \equiv \mathrm{E}\left[e_n^2\right] = h^2\, \mathrm{E}\left[x_n^2\right] - 2h\, \mathrm{E}\left[y_n x_n\right] + \mathrm{E}\left[y_n^2\right] \equiv h^2 \sigma_x^2 - 2h\, \sigma_{yx} + \sigma_y^2. \tag{10.4}$$

Setting to zero the derivative of (10.4) with respect to h,

$$\partial \mathcal{E}_h / \partial h = 2h\, \mathrm{E}\left[x_n^2\right] - 2\, \mathrm{E}\left[y_n x_n\right] = -2\, \mathrm{E}\left[e_n x_n\right] = 0,$$

we find the optimal coefficient and the MSE:

$$h_{\text{opt}} = \sigma_{yx}/\sigma_x^2, \quad \mathcal{E}_{\min} = \sigma_y^2 - \sigma_{yx}^2/\sigma_x^2. \tag{10.5}$$

An adaptive implementation is found by solving the equation $\partial\mathcal{E}_h/\partial h = 0$ iteratively using a gradient descent method, as described in section 4.5: the update equation is $h^{(n+1)} = h^{(n)} - \mu\partial\mathcal{E}_h/\partial h$, which upon using $\partial\mathcal{E}_h/\partial h = 2\sigma_x^2 h - 2\sigma_{yx}$ becomes

$$h^{(n+1)} = (1 - 2\mu\sigma_x^2)h^{(n)} + 2\mu\sigma_{yx}. \tag{10.6}$$

This is a simple inhomogeneous difference equation whose exact solution is

$$\begin{aligned} h^{(n)} &= \frac{\sigma_{yx}}{\sigma_x^2} + (1 - 2\mu\sigma_x^2)^n \left(h^{(0)} - \frac{\sigma_{yx}}{\sigma_x^2}\right) \\ &= h_{\text{opt}} + (1 - 2\mu\sigma_x^2)^n \left(h^{(0)} - h_{\text{opt}}\right), \end{aligned} \tag{10.7}$$

which shows that $h^{(n)} \to h_{\text{opt}}$ as $n \to \infty$ provided that $|1 - 2\mu\sigma_x^2| < 1$. To ensure convergence we must choose a learning rate $0 < \mu < 1/\sigma_x^2$. Note that convergence is guaranteed by smaller values of μ, i.e., values close to 0 (validating the Taylor expansion), but these values slow the convergence speed since $|1 - 2\mu\sigma_x^2|$ is then closer to 1.

10.2 The LMS Algorithm

The Widrow-Huff *least mean square* (LMS) algorithm is an adaptive method that requires no knowledge of the variance or the cross-correlation values needed to build the optimal solution [125]. The algorithm is based on removing the expectation operator that defines the expected error $\mathcal{E}_h$ when computing the derivative $\partial\mathcal{E}_h/\partial h$, i.e., making the following replacement known as the stochastic approximation:

$$\partial\mathcal{E}_h/\partial h = -2\mathrm{E}\left[e_n x_n\right] \;\to\; \partial\mathcal{E}_h/\partial h = -2e_n x_n. \tag{10.8}$$

The Widrow-Huff LMS algorithm for real, single reference channel, and filter with one element takes the following iterative form starting with zero initial condition $h^{(0)} = 0$:

$$\textbf{1. } \hat{y}_n = h^{(n)}x_n, \quad \textbf{2. } e_n = y_n - \hat{y}_n, \quad \textbf{3. } h^{(n+1)} = h^{(n)} + 2\mu\, e_n x_n.$$

Thus, the error at every step is fed back to control the filter adaptation while the filter is used to decorrelate the reference channel and the error output, i.e., $h^{(n)} \to h_{\text{opt}}$ while $e_n x_n \to 0$. In practice, the filter never reaches the theoretical optimal value but fluctuates about it. The statistical properties of the filter are impossible to calculate unless quite restrictive conditions are imposed, including requirements such as $\mathrm{E}[h^{(n)}x_n^2] = \mathrm{E}[h^{(n)}]\mathrm{E}[x_n^2]$, which are used to write

$$\mathrm{E}\left[h^{(n+1)}\right] = \mathrm{E}\big[h^{(n)}\big] + 2\mu\,\mathrm{E}[e_n x_n] = (1 - 2\mu\sigma_x^2)\mathrm{E}\left[h^{(n)}\right] + 2\mu\sigma_{yx}, \tag{10.9}$$

whose solution and the corresponding MSE are

$$\mathrm{E}\left[h^{(n)}\right] = \sigma_{yx}/\sigma_x^2 = h_{\text{opt}}, \quad \mathrm{E}\big[e_n^2\big] = \mathcal{E}_{\min} + \mathrm{E}\big[(h^{(n)} - h_{\text{opt}})^2\big]\sigma_x^2, \tag{10.10}$$

where we assumed that the filter and x are statistically independent. The limit $n \to \infty$ of the second term on the right hand side of the MSE is known as the *excess MSE* and is easily calculated to equal to $\mu\mathcal{E}_{\text{min}}\sigma_x^2$. Thus, in addition to controlling the convergence properties of the solution, μ also controls the size of the fluctuations about the optimal solution, i.e., the excess MSE, whose ratio to $\mathcal{E}_{\text{min}}$ is termed the filter *misadjustment*.

As an example consider zero-mean random process $x_n \sim \mathcal{N}(0,1)$, zero-mean random noise $\nu_n \sim \mathcal{N}(0,1)$, and the zero-mean process $y_n = -3x_n + \nu_n$. Multiplying both sides of the defining equation with x_n and taking expectation values, we obtain $\sigma_{yx} = -3\sigma_x^2$ (since ν_n and x_n are assumed uncorrelated). Using equation (10.10) the optimal Wiener filter with 1 element is $h_{\text{opt}} = -3$, (an obvious solution by simply inspecting the model equation). Figure 10.2 shows results of a simulation with 1000 data samples: as expected both convergence rates and fluctuations about the optimal filter value increase with increasing μ.

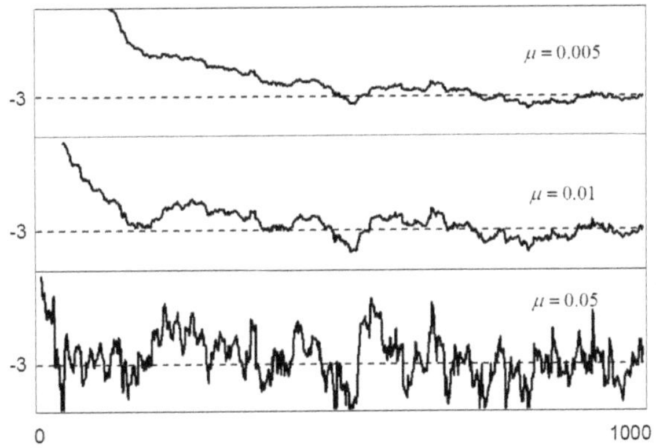

Figure 10.2: LMS filter example convergence rates and fluctuations.

For a real filter with $M > 1$, the increment $\Delta\boldsymbol{h}$ used in the gradient descent algorithm is

$$\Delta\boldsymbol{h} = -\mu\,\nabla\mathcal{E}_{\boldsymbol{h}}, \quad \mu > 0,$$

where the positive learning rate μ must be small enough to justify the two-term Taylor approximation to the objective function [see section 4.5] and $\mu < 2/\lambda_{\text{max}}$, where λ_{max} is the largest eigenvalue of the auto-correlation matrix $\boldsymbol{R}_{xx}$. The inequality is often written as $0 < \mu < 2/\boldsymbol{R}_{xx}[0]$ since $M\lambda_{\text{max}} > \textbf{Trace}(\boldsymbol{R}_{xx}) = M\boldsymbol{R}_{xx}[0]$. The corresponding LMS algorithm is:

$$\textbf{1.}\ \hat{y}_n = \boldsymbol{h}^{(n)T}\boldsymbol{x}_n, \quad \textbf{2.}\ e_n = y_n - \hat{y}_n, \quad \textbf{3.}\ \boldsymbol{h}^{(n+1)} = \boldsymbol{h}^{(n)} + 2\mu\, e_n\boldsymbol{x}_n,$$

where $\boldsymbol{x}_n \equiv \left[x_n, x_{n-1}, \ldots, x_{n-(M-1)}\right]^T$ is the time reversed data vector.

A useful application of correlation cancellation is when the primary channel y_n is available without any reference channels and the primary y_n is the sum of two independent components: a periodic signal and a broadband signal. When the periodic signal is of interest we seek a *self-tuning* LMS filter by using a delayed version of the primary channel as a reference. The periodic component has an auto-correlation function that is wide, while the broadband component has a narrow auto-correlation function. If we denote the effective correlation widths of the broadband and the periodic components by Δ_{BB} and Δ_{NB}, respectively, then the delayed signal y_{n-K} will be uncorrelated with the broadband component of the primary channel if we choose $\Delta_{BB} < K < \Delta_{NB}$. Thus, an LMS filter using the delayed primary will cancel the periodic component, with an output that is very close to the broadband signal; the periodic component is found by simply subtracting the filtered output from y_n. The self-tuning ability of the LMS filter allows for detecting extremely low amplitude sinusoidal signals in white noise and in such circumstances the LMS filter is known as an *adaptive line enhancer* [125]. The adaptive line enhancer is applicable to input signals with a variety of signal and noise parameters with little or no a priori information; it can estimate and track instantaneous frequencies and is useful when the signal is frequency modulated.

10.3 Complex LMS

In this section we generalize the real LMS algorithm to complex zero-mean data with a single reference channel and a filter with more than 1 element [126]. Defining the reference time reversed data vector $\boldsymbol{x}_n \equiv \left[x_n, \ldots, x_{n-(M-1)}\right]^T$, the primary channel estimate is $\hat{y}_n = \boldsymbol{h}^T \boldsymbol{x}_n$. The MSE $\mathcal{E} = \mathrm{E}[|e_n|^2]$ is now a function of both $\boldsymbol{h}$ and its complex conjugate $\boldsymbol{h}^*$. Using equation (4.45) and results of section 4.5, the gradient descent choice is $\Delta h_m = -2\mu\, \partial\mathcal{E}/\partial h_m^*$, $\mu > 0$, for which

$$\mathcal{E}\left(\boldsymbol{h} + \Delta\boldsymbol{h}, \boldsymbol{h}^* + \Delta\boldsymbol{h}^*\right) \approx \mathcal{E}\left(\boldsymbol{h}, \boldsymbol{h}^*\right) - 4\mu \left|\partial\mathcal{E}/\partial\boldsymbol{h}\right|^2 \le \mathcal{E}\left(\boldsymbol{h}, \boldsymbol{h}^*\right). \tag{10.11}$$

Using the stochastic approximation (removing the expectation operator)

$$\partial\mathcal{E}/\partial h_m^* = -E\left[e_n x_{n-m}^*\right] \quad \rightarrow \quad -\ e_n x_{n-m}^*, \ \ 0 \le m \le M-1, \tag{10.12}$$

we arrive at the complex LMS (single reference) algorithm:

$$\mathbf{1.}\ \hat{y}_n = \boldsymbol{h}^{(n)T}\boldsymbol{x}_n, \ \ \mathbf{2.}\ e_n = y_n - \hat{y}_n, \ \ \mathbf{3.}\ \boldsymbol{h}^{(n+1)} = \boldsymbol{h}^{(n)} + 2\mu\, e_n \boldsymbol{x}_n^*.$$

To see the convergence properties we start with

$$\begin{aligned}\mathcal{E}\left(\boldsymbol{h}, \boldsymbol{h}^*\right) = \mathrm{E}\left[|e_n|^2\right] &= \sum_{j,k=0}^{M-1} h_j^* h_k \mathrm{E}\left[x_{n-k} x_{n-j}^*\right] - \sum_{k=0}^{M-1} h_k^* \mathrm{E}\left[y_n x_{n-k}^*\right] - \\ &\sum_{k=0}^{M-1} h_k \mathrm{E}\left[y_n^* x_{n-k}\right] + \mathrm{E}\left[|y_n|^2\right],\end{aligned} \tag{10.13}$$

which together with $\mathrm{E}\left[x_{n-k} x_{n-j}^*\right] = R_{xx}\left[j-k\right]$, and $\mathrm{E}[y_n x_{n-k}^*] = r_{yx}\left[k\right]$, and the symmetry relation $R_{xx}\left[-k\right] = R_{xx}^*\left[k\right]$ will give

$$\frac{\partial\mathcal{E}}{\partial h_m} = \sum_{k=0}^{M-1} h_k^* R_{xx}\left[k-m\right] - r_{yx}^*\left[m\right], \quad \frac{\partial\mathcal{E}}{\partial h_m^*} = \sum_{k=0}^{M-1} R_{xx}\left[m-k\right] h_k - r_{yx}\left[m\right].$$

The update equation for the filter vector is an inhomogeneous difference equation

$$\boldsymbol{h}^{(n+1)} = \boldsymbol{h}^{(n)} + \Delta\boldsymbol{h} = \left(\boldsymbol{I} - 2\mu\boldsymbol{R}_{xx}\right)\boldsymbol{h}^{(n)} + 2\mu\boldsymbol{r}_{yx}, \tag{10.14}$$

whose solution is

$$\boldsymbol{h}^{(n)} = \boldsymbol{h}_{\text{opt}} + \left(\boldsymbol{I} - 2\mu\boldsymbol{R}_{xx}\right)^n \left(\boldsymbol{h}^{(0)} - \boldsymbol{h}_{\text{opt}}\right), \quad \boldsymbol{h}_{\text{opt}} = \boldsymbol{R}_{xx}^{-1}\boldsymbol{r}_{yx}. \tag{10.15}$$

If we denote by λ_m all the (real and positive) eigenvalues of the auto-correlation matrix $\boldsymbol{R}_{xx}$, then convergence to the optimal solution is theoretically assured if $|1 - 2\mu\lambda_m| < 1$, $0 \le m \le M-1$, which is guaranteed if we choose $0 < \mu < 1/\mathbf{Max}\,\{\lambda_m\}$. However, convergence rates for different components of the filter vector can differ significantly; the difference between convergence rates among different eigenvalues is large for signals that are highly self-correlated. For instance, consider the 2×2 real, symmetric, and Toeplitz auto-correlation matrix defined by lags 0 and 1 auto-correlation sequence values R_0 and R_1. The eigenvalues are $\lambda_0 = R_0 - |R_1|$ and $\lambda_1 = R_0 + |R_1|$. If the data is highly self-correlated, then $|R_1| < R_0$ and $|R_1| \cong R_0$ which means that λ_1/λ_0 is very large. Methods to equalize the convergence rates will be discussed in sections 10.6 and 10.7.

When the reference channel is a complex exponential at some frequency ω_0, the complex LMS filter acts as a pure notch filter at that frequency; to see this, we assume a filter with M elements and suppose that the reference channel is a pure tone $x_n = A\exp(in\omega_0)$. Then $x^*_{n-m} = A^*\exp(-i\omega_0(n-m))$ and the update equations for the filter components are

$$h_m^{(n+1)} = h_m^{(n)} + 2\mu e_n A^* e^{-i\omega_0(n-m)}, \;\; 0 \le m \le M-1. \tag{10.16}$$

The index m from the exponent can be eliminated by $h_m^{(n)} = g_m^{(n)}\exp(-i\omega_0(n-m))$

$$e^{-i\omega_0} g_m^{(n+1)} = g_m^{(n)} + 2\mu e_n A^*, \;\; 0 \le m \le M-1, \tag{10.17}$$

which is now a difference equation in n. Taking the Z-transform of both sides of (10.17), i.e., multiplying by z^{-n} and summing over all n gives

$$e^{-i\omega_0} z G_m(z) = G_m(z) + 2\mu A^* E(z), \;\; 0 \le m \le M-1. \tag{10.18}$$

Thus,

$$G_m(z) = E(z)\frac{2\mu A^* e^{i\omega_0}}{z - e^{i\omega_0}}, \;\; 0 \le m \le M-1. \tag{10.19}$$

Taking **Z** transforms of the estimation equation

$$\hat{y}_n = \sum_{m=0}^{M-1} h_m^{(n)} x_{n-m} = \sum_{m=0}^{M-1} g_m^{(n)} A, \tag{10.20}$$

and the error equation $e_n = y_n - \hat{y}_n$, we find

$$\hat{Y}(z) = \sum_{m=0}^{M-1} A G_m(z) = \frac{2\mu M|A|^2 e^{i\omega_0}}{z - e^{i\omega_0}} E(z), \;\; E(z) = Y(z) - \hat{Y}(z). \tag{10.21}$$

The transfer function is

$$\frac{E(z)}{Y(z)} = \frac{1 - e^{i\omega_0}z^{-1}}{1 - e^{i\omega_0}\left(1 - 2\mu M|A|^2\right)z^{-1}}, \tag{10.22}$$

which has a zero at $z = \exp(i\omega_0)$, i.e., a notch at that frequency. The pole in this filter is at $z = \exp(i\omega_0)\left(1 - 2\mu M|A|^2\right)$, which for sufficiently small values of μ and $|A|$ will always be inside the unit circle, ensuring a stable filter. Thus, a sinusoidal component at the frequency ω_0 will be canceled from the primary even if $|A| \ll 1$ [127].

10.4 Sign Adaptive LMS Algorithms

For real data, instead of using MSE $\mathrm{E}[e_n^2]$ as a basis for an adaptive filter algorithm, we can use the *mean absolute error* criterion, i.e., minimize $\mathrm{E}[|e_n|]$. With $\mathrm{E}[|e_n|] = \mathrm{E}\big[|y_n - \boldsymbol{h}^T \cdot \boldsymbol{x}_n|\big]$, we have $\partial\mathrm{E}[|e_n|]/\partial\boldsymbol{h} = -\mathrm{E}\big[\boldsymbol{x}_n\mathbf{sign}(e_n)\big]$, where $\mathbf{sign}(e_n)$ is equal to $+1, -1, 0$ for $e_n > 0$, $e_n < 0$, and $e_n = 0$, respectively; no adaptation is required, of course, when $e_n = 0$. This leads to the *signed error* LMS filter update equation

$$\boldsymbol{h}^{(n+1)} = \boldsymbol{h}^{(n)} + \mu\, \boldsymbol{x}_n\, \mathbf{sign}(e_n), \tag{10.23}$$

which has the advantage of reducing the computational complexity by avoiding M multiplications $e_n\boldsymbol{x}_n$ in the usual LMS algorithm. The step size μ must be in the range

$$0 < \mu < 2|e_n|/|\boldsymbol{x}_n^+\boldsymbol{x}_n|\,.$$

An alternative is to replace $\boldsymbol{x}_n$ with its sign to obtain the *signed data* LMS filter update equation

$$\boldsymbol{h}^{(n+1)} = \boldsymbol{h}^{(n)} + \mu\, \mathbf{sign}\,(\boldsymbol{x}_n)\, e_n, \tag{10.24}$$

or use the sign of both the data and error, producing the *signed error and data* LMS filter update equation

$$\boldsymbol{h}_{n+1} = \boldsymbol{h}_n + \mu_s\, \mathbf{sign}(\boldsymbol{x}_n)\, \mathbf{sign}(e_n). \tag{10.25}$$

If the step size in (10.25) is a power of two (computable with a bit shift), this algorithm requires no multiplications; it is used in CCITT ADPCM standard for 32000 bps voice digitization systems. The lower computational complexity comes at a price: typically the rate of convergence is lower than the LMS, and their excess MSE is higher.

A related algorithm that has been shown to achieve better performance than the LMS and the sign adaptive algorithm for some signal and noise statistics is the least mean K^{th} power adaptive filter that minimizes $\mathrm{E}\big[|e_n|^K\big]$ for some positive integer $K > 1$, with the update equation [128]

$$\boldsymbol{h}^{(n+1)} = \boldsymbol{h}^{(n)} + \mu\, K\big|e_n\big|^{K-1}\, \mathbf{sign}(e_n)\, \boldsymbol{x}_n. \tag{10.26}$$

10.5 Normalized LMS Algorithm

In the standard complex LMS algorithm filter update $\boldsymbol{h}^{(n+1)} = \boldsymbol{h}^{(n)} + \mu e_n \boldsymbol{x}_n^*$, the correction $\mu\boldsymbol{x}_n^* e_n$ is directly proportional to the input vector; if $|\boldsymbol{x}_n|$ happens to be large, the LMS filter could take a large step, resulting in what is termed *gradient noise amplification*. To correct this problem, a normalized LMS algorithm can be used, in which the step size is normalized by $||\boldsymbol{x}_n||^2$.

Rather than simply expressing the solution, we pose this as an optimization problem. Given the input vector $\boldsymbol{x}_n$ (the time reversed data vector) and a desired response y_n, the new filter vector $\boldsymbol{h}^{(n+1)}$ is to be determined to minimize the squared norm of the change in the weight vector, $||\boldsymbol{h}^{(n+1)} - \boldsymbol{h}^{(n)}||^2$, subject to the constraint that the updated weight vector provides a filter output equal to the desired signal y_n. Thus, we have the constrained minimization problem

$$\underset{\boldsymbol{h}^{(n+1)}}{\arg\min}\, \big|\big|\boldsymbol{h}^{(n+1)} - \boldsymbol{h}^{(n)}\big|\big|^2 \quad \text{subject to} \quad \boldsymbol{h}^{(n+1)T}\boldsymbol{x}_n = \boldsymbol{x}_n^T\boldsymbol{h}^{(n+1)} = y_n. \tag{10.27}$$

Introducing a Lagrange multiplier λ, we differentiate

$$||\boldsymbol{h}^{(n+1)} - \boldsymbol{h}^{(n)}||^2 + \lambda\left(y_n - \boldsymbol{x}_n^T\boldsymbol{h}^{(n+1)}\right) \tag{10.28}$$

with respect to $\boldsymbol{h}^{(n+1)}$ and set the result to zero to obtain

$$\boldsymbol{h}^{(n+1)+} - \boldsymbol{h}^{(n)+} - \lambda\, \boldsymbol{x}_n^T = 0 \;\Rightarrow\; \boldsymbol{h}^{(n+1)} = \boldsymbol{h}^{(n)} + \lambda\, \boldsymbol{x}_n^*, \tag{10.29}$$

which when substituted into the constraint $y_n = \boldsymbol{x}_n^T \boldsymbol{h}^{(n+1)}$ gives

$$\lambda = \frac{y_n - \boldsymbol{x}_n^T \boldsymbol{h}^{(n)}}{\boldsymbol{x}_n^T \boldsymbol{x}_n^*} = \frac{e_n}{\|\boldsymbol{x}_n\|^2}. \tag{10.30}$$

Substituting for λ into the filter update equation (10.29) gives

$$\boldsymbol{h}^{(n+1)} = \boldsymbol{h}^{(n)} + \frac{e_n}{\|\boldsymbol{x}_n\|^2}\, \boldsymbol{x}_n^*\,. \tag{10.31}$$

Note that the adaptation constant μ here is dimensionless whereas in the standard LMS it has the dimension of power. A more robust formula is

$$\boldsymbol{h}^{(n+1)} = \boldsymbol{h}^{(n)} + \frac{\mu\, e_n}{\|\boldsymbol{x}_n\|^2 + \varepsilon}\, \boldsymbol{x}_n^*, \tag{10.32}$$

where $0 < \mu < 2$ and $0 < \varepsilon \ll 1$ is a constant to ensure numerical stability.

Some other variants of the LMS algorithm address problems associated with large fluctuations in direction towards the minimum during the adaptation process. For instance, the update can be replaced by an average of the present and the previous L updates, or we can use a first order IIR low pass filter when the filtered gradient $\tilde{g}$ satisfies the equation $\tilde{\boldsymbol{g}}^{(n+1)} = \beta\tilde{\boldsymbol{g}}^{(n)} + (1-\beta)\boldsymbol{g}^{(n)}$ where $\boldsymbol{g}^{(n)} \equiv -2e_n \boldsymbol{x}_n^*$; this method is equivalent to introducing a momentum term in the gradient descent algorithm and will be discussed in section 12.6 in the context of neural networks. When impulsive processes in the primary or the reference are present, the LMS algorithm often becomes unstable. A possible solution is to use a nonlinear smoothing filter, such as a median filter, to reduce the degrading effects on the filter update, but the performance of such a filter is highly data dependent.

10.6 Equalizing LMS Convergence Rates

As discussed in section 10.3 the learning parameter μ controls the convergence rates of all filter coefficients and fluctuations around the optimal solution but different components of the filter vector converge at different rates. Equalization of the convergence rates among all filter vector components and increasing the rate of convergence can be investigated using the formulation of section 10.3. Suppose that instead of the single parameter μ we use multiple parameters in the form of a Hermitian and positive definite $M \times M$ matrix $\boldsymbol{W}$ and choose the filter vector increments

$$\Delta \boldsymbol{h}^* = -\, \boldsymbol{W}\, \partial\mathcal{E}/\partial \boldsymbol{h}, \tag{10.33}$$

when we obtain the following Taylor expansion:

$$\mathcal{E}\left(\boldsymbol{h} + \Delta\boldsymbol{h}, \boldsymbol{h}^* + \Delta\boldsymbol{h}^*\right) \approx \mathcal{E}\left(\boldsymbol{h}, \boldsymbol{h}^*\right) - 2\,\boldsymbol{Re}\left(\partial\mathcal{E}/\partial\boldsymbol{h}^+\, \boldsymbol{W}\, \partial\mathcal{E}/\partial\boldsymbol{h}\right) \leq \mathcal{E}\left(\boldsymbol{h}, \boldsymbol{h}^*\right),$$

guaranteeing descent towards the minimum of $\mathcal{E}$ in view of the fact that the matrix $\boldsymbol{W}$ is assumed Hermitian and positive definite (but otherwise arbitrary). Using the derivatives (10.12), we arrive at the filter vector update equation

$$\boldsymbol{h}^{(n+1)} = \left(\boldsymbol{I} - \boldsymbol{W}\boldsymbol{R}_{xx}\right)\boldsymbol{h}^{(n)} + \boldsymbol{W}\boldsymbol{r}_{yx}, \tag{10.34}$$

which is again a difference equation whose solution is

$$\boldsymbol{h}^{(n)} = \boldsymbol{h}_{\text{opt}} + (\boldsymbol{I} - \boldsymbol{W}\boldsymbol{R}_{xx})^n \left(\boldsymbol{h}^{(0)} - \boldsymbol{h}_{\text{opt}}\right). \tag{10.35}$$

The choice of $\boldsymbol{W}$ can affect the convergence speed; in fact, the iteration converges in 1 step if we choose $\boldsymbol{W} = \boldsymbol{R}_{xx}^{-1}$, but if the data were truly stationary and the inverse of the auto-correlation matrix was known then we would not need an adaptive algorithm to begin with! The point of the example is that if $\boldsymbol{W} \approx \kappa\, \boldsymbol{R}_{xx}^{-1}$ then the convergence requirement is $|1-\kappa| < 1$, i.e., if we choose $0 < \kappa < 2$ then all filter vector components will converge at the same rate. The closer the matrix $\boldsymbol{W}$ is to the inverse of the auto-correlation matrix, the faster the rate of convergence. The recursive least squares algorithm (RLS) is an effective way to approach this solution without a knowledge of $\boldsymbol{R}_{xx}^{-1}$.

10.7 Recursive Least Squares (RLS)

The LMS algorithm provides an iterative method to approach the optimal Wiener solution: the adaptive filter coefficients are actually never optimal and only converge to it. In the recursive least squares (RLS) formulation of the Wiener filtering problem we replace the minimization of the statistically defined quantity $\mathcal{E} = \mathrm{E}[|e_n|^2]$ with a least-squares minimization of a time-averaged quantity that includes all estimation error terms from the initial instant to the current time index. In order to track non-stationary behavior we should give more weight to recent data and less weight to distant ones and so we introduce a forgetting-factor $0 < \lambda \le 1$ and minimize

$$\mathcal{E}_N = \sum_{n=0}^{N} \lambda^{N-n} |e_n|^2, \quad e_n = y_n - \boldsymbol{h}^T \boldsymbol{x}_n, \tag{10.36}$$

where $\boldsymbol{h}_n = \left[h_0, \ldots, h_{M-1}\right]^T$ and $\boldsymbol{x}_n \equiv \left[x_n, \ldots, x_{n-(M-1)}\right]^T$ is the time reversed data vector. Thus, we minimize

$$\mathcal{E}_N = \sum_{n=0}^{N} \lambda^{N-n} \Big(|y_n|^2 - \boldsymbol{h}^+ y_n \boldsymbol{x}_n^* - y_n^* \boldsymbol{x}_n^T \boldsymbol{h} + \boldsymbol{h}^+ \boldsymbol{x}_n^* \boldsymbol{x}_n^T \boldsymbol{h} \Big). \tag{10.37}$$

The least squares filter at step N, denoted by $\boldsymbol{h}_N$, is then found from

$$\partial \mathcal{E}_N / \partial \boldsymbol{h}^+ = 0 \;\Rightarrow\; \left[\sum_{n=0}^{N} \lambda^{N-n} \boldsymbol{x}_n^* \boldsymbol{x}_n^T \right] \boldsymbol{h}_N = \sum_{n=0}^{N} \lambda^{N-n} y_n \boldsymbol{x}_n^*. \tag{10.38}$$

Defining the Hermitian matrix $\boldsymbol{R}_N$ and the vector $\boldsymbol{r}_N$ [1],

$$\boldsymbol{R}_N \equiv \sum_{n=0}^{N} \lambda^{N-n}\, \boldsymbol{x}_n^* \boldsymbol{x}_n^T, \quad \boldsymbol{r}_N \equiv \sum_{n=0}^{N} \lambda^{N-n}\, y_n \boldsymbol{x}_n^*, \tag{10.39}$$

we find the least squares analogue of the Wiener filter:

$$\boldsymbol{R}_N \boldsymbol{h}_N = \boldsymbol{r}_N \;\Rightarrow\; \boldsymbol{h}_N = \boldsymbol{R}_N^{-1} \boldsymbol{r}_N. \tag{10.40}$$

[1] For $\lambda = 1$, $\boldsymbol{R}_N$ can be thought of as an estimate of the scaled auto-correlation matrix $(N+1)\boldsymbol{R}_{xx} = \mathrm{E}[\boldsymbol{x}\boldsymbol{x}^+]$ for the signal vector $\boldsymbol{x} = [x_{n-(M-1)}, \ldots, x_n]^T$ [see equation (6.73) and note that $\boldsymbol{x}_n = \boldsymbol{J}\boldsymbol{x}$], while $\boldsymbol{r}_N$ can be thought of as an estimate of the scaled cross-correlation vector $(N+1)\boldsymbol{r}_{yx}$.

The quantities $\boldsymbol{R}_N$ and $\boldsymbol{r}_N$ play the roles of the true auto-correlation matrix $\boldsymbol{R}_{xx}$ and cross-correlation vector $\boldsymbol{r}_{yx}$ of the Wiener filtering problem and they satisfy rank-one update equations

$$\boldsymbol{R}^*_N = \lambda \boldsymbol{R}^*_{N-1} + \boldsymbol{x}_N \boldsymbol{x}^+_N, \quad \boldsymbol{r}_N = \lambda \boldsymbol{r}_{N-1} + y_N \boldsymbol{x}^*_N, \tag{10.41}$$

which allow for a recursive computation of the matrix inverse $\boldsymbol{R}^{-1}_N$ as described in theorem 50, also known as the *matrix inversion lemma*.

Theorem 50. *Consider two $N \times N$ non-singular matrices $\boldsymbol{A}$ and $\boldsymbol{B}$ and*

$$\boldsymbol{B} = \boldsymbol{A} + \boldsymbol{C}\boldsymbol{Q}\boldsymbol{D}, \tag{10.42}$$

where $\boldsymbol{C}$ is $N \times M$, $\boldsymbol{D}$ is $M \times N$, $\boldsymbol{Q}$ is $M \times M$, and $M < N$. Then

$$\boldsymbol{B}^{-1} = \boldsymbol{A}^{-1} - \boldsymbol{A}^{-1}\boldsymbol{C}\left(\boldsymbol{Q}^{-1} + \boldsymbol{D}\boldsymbol{A}^{-1}\boldsymbol{C}\right)^{-1}\boldsymbol{D}\boldsymbol{A}^{-1}. \tag{10.43}$$

The case of interest to us is the rank-one update when $\boldsymbol{Q} = \boldsymbol{I}$ and $M = 1$ when

$$\boldsymbol{B}^{-1} = \left(\boldsymbol{A} + \boldsymbol{c}\boldsymbol{d}^+\right)^{-1} = \boldsymbol{A}^{-1} - \frac{\boldsymbol{A}^{-1}\boldsymbol{c}\boldsymbol{d}^+\boldsymbol{A}^{-1}}{1 + \boldsymbol{d}^+\boldsymbol{A}^{-1}\boldsymbol{c}}, \tag{10.44}$$

and so

$$\boldsymbol{R}^{-1*}_N = \frac{1}{\lambda}\boldsymbol{R}^{-1*}_{N-1} - \frac{\boldsymbol{R}^{-1*}_{N-1}\,\boldsymbol{x}_N \boldsymbol{x}^+_N\,\boldsymbol{R}^{-1*}_{N-1}}{\lambda^2 + \lambda \boldsymbol{x}^+_N\,\boldsymbol{R}^{-1*}_{N-1}\,\boldsymbol{x}_N}. \tag{10.45}$$

We denote $\boldsymbol{R}^{-1}_N$ by $\boldsymbol{P}_N$ and define the *a priori Kalman gain vector* as the complex conjugate of the inverse matrix from the previous step $N-1$ acting on the present step N (time reversed) data vector $\boldsymbol{x}_N$,

$$\boldsymbol{K}_{N|N-1} \equiv \lambda^{-1}\,\boldsymbol{R}^{-1*}_{N-1}\boldsymbol{x}_N = \lambda^{-1}\,\boldsymbol{P}^*_{N-1}\boldsymbol{x}_N, \tag{10.46}$$

and define a real and positive number ν_N (the "likelihood variable") by

$$\nu_N \equiv \lambda^{-1}\,\boldsymbol{x}^+_N\,\boldsymbol{R}^{-1*}_{N-1}\,\boldsymbol{x}_N = \boldsymbol{x}^+_N\,\boldsymbol{K}_{N|N-1} = \boldsymbol{K}^+_{N|N-1}\boldsymbol{x}_N. \tag{10.47}$$

Then we have

$$\boldsymbol{P}^*_N = \frac{1}{\lambda}\,\boldsymbol{P}^*_{N-1} - \frac{1}{\lambda^2 + \lambda\nu_N}\,\boldsymbol{K}_{N|N-1}\,\boldsymbol{K}^+_{N|N-1}. \tag{10.48}$$

We define the *a posteriori Kalman gain vector*

$$\boldsymbol{K}_N \equiv \boldsymbol{R}^{-1*}_N\boldsymbol{x}_N = \boldsymbol{P}^*_N\boldsymbol{x}_N, \tag{10.49}$$

which can be written in terms of the a priori Kalman gain vector $\boldsymbol{K}_{N|N-1}$; to see this we start with the rank-one update equation for $\boldsymbol{R}^*_N$, multiply on the left by $\boldsymbol{R}^{-1*}_N$ and on the right by $\boldsymbol{R}^{-1*}_{N-1}$ to find

$$\boldsymbol{R}^{-1*}_{N-1} = \lambda\boldsymbol{R}^{-1*}_N + \boldsymbol{R}^{-1*}_N\boldsymbol{x}_N\boldsymbol{x}^+_N\boldsymbol{R}^{-1*}_{N-1}, \tag{10.50}$$

which upon acting on $\boldsymbol{x}_N$ and using the definition of the a priori Kalman gain vector gives

$$\boldsymbol{K}_{N|N-1} = \boldsymbol{R}^{-1*}_N\boldsymbol{x}_N + \lambda^{-1}\,\boldsymbol{R}^{-1*}_N\boldsymbol{x}_N\boldsymbol{x}^+_N\boldsymbol{R}^{-1*}_{N-1}\boldsymbol{x}_N = (1 + \nu_N)\boldsymbol{K}_N. \tag{10.51}$$

Thus,

$$\boldsymbol{K}_N = \frac{1}{1+\nu_N}\boldsymbol{K}_{N|N-1}. \tag{10.52}$$

Now we proceed to find an update equation for the filter $\boldsymbol{h}_{N-1}$ by considering the quantity $\boldsymbol{R}_N\boldsymbol{h}_N$ and using the matrix rank-one update equation (10.41)

$$\boldsymbol{R}_N\boldsymbol{h}_{N-1} = \lambda\,\boldsymbol{R}_{N-1}\boldsymbol{h}_{N-1} + (\boldsymbol{x}_N\boldsymbol{x}_N^+)^*\boldsymbol{h}_{N-1} = \lambda\boldsymbol{r}_{N-1} + \boldsymbol{x}_N^*\hat{y}_{N|N-1}, \tag{10.53}$$

where we have defined the *a priori estimate* $\hat{y}_{N|N-1}$ as the filter at previous step $N-1$ acting on present data $\boldsymbol{x}_N$,

$$\hat{y}_{N|N-1} \equiv \boldsymbol{h}_{N-1}^T\boldsymbol{x}_N = \boldsymbol{x}_N^T\boldsymbol{h}_{N-1}. \tag{10.54}$$

Next we use the rank-one update equation (10.41) for $\boldsymbol{r}_N$ to replace $\lambda\boldsymbol{r}_{N-1}$,

$$\boldsymbol{R}_N\boldsymbol{h}_{N-1} = \boldsymbol{r}_N - (y_N - \hat{y}_{N|N-1})\boldsymbol{x}_N^* = \boldsymbol{r}_N - e_{N|N-1}\boldsymbol{x}_N^*, \tag{10.55}$$

where we have defined the *a priori estimation error*:

$$e_{N|N-1} \equiv y_N - \hat{y}_{N|N-1}. \tag{10.56}$$

Thus,

$$\boldsymbol{h}_{N-1} = \boldsymbol{R}_N^{-1}\boldsymbol{r}_N - e_{N|N-1}\boldsymbol{R}_N^{-1}\boldsymbol{x}_N^* = \boldsymbol{h}_N - e_{N|N-1}\boldsymbol{K}_N^*, \tag{10.57}$$

and so

$$\boldsymbol{h}_N = \boldsymbol{h}_{N-1} + e_{N|N-1}\boldsymbol{K}_N^* = \boldsymbol{h}_{N-1} + \frac{1}{1+\nu_N}e_{N|N-1}\boldsymbol{K}_{N|N-1}^*. \tag{10.58}$$

The estimation error at step N is

$$\begin{aligned} e_N &= y_N - \hat{y}_N = y_N - \boldsymbol{h}_N^T\boldsymbol{x}_N = y_N - \left(\boldsymbol{h}_{N-1}^T + e_{N|N-1}\boldsymbol{K}_N^+\right)\boldsymbol{x}_N \\ &= e_{N|N-1}\left(1 - \boldsymbol{K}_N^+\boldsymbol{x}_N\right). \end{aligned} \tag{10.59}$$

Next we use

$$\boldsymbol{K}_N^+\boldsymbol{x}_N = \frac{1}{1+\nu_N}\boldsymbol{K}_{N|N-1}^*\boldsymbol{x}_N = \frac{\nu_N}{1+\nu_N}, \tag{10.60}$$

in (10.59) to obtain

$$e_N = e_{N|N-1}\left(1 - \frac{\nu_N}{1+\nu_N}\right) = \frac{1}{1+\nu_N}e_{N|N-1}. \tag{10.61}$$

Finally, the inverse matrix is updated:

$$\boldsymbol{P}_N^* = \frac{1}{\lambda}\boldsymbol{P}_{N-1}^* - \frac{1}{1+\nu_N}\boldsymbol{K}_{N|N-1}\boldsymbol{K}_{N|N-1}^+ = \frac{1}{\lambda}\boldsymbol{P}_{N-1}^* - \boldsymbol{K}_N\boldsymbol{K}_{N|N-1}^+. \tag{10.62}$$

The initializations are $\boldsymbol{h}_0 = \boldsymbol{0}$ and $\boldsymbol{P}_0 = \kappa\boldsymbol{I}$, where κ is a very large positive number that in practice is chosen to be at least an order of magnitude larger than the inverse standard deviation of the first data segment $\boldsymbol{x}_1$. The resulting method is known as the *direct form algorithm* and is summarized below for $N \geq 1$.

- $\boldsymbol{K}_{N|N-1} = \lambda^{-1}\boldsymbol{P}_{N-1}^*\boldsymbol{x}_N,\ \ \nu_N = \boldsymbol{K}_{N|N-1}^+\boldsymbol{x}_N,\ \ \boldsymbol{K}_N = (1+\nu_N)^{-1}\boldsymbol{K}_{N|N-1}$

- $\hat{y}_{N|N-1} = \boldsymbol{h}^T_{N-1}\boldsymbol{x}_N$, $e_{N|N-1} = y_N - \hat{y}_{N|N-1}$, $e_N = (1+\nu_N)^{-1} e_{N|N-1}$
- $\boldsymbol{h}_N = \boldsymbol{h}_{N-1} + e_{N|N-1}\boldsymbol{K}^*_N$, $\boldsymbol{P}^*_N = \lambda^{-1}\boldsymbol{P}^*_{N-1} - \boldsymbol{K}_N\boldsymbol{K}^+_{N|N-1}$

The effective memory in $\mathcal{E}_N$ is given by

$$\mathcal{M}_\lambda = \sum_{n=0}^{\infty} n\lambda^n \Big/ \sum_{n=0}^{\infty} \lambda^n = \lambda/(1-\lambda), \tag{10.63}$$

with $\lambda = 1$ corresponding to the stationary case (infinite memory).

As an example, consider $2,000$ samples of an AR(2) process (the reference channel) defined by $x_n = 0.8x_{n-1} - 0.65x_{n-2} + \nu_n$ where $\nu_n \sim \mathcal{N}(0,1)$, and $2,000$ samples of the primary channel defined by $y_n = x_n + 1.2x_{n-1} - 3x_{n-2} + \epsilon_n$ with $\epsilon_n \sim \mathcal{N}(0,1)$. The optimal Wiener filter is clearly given by $\boldsymbol{h}_{\text{opt}} = [1, 1.2, -3]^T$. Using $\lambda = 1$ for stationary data, we find all three filter coefficients converging after $1,000$ steps with a mean value of $[0.97, 1.24, -3.07]^T$ and a standard deviation (calculated between steps $1,000$ and $2,000$) of 0.006.

10.8 RLS Implementation

The direct form RLS algorithm at the end of section 10.7 is susceptible to round-off error in the calculation of the inverse matrix $\boldsymbol{P}_N$ in equation (10.62). The simplest solution is to use upper triangular Cholesky factors of $\boldsymbol{P}_N$ and then to update the square root matrix $\boldsymbol{U}_N$ directly, after which the corresponding filter equation (10.40) can be solved using the upper triangular form of the Cholesky factor. In particular we write

$$\boldsymbol{P}_N = \boldsymbol{U}^*_N \boldsymbol{D}_N \boldsymbol{U}^T_N \tag{10.64}$$

where $\boldsymbol{U}_N$ is an upper triangular matrix with 1s along the diagonal and $\boldsymbol{D}_N$ is a diagonal matrix with real and positive diagonal elements. We also define two vectors

$$\boldsymbol{f}_{N-1} = \boldsymbol{U}^T_{N-1}\boldsymbol{x}^*_N \text{ and } \boldsymbol{v}_{N-1} = \boldsymbol{D}_{N-1}\boldsymbol{f}_{N-1}, \tag{10.65}$$

and a scalar $\alpha = \boldsymbol{x}^T_N \boldsymbol{P}_{N-1}\boldsymbol{x}^*_N$, in terms of which the update equation for $\boldsymbol{P}_N$ becomes

$$\boldsymbol{U}^*_N\boldsymbol{D}_N\boldsymbol{U}^T_N = \frac{1}{\lambda}\boldsymbol{U}^*_{N-1}\Big(\boldsymbol{D}_{N-1} - \frac{1}{\alpha}\boldsymbol{v}_{N-1}\boldsymbol{v}^+_{N-1}\Big)\boldsymbol{U}^T_{N-1}. \tag{10.66}$$

Now using the upper triangular Cholesky factorization of the term inside the parentheses, namely,

$$\boldsymbol{D}_{N-1} - \frac{1}{\alpha}\boldsymbol{v}_{N-1}\boldsymbol{v}^+_{N-1} = \boldsymbol{U}'^*_{N-1}\boldsymbol{D}'_{N-1}\boldsymbol{U}'^T_{N-1}, \tag{10.67}$$

we find the update equations for the Cholesky factors of $\boldsymbol{P}_N$,

$$\boldsymbol{U}^*_N = \boldsymbol{U}^*_{N-1}\boldsymbol{U}'_{N-1} \text{ and } \boldsymbol{D}_N = \boldsymbol{U}'_{N-1}/\lambda, \tag{10.68}$$

which form the basis for the *square root RLS algorithm* [129].

A different approach often used in dynamically updated adaptive beamforming [see section 11.7], is to use QR factors of the Cholesky factor of the matrix $\boldsymbol{R}_N$. Our discussion here is in the context of adaptive beamforming so the $M \times 1$ vector $\boldsymbol{x}_n$ refers to a snapshot at time step n of data recorded at a spatial linear array of M sensors; the $M \times M$ "correlation matrix" at each time step satisfies the rank-1 update equation with a forgetting factor $0 < \lambda \leq 1$,

$$\boldsymbol{R}_N = \lambda\, \boldsymbol{R}_{N-1} + \boldsymbol{x}_N \boldsymbol{x}_N^+, \tag{10.69}$$

with no "cross-correlation vector". The output of the adaptive beamformer at each time step n is a weighted sum $\boldsymbol{w}^+\boldsymbol{x}_n$; the definition of the weight vector here includes a complex conjugation, and since in this context $\boldsymbol{x}_n$ is not a time reversed version of the data vector, no complex conjugation is required [as was the case in (10.41)]. The problem here is to find the optimal weight vector including a linear constraint of the form $\boldsymbol{A}^+\boldsymbol{w} = \boldsymbol{b}$; the optimal solution leads to the adaptive beamformer output $\boldsymbol{b}^+(\boldsymbol{A}^+\boldsymbol{R}^{-1}\boldsymbol{A})^{-1}\boldsymbol{b}$. Thus, the RLS algorithm in this case provides a means of calculating the adaptive beamformer output without having to invert the correlation matrix at each step by using an update equation for the $M \times M$ Cholesky square root factor of the correlation matrix [130],

$$\boldsymbol{R}_n = \boldsymbol{L}_n \boldsymbol{L}_n^+. \tag{10.70}$$

Using the Cholesky factors, the correlation matrix update equation (10.69) is

$$\boldsymbol{L}_N \boldsymbol{L}_N^+ = \lambda\, \boldsymbol{L}_{N-1} \boldsymbol{L}_{N-1}^+ + \boldsymbol{x}_N \boldsymbol{x}_N^+. \tag{10.71}$$

Defining an augmented $M \times (M+1)$ matrix $\boldsymbol{L}'$, we have

$$\boldsymbol{L}' \equiv \left[\sqrt{\lambda}\, \boldsymbol{L}_{N-1} \,|\, \boldsymbol{x}_N\right] \quad \rightarrow \quad \boldsymbol{R}_N = \boldsymbol{L}_N \boldsymbol{L}_N^+ = \boldsymbol{L}'\boldsymbol{L}'^+. \tag{10.72}$$

Now we perform a "square" QR factorization [see figure 2.1] of $\boldsymbol{L}'^+$ in the form

$$\boldsymbol{L}'^+ = \boldsymbol{Q}\boldsymbol{V}, \tag{10.73}$$

where $\boldsymbol{Q}$ is $(M+1) \times (M+1)$ and unitary, $\boldsymbol{Q}^+\boldsymbol{Q} = \boldsymbol{I}$, and $\boldsymbol{V}$ is $(M+1) \times M$ and has a zero last row while the rest of it, denoted by $\boldsymbol{U}$, is $M \times M$ and upper triangular, as depicted in figure 10.3.

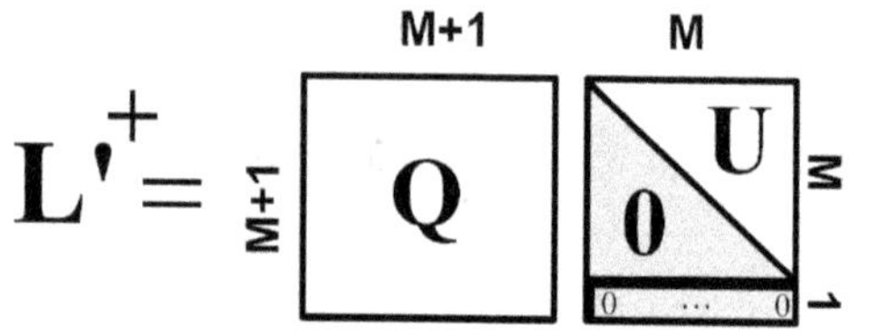

Figure 10.3: QR factors of $\boldsymbol{L}'^+$.

Using the QR factors (10.73) in the update equation (10.72) we have

$$\boldsymbol{L}_N \boldsymbol{L}_N^+ = \boldsymbol{L}'\boldsymbol{L}'^+ = \boldsymbol{V}^+\boldsymbol{Q}^+\boldsymbol{Q}\boldsymbol{V} = \boldsymbol{U}^+\boldsymbol{U}, \tag{10.74}$$

which shows that $\boldsymbol{L}_N = \boldsymbol{U}^+$. Thus, starting at step $N-1$ we construct $\boldsymbol{L}'$ using equation (10.72), find the QR factors of its Hermitian conjugate using (10.73), and extract the upper triangular sub-matrix of $\boldsymbol{V}$, namely, $\boldsymbol{U}$; the updated lower triangular matrix $\boldsymbol{L}_N$ is then given by $\boldsymbol{U}^+$.

CHAPTER 11

Optimal Processing of Linear Arrays

11.1 Uniform Linear Array (ULA)

A uniform linear array (ULA) was introduced at the end of section 1.9 as an array of M omni-directional receivers on the x-axis and whose Cartesian position coordinates are $(md, 0, 0)$, $0 \leq m \leq M-1$. Figure 11.1(a) shows a cone with its apex at the origin and vertical height h to the circular base that is orthogonal to the x-axis. Consider a point at distance s from the origin and on the circular base; its position vector in Cartesian coordinates is $\boldsymbol{S} \leftrightarrow (s\cos\phi\sin\theta = h, s\sin\phi\sin\theta, s\cos\theta)$, where (θ, ϕ) are the usual polar and azimuthal angles in spherical polar coordinates. The angle ψ between $\boldsymbol{S}$ and the array (the x-axis), is the cone angle and is found by taking the dot product between two unit vectors $\boldsymbol{S}/s$ and $[1, 0, 0]$; thus, $\cos\psi = \cos\phi\sin\theta$. The ULA response to plane waves with direction vector parallel to $\boldsymbol{S}$ is independent of (θ, ϕ) and only depends on the cone angle ψ.

The relation between ψ and the spherical angles cannot be inverted to obtain a unique pair (θ, ϕ) for a given ψ; thus, the cone is known as the cone of ambiguity. In many underwater acoustic applications, for instance, a sonar line array parallel to the plane of the ocean surface, θ is nearly $90°$, and in this case, the cone angle ψ is the same as the azimuthal angle ϕ. Since our discussion in this chapter will depend on the cone angle alone, we choose to work in any plane containing the x-axis; in underwater applications, it is natural to choose the xy-plane, which is parallel to the ocean surface; figure 11.1(b) shows such a plane containing a plane wave with wave vector $\boldsymbol{k}$ incident on the array. Reflection symmetry about the array means that we can take the wave direction ψ in the range $[0°, 180°]$.

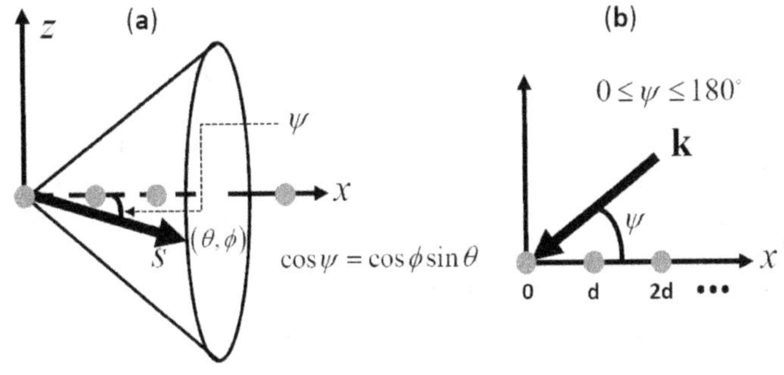

Figure 11.1: ULA cone of ambiguity and the propagation plane.

In the example following theorem 6, we described the application of the sampling theorem to the normalized spatial frequency $\tilde{f} = d\cos\psi/\lambda$ of plane waves from a far-field source with linear frequency $f = \omega/2\pi$ leading to equation (1.42) that is satisfied by half-wavelength spacing of array elements. In addition, we showed in section 7.10 that isotropic noise sampled on sensors of a linear half-wavelength array are independent. We will, therefore, assume half-wavelength arrays with $d = \lambda/2$ in this chapter.

11.2 The Signal Model on a ULA

The received narrow-band signal undergoes a process of *base-banding* consisting of demodulation of the propagation frequency ω_0 and leaving a complex base-band signal $s(t)$ whose bandwidth we denote by B. The narrow-band assumption is often taken to mean that the bandwidth B is small compared to the high propagation frequency f_0; a more accurate criterion is $TB \ll 1$, where T is the propagation time across the array $T = (M-1)d/c$. For example, for $M = 30$, $f_0 = 1000$ Hz and $c = 1500$ m/s, $d = \lambda_0/2 = 0.75$ m and $T = 0.01$ seconds; thus, $TB \ll 100$, and $B \sim 1$ Hz clearly satisfies the narrow-band assumption.

Let us denote by $x_m(t)$ the base-banded received signal from direction ψ_s at receiver m with position $(md, 0, 0)$ at time t. Neglecting any interference and noise, $x_m(t) = s(t - \tau_m)$ where $\tau_m(\psi_s)$ is the propagation time difference (positive or negative depending on the value of ψ_s) between receiver m and receiver 0 at the origin. The narrow-band assumption leads to a description of the time shift in terms of a phase difference, just as in the plane wave example of equation (1.39). Thus, the base-banded received signal model at each receiver without noise is

$$x_m(t) = s(t)\exp(-i\omega_0\tau_m), \quad \tau_m \equiv \tau_m(\psi_s), \tag{11.1}$$

where $\tau_m = md\cos\psi_s/c$ is the time delay (or advance) at receiver m (relative to the receiver at the origin) for a signal whose propagation vector makes an angle ψ_s with the x-axis; $s(t)$ is a zero-mean WSS complex random process with power $\mathcal{P}_s = \mathrm{E}[|s(t)|^2]$. Using $c = f_0\lambda_0$ and the definition of the normalized spatial frequency (1.41) (the subscript s on $\tilde{f}$ and ψ indicate the signal), we have

$$x_m(t) = s(t)e^{-im\tilde{\omega}_s} = s(t)e^{-2i\pi m\tilde{f}_s}, \quad \frac{\tilde{\omega}_s}{2\pi} = \tilde{f}_s \equiv \frac{d}{\lambda_s}\cos\psi_s\ . \tag{11.2}$$

Putting the data from all the receivers at time t together as an $M \times 1$ vector $\boldsymbol{x}(t) = [x_0(t), \dots, x_{M-1}(t)]^T$, we have

$$\boldsymbol{x}(t) \equiv s(t)\,\boldsymbol{d}_s, \tag{11.3}$$

where we define the *signal direction vector* [131] [c.f. the signal vector (6.30) in section 6.2]

$$\boldsymbol{d}_s \equiv [1, e^{-2i\pi\tilde{f}_s}, \dots, e^{-2i\pi(M-1)\tilde{f}_s}]^T, \quad \boldsymbol{d}_s^+\boldsymbol{d}_s = M. \tag{11.4}$$

In addition to the signal $s(t)$, we assume the presence of undesired signal(s), or interference(s), denoted by $u(t)$, and receiver thermal white noise [1] $\nu(t)$; in sonar the in-band white noise is due to flow and it is partially eliminated by averaging several closely spaced hydrophones that compose one sensor element. The interference model is identical in form to the signal model except for different power(s) and direction vector(s). The signal, interference, and thermal noise are assumed to be mutually uncorrelated. Thus, we have the full *signal + interference + noise* model [2]

$$\boldsymbol{x}(t) = s(t)\,\boldsymbol{d}_s + \boldsymbol{u}(t) + \boldsymbol{\nu}(t) = s(t)\,\boldsymbol{d}_s + u(t)\,\boldsymbol{d}_u + \boldsymbol{\nu}(t), \tag{11.5}$$

[1] Thermal white noise power is given by $\sigma_\nu^2 = kTB$ where $k = 1.38 \times 10^{-23}$ J/°K is Boltzmann's constant, T is the equivalent noise temperature in °K, and B is the receiver bandwidth in Hz.

[2] We neglect cable strum, which is a source of interference in sonar and has a different dispersion relation from acoustic waves.

where $\boldsymbol{u}(t)$ has a direction vector $\boldsymbol{d}_u$ associated with the interference arrival angle ψ_u, and $\boldsymbol{\nu}(t) = [\nu_0(t), \ldots, \nu_{M-1}(t)]^T$ is the vector composed of independent and identically distributed Gaussian white noises $\nu_m(t)$ on the receivers.

When the source signal is wide-band, we compute the STFT of each receiver time-series $x_m(t)$, $0 \leq t \leq T$ for L, with possibly overlapped subsections, to obtain (neglecting interference and noise terms)

$$X_m^{(l)}(\omega) = e^{-i\omega\tau_m(\psi_s)} S_m^{(l)}(\omega), \quad 1 \leq l \leq L, \quad 0 \leq m \leq M-1, \tag{11.6}$$

where $S_m^{(l)}(\omega)$ is the Fourier transform of the windowed signal (sub-section l) on receiver m. Defining the vectors

$$\boldsymbol{X}^{(l)}(\omega) = [X_0^{(l)}(\omega), \ldots, X_{M-1}^{(l)}(\omega)]^T, \quad 1 \leq l \leq L, \tag{11.7}$$

leads to the signal model (neglecting interference and noise terms) [132]

$$\boldsymbol{X}^{(l)}(\omega) = \boldsymbol{d}_s(\omega) \otimes \boldsymbol{S}^{(l)}(\omega), \quad \boldsymbol{S}^{(l)}(\omega) = [S_0^{(l)}(\omega), \ldots, S_{M-1}^{(l)}(\omega)]^T, \tag{11.8}$$

which is now amenable to narrow-band techniques applied to each frequency ω independently of each other.

Two topics of interest in array signal processing are spatial discrimination or filtering, also referred to as beamforming, and bearing estimation or determination of direction of arrival (DOA) of a signal; both topics are intimately related. Beamforming is the process of linearly combining all received data from all sensors so as to obtain a narrow beam pointing to a specific direction; this is equivalent to a spatial filter that emphasizes incoming signals from that direction while attenuating those coming in from other directions. DOA estimation, on the other hand, can be thought of as calculating beams for a large number of angles and picking the signal direction based on some optimality criterion such as maximum likelihood; the exception to this picture is DOA estimation based on the model of multiple complex sinusoids in white noise when methods such as MUSIC and ESPRIT are used to estimate the signal spatial frequencies, and hence their direction.

Beamforming is performed by calculating a set of weights (coefficients), and the resulting beam is the digital equivalent of a mechanically steered antenna. Beamformers generally steer the array to the direction of the signal of interest, assuming that the direction is known. Bearing or angle estimation refers to finding the arrival angle of a propagating signal incident on an array of sensors. Spectral estimation techniques to find PSD peaks of zero-mean WSS random time series can be used to find peaks in spatial spectra as a function of spatial frequencies which, using the equation for normalized spatial frequency (11.2), determine the signal direction of arrival ψ_s. Narrow-band techniques in both topics are based on estimation of the data's spatial auto-correlation matrix,

$$\boldsymbol{R}_{xx} = \mathrm{E}[\boldsymbol{x}\boldsymbol{x}^+], \tag{11.9}$$

which, using the signal model (11.5) and defining the interference power $\mathcal{P}_u = \mathrm{E}[|u|^2]$ (signal power was defined earlier $\mathcal{P}_s = \mathrm{E}[|s|^2]$), can be written as

$$\boldsymbol{R}_{xx} = \mathcal{P}_s \boldsymbol{d}_s \boldsymbol{d}_s^+ + \mathcal{P}_u \boldsymbol{d}_u \boldsymbol{d}_u^+ + \sigma_\nu^2 \boldsymbol{I}. \tag{11.10}$$

This is the spatial analogue of the temporal auto-correlation of a sum of complex sinusoids in zero-mean white noise (6.38). In practice, the spatial auto-correlation matrix must be estimated using a number of available snapshots, $\boldsymbol{x}(t_1), \ldots, \boldsymbol{x}(t_N)$. Wide-band techniques are based on estimating the power spectral matrix at each frequency using

$$\boldsymbol{C}(\omega) = \frac{1}{N} \sum_{l=1}^{L} \boldsymbol{X}^{(l)}(\omega) \left[\boldsymbol{X}^{(l)}(\omega)\right]^+, \tag{11.11}$$

where $\boldsymbol{X}^{(l)}(\omega)$ is defined by (11.8). Narrow-band estimation techniques are then applied to each frequency component independently. In wide-band estimation, all narrow-band DOAs are averaged [133].

11.3 Beamforming

Beamforming is the process of calculating a weighted average of the snapshot vector $\boldsymbol{x}(t)$ (for a fixed t) using a weighting vector $\boldsymbol{w}$:

$$y(t) = \boldsymbol{w}^{+}\boldsymbol{x}(t), \quad \boldsymbol{w} = [w_0, \dots, w_{M-1}]^T, \tag{11.12}$$

in order to produce the digital equivalent of a narrow beam in a specific direction to enhance the signal and reduce noise. As the signal model (11.3) shows, the simplest weight vector is the one that cancels the phase shifts included in the direction vector $\boldsymbol{d}_s$; since $\boldsymbol{d}_s^{+}\boldsymbol{d}_s = M$, the cancellation occurs for $\boldsymbol{w} \propto \boldsymbol{d}_s$, and this beamformer is known as the *conventional beamformer* (CBF).

The *beam response* b_ψ of a ULA is the spatial frequency response of the beamformer and is defined for a fixed weight vector $\boldsymbol{w}$ as a function of the normalized spatial frequency (11.2) and the *beam steering vector* $\boldsymbol{\psi}$, which depends on the *steering angle* ψ to which the beam is steered, and

$$b_\psi \equiv \frac{1}{M}\boldsymbol{w}^{+}\boldsymbol{\psi}, \quad \boldsymbol{\psi} \equiv [1, e^{-2i\pi\tilde{f}_\psi}, \dots, e^{-2i\pi(M-1)\tilde{f}_\psi}]^T, \quad \boldsymbol{\psi}^{+}\boldsymbol{\psi} = M. \tag{11.13}$$

Division by M in the equation for b_ψ ensures that the beam response of the spatial matched filter [discussed later in the paragraph leading to equation (11.20)] is unity in the signal direction. When calculating the beam response, the weight vector is fixed while the steering vector $\boldsymbol{\psi}$ is varied over all possible angles ψ. Thus, when a CBF weight vector matches the signal direction vector, i.e., $\boldsymbol{w} \propto \boldsymbol{d}_s$, its output is high when $\psi = \psi_s$, and low when $\psi \neq \psi_s$. The *beam pattern* B_ψ is the squared magnitude of the beam response; equation (11.13) shows that the beam response is the Fourier transform of the complex conjugated weight vector. Parseval's relation (3.59) then gives

$$\int_{-0.5}^{0.5} |b_\psi|^2 \, d\tilde{f}_\psi = \int_{-0.5}^{0.5} B_\psi \, d\tilde{f}_\psi = \frac{1}{M^2}|\boldsymbol{w}|^2. \tag{11.14}$$

An example of a beam pattern is a ULA with 16 elements and uniform weights $w_m = 1$, i.e., the CBF is matched to a signal from 90°),

$$B_\psi \equiv |b_\psi|^2 = \frac{1}{M^2}\left|\frac{\sin(\pi M\tilde{f}_\psi)}{\sin(\pi\tilde{f}_\psi)}\right|^2, \tag{11.15}$$

as illustrated in figure 11.2 for two arrays: on the left is the beam pattern for a half-wavelength array, and on the right is the beam pattern for a double-wavelength array. Equation (11.15) can be used to show that the first side-lobe level is approximately 13 dB below the main-lobe, independent of M. The side-lobe level can be lowered at the expense of increasing the main-lobe width by multiplying each snapshot $\boldsymbol{x}$ with a window function; this is known as *tapered* or *shaded* beamforming. The usual taper is the Dolph-Chebyshev window, which minimizes the width of the main lobe subject to a specified side-lobe level; it can be computed as the inverse DFT of

$$\cos\left\{M\cos^{-1}\left[\beta\cos\left(m\pi/M\right)\right]\right\} / \cosh\left[M\cosh^{-1}(\beta)\right], \quad 0 \le m \le M-1,$$

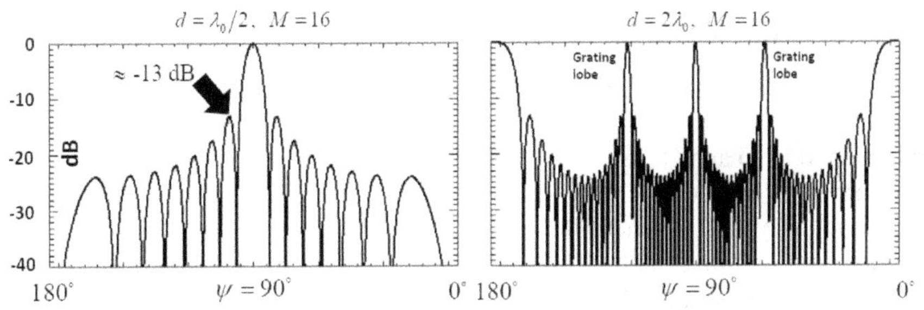

Figure 11.2: Beam pattern for a signal direction of arrival 90° with 16 elements: half-wavelength array (left) and double-wavelength array (right).

where the side-lobe level in dB is given by -20α, and $\beta = \cosh\left[M^{-1}\cosh^{-1}\left(10^{\alpha}\right)\right]$. The double-wavelength array clearly demonstrates the grating lobes that cause ambiguity in direction of arrival when the sampling theorem requirement (1.42) is not met.

The array aperture for a half-wavelength array is given by $L = (M-1)\lambda_{0}/2$; the larger the aperture, the finer the angular resolution of the array. The angular resolution $\Delta\psi$ is defined similarly to frequency resolution as the -3 dB width of the main lobe (see figure 8.3), which, for a half-wavelength array, is given approximately by

$$\Delta\psi \approx \lambda_{0}/L = 2/(M-1). \tag{11.16}$$

Figure 11.3 shows dramatically different angular resolutions for the array response for two half-wavelength arrays with 4 and 64 elements.

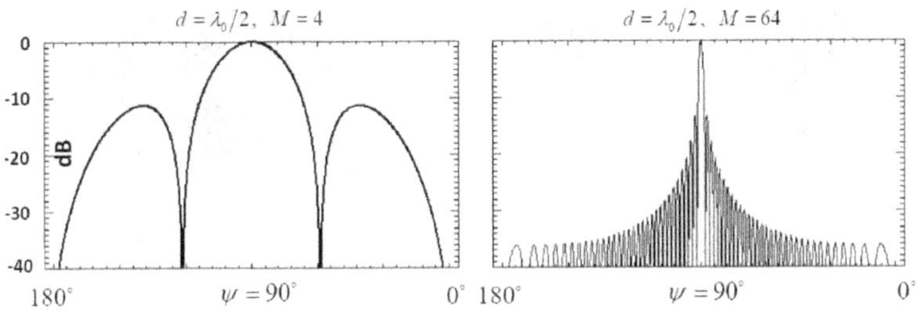

Figure 11.3: Beam patterns for two uniformly weighted ULAs: 4 and 64 elements.

Consider a single narrow-band signal $s(t)$ arriving from direction ψ_{s} at a half-wavelength ULA with M receivers. The model of the received *signal* + *uncorrelated white noise* is given by equation (11.5) (without the interference term). The beamformer output is

$$y(t) = \boldsymbol{w}^{+}\boldsymbol{d}_{s}s(t) + \boldsymbol{w}^{+}\boldsymbol{\nu}(t). \tag{11.17}$$

The signal and noise powers of the beamformer output are

$$\mathrm{E}\left[\left|\boldsymbol{w}^{+}\boldsymbol{d}_{s}s(t)\right|^{2}\right] = \mathcal{P}_{s}\left|\boldsymbol{w}^{+}\boldsymbol{d}_{s}\right|^{2},\ \ \mathrm{E}\left[\left|\boldsymbol{w}^{+}\boldsymbol{\nu}\right|^{2}\right] = |\boldsymbol{w}|^{2}\sigma_{\nu}^{2}, \tag{11.18}$$

giving the beamformer SNR (also known as the array gain [3])

$$\text{SNR} = \left(|\boldsymbol{w}^{+}\boldsymbol{d}_{s}|^{2}/|\boldsymbol{w}|^{2}\right)\left(\mathcal{P}_{s}/\sigma_{\nu}^{2}\right). \tag{11.19}$$

The signal model [(11.1) and (11.5)] clearly shows that in the presence of uncorrelated noise, gain can be achieved by correcting for the time delays (or advances) and summing all the phase corrected element data. Since the time delays (or advances) are represented by the signal direction vector $\boldsymbol{d}_s$, this process of phase correction and summing can be applied using the *spatial matched filter* $\boldsymbol{w}_{\text{smf}} = \boldsymbol{d}_s$, whose square norm is $\|\boldsymbol{w}\|^2 = M$. The associated beamformer is the CBF, with unity beam response in the signal direction and SNR,

$$\text{SNR}_{\text{smf}} = M\,\mathcal{P}_{s}/\sigma_{\nu}^{2}, \tag{11.20}$$

showing a gain (with respect to the SNR at each element) equal to the number of elements M of the array (the gain is usually written in dB scale as $10\log_{10} M$). The spatial matched filter SNR in (11.20) maximizes the SNR in the presence of spatially uncorrelated white noise (sensor thermal noise). If, in addition to spatially uncorrelated white sensor noise, spatially correlated interference at the signal frequency is present, the CBF is no longer optimal; the optimal beamformer in this case will be discussed in section 11.4.

Two plots on the left of figure 11.4 show the real and imaginary parts of a complex sinusoid at 1.7 Hz with a Gaussian envelope and unit amplitude arriving from $\psi = 65°$ at a half-wavelength ULA with 50 elements. The image on the right of figure 11.4 shows the squared magnitude (in dB) of the data for all 50 elements. Data were 100 seconds, sampled at 10 Hz, and included complex zero-mean and unit-variance white noise added to each element. The signal is centered at 50 seconds into the time series at element 0; the element-level SNR of 0 dB agrees with the SNR calculated from the image in 11.4. The CBF beams,

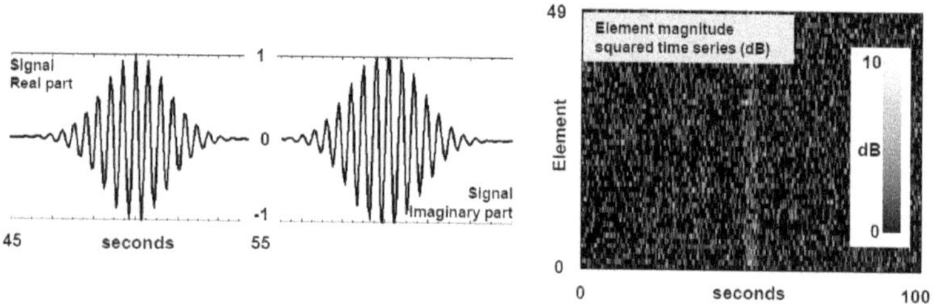

Figure 11.4: Real and imaginary parts of the signal (a), and squared magnitude (dB) of element data (b).

using a weight vector matched to $\psi = 65°$, were calculated for all angles between 0° and 180°, and the squared magnitude of all 100 seconds for all beams are shown in figure 11.5(b); the signal is clearly seen at beam angle 65° and time 50 seconds, and the SNR for the image is ≈ 17 dB; this agrees with the expected SNR gain with respect to the element level SNR of $10\log_{10} 50 = 16.98$ dB. Figure 11.5(a) shows the power in each beam calculated in the signal band 1.7 ± 1.5 Hz for all angles; the peak, ≈ 7 dB above the background, is at the signal direction (65°).

[3]The general definition of array gain for non-isotropic signal and noise with powers $\mathcal{P}_s(\theta,\phi)$ and $\mathcal{N}(\theta,\phi)$, respectively, for an array with beam pattern $B(\theta,\phi)$ is $\left[\int_{4\pi} \mathcal{P}_s(\theta,\phi)B(\theta,\phi)d\Omega / \int_{4\pi} \mathcal{N}(\theta,\phi)B(\theta,\phi)d\Omega\right] / \left[\int_{4\pi} \mathcal{P}_s(\theta,\phi)d\Omega / \int_{4\pi} \mathcal{N}(\theta,\phi)d\Omega\right]$.

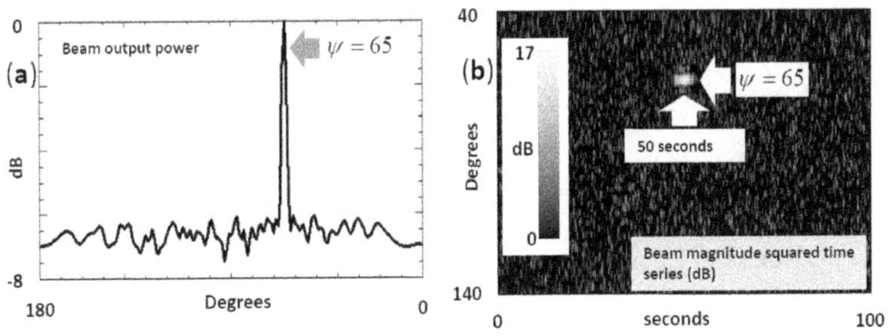

Figure 11.5: (a) In-band power for all beam angles, (b) squared magnitude (dB) of matched filter beam data.

11.4 Optimal Beamforming

Using the *signal* + *interference* + *noise* model of equation (11.5) and defining the *interference* + *noise* correlation matrix $\boldsymbol{R}$ as

$$\boldsymbol{R} \equiv \mathrm{E}[(\boldsymbol{u}+\boldsymbol{\nu})(\boldsymbol{u}+\boldsymbol{\nu})^{+}] = \boldsymbol{R}_{\mathrm{uu}} + \sigma_{\nu}^{2}\boldsymbol{I}, \tag{11.21}$$

the data auto-correlation matrix of equation (11.9) is

$$\boldsymbol{R}_{xx} = \mathcal{P}_{s}\boldsymbol{d}_{s}\boldsymbol{d}_{s}^{+} + \boldsymbol{R}_{\mathrm{uu}} + \sigma_{\nu}^{2}\boldsymbol{I} = \mathcal{P}_{s}\boldsymbol{d}_{s}\boldsymbol{d}_{s}^{+} + \boldsymbol{R}\,. \tag{11.22}$$

Equation (10.44) of theorem 50 provides a useful result for the inverse of the data auto-correlation matrix in terms of the inverse of the *interference* + *noise* correlation matrix:

$$\boldsymbol{R}_{xx}^{-1} = \boldsymbol{R}^{-1} - \frac{\boldsymbol{R}^{-1}\boldsymbol{d}_{s}\boldsymbol{d}_{s}^{+}\boldsymbol{R}^{-1}}{\boldsymbol{d}_{s}^{+}\boldsymbol{R}^{-1}\boldsymbol{d}_{s} + 1/\mathcal{P}_{s}}\,. \tag{11.23}$$

For a known desired signal $s(t)$, the optimal minimum MSE (MMSE) weight vector is found by minimizing $\mathrm{E}[|e(t)|^{2}]$, where

$$e(t) = s(t) - \boldsymbol{w}^{+}\boldsymbol{x}(t). \tag{11.24}$$

The MMSE solution is the *Wiener filter* [4],

$$\boldsymbol{w}_{\mathrm{MMSE}} = \boldsymbol{R}_{xx}^{-1}\cdot\boldsymbol{r}_{xs},\ \ \boldsymbol{r}_{xs} \equiv \mathrm{E}[\boldsymbol{x}s^{*}], \tag{11.25}$$

which, in view of the relation $\mathrm{E}[\boldsymbol{x}s^{*}] = \alpha\,\boldsymbol{d}_{s}$ for some constant α and equation (11.23), is of the form

$$\begin{aligned}\boldsymbol{w}_{\mathrm{MMSE}} &= \alpha\boldsymbol{R}_{xx}^{-1}\boldsymbol{d}_{s} = \boldsymbol{R}^{-1}\boldsymbol{d}_{s}\Big(1 - \frac{\boldsymbol{d}_{s}^{+}\boldsymbol{R}^{-1}\boldsymbol{d}_{s}}{\boldsymbol{d}_{s}^{+}\boldsymbol{R}^{-1}\boldsymbol{d}_{s} + 1/\mathcal{P}_{s}}\Big)\\ &= \alpha\,\boldsymbol{R}^{-1}\,\boldsymbol{d}_{s},\end{aligned} \tag{11.26}$$

where $\boldsymbol{R}$ is the *interference* + *noise* correlation matrix (11.21).

[4]Although the equations are formally the same as the Wiener filter equation of section 9.2, there is a subtle difference, often overlooked, between a weight vector $\boldsymbol{w}$ and a filter vector $\boldsymbol{h}$. Here the MMSE equation for a weight vector has a term of the form $\boldsymbol{w}^{+}\boldsymbol{x}$, with $\boldsymbol{x} = [x_0, \ldots]^T$, whereas the MMSE criterion in the context of a Wiener filter contains the term $\boldsymbol{h}^T\boldsymbol{x}_{\mathrm{tr}}$, with the subscript "tr" indicating time reversal, $\boldsymbol{x}_{\mathrm{tr}} = [\ldots, x_0]^T$; the difference is discussed in section 6.6.

A different optimality criterion, namely maximizing the output signal to *interference* + *noise* ratio (SINR) with respect to the weight vector, will give the same optimal weight vector with a specific normalization. The SINR to be maximized is

$$\mathrm{SINR} = \frac{\mathcal{P}_s\, |\boldsymbol{w}^+\boldsymbol{d}_s|^2}{\mathrm{E}[|\boldsymbol{w}^+(\boldsymbol{u}+\boldsymbol{\nu})|^2]} = \frac{\mathcal{P}_s\, |\boldsymbol{w}^+\boldsymbol{d}_s|^2}{\boldsymbol{w}^+\boldsymbol{R}\boldsymbol{w}}. \tag{11.27}$$

The simplest way to maximize (11.27) is to minimize the denominator and constrain the numerator to be a constant

$$\arg\min_{\boldsymbol{w}} \left(\boldsymbol{w}^+\boldsymbol{R}\boldsymbol{w}\right), \quad \text{subject to } \boldsymbol{w}^+\boldsymbol{d}_s = 1, \tag{11.28}$$

with the solution

$$\boldsymbol{w}_{\mathrm{MVDR}} = \frac{\boldsymbol{R}^{-1}\boldsymbol{d}_s}{\boldsymbol{d}_s^+\boldsymbol{R}^{-1}\boldsymbol{d}_s}. \tag{11.29}$$

This is the *minimum variance distortionless response (MVDR)* (MVDR) beamformer (the spatial frequency analogue of the MVDSE of section 8.9), also known as the Capon beamformer [91] or the adaptive beamformer. Using (11.29) in (11.27) gives the maximum SINR for the MVDR beamformer:

$$\mathrm{SINR}_{\mathrm{MVDR}} = \mathcal{P}_s\, \boldsymbol{d}_s^+\boldsymbol{R}^{-1}\boldsymbol{d}_s. \tag{11.30}$$

Another commonly used optimal weight vector is based on the maximization of SINR but with a different constraint from the MVDR distortionless condition; it is known as the unit noise gain beamformer and is defined by

$$\arg\max_{\boldsymbol{w}} \mathrm{SINR}, \quad \text{subject to } |\boldsymbol{w}|^2 = 1, \tag{11.31}$$

with the solution

$$\boldsymbol{w}_{\mathrm{UNG}} = \frac{\boldsymbol{R}^{-1}\boldsymbol{d}_s}{\sqrt{\boldsymbol{d}_s^+\boldsymbol{R}^{-2}\boldsymbol{d}_s}}. \tag{11.32}$$

Yet another constrained maximization of SINR is based on the requirement that the *interference* + *noise* has unit power; this is known as the *adaptive matched filter* [134] and is used in constant false alarm rate (CFAR) detection

$$\arg\max_{\boldsymbol{w}} \mathrm{SINR}, \quad \text{subject to } \boldsymbol{w}^+\boldsymbol{R}\boldsymbol{w} = 1, \tag{11.33}$$

with the solution

$$\boldsymbol{w}_{\mathrm{AMF}} = \frac{\boldsymbol{R}^{-1}\boldsymbol{d}_s}{\sqrt{\boldsymbol{d}_s^+\boldsymbol{R}^{-1}\boldsymbol{d}_s}}. \tag{11.34}$$

All optimal beamformer weight vectors discussed above are a constant multiple of $\boldsymbol{R}^{-1}\boldsymbol{d}_s$. The actual constant in each case depends on the constraint, which is motivated by the particular application. The maximum SINR in all cases is the same as the result for the MVDR, i.e., equation (11.30). Since the signal direction vector is $\boldsymbol{d}_s$, all the above *steered beamformers* pass the signal from direction ψ_s undistorted (but possibly gained) while attenuating "loud" interferences from all other directions that could produce large beam outputs through the side-lobes.

To see the effect of optimal beamformers on loud interferences, we will look at the eigenstructure of the *interference* + *noise* correlation matrix using the results of section 6.3; the only difference is that we are using spatial frequencies here instead of temporal frequencies. Suppose there are K interferences with $K < M$; these are the "signals" that span the signal subspace of the range space of the $M \times M$ Hermitian and positive definite *interference* + *noise* correlation matrix $\boldsymbol{R}$. The range space is the direct sum of two orthogonal subspaces: the signal subspace spanned by the "signal eigenvectors" and the noise subspace spanned by the "noise eigenvectors." The $M \times M$ correlation matrix and its inverse have the following eigen expansions:

$$\boldsymbol{R} = \sum_{m=1}^{M} \lambda_{m} \boldsymbol{v}_{m} \boldsymbol{v}_{m}^{+}, \quad \boldsymbol{R}^{-1} = \sum_{m=1}^{M} \lambda_{m}^{-1} \boldsymbol{v}_{m} \boldsymbol{v}_{m}^{+}, \tag{11.35}$$

where $\lambda_1, \ldots, \lambda_K$ are the "signal eigenvalues" with associated signal eigenvectors $\boldsymbol{v}_1, \ldots, \boldsymbol{v}_K$, and $\lambda_{K+1} = \ldots = \lambda_M = \sigma_\nu^2$ are the noise eigenvalues with associated noise eigenvectors $\boldsymbol{v}_{K+1}, \ldots, \boldsymbol{v}_M$; the eigen-expansion of the correlation matrix is the starting point for many important approximations to the MVDR and will be discussed in section 11.8. Thus,

$$\boldsymbol{R}^{-1}\boldsymbol{d}_s = \frac{1}{\sigma_\nu^2}\left[\boldsymbol{d}_s - \sum_{m=1}^{M}\left(1 - \frac{\sigma_\nu^2}{\lambda_m}\right)\boldsymbol{v}_m^+ \cdot \boldsymbol{d}_s\, \boldsymbol{v}_m\right]. \tag{11.36}$$

Using (11.13) and (11.29), the MVDR beam response is

$$b_\psi^{\text{MVDR}} = \frac{1}{M}\boldsymbol{w}_{\text{MVDR}}^+ \boldsymbol{\psi} = \frac{\boldsymbol{d}_s^+ \cdot \boldsymbol{\psi} - \sum\limits_{m=1}^{M} \left(1 - \sigma_\nu^2/\lambda_m\right) \boldsymbol{d}_s^+ \cdot \boldsymbol{v}_m \ \boldsymbol{v}_m^+ \cdot \boldsymbol{\psi}}{M\, \sigma_\nu^2 \left(\boldsymbol{d}_s^+ \boldsymbol{R}^{-1} \boldsymbol{d}_s\right)}, \tag{11.37}$$

which shows that in the absence of interference ($K = 0$), i.e., when all eigenvalues of the correlation matrix are equal to the noise variance, the sum in the numerator vanishes, only the first term in the numerator $\boldsymbol{d}_s^+ \cdot \boldsymbol{\psi}$ survives, and the MVDR beam response reduces to that of the spatial matched filter [see the discussion before equation (11.20)].

When K interferences are present ($1 \le K < M$) and there is at least one eigenvalue greater than the noise variance, the first term in the numerator of equation (11.37) is known as the *quiescent response* of the MVDR beamformer. The last part of the second term in the numerator, namely, $\boldsymbol{v}_m^+ \cdot \boldsymbol{\psi}$, is known as an *eigenbeam*. For noise eigenvectors with indices $K+1 \le m \le M$, we have $1 - \sigma_\nu^2/\lambda_m \approx 0$, and the MVDR beam response is the same as the quiescent response. However, for a loud interference from any direction ψ_m, $1 \le m \le K$, $\lambda_m \gg \sigma_\nu^2$ and $1 - \sigma_\nu^2/\lambda_m \approx 1$, and the eigenbeam weighted by $\boldsymbol{v}_m^+ \cdot \boldsymbol{\psi}_s$ (where $\boldsymbol{\psi}_s$ denotes the steering vector in the signal direction ψ_s) is subtracted from the quiescent response, reducing the MVDR beam response significantly (depending on how loud the interference is) in the direction of the corresponding interference. The cancellation of a loud interference in the beam response appears as a deep "null" in the direction of the interference [135]. To study the depth of the null in the beam response in the interference direction ψ_1 (without loss of generality we choose the $k = 1$ interference to be the direction of interest here), we consider the beam response at ψ_1 using the MVDR weight vector (11.29),

$$b_1^{\text{MVDR}} = \frac{1}{M}\,\boldsymbol{w}_{\text{MVD}}^+ \boldsymbol{\psi}_1 = \frac{1}{M}\,\frac{\boldsymbol{d}_s^+ \boldsymbol{R}^{-1} \boldsymbol{\psi}_1}{\boldsymbol{d}_s^+ \boldsymbol{R}^{-1} \boldsymbol{d}_s}, \tag{11.38}$$

where $\boldsymbol{\psi}_1$ is the steering vector in the interference direction ψ_1.

In analogy to the signal model (11.3), we have the *interference + noise* model

$$\boldsymbol{q}(t) = \sum_{k=1}^{K} \boldsymbol{d}_k u_k(t) + \boldsymbol{\nu}(t), \tag{11.39}$$

with $K < M$ interferers arriving from directions ψ_k with powers $\mathcal{P}_k$, $1 \le k \le K$. The $M \times M$ *interference + noise* correlation matrix is then found by generalizing equation (11.10) to K interferences,

$$\boldsymbol{R} = \sum_{k=1}^{K} \mathcal{P}_k \boldsymbol{d}_k \boldsymbol{d}_k^+ + \sigma_\nu^2 \boldsymbol{I}, \tag{11.40}$$

where $\boldsymbol{d}_k$ is the interference direction vector in the direction ψ_k. This can be written as

$$\boldsymbol{R} = \mathcal{P}_1 \boldsymbol{d}_1 \boldsymbol{d}_1^+ + \sum_{k=2}^{K} \mathcal{P}_k \boldsymbol{d}_k \boldsymbol{d}_k^+ + \sigma_\nu^2 \boldsymbol{I} \equiv \mathcal{P}_1 \boldsymbol{d}_1 \boldsymbol{d}_1^+ + \boldsymbol{R}', \tag{11.41}$$

where $\boldsymbol{R}'$ is the *interference + noise* correlation matrix in the absence of the $k = 1$ interference. Using equation (11.23), we have

$$\boldsymbol{R}^{-1} = \boldsymbol{R}'^{-1} - \mathcal{P}_1 \frac{\boldsymbol{R}'^{-1} \boldsymbol{d}_1 \boldsymbol{d}_1^+ \boldsymbol{R}'^{-1}}{1 + \mathcal{P}_1 \boldsymbol{d}_1^+ \boldsymbol{R}'^{-1} \boldsymbol{d}_1}, \tag{11.42}$$

which, when substituted into the beam response equation (11.38) for the MVDR beamformer matched to the interference direction ψ_1, gives

$$b_1^{\text{MVDR}} = \frac{1}{M} \frac{(\boldsymbol{d}_s^+ \boldsymbol{R}'^{-1} \boldsymbol{d}_1 / \boldsymbol{d}_s^+ \boldsymbol{R}^{-1} \boldsymbol{d}_s)}{1 + \mathcal{P}_1 \boldsymbol{\psi}_1^+ \boldsymbol{R}'^{-1} \boldsymbol{\psi}_1}. \tag{11.43}$$

The numerator in the beam response (11.43) is the optimal beam response of the MVDR beamformer steered in the direction ψ_1 and in the absence of the $k = 1$ interferer, while the denominator is the amount by which this response is reduced in proportion to $\mathcal{P}_1^2$ because of the $k = 1$ interference. The stronger the interference, the greater the suppression of the beam pattern in that interference direction. Weak interfering sources are, of course, ignored by the MVDR beamformer.

As an example, consider a 50-element half-wavelength sonar array with $d = 7.5$ m corresponding to $f_0 = 100$ Hz and $c = 1500$ m. We use a model of a plane wave signal with unit power incident at 65°, together with two loud interferences at the same frequency and equal powers of 50^2 arriving from 105° and 122°, and complex zero-mean unit-variance white noise at each receiver. We use equation (6.38) to calculate the *interference + noise* correlation matrix (only the two interferences are used to construct the "signal" matrix $\boldsymbol{S}$) and equation (6.42) to calculate its inverse. The correlation matrix has two large eigenvalues ($\approx 1.2 \times 10^7$) corresponding to the two interferences and 48 small eigenvalues (≈ 1) corresponding to the noise. Figure 11.6 shows the beam patterns for CBF (dashed) and MVDR (solid), with a maximum value of 0 dB at the signal direction 65° for both beamformers. The MVDR has placed nulls nearly 100 dB below the side-lobe level of about 40 dB, at both interference directions of 105° and 122° with no prior knowledge of the interference directions! It is important to note that for an array with M elements, only $M - 1$ loud interferences can be suppressed.

Now consider the effect of optimal beamforming on white noise that exists on each array element. The noise component in the MVDR beamformer output is $\boldsymbol{w}_{\text{MVDR}}^+ \boldsymbol{\nu}$, and its power is

$$\text{E}[|\boldsymbol{w}_{\text{MVDR}}^+ \boldsymbol{\nu}|^2] = \sigma_\nu^2 \, |\boldsymbol{w}_{\text{MVDR}}|^2, \tag{11.44}$$

showing that noise is enhanced by the total energy of the optimal weight vector which, according to Parseval's relation (11.14), is the area under the beam pattern. Thus, high side-lobes would enhance the noise, and the optimal beamformer (MVDR, etc.) attempts to lower the side-lobe levels to maximize the SINR.

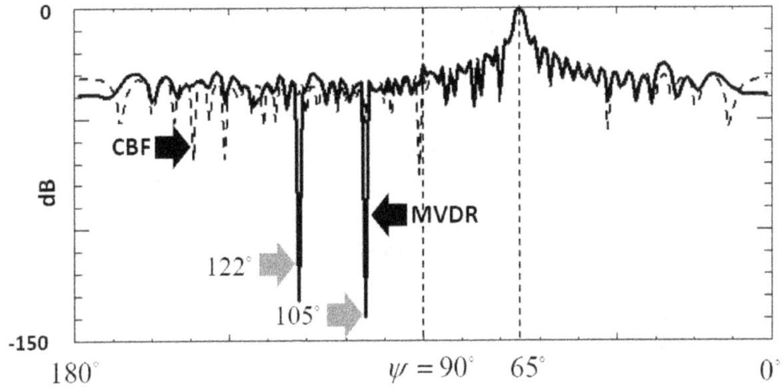

Figure 11.6: Beam patterns (in dB) for CBF and MVDR beamformer.

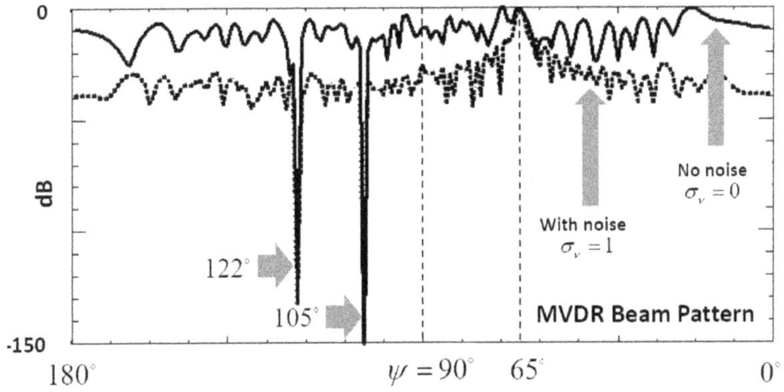

Figure 11.7: Beam patterns (dB) for MVDR beamformer with and without white noise.

Figure 11.7 illustrates the same MVDR beam pattern as in example of figure 11.6, with and without noise; the beam pattern without noise shows the usual suppression of the two interferences but with nearly 30 dB higher side-lobe levels, some as high as the response at the signal direction.

The optimal beamformer weights discussed in this section have been the solutions to an optimization problem with a single constraint. If we have additional constraints, for instance when we know the arrival angle for a specific interference and wish to place a null in that direction, then we obtain the linearly constrained minimum variance (LCMV) beamformer [136], which is the solution to the linearly constrained optimization problem

$$\arg\min_{\boldsymbol{w}} \left(\boldsymbol{w}^{+}\boldsymbol{R}\boldsymbol{w}\right), \quad \text{subject to } \boldsymbol{A}^{+}\boldsymbol{w} = \boldsymbol{b}, \tag{11.45}$$

where $\boldsymbol{A}$ is a $K \times M$ full rank complex matrix and $\boldsymbol{b}$ is a $K \times 1$ vector. The solution to this problem is shown in equation (4.12) of section 4.2. Thus, the LCMV beamformer weight vector is

$$\boldsymbol{w}_{\text{LCMV}} = \boldsymbol{R}^{-1}\boldsymbol{A}\left(\boldsymbol{A}^{+}\boldsymbol{R}^{-1}\boldsymbol{A}\right)^{-1}\boldsymbol{b}, \tag{11.46}$$

with array output power

$$\mathrm{E}\left[\left|\boldsymbol{w}_{\text{LCMV}}^{+}\boldsymbol{x}\right|^{2}\right] = \boldsymbol{w}_{\text{LCMV}}^{+}\boldsymbol{R}\boldsymbol{w}_{\text{LCMV}} = \boldsymbol{b}^{+}\left(\boldsymbol{A}^{+}\boldsymbol{R}^{-1}\boldsymbol{A}\right)^{-1}\boldsymbol{b}. \tag{11.47}$$

As observed in section 4.2, equation (4.14), the projection matrix $\boldsymbol{P}$ projects the solution orthogonally onto the range space of $\boldsymbol{A}$, and the projection $\boldsymbol{A}(\boldsymbol{A}^+\boldsymbol{A})^{-1}\boldsymbol{b}$, being independent of the *interference + noise* correlation matrix, is therefore the non-adaptive component of the LCMV weight vector (it is the only remaining component when the correlation matrix is the identity matrix).

The decomposition of the LCMV into two orthogonal components has applications to the LMS adaptive solution of LCMV beamformers [137] and is the basis for the generalized side-lobe canceler [138]; it is depicted in figure 11.8 for the MVDR whose single linear constraint is $\boldsymbol{d}_s^+\boldsymbol{w} = 1$. The upper branch is the non-adaptive portion of the optimal beamformer and is a matched filter, while the lower branch is the adaptive portion that will solve an unconstrained minimization problem; it consists of a blocking matrix $\boldsymbol{B}$ to prevent the signal from direction ψ_s to pass through. Thus, the blocked data consists of interferences from directions other than the signal's direction, which are, therefore, correlated with the interferences on the upper branch. The correlated interferences can be canceled using a Wiener filter $\boldsymbol{w}$ which is the solution to the unconstrained minimization problem:

$$\arg\min_{\boldsymbol{w}} \mathrm{E}\left[\left|\boldsymbol{d}_s^+\boldsymbol{x} - \boldsymbol{w}^+\boldsymbol{B}^+\boldsymbol{x}\right|^2\right] = (\boldsymbol{d}_s - \boldsymbol{B}\boldsymbol{w})^+\boldsymbol{R}(\boldsymbol{d}_s - \boldsymbol{B}\boldsymbol{w}). \tag{11.48}$$

The Wiener solution to (11.48) is given by $\boldsymbol{R}^{-1}\boldsymbol{r}$, where $\boldsymbol{R} = \mathrm{E}\left[\left|\boldsymbol{B}^+\boldsymbol{x}\right|^2\right]$ and $\boldsymbol{r} = \mathrm{E}\left[\boldsymbol{B}^+\boldsymbol{x}\boldsymbol{x}^+\boldsymbol{d}_s\right]$, namely,

$$\boldsymbol{w} = (\boldsymbol{B}^+\boldsymbol{R}\boldsymbol{B})^{-1}\boldsymbol{B}^+\boldsymbol{R}\boldsymbol{d}_s. \tag{11.49}$$

The output $y(t)$ in figure 11.8 must equal the output of the LCMV beamformer using the LCMV weight

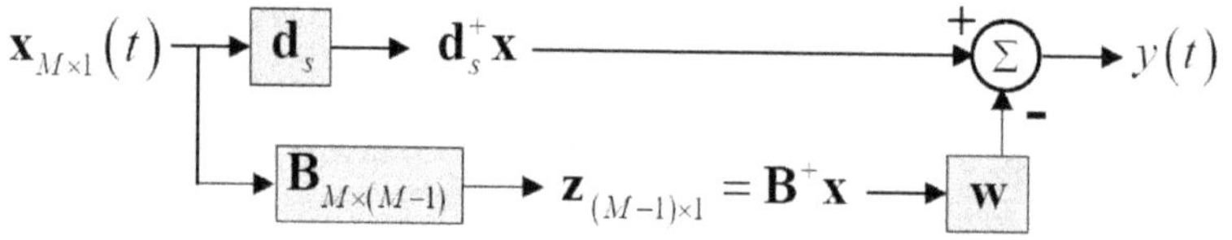

Figure 11.8: Generalized side-lobe canceler representation of MVDR constraint $\boldsymbol{d}_s^+\boldsymbol{w} = 1$, with $M\times(M-1)$ blocking matrix $\boldsymbol{B}$.

vector (11.46). Thus,

$$\left(\boldsymbol{I} - \boldsymbol{B}(\boldsymbol{B}^+\boldsymbol{R}\boldsymbol{B})^{-1}\boldsymbol{B}^+\boldsymbol{R}\right)\boldsymbol{d}_s = \boldsymbol{R}^{-1}\boldsymbol{A}(\boldsymbol{A}^+\boldsymbol{R}^{-1}\boldsymbol{A})^{-1}\boldsymbol{b}, \tag{11.50}$$

which, upon multiplying on the left by $\boldsymbol{B}^+\boldsymbol{R}$, gives

$$\boldsymbol{B}^+\boldsymbol{A}(\boldsymbol{A}^+\boldsymbol{R}^{-1}\boldsymbol{A})^{-1}\boldsymbol{b} = \boldsymbol{0}. \tag{11.51}$$

Equation (11.51) is satisfied by a blocking matrix such that $\boldsymbol{B}^+\boldsymbol{A} = \boldsymbol{0}$. For a MVDR when the linear constraint is $\boldsymbol{d}_s^+\boldsymbol{w} = 1$, the blocking matrix must be such that $\boldsymbol{B}^+\boldsymbol{d}_s = \boldsymbol{0}$; this ensures that the $M-1$ channels on the lower branch of figure 11.8 contain no signal from direction ψ_s. An example $M\times(M-1)$ blocking matrix $\boldsymbol{B}$ in this case is one whose columns are

$$\boldsymbol{b}_m \equiv [1, e^{-2i\pi\tilde{f}_m}, \ldots, e^{-2i\pi(M-1)\tilde{f}_m}]^T, \quad m = 1, \ldots, M-1, \quad \tilde{f}_m \equiv \tilde{f}_s + m/M,$$

where $\tilde{f}_s = d\cos\psi_s/\lambda_0$ [see equation (11.2)]. The resulting blocking matrix satisfies $\boldsymbol{B}^+\boldsymbol{d}_s = \boldsymbol{0}$, since $\boldsymbol{B} = [\boldsymbol{b}_1, \ldots, \boldsymbol{b}_m]$ and

$$\boldsymbol{b}_m^+\boldsymbol{d}_s = \sum_{k=0}^{M-1} e^{2ikm\pi/M} = \frac{1 - e^{2im\pi}}{1 - e^{2im\pi/M}} = 0, \quad m = 1\ldots, M-1.$$

Although the general side-lobe canceler is a powerful alternative formulation of the optimal beamformer, it has a number of differences with the optimal beamformers discussed in this section that render it of more theoretical than practical value. Optimal beamformers use a single *interference + noise* correlation matrix to substantially attenuate loud interferences from directions other than that of the signal's direction. The general side-lobe canceler, on the other hand, uses correlation cancellation to reduce interferences from other than signal directions, and in doing so it must use a new correlation matrix and blocking matrix for every interference direction.

11.5 Performance of the Optimal Beamformer

A measure of performance of the optimal beamformer is the SINR loss $\mathcal{L}_{\text{SINR}}$, defined as the ratio between the optimal SINR, equation (11.30), and the spatial matched filter SNR, equation (11.20); a positive number less than 1,

$$\mathcal{L}_{\text{int}} = \frac{\text{SINR}_{\text{MVDR}}}{\text{SNR}_{\text{smf}}} = \frac{\mathcal{P}_s\, \boldsymbol{d}_s^+ \boldsymbol{R}^{-1} \boldsymbol{d}_s}{M\, \mathcal{P}_s / \sigma_\nu^2} = \frac{\sigma_\nu^2}{M} \boldsymbol{d}_s^+ \boldsymbol{R}^{-1} \boldsymbol{d}_s . \tag{11.52}$$

The optimal beamformer performance is adversely affected by two issues: the breakdown of the narrow-band assumption and mismatch between the actual signal and the signal model. If the narrow-band assumption is valid, then the correlation matrix of a single interference in direction ψ_k, namely,

$$\boldsymbol{R}_k = \mathcal{P}_k \boldsymbol{d}_k \boldsymbol{d}_k^+ ,$$

will have as its (l, n) element a complex exponential $\propto \exp\left[2i\pi f_0 (l-n) d \cos\psi_k / c\right]$. If the narrow-band assumption is violated, then the (l, n) element of the above correlation matrix is the integral of the complex exponential over the non-zero bandwidth, producing a "dispersion" where the complex exponential (as a function of frequencies in the bandwidth) is now multiplied by the signal PSD and then integrated over the bandwidth [139]; this dispersion represents de-correlation of the signal across the array for any non-zero arrival angle. A technique to correct for dispersion is time-delay steering, which effectively focuses the dispersed signal from a given look direction, but that can increase dispersion in other directions.

To study the effect of signal mismatch on the performance of the optimal beamformer, consider a steering vector $\boldsymbol{\psi}$ and signal direction vector $\boldsymbol{d}_s \neq \boldsymbol{\psi}$. The MVDR beam response steered to a direction different from the actual signal direction is known as the side-lobe level (SLL) of the optimal beamformer and is given by the MVDR beam pattern

$$\text{SLL}_{\text{MVDR}}(\boldsymbol{\psi}) = \left| \frac{\boldsymbol{\psi}^+ \boldsymbol{R}^{-1} \boldsymbol{d}_s}{\boldsymbol{\psi}^+ \boldsymbol{R}^{-1} \boldsymbol{\psi}} \right|^2 . \tag{11.53}$$

The SINR for the mismatched MVDR beamformer is found by using the MVDR weight vector of equation (11.29) with the assumed mismatched direction vector [5], i.e., $\boldsymbol{w}_{\text{MVDR}} = \alpha \boldsymbol{R}^{-1} \boldsymbol{\psi}$ and the signal $\boldsymbol{s} = \sqrt{\mathcal{P}_s}\, \boldsymbol{d}_s$,

$$\begin{aligned} \text{SINR} &= \frac{\mathrm{E}\left[|\boldsymbol{w}_{\text{MVDR}}^+ \boldsymbol{s}|^2\right]}{\left(\boldsymbol{w}_{\text{MVDR}}^+ \boldsymbol{R} \boldsymbol{w}_{\text{MVDR}}\right)} = \mathcal{P}_s \frac{|\boldsymbol{\psi}^+ \boldsymbol{R}^{-1} \boldsymbol{d}_s|^2}{\boldsymbol{\psi}^+ \boldsymbol{R}^{-1} \boldsymbol{\psi}} \\ &= \mathcal{P}_s \left(\boldsymbol{d}_s^+ \boldsymbol{R}^{-1} \boldsymbol{d}_s\right) \frac{|\boldsymbol{\psi}^+ \boldsymbol{R}^{-1} \boldsymbol{d}_s|^2}{\left(\boldsymbol{\psi}^+ \boldsymbol{R}^{-1} \boldsymbol{\psi}\right)\left(\boldsymbol{d}_s^+ \boldsymbol{R}^{-1} \boldsymbol{d}_s\right)} \equiv \mathcal{P}_s \left(\boldsymbol{d}_s^+ \boldsymbol{R}^{-1} \boldsymbol{d}_s\right) \times \mathcal{L}_{\text{sigmis}} , \end{aligned} \tag{11.54}$$

[5]This analysis gets more complicated when the correlation matrix includes the signal [131].

which defines the loss due to signal mismatch $\mathcal{L}_{\text{sigmis}}$. It is natural to use equation (1.8) and the associated norm together with the Cauchy-Schwartz inequality (1.9) to define [131]

$$\cos^2(\boldsymbol{\psi}, \boldsymbol{d}; \boldsymbol{W}) \equiv \frac{|\boldsymbol{\psi}^+ \boldsymbol{W} \boldsymbol{d}|^2}{(\boldsymbol{d}^+ \boldsymbol{W} \boldsymbol{d})(\boldsymbol{\psi}^+ \boldsymbol{W} \boldsymbol{\psi})}, \tag{11.55}$$

in terms of which we have

$$0 \le \mathcal{L}_{\text{sigmis}} = \cos^2(\boldsymbol{\psi}, \boldsymbol{d}_s; \boldsymbol{R}^{-1}) \le 1. \tag{11.56}$$

Thus, the optimal mismatched MVDR SINR is reduced from its maximum value (11.30) by the signal mismatch loss factor. We can further use the matched filter SNR (11.20) and the SINR loss due to the presence of interference (11.52) to write

$$\text{SINR} = \text{SNR}_{\text{mf}} \times \mathcal{L}_{\text{int}} \times \mathcal{L}_{\text{sigmis}}, \tag{11.57}$$

i.e., the SINR of the mismatched optimal beamformer is equal to the matched filter SNR (maximum attainable SNR in the absence of interference) multiplied by the product of two losses: loss due to the presence of the interference, and loss due to signal mismatch in the MVDR beamformer.

11.6 Optimal Beamforming in Practice

The direct method to calculate the optimal beamformer weight vector is to estimate the *interference + noise* correlation matrix and then to invert it; this is known as sample matrix inversion (SMI) beamformer. The MLE of the *interference + noise* correlation matrix, equation (11.9), is given by

$$\hat{\boldsymbol{R}} = N^{-1} \sum_{n=1}^{N} \boldsymbol{x}_n \boldsymbol{x}_n^+, \tag{11.58}$$

where $\boldsymbol{x}_n$ is the snapshot vector at time step t_n, $1 \le n \le N$. The SINR of the SMI beamformer is found by using the SMI weight vector

$$\boldsymbol{w}_{\text{SMI}} = \frac{\hat{\boldsymbol{R}}^{-1} \boldsymbol{d}_s}{\boldsymbol{d}_s^+ \hat{\boldsymbol{R}}^{-1} \boldsymbol{d}_s} \tag{11.59}$$

in the SINR equation (11.27) and taking its ratio to the "true" (i.e., when using the exact correlation matrix $\boldsymbol{R}$) MVDR SINR (11.30) to obtain the normalized SMI SINR:

$$\rho_{\text{SMI}} = \frac{|\boldsymbol{d}_s^+ \hat{\boldsymbol{R}}^{-1} \boldsymbol{d}_s|^2}{\boldsymbol{d}_s^+ \hat{\boldsymbol{R}}^{-1} \boldsymbol{R} \hat{\boldsymbol{R}}^{-1} \boldsymbol{d}_s} \times \frac{1}{\boldsymbol{d}_s^+ \boldsymbol{R}^{-1} \boldsymbol{d}_s}. \tag{11.60}$$

The SMI beamformer suffers from signal contamination of the *interference + noise* correlation matrix; if the signal has relatively large amplitude, the beamformer will try to maximize the SINR and at the same time to suppress the signal! Radar systems overcome the signal contamination problem by using a guard range to eliminate the desired signal, while in communication systems, the "desired user" must stay silent during correlation estimation. Passive sonar systems have no way to exclude the signal from estimation and consequently have sub-optimal performance. The loss in SINR due to estimation of the correlation matrix is represented by the normalized SMI SINR of equation (11.60); this loss can be shown [140] to have a beta distribution with parameters $\alpha = N + 2 - M$ and $\beta = M - 1$, with mean $\alpha/(\alpha + \beta)$,

$$f_{\rho}(\rho) = \frac{N!}{(M-2)!\,(N+1-M)!}\,(1-\rho)^{M-2}\rho^{N+1-M}. \tag{11.61}$$

Thus, the mean of the normalized SINR is

$$\mathrm{E}[\rho_{\text{SMI}}] = \frac{N+2-M}{N+1}. \tag{11.62}$$

Solving equation (11.62) for the mean value of 0.5, i.e., when the beamformer output is within 3 dB of the optimal value, gives $N = 2M - 3$; this is often written as $N \approx 2M$ for good beamformer performance. Solving (11.62) for $N = M$ gives $\mathrm{E}(\rho) = 2/(M + 1)$, which decreases rapidly for increasing M. Thus, $N = M$ (i.e., *low sample support*) is not considered to produce adequate performance in general. Other beamformers, however, such as the multi-stage Wiener filter [141], have been shown to provide significantly improved performance for low sample support values.

In addition to SINR loss, the SMI beamformer suffers from losses due to its side-lobe levels, SLL_{SMI}. Using the SMI weight vector in equation (11.53) we find

$$\text{SLL}_{\text{SMI}} = \left| \frac{\boldsymbol{\psi}^{+}\hat{\boldsymbol{R}}^{-1}\boldsymbol{d}_{s}}{\boldsymbol{\psi}^{+}\hat{\boldsymbol{R}}^{-1}\boldsymbol{\psi}} \right|^{2}, \tag{11.63}$$

and it can be shown [142, 143] that ρ_{SMI} in equation (11.60) (SINR loss for a signal with direction vector $\boldsymbol{d}_s$ arriving into a beam with steering vector $\boldsymbol{\psi}$) has an expected value given by

$$\mathrm{E}\left[\rho_{\text{SMI}}\right] = \frac{1 + (N+1-M)\cos^{2}\left(\boldsymbol{\psi}, \boldsymbol{d}_{s}; \boldsymbol{R}^{-1}\right)}{N+1}. \tag{11.64}$$

The expected loss reduces to equation (11.62) when the signal direction vector matches the steered direction, i.e., $\boldsymbol{d}_s = \boldsymbol{\psi}$. If the signal arrives at an angle orthogonal to the steering direction, i.e., when $\boldsymbol{\psi}^{+}\hat{\boldsymbol{R}}^{-1}\boldsymbol{d}_s = 0$, then the expected loss is equal to $1/(N+1)$ (or $-10\ \log_{10}(N+1)$ in dB), which serves as a theoretical lower bound on SLL for an SMI beamformer. For instance, 1000 snapshots should be adequate to have an SMI SLL of -30 dB. As an example, consider a 64-element half-wavelength array with a signal arriving at 65°, an interference with power 40 dB arriving from 55°, and $\sigma_{\nu}^{2} = 1$. Figure 11.9 shows the beam patterns for SMI MVDR beamformers matched to the signal direction vector using 100 and 1000 snapshots with SLL levels of -20 and -30 dB, respectively, and both exhibit deep nulls at 55°.

Using 100 snapshots to estimate the *interference + noise* correlation matrix, we find 1 large eigenvalue corresponding to the interference and 63 small eigenvalues corresponding to white noise. The mean and standard deviation of the 63 "noise" eigenvalues are 0.98 and 0.78, respectively; this large variation in the noise eigenvalues adversely affects the beam response, as is theoretically evident from the eigenbeam expansion (11.37). In order to reduce the variation of the noise eigenvalues, the diagonal elements of the estimated correlation matrix are "loaded" by a multiple of the white noise variance [144], where the loading factor is usually chosen between 1 and 10, i.e.,

$$\hat{\boldsymbol{R}} \;\rightarrow\; \hat{\boldsymbol{R}} + \kappa\sigma_{\nu}^{2}\boldsymbol{I}, \quad 1 \leq \kappa \leq 10. \tag{11.65}$$

Figure 11.10 relates to the same example of figure 11.9 for the case of 100-snapshot estimation; the dashed line is the beam pattern without diagonal loading, and the solid line is the beam pattern for diagonal loading equal to $10\sigma_{\nu}^{2} = 10$, showing a reduction in SLL by nearly 15 dB.

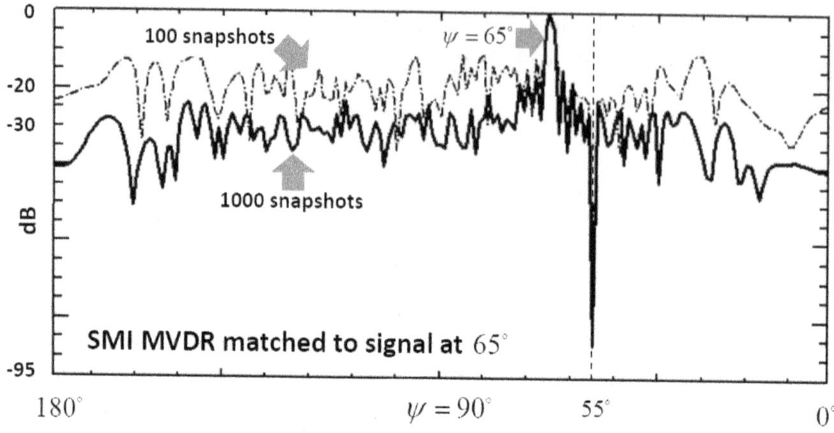

Figure 11.9: SMI MVDR (matched to signal at 25°) beam patterns with correlation matrix estimated from 100 (dashed) and 1000 snapshot (solid).

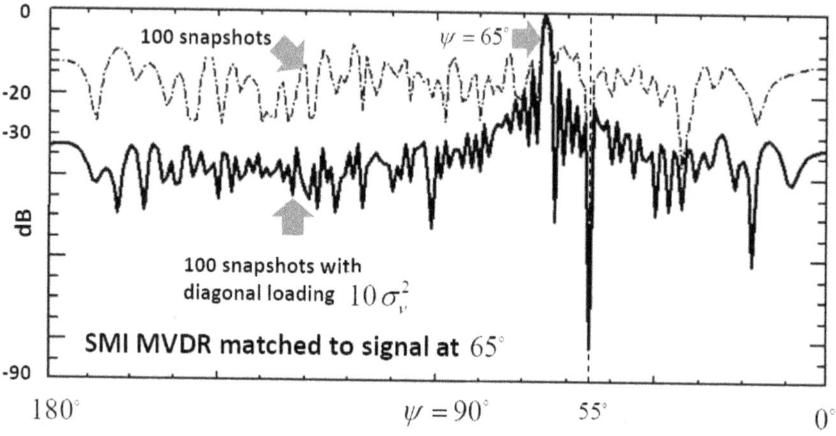

Figure 11.10: SMI MVDR beam patterns with correlation matrix estimated from 100 snapshots: no diagonal loading (dashed) and with diagonal loading $10\sigma_\nu^2 = 10$ (solid).

Estimation of the correlation matrix for the SMI beamformer and the diagonal loading can be combined into a single matrix multiplication by using the $M \times N$ array-snapshot data matrix,

$$\boldsymbol{X} \equiv [\boldsymbol{x}_1, \ldots, \boldsymbol{x}_N], \quad \boldsymbol{x}_n = [x_1[t_n], \ldots, x_M[t_n]]^T, \tag{11.66}$$

to construct an augmented $M \times (N + M)$ matrix $\boldsymbol{X}'$ to estimate the diagonally loaded correlation matrix

$$\boldsymbol{X}' \equiv [\boldsymbol{X} \mid \sqrt{N\kappa}\,\sigma_\nu \boldsymbol{I}] \Rightarrow \hat{\boldsymbol{R}} = \frac{1}{N}\boldsymbol{X}'\boldsymbol{X}'^{+}. \tag{11.67}$$

To solve for the optimal weight vector $\boldsymbol{w}_{\text{SMI}} = \hat{\boldsymbol{R}}^{-1}\boldsymbol{d}_s / \boldsymbol{d}_s^{+}\hat{\boldsymbol{R}}^{-1}\boldsymbol{d}_s$ we use the Cholesky factorization of the estimated correlation matrix (see section 2.6), namely, $\hat{\boldsymbol{R}} = \boldsymbol{L}\boldsymbol{L}^{+}$ and solve two triangular linear equations

[130]: first we solve $\boldsymbol{L}\boldsymbol{p} = \boldsymbol{d}_s$, and then we solve $\boldsymbol{L}^{+}\boldsymbol{q} = \boldsymbol{p}$. Using the identity $\boldsymbol{d}_s^{+}\hat{\boldsymbol{R}}^{-1}\boldsymbol{d}_s = \boldsymbol{p}^{+}\boldsymbol{p} = |\boldsymbol{p}|^2$ the SMI MVDR weight vector is $\boldsymbol{w}_{\text{SMI}} = \boldsymbol{q}/|\boldsymbol{p}|^2$.

Figure 11.11 shows a comparison between an unshaded CBF and an SMI MVDR beamformer (weights updated every 5 seconds at 250 Hz sampling frequency) for a dynamic model consisting of a stationary target in the presence of strong non-stationary interferers, three of which emit broadside noise continuously and one that emits a high amplitude short duration impulse at 60 seconds. The target continuously emits a narrow-band tone at 100 Hz and remains at the same azimuth relative to the ULA comprising 32 elements 7.5 m apart (half-wavelength array at 100 Hz design frequency and sound speed of 1500 m/s); this is typical target behavior in many detection applications. One of the broadside interferers is at short range moving parallel to the array, traversing the beams rapidly (this is referred to as a contact with high beam rate), and is poorly resolved in azimuth when arriving at or near endfire CBF beams. The MVDR beamformer skews the adjacent beams so as to minimize the output power, thus increasing resolution of those beams. As expected, the signal is preserved without distortion on the beam that corresponds to the signal direction of arrival. The MVDR has similarly resolved two contacts near broadside that are unresolved in the CBF beam time record. Thus, the SMI MVDR beamformer has uncovered the stationary target in the presence of a meandering interferer with random azimuthal motion. The impulse interference has leaked through the sidelobes of the CBF beams and is undetectable, whereas the adaptive beamformer has eliminated the sidelobe contamination by placing a deep null in the impulsive contact direction.

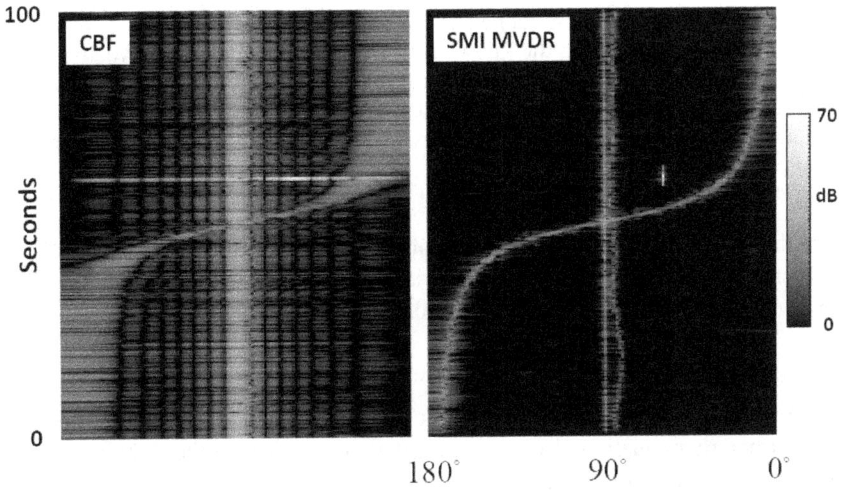

Figure 11.11: SMI MVDR beamformer vs CBF.

11.7 Recursive Methods in SMI Beamforming

Recursive solutions to the SMI MVDR weight vector, also known as dynamical updating adaptive algorithms, include the RLS method discussed in chapter 10. The main equation to be solved is, of course, $\boldsymbol{R}'_N \boldsymbol{w}^{(N)} = N\boldsymbol{d}_s$, where $\boldsymbol{R}'_N = N\hat{\boldsymbol{R}}$ satisfies a rank-1 update equation [see (10.41)]; the SMI MVDR weight vector is then a normalized version of the solution. The rank-1 update equation for the (scaled) correlation matrix is found by writing (a forgetting factor can be included as in section 10.7)

$$\boldsymbol{R'}_N \boldsymbol{w}^{(N)} = (\boldsymbol{R'}_{N-1} + \boldsymbol{x}_N \boldsymbol{x}_N^+)\boldsymbol{w}^{(N)} = N\boldsymbol{d}_s. \tag{11.68}$$

Thus, the RLS method of section 10.8 using QR factorization of the Cholesky square root factor of $\boldsymbol{R'}_N$, described by equations (10.70)–(10.74), is directly applicable to finding the SMI MVDR weight vector, which is a normalized version of the solution to the RLS recursion. Starting at step N, we construct $\boldsymbol{L'}$ using equation (10.72), which includes the previous step Cholesky factor $\boldsymbol{L}_{N-1}$ and the present data vector $\boldsymbol{x}_N$, find the QR factors of its Hermitian conjugate using (10.73), and extract the upper triangular $\boldsymbol{U}$ sub-matrix of $\boldsymbol{V}$; the updated lower triangular matrix $\boldsymbol{L}_N$ is then given by $\boldsymbol{U}^+$.

Once the updated matrix $\boldsymbol{L}_N$ is calculated, the dynamically updated array output power, equation (11.47), at step N becomes

$$\boldsymbol{b}^+\left[(\boldsymbol{L}_N^{-1}\boldsymbol{A})^+(\boldsymbol{L}_N^{-1}\boldsymbol{A})\right]^{-1}\boldsymbol{b} = \boldsymbol{b}^+(\boldsymbol{C}^+\boldsymbol{C})^{-1}\boldsymbol{b}, \tag{11.69}$$

where the matrix $\boldsymbol{C}$ is the solution to the equation $\boldsymbol{L}_N\boldsymbol{C} = \boldsymbol{A}$. The latter is easily solved using the lower triangular property of $\boldsymbol{L}_N$, and so the array output power can be calculated without the inversion of the Cholesky factor $\boldsymbol{L}_N$.

The LMS method, however, is not directly applicable, since the weight vector update equation [6]

$$\boldsymbol{w}^{(n+1)} = \boldsymbol{w}^{(n)} + 2\mu\, e_n \boldsymbol{x}_n^*$$

includes the "desired signal" (the error sequence e_n cannot be calculated without it) which is unknown in the context of optimal array processing; the known quantity here is the direction vector for the signal and the algorithm must be altered so as to use the known information. The linearly constrained minimization problem is given in (11.45), and the idea is to find a steepest descent method to minimize a single objective function by including a Lagrange multiplier and arriving at the following weight vector update equation [137]:

$$\boldsymbol{w}^{(n+1)} = \boldsymbol{w}^{(n)} - \mu\left[\boldsymbol{I} - \boldsymbol{A}(\boldsymbol{A}^+\boldsymbol{A})^{-1}\boldsymbol{A}^+\right]\boldsymbol{R}\boldsymbol{w}^{(n)} + \boldsymbol{A}(\boldsymbol{A}^+\boldsymbol{A})^{-1}\left(\boldsymbol{b} - \boldsymbol{A}^+\boldsymbol{w}^{(n)}\right). \tag{11.70}$$

The last term on the right hand side is the equation of the constraint; it is zero only for the actual known values, and its inclusion in the update equation provides numerical stability to the algorithm. Using the projection operator defined in (4.14), we write the update equation in the form

$$\boldsymbol{w}^{(n+1)} = \boldsymbol{A}(\boldsymbol{A}^+\boldsymbol{A})^{-1}\boldsymbol{b} + \boldsymbol{Q}(\boldsymbol{I} - \mu\boldsymbol{R})\boldsymbol{w}^{(n)}, \quad \boldsymbol{Q} \equiv \boldsymbol{I} - \boldsymbol{P}, \tag{11.71}$$

where the projection matrix $\boldsymbol{Q}$ projects the solution orthogonally onto the null space of $\boldsymbol{A}$ [as observed for the LCMV beamformer after equation (11.47)]. Thus, the second term on the right hand side of (11.71) is the only part of the update equation that depends on the data and can be adapted, while the first term does not depend on the data and is unaffected by the update equation. Replacing the matrix $\boldsymbol{R}$ by its estimate, namely, $\boldsymbol{x}_n\boldsymbol{x}_n^+$, and using the relation $\boldsymbol{x}^+\boldsymbol{w} = \boldsymbol{w}^T\boldsymbol{x}^* = (\boldsymbol{w}^+\boldsymbol{x})^*$, we arrive at the LMS adaptive beamformer weight vector update equation [137]:

$$\boldsymbol{w}^{(n+1)} = \boldsymbol{A}(\boldsymbol{A}^+\boldsymbol{A})^{-1}\boldsymbol{b} + \boldsymbol{Q}\left[\boldsymbol{w}^{(n)} - \mu\left(\boldsymbol{w}^{(n)+}\boldsymbol{x}_n\right)^*\boldsymbol{x}_n\right], \quad \boldsymbol{Q} \equiv \boldsymbol{I} - \boldsymbol{P}. \tag{11.72}$$

Taking expectation values of (11.71), we note that $\boldsymbol{w}^{(n)} \to \boldsymbol{A}(\boldsymbol{A}^+\boldsymbol{A})^{-1}\boldsymbol{b}$ as $n \to \infty$ if the eigenvalues of the Hermitian matrix $\boldsymbol{Q}(\boldsymbol{I} - \mu\boldsymbol{R})$ have magnitudes less than 1; the eigenvalue equation is

$$\left[\boldsymbol{Q}(\boldsymbol{I} - \mu\boldsymbol{R})\right]\boldsymbol{q} = \kappa\boldsymbol{q}, \tag{11.73}$$

[6] The LMS "filtering" equation is of the form $\boldsymbol{h}^T\boldsymbol{x}_n$ whereas in the context of array processing we have defined the "weighting" equation as $\boldsymbol{w}^+\boldsymbol{x}_n$. So the "filter update" equation must be complex conjugated to get the "weight vector update" equation.

which, on setting $\boldsymbol{q} = \boldsymbol{Qp}$ and using the projection property $\boldsymbol{Q}^2 = \boldsymbol{Q}$, can be written as

$$-\mu\, \boldsymbol{QRQp} = (\kappa - 1)\, \boldsymbol{Qp}. \tag{11.74}$$

Now consider the eigenvalue equation for the matrix $\boldsymbol{QRQ}$, namely,

$$\boldsymbol{QRQ}\, \boldsymbol{q}' = \lambda\, \boldsymbol{q}'. \tag{11.75}$$

Multiplying on the left by $\boldsymbol{Q}$ and using the projection property, we find

$$-\mu\, \boldsymbol{QRQ}\, \boldsymbol{q}' = -\mu\lambda\, \boldsymbol{Q}\, \boldsymbol{q}', \tag{11.76}$$

which is equivalent to the eigenvalue equation 11.74. Thus, $\kappa - 1 = -\mu\lambda$ and so $\kappa = 1 - \mu\lambda$, and the LMS convergence condition becomes

$$\left|1 - \mu\lambda\right| < 1, \quad \text{or equivalently} \;\; 0 < \mu < 1/\lambda_{\mathbf{max}}, \tag{11.77}$$

where $\lambda_{\mathbf{max}}$ is the maximum eigenvalue of the matrix $\boldsymbol{QRQ}$. The latter has exactly K zero eigenvalues; to see this we use the eigenvalue-eigenvector expansion for $\boldsymbol{R}$ to write

$$\boldsymbol{QRQ} = \sum_{m=1}^{M} \lambda_m \boldsymbol{Qv}_m \left(\boldsymbol{Q}^+ \boldsymbol{v}_m\right)^+. \tag{11.78}$$

The M vectors $\boldsymbol{Qv}_m$ are in the null space of $K \times M$ matrix $\boldsymbol{A}^+$ which has dimension $M - K$ (the range space of $\boldsymbol{A}^+$ has dimension K); thus, only $M - K$ of them are independent, and so there are exactly K zero eigenvalues of the $M \times M$ matrix $\boldsymbol{QRQ}$.

11.8 PCA and Dominant Mode Rejection (DMR) Beamforming

The discussion following equation (11.62) indicated that the required number of snapshots for the MVDR adaptive beamformer is twice the number of array elements for the mean output SINR to be within 3 dB of the optimal value; this requirement is difficult to meet for non-stationary environments and large arrays when the changes happen on a scale shorter than the required number of snapshots. In such circumstances, the beamformer *degrees of freedom* must be reduced so as to achieve faster convergence; i.e., optimization is performed in a dimensionally reduced space [145] by rank reduction of the correlation matrix.

Using the eigen-expansion (11.35) of the correlation matrix, we can form an eigenspace beam by choosing a subset of the eigenvectors to represent the reduced rank correlation matrix; this is done by simply setting to zero the eigenvalues corresponding to eigenvectors we wish to discard. Since the eigenvectors are orthogonal, we are essentially projecting the correlation matrix onto a subspace formed by the remaining eigenvectors. For instance, the PCA beamformer assumes that there is one signal and $K - 1$ interferences whose eigenvalues are much larger than the eigenvalues of white noise σ_ν^2. The PCA reduced-rank correlation matrix and its inverse are based on discarding the $M - K$ noise eigenvectors, i.e.,

$$\tilde{\boldsymbol{R}}_{\text{PCA}} = \sum_{k=1}^{K} \lambda_k \boldsymbol{v}_k \boldsymbol{v}_k^+, \quad \tilde{\boldsymbol{R}}_{\text{PCA}}^{-1} = \sum_{k=1}^{K} \frac{1}{\lambda_k} \boldsymbol{v}_k \boldsymbol{v}_k^+. \tag{11.79}$$

The PCA weight vector is the MVDR weight vector of equation (11.29) but using the reduced-rank correlation matrix inverse (11.79); in practice, this is equivalent to using the full eigen-expansion of the correlation matrix but assuming that the signal direction vector $\boldsymbol{d}_s$ is orthogonal to the noise eigenspace, i.e., $\boldsymbol{v}_k^+ \cdot \boldsymbol{d}_s = 0,\ K+1 \le k \le M$,

$$\boldsymbol{w}_{\text{PCA}} = \frac{\sum\limits_{k=1}^{K} \lambda_k^{-1} (\boldsymbol{v}_k^+ \cdot \boldsymbol{d}_s)\, \boldsymbol{v}_k}{\sum\limits_{k=1}^{K} |\lambda_k^{-1} \boldsymbol{d}_s^+ \cdot \boldsymbol{v}_k|^2}. \tag{11.80}$$

Instead of projecting into a *signal* + *interference* subspace (the PCA beamformer), the dominant mode rejection (DMR) beamformer, also known as the enhanced minimum variance distortionless response (EMVDR) beamformer, uses the subspace of the dominant interferences it should reject so that it can detect signals that are less "loud"; in the eigen-expansion (11.35), we keep the K largest eigenvalues and replace the "noise eigenvalues" by their average, namely,

$$\sigma_\nu^2 \equiv \frac{1}{M-K} \sum_{k=K+1}^{M} \lambda_k = \frac{1}{M-K}\left[\textbf{Trace}(\boldsymbol{R}) - \sum_{k=1}^{K} \lambda_k\right]. \tag{11.81}$$

Using the completeness relation for the range space of the correlation matrix,

$$\sum_{m=1}^{M} \lambda_m \boldsymbol{v}_m \boldsymbol{v}_m^+ = \sum_{k=1}^{K} \lambda_k \boldsymbol{v}_k \boldsymbol{v}_k^+ + \sum_{k=K+1}^{M} \lambda_k \boldsymbol{v}_k \boldsymbol{v}_k^+ = \boldsymbol{I}, \tag{11.82}$$

we arrive at the DMR approximation to the correlation matrix [146, 147]:

$$\tilde{\boldsymbol{R}} = \sigma_\nu^2 \left[\boldsymbol{I} + \sum_{k=1}^{K} (\lambda_k/\sigma_\nu^2 - 1) \boldsymbol{v}_k \boldsymbol{v}_k^+\right]. \tag{11.83}$$

Although this approximation reduces the degrees of freedom from M eigenvalues to $K < M$, thus reducing the number of required snapshots to $\approx 2K$, it is a full-rank approximation to the correlation matrix. The inverse of the DMR approximation to the correlation matrix is

$$\tilde{\boldsymbol{R}}^{-1} = \frac{1}{\sigma_\nu^2} \left[\boldsymbol{I} - \sum_{k=1}^{K} (1 - \sigma_\nu^2/\lambda_k) \boldsymbol{v}_k \boldsymbol{v}_k^+\right], \tag{11.84}$$

and as $\sigma_\nu^2/\lambda_k \to 0$

$$\tilde{\boldsymbol{R}}^{-1} \to \frac{1}{\sigma_\nu^2} \left[\boldsymbol{I} - \sum_{k=1}^{K} \boldsymbol{v}_k \boldsymbol{v}_k^+\right], \tag{11.85}$$

which is the projection matrix onto the noise subspace. Equation (11.84) shows that the DMR beamformer places a null of depth σ_ν^2/λ_k at the k-th eigenvector, i.e., in the direction of the k-th interference. If we load the diagonals, we have

$$[\tilde{\boldsymbol{R}} + \epsilon\, \boldsymbol{I}]^{-1} = \frac{1}{\sigma_\nu^2 + \epsilon} \left[\boldsymbol{I} - \sum_{k=1}^{K} \frac{\lambda_k - \sigma_\nu^2}{\lambda_k + \epsilon} \boldsymbol{v}_k \boldsymbol{v}_k^+\right], \tag{11.86}$$

which shows a decrease in the null depth, as expected.

Substituting the diagonally loaded form (11.85) in (11.29) we find the DMR approximation to the MVDR weight vector, namely,

$$\boldsymbol{w}_{\text{DMR}} = \frac{\boldsymbol{d}_s - \sum_{k=1}^{K} \gamma_k \left(\boldsymbol{v}_k^+ \cdot \boldsymbol{d}_s\right) \boldsymbol{v}_k}{|\boldsymbol{d}_s|^2 - \sum_{k=1}^{K} \gamma_k \, |\boldsymbol{v}_k^+ \cdot \boldsymbol{d}_s|^2}, \quad \gamma_k \equiv \frac{\lambda_k - \sigma_\nu^2}{\lambda_k + \epsilon}. \tag{11.87}$$

The DMR beamformer is particularly useful when data non-stationarities prevent obtaining the required number of snapshots for a successful adaptive MVDR application; it achieves faster convergence by reducing the adaptive degrees of freedom. Snapshot performance studies with a single interference indicate that the DMR beamformer mean performance is determined by the eigenvectors and is unaffected by the eigenvalues [148].

11.9 Direction of Arrival (DOA) Estimation

Direction of Arrival (DOA) or angle estimation is the problem of finding the direction of a spatially propagating signal using an array, in particular a ULA. When discussing beamformers we have always steered the array in the direction of signal of interest that is assumed known. When the direction is unknown or is known with some uncertainty, then we must estimate the actual direction from data. The most straightforward method is to use the MVDR beamformer (or any of its variants) and form beams for a sufficiently dense set of directions ψ and then pick as the estimated signal direction the angle ψ_s with maximum beam power. A justification for this method can be sought by finding the MLE of the log-likelihood function by assuming that the interference and noise in (11.5) form a zero-mean process with a complex Gaussian density function. If $\mathcal{P}$ denotes the unknown signal power, then the log-likelihood function for a half-wavelength ULA with M elements is

$$\begin{aligned} \ln f(\boldsymbol{x}; \mathcal{P}, \psi) &= -\ln \pi^M |\boldsymbol{R}| - \left(\boldsymbol{x} - \sqrt{\mathcal{P}}\boldsymbol{d}_\psi\right)^+ \boldsymbol{R}^{-1} \left(\boldsymbol{x} - \sqrt{\mathcal{P}}\boldsymbol{d}_\psi\right) \\ &\approx -\boldsymbol{x}^+ \boldsymbol{R}^{-1} \boldsymbol{x} + \sqrt{\mathcal{P}}\left(\boldsymbol{x}^+ \boldsymbol{R}^{-1} \boldsymbol{d}_\psi + \boldsymbol{d}_\psi^+ \boldsymbol{R}^{-1} \boldsymbol{x}\right) - \mathcal{P} \boldsymbol{d}_\psi^+ \boldsymbol{R}^{-1} \boldsymbol{d}_\psi, \end{aligned} \tag{11.88}$$

where $\boldsymbol{R}$ is the interference plus noise correlation matrix defined in (11.21), $\boldsymbol{d}_\psi$ is a function of ψ [equation (11.4)] through its dependence on spatial frequency $\tilde{f}_s = d \cos \psi / \lambda$, and we have dropped the constant term in the last equation. The ML DOA estimate is the angle ψ_s that maximizes the right hand side of (11.88) as a function of ψ; the solution is found by setting the derivative of $\ln f(\boldsymbol{x}; \mathcal{P}, \psi)$ to zero,

$$\frac{\partial \ln f(\boldsymbol{x}; \mathcal{P}, \psi)}{\partial \psi} = \boldsymbol{Re}\left[\left(\boldsymbol{x} - \sqrt{\mathcal{P}}\, \boldsymbol{d}_\psi\right)^+ \boldsymbol{R}^{-1} \sqrt{\mathcal{P}}\, \frac{\partial \boldsymbol{d}_\psi}{\partial \psi}\right] = 0. \tag{11.89}$$

The solution is the same as finding the angle that maximizes the optimal beamformer output power, i.e., the MVDR beamformer, but with the adaptive matched filter normalization given in (11.33) and (11.34) [134]. To find the CRLB on the angle estimate we use the Fisher information matrix inverse (5.172) for a complex exponential in WGN $\sim \mathcal{N}\left(\boldsymbol{0}, \sigma_\nu^2 \boldsymbol{I}\right)$, where the parameter vector is $\boldsymbol{\theta} = [A, \omega]^T$. Our model for the signal on the ULA is (11.2), and so we have the parameter vector $\boldsymbol{\theta} = [A, \tilde{\omega}_s]^T$. We are, however, interested in the transformed parameter vector $\boldsymbol{\eta} = [A, \psi_s]^T$, where $\omega = d \cos \psi_s / 2\pi\lambda$ with the Jacobian transformation matrix

$$\mathbf{J} = \frac{\partial \boldsymbol{\theta}}{\partial \boldsymbol{\eta}} = \begin{bmatrix} 1 & 0 \\ 0 & -\frac{2\pi\lambda}{d \, \sin \psi_s} \end{bmatrix}.$$

Using the result for the CRLB of a transformation of a parameter vector [see theorem 23] and the inverse Fisher information matrix (5.172), the variance of $\hat{\psi}_s$ for a half-wavelength ULA with M elements is bounded by the (2, 2) elements of the matrix $\mathbf{J}\mathbf{I}_{\theta\theta}^{-1}\mathbf{J}^+$, i.e.,

$$\sigma_{\hat{\psi}}^2 \geq \frac{\pi^2\lambda^2}{d^2 \sin^2 \psi_s} \frac{\sigma_\nu^2}{A^2} \frac{48}{N(N-1)(5N-1)} = \frac{48\pi^2(N-1)}{N(5N-1)} \frac{\sigma_\nu^2}{A^2} \left(\frac{\lambda}{L}\right)^2, \tag{11.90}$$

where $L \equiv (N-1)d$ is the array length.

If we use the signal model of complex sinusoids in complex WGN, then all frequency estimation techniques for complex exponential time series in white noise, such as Pisarenko, MUSIC, and ESPRIT (sections 6.4, 6.3, and 6.5), can be applied to find all signal direction(s) using the estimated spatial correlation matrix once we make the correspondence between the signal vector $\boldsymbol{s}_k$ of equation (6.30), which is a function of frequency, and the signal direction vector $\boldsymbol{d}_k$ of equation (11.4), which is a function of spatial frequency $\tilde{f}_k = d\cos\psi_k/\lambda$.

Figure 11.12(a) shows model-based results for an example of a half-wavelength ULA composed of 20 elements, with two unit amplitude complex plane wave signals arriving at 55° and 60° degrees in the presence of white noise with unit power. The array is too short to allow resolving the two directions, as seen in the spatial spectrum (CBF). The MVDR and the MUSIC spatial spectra, computed at 181 angles from 0° to 180° and using 1000 snapshots, clearly resolve both directions. When the white noise power is increased to 10, both the CBF and MVDR fail to resolve the directions while the MUSIC spatial spectrum still resolves both, as shown in 11.12(b).

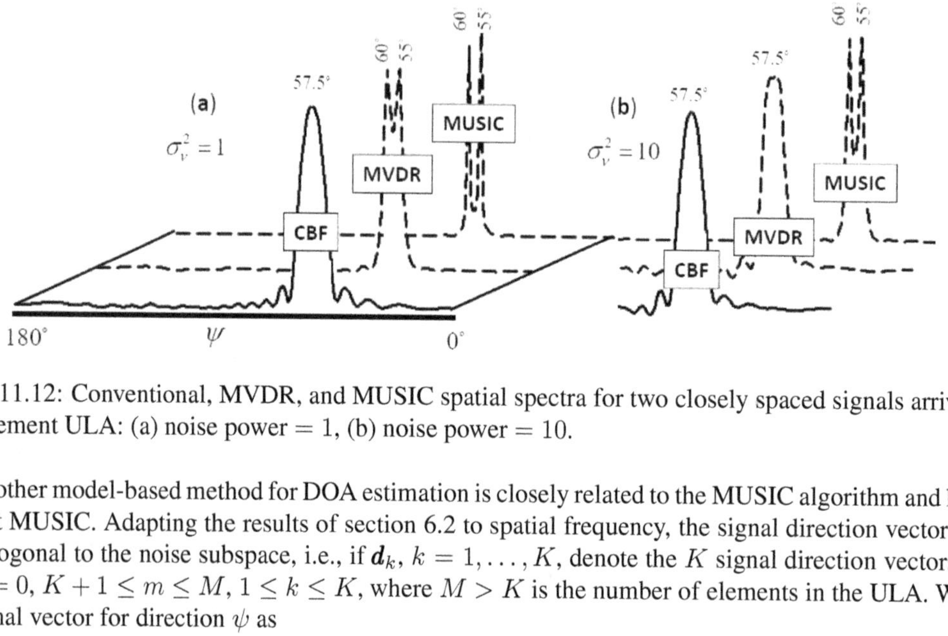

Figure 11.12: Conventional, MVDR, and MUSIC spatial spectra for two closely spaced signals arriving at a 20-element ULA: (a) noise power = 1, (b) noise power = 10.

Another model-based method for DOA estimation is closely related to the MUSIC algorithm and known as Root MUSIC. Adapting the results of section 6.2 to spatial frequency, the signal direction vectors must be orthogonal to the noise subspace, i.e., if $\boldsymbol{d}_k$, $k = 1, \ldots, K$, denote the K signal direction vectors, then $\boldsymbol{v}_m^+ \boldsymbol{d}_k = 0$, $K+1 \leq m \leq M$, $1 \leq k \leq K$, where $M > K$ is the number of elements in the ULA. Writing the signal vector for direction ψ as

$$\boldsymbol{d}_\psi = \left[1, z^{-1}, \ldots, z^{-(M-1)}\right]^T, \quad z \equiv e^{2i\pi d\cos\psi/\lambda}, \tag{11.91}$$

the orthogonality of the signal direction vectors and the noise subspace can be expressed as a polynomial equation in z^{-1}

$$P(z) \equiv \left|\boldsymbol{V}_n^+ \boldsymbol{d}_\psi\right|^2 = \boldsymbol{d}_\psi^+ \boldsymbol{V}_n \boldsymbol{V}_n^+ \boldsymbol{d}_\psi = 0, \tag{11.92}$$

where $\boldsymbol{V}_n$ is the noise eigenvector matrix composed of the $M-K$ noise eigenvectors $\boldsymbol{v}_k$, $k=K=1,\ldots,M$. The roots (zeros) of $P(z)$ contain directional information: if z_k is a root, $P(z_k)=0$, then the associated signal direction is found from

$$\psi_k = \cos^{-1}\left[\frac{\lambda\ \arg(z_k)}{2\pi d}\right]. \tag{11.93}$$

The roots are, in practice, displaced somewhat from the unit circle due to the presence of noise; the K closest roots to the unit circle provide the estimated signal directions. The Root MUSIC estimates for the example of figure 11.12 are $54.9°, 60°$ for noise power 1, and $54.8°, 60°$ for noise power 10.

It is easy to generalize the complex sinusoid in a white noise model to a uniform planar array. Consider $M\times N$ sensors placed on a regular rectangular grid with element position coordinates $(md, nd, 0)$, $0\le m\le M-1, 0\le n\le N-1$, and K signals arriving from directions defined by spherical polar coordinates (ϕ_k,θ_k). The angles between each signal direction and the x axis is ψ_k, while the angle with the y axis is χ_k, as illustrated in figure 11.13.

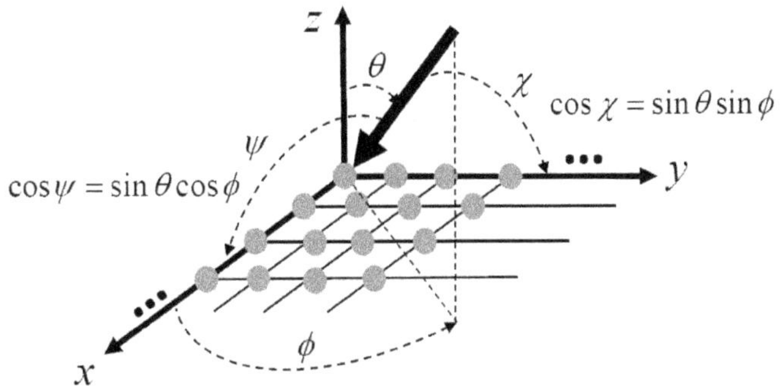

Figure 11.13: Uniform planar array.

An incoming plane wave signal has the form

$$s(t)\, e^{-i\omega t - i\boldsymbol{k}^T\boldsymbol{r}},\quad \boldsymbol{k}^T\boldsymbol{r} = \frac{2\pi}{\lambda}\,(x\cos\psi + y\cos\chi),$$

and for K incoming signals the recorded data (excluding noise) on element (m,n) with coordinate vector $[md, nd, 0]$ is

$$x_{mn}(t) = \sum_{k=1}^{K} s_k(t) e^{-2i\pi\left(m\tilde{f}_k + n\tilde{g}_k\right)},\quad \tilde{f}_k = \frac{d\cos\psi_k}{\lambda},\ \tilde{g}_k = \frac{d\cos\chi_k}{\lambda}. \tag{11.94}$$

The recorded data on all elements can be written as a single vector

$$\begin{aligned}\boldsymbol{x}_s &\equiv \sum_{k=1}^{K}\left[\boldsymbol{f}_k^T, e^{-2i\pi\tilde{g}_k}\boldsymbol{f}_k^T,\ldots, e^{-2i(N-1)\pi\tilde{g}_k}\boldsymbol{f}_k^T\right]^T s_k(t) \\ &= \sum_{k=1}^{K} s_k(t)\,\boldsymbol{g}_k\otimes\boldsymbol{f}_k \equiv \sum_{k=1}^{K} s_k(t)\,\boldsymbol{d}_k,\end{aligned} \tag{11.95}$$

defining the $MN \times 1$ signal direction vector $\boldsymbol{d}_k(\chi_k, \psi_k)$ as the Kronecker product [7] $\boldsymbol{g}_k \otimes \boldsymbol{f}_k$ between the direction vectors associated with χ_k and ψ_k

$$\boldsymbol{g}_k = \left[1, e^{-2i\pi\tilde{g}_k}, \ldots, e^{-2i(N-1)\pi\tilde{g}_k}\right]^T, \quad \tilde{g}_k = \frac{d\cos\chi_k}{\lambda},$$

$$\boldsymbol{f}_k = \left[1, e^{-2i\pi\tilde{f}_k}, \ldots, e^{-2i(M-1)\pi\tilde{f}_k}\right]^T, \quad \tilde{f}_k = \frac{d\cos\psi_k}{\lambda}. \tag{11.96}$$

The received signal vector at time t including noise $\boldsymbol{x}(t) = \boldsymbol{x}_s(t) + \boldsymbol{\nu}(t)$ has a theoretical correlation matrix

$$\boldsymbol{R}_{xx} = \sum_{k=1}^{K} \mathcal{P}_k\, \boldsymbol{d}_k \boldsymbol{d}_k^+ + \sigma_\nu^2 \boldsymbol{I}, \tag{11.97}$$

to which MUSIC, Root MUSIC, and ESPRIT algorithms are applicable. Forming the estimated correlation matrix from L snapshots,

$$\hat{\boldsymbol{R}}_{xx} = \frac{1}{L}\sum_{l=1}^{L} \boldsymbol{x}(t_l)\boldsymbol{x}^+(t_l), \tag{11.98}$$

we estimate the number of signals K and directions (ψ_k, χ_k) from which we calculate the spherical polar angles (ϕ_k, θ_k) using the equations

$$\phi_k = \tan^{-1}\left[\cos\chi_k / \cos\psi_k\right], \quad \theta_k = \sin^{-1}\left[\sqrt{\cos^2\psi_k + \cos^2\chi_k}\right]. \tag{11.99}$$

[7] The Kronecker product [149] of two column vectors $\boldsymbol{a} = [a_1, \ldots, a_M]^T$ and $\boldsymbol{b} = [b_1, \ldots, b_N]^T$ is here defined by the $MN \times 1$ column vector $\boldsymbol{a} \otimes \boldsymbol{b} = [a_1\boldsymbol{b}, \ldots, a_M\boldsymbol{b}]^T$, which in this case is also known as the Khatri-Rao product between the two column vectors.

CHAPTER 12

Neural Networks

AMIR-HOMAYOON NAJMI AND PATRICK EMMANUEL—JHU APL
TODD K. MOON—UTAH STATE UNIVERSITY

12.1 Introduction

Neural networks have become a major field of research in machine learning, with applications to complex problems such as optimization, pattern recognition, and system identification. The most important among many reasons to use the human nervous system (a network of nearly 90 billion interconnected neurons) as a model is the fact that the human brain is able to successfully deal with complex problems such as face recognition. Conversely, better neural network models and a fundamental understanding of their inner workings will hopefully enable us to understand the human brain better; if our models learn to recognize, generalize, and discriminate complex patterns, perhaps they will reveal how the brain uses the same identified mechanisms in its processes. The potential to apply useful ideas from a model of the human nervous system to difficult problems is perhaps the most immediate reason neural networks have gained such prominence in modern computer applications.

The most fundamental structure of a neural network is based on a single neuron, illustrated in figure 12.1, consisting of a cell body called the soma, several extensions of the soma called dendrites, and the axon, which is a single nerve fiber connecting the soma to thousands of other neurons. The axon and dendrites can be thought of as insulated conductors with different impedances that transmit electrical signals to the neuron. The connection between neurons can occur on the soma or on junctions on the dendrites known as synapses that regulate signals between neurons. The totality of the neurons with their dendrites and axon connections and synapses form a neural network.

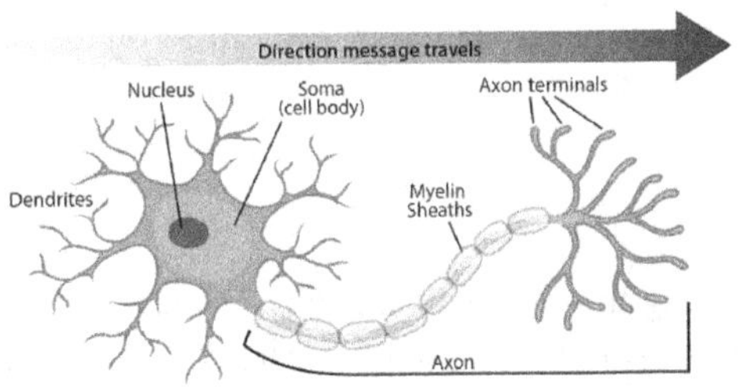

Figure 12.1: Anatomy of a neuron (by permission from ASU's Ask A Biologist `https://askabiologist.asu.edu`).

Success of models of the human neural network based on the perceptron with activation function σ, as described in the next section 12.2, is closely connected with theorem 51 which is known as the universal approximation theorem of neural networks [150], and is illustrated in figure 12.2 for the one dimensional space $\mathbb{R}$.

Theorem 51. *If $f(\boldsymbol{x})$ is a continuous function defined on a compact subset of $\mathbb{R}^D$, then for any $\epsilon > 0$ there exists a positive integer N, real numbers $b_n \in \mathbb{R}$, real vectors $\boldsymbol{w}_n \in \mathbb{R}^D$, and real numbers $\alpha_n \in \mathbb{R}$, for which*

$$|f(\boldsymbol{x}) - h(\boldsymbol{x})| < \epsilon, \;\; h(\boldsymbol{x}) = \sum_{n=1}^{N} \alpha_n \sigma\Big(\boldsymbol{w}_n^T \boldsymbol{x} + b_n\Big), \tag{12.1}$$

where the function σ satisfies the following conditions:

- *σ is sigmoidal, i.e., $\sigma(v) \to 1$ as $v \to \infty$, and $\sigma(v) \to 0$ as $v \to -\infty$.*
- *σ is discriminatory, i.e., it cannot map the linear variety $\boldsymbol{w}_n^T \boldsymbol{x} + b_n$ to a set of measure 0.*

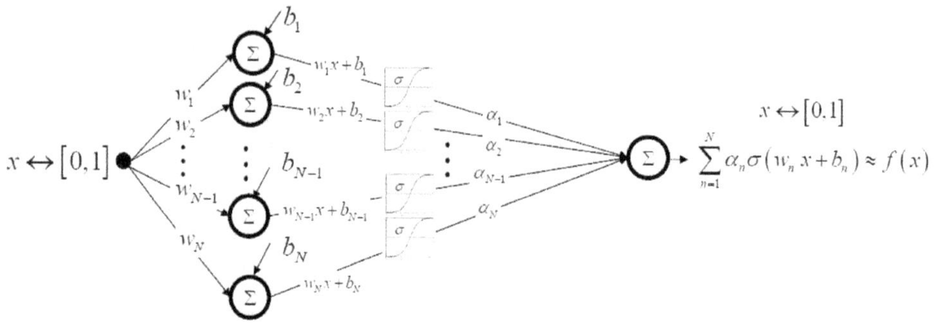

Figure 12.2: The universal approximation theorem 51: $x \in [0, 1] \subset \mathbb{R}$.

The second condition in theorem 51 on the function σ ensures that the inverse function $\sigma^{-1}(.)$ is well-defined and continuous on $[0, 1]$. Both conditions on the function σ are satisfied by the sigmoid function

$$\sigma(v) = 1/(1 + e^{-v}). \tag{12.2}$$

Using the sigmoid function, it is not difficult to visualize the result of the theorem for an arbitrary continuous function $f(x)$ of a single variable $x \in [a, b] \subset \mathbb{R}$. We note that $\sigma(wx + b)$ is equal to $1/2$ at $x = -b/w$ at which point its derivative is $w/4$. Changing the ratio $-b/w$ simply shifts the sigmoid function by that amount while changing w changes the gradient at that point. Thus, instead of the pair of parameters (w, b) it is more convenient to work with parameters (w, s) where $s \equiv -b/w$, and so we consider the function $\sigma(w(x - s))$. Figure 12.3 shows the function $\sigma(w(x - 0.499)) - \sigma(w(x - 0.501))$ for three different values of w with x ranging in $[0.48, 0.52]$. Clearly, a pair of sigmoids are required to produce (any) one of the functions depicted in figure 12.3 that, when appropriately shifted and scaled, can produce an approximation to any given continuous function defined on a compact subset of $\mathbb{R}$ to any degree of accuracy. The approximation improves as the number of shifts s_n are increased, thus increasing the number of nearly rectangular bumps centered on s_n and covering the interval $[0, 1]$.

The application of theorem 51 to a neural network with a single hidden layer requires the insertion of a final sigmoid function to obtain the network output. We will introduce the neuron model in section 12.2 and neural networks in section 12.3.

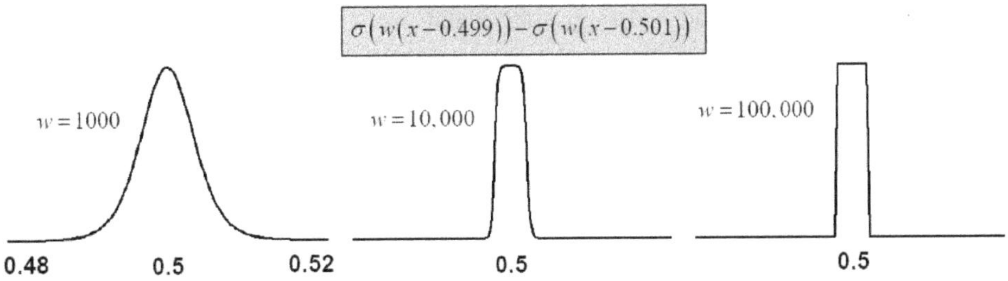

Figure 12.3: The difference between two sigmoid functions $\sigma(w(x - s + 0.001)) - \sigma(w(x - s - 0.001))$ for $s = 0.5$, $w = 1000, 10,000, 100,000$, for $x \in [0.48, 0.52]$.

Although it appears that a sigmoidal function is necessary for the approximation theorem to hold, current practice has replaced the sigmoid [151] with other activation functions such as the ReLU (see section 12.2 and figure 12.6). The practical challenge is how to learn the parameters of a neural network that can produce arbitrarily close approximations to any continuous function.

12.2 The Perceptron

The simplest model of a neuron is the classical perceptron [152] as shown in figure 12.4. Signals on dendrites are modeled by a real vector $\boldsymbol{x} \equiv [x_1, \ldots, x_N]^T$, whose N elements are linearly combined using N elements of a real weight vector $\boldsymbol{w} \equiv [w_1, \ldots, w_N]^T$, producing a unity output when that linear combination is greater than or equal to an internal threshold value $-b$. This threshold computing unit is known as a perceptron; it divides the N dimensional space $\mathbb{R}^N$ into two regions separated by a decision boundary given by the hyperplane $\boldsymbol{w}^T\boldsymbol{x} + b = 0$.

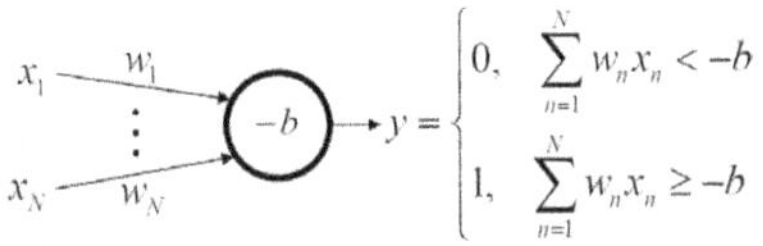

Figure 12.4: A threshold computing unit model of a neuron.

By defining an additional input equal to b, we set the internal threshold to 0; the input b is equivalent to an input of 1 multiplied by a weight b; thus, the output of the classical perceptron is $y = b + \boldsymbol{w}^T\boldsymbol{x}$. In the modern development of neural networks, the output of a classical perceptron is passed through a possibly nonlinear, or piecewise linear, activation function σ; the resulting threshold computing unit is known as an artificial neuron (often simply referred to as a neuron), and is illustrated in figure 12.5.

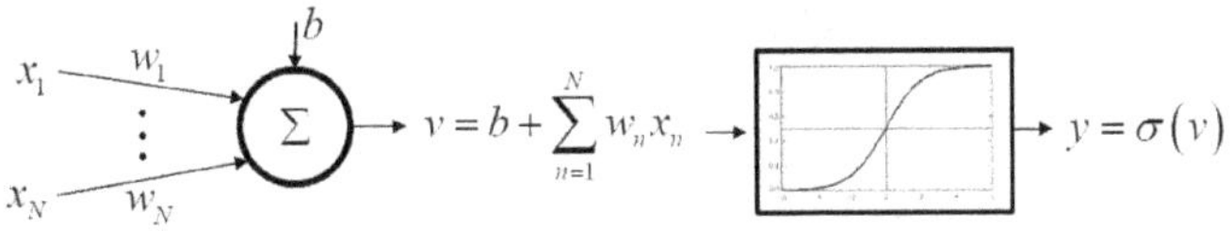

Figure 12.5: An artificial neuron.

Some common activation functions [153] illustrated in figure 12.6 are:

- The sigmoid (or logistic) function $\sigma(v) = 1/(1 + e^{-v})$ (this is used in figure 12.5).
- The hyperbolic tangent $\sigma(v) = (e^v - e^{-v})/(e^v + e^{-v})$.
- The Rectified linear unit (ReLU) $\sigma(v) = \text{Max}(0, v)$.
- The softplus function, a smooth approximation to ReLU, $\sigma(v) = \ln(1 + e^v)$. Note that the derivative of the softplus is the sigmoid function.
- Leaky Rectified linear unit (LReLU) is much the same as ReLU except that it allows for a small gradient for negative input, $\sigma(v) = v$ for $v \geq 0$, and αv for $v < 0$.
- Exponential linear unit (ELU), $\sigma(v) = v$ for $v \geq 0$, and $\alpha(e^v - 1)$ for $v < 0$.

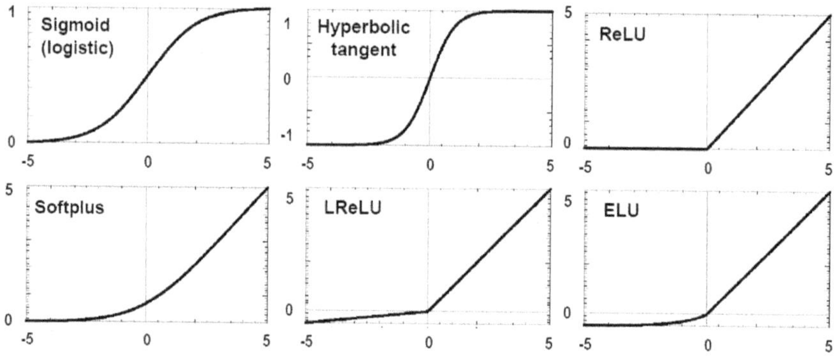

Figure 12.6: Some activation functions.

A single neuron can be used as a linear classifier. For instance, using the sigmoid function, the output of a single neuron can be interpreted as the probability of class membership in a binary classification problem. Suppose that the outcome of a Bernoulli trial is a discrete random variable y taking the values 1 or 0 with probabilities p and $1 - p$, where p is a function of the input variable x. Then, in N independent trials the conditional likelihood function is

$$\prod_{n=1}^{N} \Pr(y = y_n | x = x_n) = \prod_{n=1}^{N} p(x_n; \theta)^{y_n} (1 - p(x_n; \theta))^{1-y_n}, \tag{12.3}$$

for some parameter θ. The logistic regression model is given by

$$\ln \frac{p}{1-p} = b + \boldsymbol{w}^T \boldsymbol{x}, \tag{12.4}$$

whose solution

$$p = \frac{1}{1 + e^{-b-\boldsymbol{w}^T\boldsymbol{x}}} \tag{12.5}$$

is the nonlinear output of the sigmoid function for a neuron (this is also the Boltzmann distribution for a two-state system whose states differ in energy by $b + \boldsymbol{w}^T\boldsymbol{x}$). The mis-classification error is minimized when we actually predict $y = 1$ for $p \geq 0.5$, and $y = 0$ for $p < 0.5$; i.e., $y = 1$ when $b + \boldsymbol{w}^T\boldsymbol{x} \geq 0$ and $y = 0$ otherwise. The decision boundary is the solution to $b + \boldsymbol{w}^T\boldsymbol{x} = 0$ and so logistic regression is a linear classifier. Note that the distance of any point $\boldsymbol{x}$ to the boundary is given by $\left|b + \boldsymbol{w}^T\boldsymbol{x}\right| / \sqrt{b^2 + \boldsymbol{w}^T\boldsymbol{w}}$, which also defines the class probability. Substituting the logistic model into the likelihood function we can proceed to determine the optimal parameters b and $\boldsymbol{w}$ by maximizing the likelihood function using numerical techniques.

Classification problems with nonlinear decision boundaries, however, require a network of neurons. For instance, figure 12.7 shows two patterns: pattern (a) can be separated by a straight line, but pattern (b) cannot.

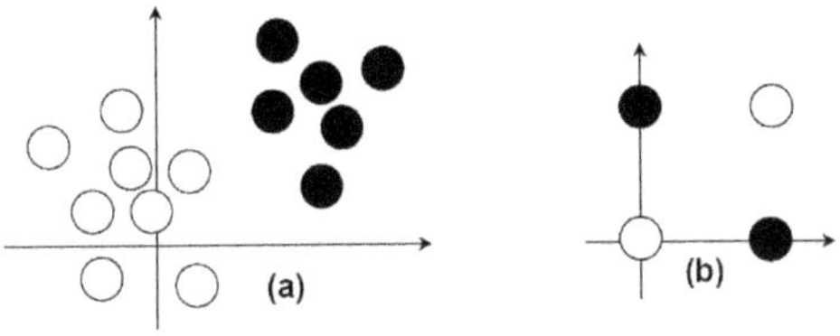

Figure 12.7: Pattern (a) shows linear separation; no straight line can separate the pattern in (b).

The procedure commonly used to train many neural networks is based on gradient descent [154] and known as backpropagation [155]. Recall that the gradient descent algorithm is an iterative method to reach a local minimum of an error surface that is a positive function of a vector $\boldsymbol{w} \in \mathbb{R}^N$ (we take the vector to be real in this chapter), namely, $\mathcal{E}(\boldsymbol{w})$ (if the surface is convex, then gradient descent finds the global minimum). We wish to choose the increment $\Delta\boldsymbol{w}$ so that we descend in the direction of maximum change of the error surface. Proceeding with a Taylor series approximation [as in deriving equation (4.44)]

$$\mathcal{E}(\boldsymbol{w} + \Delta\boldsymbol{w}) \approx \mathcal{E}(\boldsymbol{w}) + \Delta\boldsymbol{w}^T\,\nabla_{\boldsymbol{w}}\mathcal{E}, \tag{12.6}$$

we choose the increment

$$\Delta\boldsymbol{w} = -\mu\nabla_{\boldsymbol{w}}\mathcal{E}, \tag{12.7}$$

where $\mu > 0$ is the learning rate (see section 4.5), leading to a decrease in the error, i.e.,

$$\mathcal{E}(\boldsymbol{w} + \Delta\boldsymbol{w}) \approx \mathcal{E}(\boldsymbol{w}) - \mu\left|\nabla_{\boldsymbol{w}}\mathcal{E}\right|^2 < \mathcal{E}(\boldsymbol{w}). \tag{12.8}$$

There is no guarantee, of course, that the global minimum can be achieved for a non-convex error surface. Nevertheless, gradient descent, or modifications to it, are used in training most artificial neural networks that use supervised learning. We illustrate this algorithm in the training of a single neuron to approximate a specific value c for an input data vector $\boldsymbol{x}$. In practice, we would have a collection of input vectors $\boldsymbol{x}^{[k]}$ forming our training data with a weight vector $\boldsymbol{w}$, but for now we consider only a single instance and drop the superscript k. Thus, our aim is to start with a random set of elements for the weight $\boldsymbol{w}$ and the bias $\boldsymbol{b}$ vectors (see section 12.9 for more details), and to determine the optimal values that produce a final error less than some prescribed $\epsilon > 0$ using iterative techniques.

Let us define $w_0 \equiv b$, $x_0 \equiv 1$, a new input vector $\boldsymbol{x} = [x_0, \ldots, x_N]^T$ and a new weight vector $\boldsymbol{w} = [w_0, \ldots, w_N]^T$, each with $N + 1$ elements, and a loss function (an error surface):

$$\mathcal{E}(\boldsymbol{w}) \equiv (y - c)^2 = (\sigma(\boldsymbol{w}^T\boldsymbol{x}) - c)^2. \tag{12.9}$$

The iterative algorithm should stop at the final weight vector $\boldsymbol{w}_f$ when $\mathcal{E}(\boldsymbol{w}_f) \leq \epsilon$. The gradient descent algorithm is defined by

$$\boldsymbol{w} \leftarrow \boldsymbol{w} - \mu \frac{\partial \mathcal{E}(\boldsymbol{w})}{\partial \boldsymbol{w}}. \tag{12.10}$$

The derivative of the loss function $\mathcal{E}(\boldsymbol{w})$ with respect to the weight vector is

$$\frac{\partial \mathcal{E}(\boldsymbol{w})}{\partial \boldsymbol{w}} = 2\big(y(\boldsymbol{w}) - c\big)\sigma'\boldsymbol{x}, \ \ \sigma' \equiv \frac{d\sigma(v)}{dv}, \ v = \boldsymbol{w}^T\boldsymbol{x}, \tag{12.11}$$

where $y(\boldsymbol{w})$ indicates the dependence of the neuron output on the weight vector. This gives the weight vector update:

$$\boldsymbol{w} \leftarrow \boldsymbol{w} - 2\mu\Big(y(\boldsymbol{w}) - c\Big)\sigma'\boldsymbol{x}. \tag{12.12}$$

When training data consists of K vectors $\boldsymbol{x}_k$, $k = 1, \ldots, K$, the average squared error is minimized:

$$\mathcal{E} = \frac{1}{K}\sum_{k=1}^{K} \mathcal{E}_k, \ \ \mathcal{E}_k \equiv (y_k - c)^2, \tag{12.13}$$

and the weight vector update is

$$\boldsymbol{w} \leftarrow \boldsymbol{w} - 2\mu\frac{1}{K}\sum_{k=1}^{K}\Big(y_k(\boldsymbol{w}) - c\Big)\sigma'\boldsymbol{x}_k. \tag{12.14}$$

In section 12.3 we will study artificial neural networks, i.e., connected networks of individual neurons. The backpropagation of gradient function follows a similar form when the gradient is calculated at an output layer, but will be different if the gradient is calculated at a hidden (not the output) layer. Different forms of loss functions for neural networks will be discussed in section 12.5.

12.3 Fully Connected Feed Forward Neural Networks

A feed forward artificial neural network can be constructed by attaching layers of neurons whose outputs become inputs to the neurons in the next layer [156]. A feed forward network is fully connected when every neuron in one layer is connected to every neuron in the next layer. Figure 12.8 shows one such network whose input layer has N elements (data samples x_1 through x_N), three hidden layers with 3, 2, and 3 neurons, respectively, and an output layer with two neurons. The entire network is denoted by the equation $\boldsymbol{y} = h(\boldsymbol{x})$ where the input $\boldsymbol{x}$ is an $N \times 1$ vector and the output $\boldsymbol{y}$ is a 2×1 vector. The upper part of figure 12.8 shows the network connections with inputs and outputs; a square box represents a neuron as shown in the lower part of figure 12.8. Layer $l \geq 1$ has N_l neurons, each of which has N_{l-1} inputs and a single output. Each connection has a weight $w_{ij}^{(l)}$ that multiplies the output of neuron $i \geq 1$ in the previous layer $l-1$ to neuron $j \geq 1$ in layer l; the latter neuron has a bias $b_j^{(l)}$ which is equivalently defined by $w_{0j}^{(l)}$. In this example layer $l = 0$ is the input layer, $l = 4$ is the output layer, and $l = 1, 2, 3$ are the three hidden layers. Inputs to layer 1 are the N data samples of $\boldsymbol{x}$. The output of neuron j of the hidden layer $l = 1, 2, 3$ is denoted by $y_j^{(l)}$ while the final output of neuron j in the output layer 4 is denoted by y_j, all of which are denoted by the vector $\boldsymbol{y}$ with dimension $M \times 1$ with M denoting the number of final outputs. Thus, for a

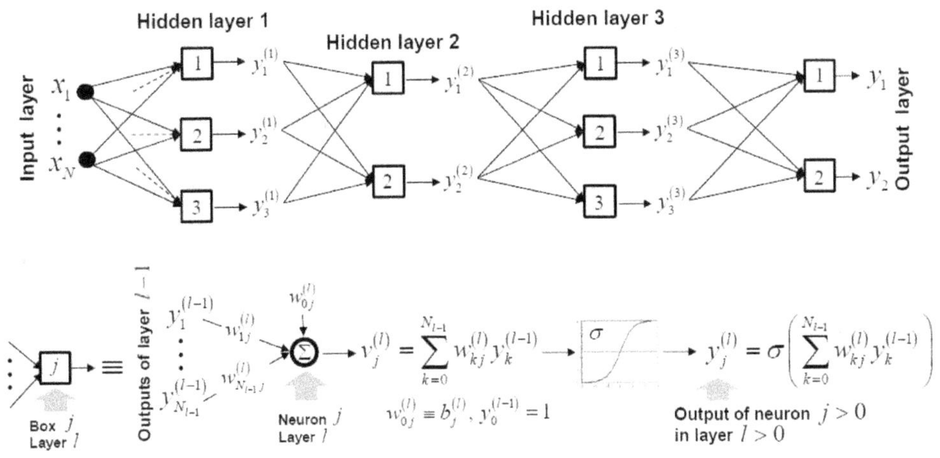

Figure 12.8: An artificial neural network $\boldsymbol{y} = h(\boldsymbol{x})$ whose input layer ($l = 0$) has N elements, three hidden layer ($l = 1, 2, 3$) with $3, 2, 3$ neurons, respectively, and an output layer ($l = 4$) with 2 neurons. Each square box in the network represents a neuron shown on the bottom.

network with $L + 1$ layers consisting of one input layer ($l = 0$), $L - 1$ hidden layers ($l = 1, \ldots, L - 1$), and one output layer ($l = L$), the output of neuron j in layer $l \geq 1$ is

$$y_j^{(l)} = \sigma\left(w_{0j}^{(l)} + \sum_{k=1}^{N_l} w_{kj}^{(l)} y_k^{(l-1)}\right), \quad 1 \leq l \leq L, \; 1 \leq j \leq N_l, \tag{12.15}$$

where $y_j^0 \equiv x_j$ is the data at the input layer $l = 0$, and $y_j^L \equiv y_j$ is the output of the output layer $l = L$.

The illustration of theorem 51 in figure 12.2, for a continuous function defined on the closed interval $[0, 1]$, can be readily generalized to apply to a neural network with a single hidden layer by simply including an activation function before the final output. If the output is to constitute an approximation to a function $f(x)$, then the input to this final activation function must be of the form $\sigma^{-1}[f(x)]$. The general requirements on the activation in theorem 51 ensure that the inverse function exists and is continuous in the same interval in which $f(x)$ is continuous and so the inverse function can be uniformly approximated by a linear combination of function $\sigma(w_n x + b_n)$. Thus, theorem 51 can be restated as theorem 52.

Theorem 52. *Given a continuous function $f(\boldsymbol{x})$ defined on a compact subset of $\mathbb{R}^D$, and $\epsilon > 0$, a neural network with a single hidden layer and N neurons, and a sigmoidal activation function exists whose output $h(\boldsymbol{x})$ uniformly approximates $f(\boldsymbol{x})$ with $|f(\boldsymbol{x}) - h(\boldsymbol{x})| < \epsilon$.*

12.4 The Backpropagation Algorithm

Training of most feed forward neural networks uses the steepest descent method in a backpropagation algorithm [155]. We have already seen how to compute the required derivatives with respect to the weight functions at an output layer. The backpropagation algorithm allows us to compute the derivatives with respect to the weight functions at all hidden layers, once we have the derivatives at an output layer. To derive the hidden layer derivatives we refer to figure 12.9, which shows the connection of an output of neuron k in layer $l-2$ to neuron i in layer $l-1$ whose output goes through neuron j in layer l to finally arrive at $y_j^{(l)}$.

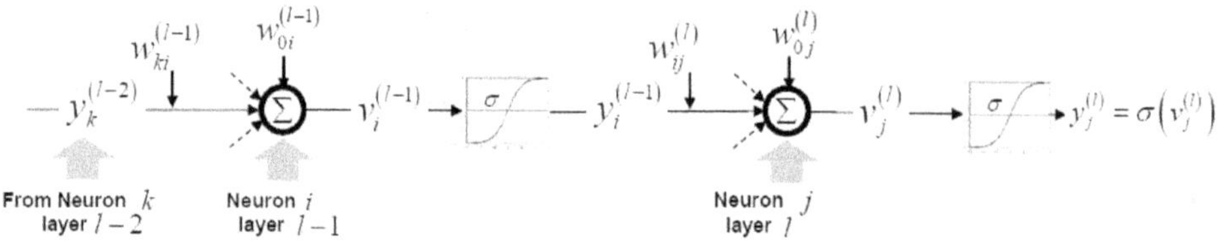

Figure 12.9: A portion of a feed forward neural network: $y_k^{(l-2)}$ is the output of neuron k in layer $l-2$, $y_i^{(l-1)}$ is the output of neuron i in layer $l-1$, and $y_j^{(l)}$ is the output of neuron j in layer l.

Using figure 12.9 we have

$$y_j^{(l)} = \sigma(v_j^{(l)}), \; v_j^{(l)} = \sum_{i=0}^{N_l} w_{ij}^{(l)} y_i^{(l-1)}, \; w_{0j}^{(l)} \equiv b_j^{(l)}, \; y_0^{(l-1)} = 1. \tag{12.16}$$

Let us assume there are M final outputs with squared errors $\mathcal{E}_m = (y_m - c_m)^2$, $1 \le m \le M$ whose sum divided by M is the average output error. Differentiating $\mathcal{E}_m$ with respect to the weights at layer $l-1$ gives

$$\frac{\partial \mathcal{E}_m}{\partial w_{ki}^{(l-1)}} = \frac{\partial \mathcal{E}_m}{\partial v_i^{(l-1)}} \frac{\partial v_i^{(l-1)}}{\partial w_{ki}^{(l-1)}} = \frac{\partial \mathcal{E}_m}{\partial v_i^{(l-1)}} y_k^{(l-2)}. \tag{12.17}$$

The last derivative on the right hand side of (12.17) is

$$\frac{\partial \mathcal{E}_m}{\partial v_i^{(l-1)}} = \sum_{j=1}^{N_l} \frac{\partial \mathcal{E}_m}{\partial v_j^{(l)}} \frac{\partial v_j^{(l)}}{\partial v_i^{(l-1)}} = \sum_{j=1}^{N_l} \frac{\partial \mathcal{E}_m}{\partial v_j^{(l)}} w_{ij}^{(l)} \sigma'\left(v_i^{(l-1)}\right), \tag{12.18}$$

which is the backpropagation formula to compute all gradients down to the first hidden layer, starting with the output layer L, and equation (12.11) for the derivative at layer L, namely,

$$\frac{\partial \mathcal{E}_m}{\partial v_j^{(L)}} = 2(y_m - c_m)\sigma'(v_m^{(L)}). \tag{12.19}$$

The gradient update at layer 1 is

$$\Delta w_{ij}^{(1)} = -\mu \frac{\partial \mathcal{E}_m}{\partial w_{ij}^{(1)}} = -\mu \, x_i \sum_{k=1}^{N_1} \frac{\partial \mathcal{E}_m}{\partial v_j^{(2)}} w_{jk}^{(2)} \sigma'\left(v_j^{(1)}\right). \tag{12.20}$$

We will study examples of neural network training and loss functions in section 12.5. For now we should emphasize the importance of the nonlinear activation in the design of an artificial neural network. In general, an activation such as the sigmoid function that has saturation levels in both directions leads to the *vanishing gradient* problem, which can permanently deactivate many of the neurons in the network, thus decreasing the capacity of the network. In addition, the output of the sigmoid is not centered at zero and this can cause jumps in gradient updates; if the incoming data is all positive, then during backpropagation gradients become all positive or all negative; for this reason, the hyperbolic tangent is preferred over the sigmoid, but it too suffers from the vanishing gradient problem [157].

The ReLU and some of its variants (Noisy ReLU, Leaky ReLU, and Exponential Linear Units) alleviate the vanishing gradient problem and as a result have, in the last few years, completely replaced the older sigmoid and hyperbolic tangent activations. The ReLU itself suffers from the *dead neuron* problem when for high learning rates some neurons never activate for the entire training data. This problem has been solved by lowering the learning rate or using other variants of the ReLU such as the Leaky ReLU when the parameter α is chosen to be a small number such as 0.01. In practice, it is best to start with ReLU with lower learning rates while monitoring the fraction of dead neurons.

Although activation functions in the hidden layers are often chosen to be the same (e.g., ReLU), the output layer activation is selected depending on the loss function $\mathcal{E}$ the neural network is designed to minimize (see section 12.5 for a discussion of loss functions), and the function $h(\boldsymbol{x})$ that the network is designed to compute. For instance, in a regression problem if the output values are in the range $[-A, +A]$, $A > 0$, then we could use the **tanh** nonlinearity, while if the output values are non-negative then the ReLU activation is appropriate. However, in regression problems a nonlinearity is often not used in the output layer; in other words, activation is the identity operation.

When the network is used as a binary classifier the sigmoid (logistic) function is the most commonly used activation at the output layer, in which case the output is the conditional probability (conditioned on the input to the network) of belonging to the positive class (typically denoted by 1 in a binary vector $[0, 1]$). When the network is used to classify among $N_c > 2$ classes of data, then the **softmax** activation function is used to minimize the cross-entropy loss function (see section 12.5),

$$\textbf{softmax}(\boldsymbol{v}) = [e^{v_1}, \ldots, e^{v_{N_c}}]^T / S, \quad S \equiv \sum_{n=1}^{N_c} e^{v_n}, \tag{12.21}$$

where $\boldsymbol{v}$ is the output before the nonlinearity as defined in figure 12.9. To avoid numerical instability the **softmax** nonlinearity is usually calculated by multiplying the numerator and the denominator by $e^{-v_{\max}}$. When $N_c = 2$, $\sigma(v_1) + \sigma(v_2) = 1$ and this reduces to the sigmoid (logistic) nonlinearity.

12.5 Loss Functions in Neural Network Training

Let $\boldsymbol{y} = h(\boldsymbol{x}; \boldsymbol{w})$ denote the network output for input $\boldsymbol{x}$ and weight vector $\boldsymbol{w}$, and consider training the network using training data $\boldsymbol{x}_k$ to reach a desired output data $\boldsymbol{d}_k$, $1 \le k \le K$. A regression problem loss function is

$$\mathcal{E} = \frac{1}{K} \sum_{k=1}^{K} \left\| h(\boldsymbol{x}_k; \boldsymbol{w}) - \boldsymbol{d}_k \right\|^2. \tag{12.22}$$

Note that this loss function requires the last activation function to be the identity, as illustrated in figure 12.10 describing a network with L hidden layers whose first $L - 1$ layers with activation σ, and the last layer L with identity activation. The input to the network is an $N \times 1$ vector $\boldsymbol{x}$ with elements $x_1, \ldots, x_N$ and the output $\boldsymbol{y} = \boldsymbol{v}^{(L)}$ has N elements.

Now consider a two-class classifier neural network with a sigmoid activation at the output layer, and training data $\boldsymbol{x}_k$ belonging to classes 0 and 1. In this case the network output determines class, e.g., we assign $\boldsymbol{x}_k$ to class 1 when $y_k \ge 0.5$ and $\boldsymbol{x}_k$ to class 0 when $y_k < 0.5$ (in practice the threshold is chosen according to some optimality criterion). Then the appropriate cost function is the binary cross-entropy

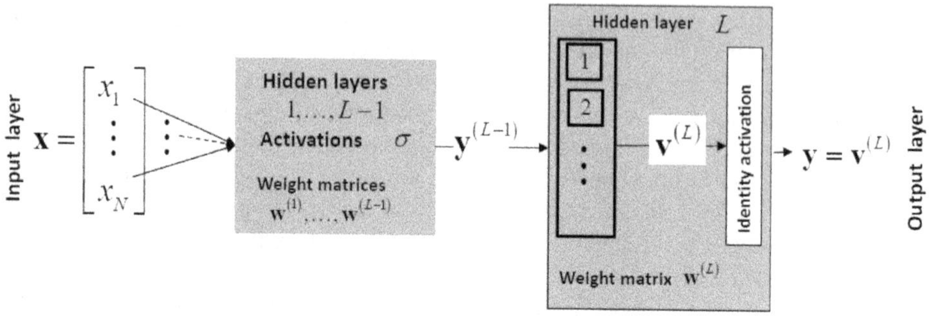

Figure 12.10: A fully connected regression neural network with L hidden layers: the first $L-1$ layers have activation σ and weight matrices $\boldsymbol{w}^{(1)}, \ldots, \boldsymbol{w}^{(L-1)}$, and the last hidden layer has weight matrix $\boldsymbol{w}^{(L)}$ with identity activation so that $\boldsymbol{y} = \boldsymbol{v}^{(L)}$.

function defined by

$$\mathcal{E} = -\frac{1}{K}\sum_{k=1}^{K}\Big(d_k \ln(y_k) + (1-d_k)\ln(1-y_k)\Big), \quad d_k = 0 \text{ or } 1. \tag{12.23}$$

When the network is used to classify data into $N_c > 2$ classes using the **softmax** activation at the output layer, the associated loss function is the cross-entropy defined by

$$\mathcal{E} = -\frac{1}{K}\sum_{k=1}^{K}\sum_{n=1}^{N_c} d_{kn} \ln y_{kn}, \quad d_{kn} = 0 \text{ or } 1. \tag{12.24}$$

For instance, consider a 3-class problem with classes A, B, and C, and a desired vector of the form [A, B, C, C, B, A]. In order to train the network we encode this vector as a *one-hot encoded matrix* whose elements are the desired values d_{kn},

$$\begin{aligned}&A \to [1,0,0] = d_{1n}, \quad B \to [0,1,0] = d_{2n}, \quad C \to [0,0,1] = d_{3n}\\ &d_{5n} = d_{2n}, \quad d_{4n} = d_{3n}, \quad d_{6n} = d_{1n}\end{aligned}$$

12.6 Gradient Descent Variants

The gradient descent algorithm

$$\boldsymbol{g}^{(n)} \equiv \nabla_{\boldsymbol{w}^{(n)}}\mathcal{E}\Big(\boldsymbol{w}^{(n)}\Big), \quad \boldsymbol{w}^{(n+1)} \leftarrow \boldsymbol{w}^{(n)} - \mu \boldsymbol{g}^{(n)}, \tag{12.25}$$

is often slow to converge. An approach to accelerate learning is momentum optimization [154] that introduces a momentum vector $\boldsymbol{m}$ to store and use previous step's gradient direction. Classical momentum algorithms accumulate a decaying sum of the previous gradients into a momentum vector $\boldsymbol{m}$ and use this in the update instead of the gradient,

$$\boldsymbol{m}^{(n+1)} \leftarrow \beta \boldsymbol{m}^{(n)} + \boldsymbol{g}^{(n)}, \quad \boldsymbol{w}^{(n+1)} \leftarrow \boldsymbol{w}^{(n)} - \mu \boldsymbol{m}^{(n+1)}. \tag{12.26}$$

Learning is accelerated in any direction along which the gradient is relatively stable across training steps, but learning is slowed in any direction along which the gradient is oscillatory; β is a friction coefficient to ensure that the momentum gradually decreases to zero.

If the momentum vector is pointing in the right direction, then it may be more accurate to evaluate the gradient a little further along than the current position. The Nesterov accelerated gradient (NAG) method is equivalent to improving the momentum vector and achieves a much better bound than standard gradient descent by evaluating the gradient at the updated value of momentum $\boldsymbol{m}^{(n+1)}$. In the momentum update (12.26), $\boldsymbol{m}^{(n)}$ does not depend on the gradient $\boldsymbol{g}^{(n)}$; Nesterov's algorithm introduces a dependence in the form

$$\boldsymbol{g}^{(n)} \equiv \boldsymbol{\nabla}_{\boldsymbol{w}^{(n)}} \mathcal{E}\left(\boldsymbol{w}^{(n)} - \beta\mu\boldsymbol{m}^{(n)}\right) \tag{12.27}$$

when using the update equations (12.26).

To address the problem of learning in a "long narrow valley" one might try to cut across the slope heading towards the global minimum, gaining progress on the variable that needs the most change (the long valley), even though it is not approaching the global minimum in the steepest direction. AdaGrad (adaptive subgradient descent) scales down the gradient in the steepest direction by its norm (so instead of going straight downhill it traverses less steep directions); it uses different learning rates on different parameters by adaptively adjusting the rates according to the "steepness" in each component [154]:

$$\begin{aligned} &\boldsymbol{g}^{(n)} \equiv \boldsymbol{\nabla}_{\boldsymbol{w}^{(n)}} \mathcal{E}\left(\boldsymbol{w}^{(n)}\right), \quad \nu^{(n+1)} \leftarrow \beta\nu^{(n)} + \left|\nu^{(n)}\right|^2, \\ &\boldsymbol{w}^{(n+1)} \leftarrow \boldsymbol{w}^{(n)} - \frac{\mu\boldsymbol{g}^{(n)}}{\sqrt{\nu^{(n+1)} + \varepsilon}}. \end{aligned} \tag{12.28}$$

This algorithm accelerates learning along directions that have changed slightly but suffers from the exploding norm problem that halts learning altogether. A simple cure is to use an exponentially weighted adaptive norm calculation for ν, known as RMSProp,

$$\nu^{(n+1)} \leftarrow \alpha\nu^{(n)} + (1-\alpha)\left|\nu^{(n)}\right|^2, \quad 0 \ll \alpha < 1.$$

A combination of RMSProp and the classical momentum method is Adam (adaptive momentum estimation) [154],

$$\begin{aligned} &\boldsymbol{g}^{(n)} \equiv \boldsymbol{\nabla}_{\boldsymbol{w}^{(n)}} \mathcal{E}\left(\boldsymbol{w}^{(n)}\right), \quad \boldsymbol{m}^{(n+1)} \leftarrow \beta\boldsymbol{m}^{(n)} + (1-\beta)\boldsymbol{g}^{(n)}, \quad \boldsymbol{m}^{(n+1)} \leftarrow \frac{\boldsymbol{m}^{(n+1)}}{(1-\beta)}, \\ &\nu^{(n+1)} \leftarrow \alpha\nu^{(n)} + (1-\alpha)\left|\nu^{(n)}\right|^2, \quad \nu^{(n+1)} \leftarrow \frac{\nu^{(n+1)}}{(1-\alpha)}, \\ &\boldsymbol{w}^{(n+1)} \leftarrow \boldsymbol{w}^{(n)} - \mu\boldsymbol{m}^{(n)}/\sqrt{\nu^{(n+1)} + \varepsilon}. \end{aligned} \tag{12.29}$$

Exponentially weighted momentum adaptive methods often use a time-dependent parameter, e.g., $\beta^{(n)} = 0.99\left(1-0.5\times0.96^{n/250}\right)$. Figure 12.11 shows a comparison between gradient descent and AdaGrad/Adam-type algorithms.

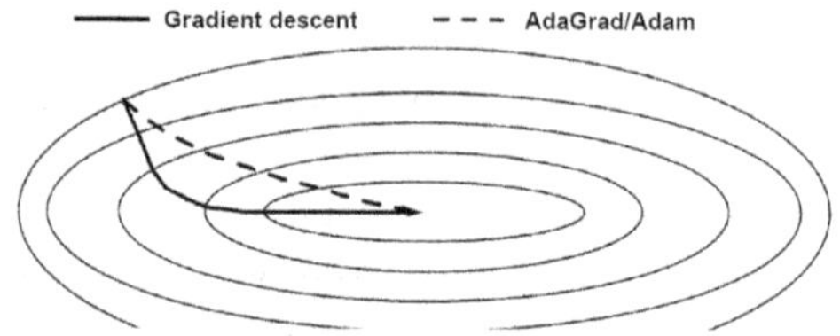

Figure 12.11: Gradient descent compared with AdaGrad/Adam.

12.7 Single Hidden Layer and Multiple Hidden Layers Neural Networks

As theorem 51 suggests, arbitrarily complicated functions can be approximated by a neural network with a single hidden layer. But deep networks with more than one hidden layer can have much higher efficiency with far fewer (possibly exponentially fewer) neurons per layer. Deep networks have been successfully used in difficult problems such as speech recognition and image classification. Development of a neural network often starts with one or two layers, and then increasing the number of hidden layers until overfitting is observed or suspected.

The number of input neurons is determined by the input data. For instance, to classify images of handwritten characters, the input images are small, say $30 \times 30 = 900$ pixels, and if no data reduction prior to the neural network is performed then 900 neurons are needed in the first layer. If the characters are classified into $26 + 26$ letters, 10 numeric digits, and a dozen punctuation marks, then 74 neurons are needed in the final output layer.

To illustrate the effect of more hidden layers in reducing the number of neurons with no loss in performance we examine the spiral data consisting of two paired sets $(\boldsymbol{x}_1, \boldsymbol{y}_1)$ and $(\boldsymbol{x}_2, \boldsymbol{y}_2)$, which we construct by

$$\begin{aligned} \boldsymbol{z} = 2\pi \frac{780}{360} \sqrt{\boldsymbol{u}_0}, \quad \boldsymbol{x}_1 &= -\boldsymbol{z}\cos(\boldsymbol{z}) + \sigma \boldsymbol{u}_1, \quad \boldsymbol{y}_1 = +\boldsymbol{z}\sin(\boldsymbol{z}) + \sigma \boldsymbol{u}_2, \\ \boldsymbol{x}_2 &= +\boldsymbol{z}\cos(\boldsymbol{z}) + \sigma \boldsymbol{u}_3, \quad \boldsymbol{y}_2 = -\boldsymbol{z}\sin(\boldsymbol{z}) + \sigma \boldsymbol{u}_4, \end{aligned}$$

where $\boldsymbol{u}_k$, $0 \le k \le 4$, are random vectors whose elements are uniformly distributed in $[0, 1]$, and σ denotes the strength of the noise (0.5 in our example); element by element multiplication and association is implied in the equations.

Figure 12.12 shows the classification boundaries for three neural networks consisting of a single hidden layer with 10 (a), 1000 (b), and 5000 (c) neurons, respectively. Clearly, increasing the number of neurons improves the network performance as expected by the universal approximation theorem 51.

Figure 12.13 shows the performance with more hidden layers but with a significantly reduced number of neurons. Image (a) shows the classification boundaries using 2 hidden layers with 10 neurons each, while image (b) shows the boundaries for 3 hidden layers with 10 neurons each; both images show similar performance to the single hidden layer network with 5000 neurons, but a reduction in the total number of neurons by a factor of approximately 150. We note that to illustrate the power of deep representations compared with shallow ones, we limited the number of epochs (see section 12.8) to 100; increasing the number of epochs does allow for better classifications using single-layer networks (even with fewer neurons) at the expense of significantly more training time. All models were optimized using Adam (see section 12.6).

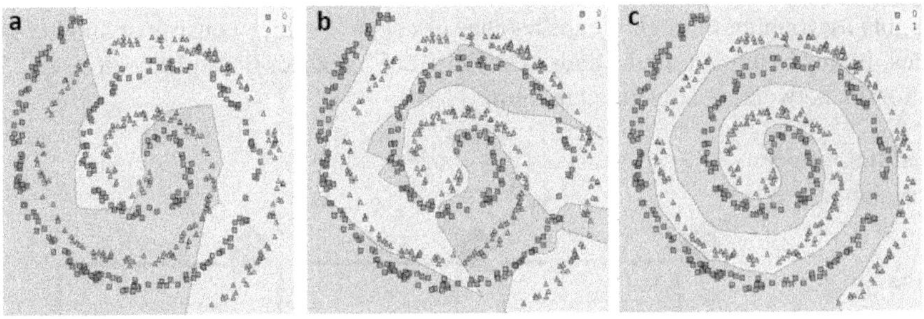

Figure 12.12: Classification boundaries for three neural networks with a single hidden layer and 10 (a), 1000 (b), and 5000 (c) neurons, respectively.

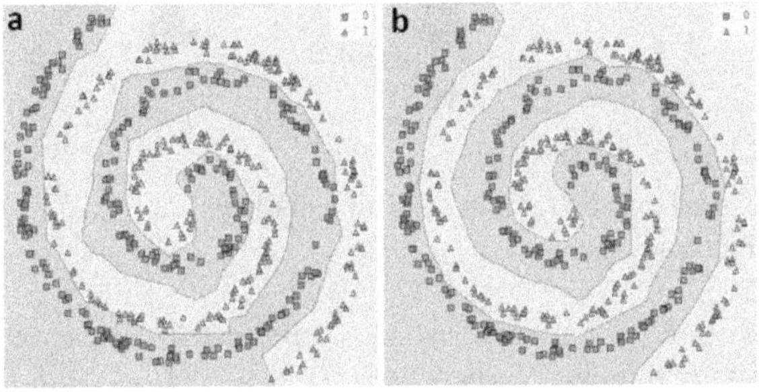

Figure 12.13: Classification boundaries for two neural networks with 2 (a) and 3 (b) hidden layers, each with 10 neurons.

12.8 Mini-Batch Training and Normalization

When performing gradient descent adaptation the average of the loss over the entire training set can be used to compute gradients for backpropagation; this is known as batch gradient descent and can be extremely time consuming and memory inefficient for large data sets. Batch training offers a more stable estimate of the error gradient which, although useful in some problems, can actually reduce performance by converging to less optimal network weights. The extreme alternative to this is to use a random ordering of the training samples, update the weights based on the loss calculated for the first training data, and use the weights as initial weights for training using the second training data and so on, to finish through the entire training set of size N_T. The process can be repeated starting with the weights from the first round, and a new random ordering of the training data. Each round is known as an epoch, by the end of which the network has seen all the training data. After completing N_E epochs, i.e., when the network has been exposed to the training data N_E times, we would have a total of $N_T \times N_E$ weight updates. This method is known as stochastic gradient descent or SGD; it can be useful in on-line learning when training can occur as new samples arrive.

The concept of stochastic mini-batch training fits somewhere between SGD and batch training. To illustrate, consider the set of indices $[1, \ldots, 20]$ that refer to 20 training samples. Choosing a mini-batch size of 4, we construct 5 mini-batches containing random indices (typically without replacement) between 1 and 20, and begin training with initial random weights $\{\boldsymbol{w}\}_0$ on mini-batch 1 resulting in updated weights $\{\boldsymbol{w}\}_1$. The weights $\{\boldsymbol{w}\}_1$ become the initial weights for training on mini-batch 2, and so on until we have the weights $\{\boldsymbol{w}\}_5$ at the end of epoch 1, which consists of 5 mini-batches. The weights $\{\boldsymbol{w}\}_5$ now become the initial weights for training through 5 mini-batches of epoch 2, which consists of another random set of training indices. In this example training ends after the set of 5 updates through epoch 3, with $\{w\}_{15}$ as the final weights for the network, as illustrated in figure 12.14.

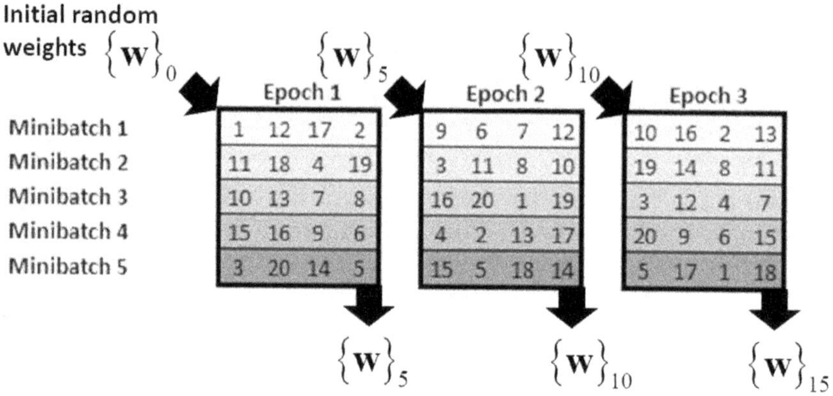

Figure 12.14: Example of mini-batch training: training indices $[1, \ldots, 20]$, mini-batch size of 4, three epochs, resulting in 15 weight updates.

Given a mini-batch size of M_{mB}, there are N_T/M_{mB} mini-batches in each epoch, and the total number of weight updates after training through N_E epochs is $N_E \times N_T/M_{mB}$. Mini-batch size is usually much smaller than the number of training samples and should divide it; smaller mini-batch sizes increase noise in the gradients which, ironically, may be useful as an implicit regularizer [158], while larger sizes have less noise in the gradients but they can train faster. Mini-batch size is often chosen to be a power of 2, e.g., $4, 8, 16, 32, \ldots$.

An important issue in the training phase of deep neural networks is the fact that the distribution of each layer's activations change as the distribution of inputs to that layer change through change in parameters of the previous layers; this change of distribution of network activations during training is known as *internal covariate shift* [159] and often slows convergence. Convergence can be accelerated by whitening the input sequences to all layers; whitening includes decorrelation as well as normalization. Currently, normalization alone (excluding decorrelation) via mini-batch statistics is the preferred method and is applied to each activation separately, i.e., each element of each activation vector is normalized by the mean and standard deviation of the set of its values within a mini-batch [159]. For instance, let $\boldsymbol{x} = [x_1, \ldots, x_d]$ denote an input vector to a fully connected network, and let $\boldsymbol{x}^{(1)}, \boldsymbol{x}^{(2)}, \ldots$, denote the training data. Let $\boldsymbol{x}^{(k_1)}, \ldots, \boldsymbol{x}^{(k_m)}$ denote the members of a mini-batch of size m, and let $\boldsymbol{y}^{(k_1)}, \ldots, \boldsymbol{y}^{(k_m)}$ denote the outputs of a layer associated with this mini-batch. Batch normalization introduces two new $d \times 1$ vector parameters $\boldsymbol{\gamma}$ and $\boldsymbol{\beta}$, for each mini-batch, that are learnt along with the rest of the network parameters, and are defined by the Batch Normalizing Transform $\mathbf{BNT}_{\gamma,\beta} : \boldsymbol{x} \to \boldsymbol{y}$:

$$\boldsymbol{\mu} \equiv \frac{1}{m}\sum_{i=1}^{m} \boldsymbol{x}^{(k_i)}, \; \boldsymbol{\sigma}^2 \equiv \frac{1}{m}\sum_{i=1}^{m} (\boldsymbol{x}^{(k_i)} - \boldsymbol{\mu})^2,$$

$$\hat{\boldsymbol{x}}^{(k_i)} = \frac{\boldsymbol{x}^{(k_i)} - \boldsymbol{\mu}}{\sqrt{\boldsymbol{\sigma}^2 + \varepsilon}}, \; 0 < \varepsilon \ll 1, \; \boldsymbol{y}^{(k_i)} = \boldsymbol{\gamma} \otimes \hat{\boldsymbol{x}}^{(k_i)} + \boldsymbol{\beta}, \; i = 1, \ldots, m, \tag{12.30}$$

where $\otimes$ indicates element by element vector multiplication. Batch normalized networks are trained using batch gradient descent or stochastic mini-batch descent with $m > 1$, and backpropagation of the usual derivatives of the loss function with respect to the weights in addition to the following derivatives with respect to the BNT variables:

$$\frac{\partial E}{\partial \hat{\boldsymbol{x}}^{(k_i)}} = \boldsymbol{\gamma} \otimes \frac{\partial E}{\partial \boldsymbol{y}^{(k_i)}}, \quad \frac{\partial E}{\partial \boldsymbol{\sigma}^2} = -\frac{1}{2\left(\boldsymbol{\sigma}^2 + \varepsilon\right)^{1.5}} \otimes \sum_{i=1}^{m} \frac{\partial E}{\partial \hat{\boldsymbol{x}}^{(k_i)}} \otimes \left(\boldsymbol{x}^{(k_i)} - \boldsymbol{\mu}\right),$$
$$\frac{\partial E}{\partial \boldsymbol{\mu}} = -\frac{2}{m}\frac{\partial E}{\partial \boldsymbol{\sigma}^2} \otimes \sum_{i=1}^{m} \left(\boldsymbol{x}^{(k_i)} - \boldsymbol{\mu}\right) - \frac{1}{2\left(\boldsymbol{\sigma}^2 + \varepsilon\right)^{0.5}} \otimes \sum_{i=1}^{m} \frac{\partial E}{\partial \hat{\boldsymbol{x}}^{(k_i)}},$$
$$\frac{\partial E}{\partial \boldsymbol{x}^{(k_i)}} = \frac{1}{\left(\boldsymbol{\sigma}^2 + \varepsilon\right)^{0.5}} \otimes \frac{\partial E}{\partial \hat{\boldsymbol{x}}^{(k_i)}} + \frac{2}{m}\frac{\partial E}{\partial \boldsymbol{\sigma}^2} \otimes \left(\boldsymbol{x}^{(k_i)} - \boldsymbol{\mu}\right) + \frac{1}{m}\frac{\partial E}{\partial \boldsymbol{\mu}},$$
$$\frac{\partial E}{\partial \boldsymbol{\gamma}} = \sum_{i=1}^{m} \frac{\partial E}{\partial \boldsymbol{y}^{(k_i)}} \otimes \hat{\boldsymbol{x}}^{(k_i)}, \quad \frac{\partial E}{\partial \boldsymbol{\beta}} = \sum_{i=1}^{m} \frac{\partial E}{\partial \boldsymbol{y}^{(k_i)}}. \tag{12.31}$$

While the normalized activations $\hat{\boldsymbol{x}}^{(k_i)}$ remain internal to the network, reducing covariate shift and accelerating network training, the BNT transformed values $\boldsymbol{y}^{(k_i)}$ are passed to other network layers. The learnt variables $\boldsymbol{\gamma}, \boldsymbol{\beta}$ applied to the normalized activations allow the BNT to represent the identity transformation and preserve network capacity.

To achieve its promise, batch normalization must be accompanied by changes in several other training parameters . Some of these changes include increasing the learning rate parameter, removing dropout (see section 12.9), reducing L_2 weight regularization, and shuffling training samples more thoroughly so the same samples do not always appear in a mini-batch together [159].

12.9 Network Initialization

An important issue in network training is the initialization of the weights [160]. Zero initial values lead to equal loss derivatives with respect to all the weights, which will lead to the same value for all the weights in every iteration during training. The neural network then essentially becomes a linear model. Hence random initial values should be the starting values. There are two possible issues with random initial values: they could lead to vanishing gradients (the vanishing gradient problem), or divergent gradients (the exploding gradient problem). One way to avert the vanishing gradient problem is to use ReLU activations.

Methods to prevent the exploding gradient problem include gradient clipping (setting a threshold value for the magnitude of the gradients that if exceeded will set them to the threshold), or modifying the parameters of random number generators (uniform or Gaussian) that are used to produce the initial values. Let us define $[-r, +r]$, $r > 0$, to be the interval in which random numbers with uniform density are generated. Gaussian random numbers are assumed to be zero-mean and are defined by their variance σ. Table 12.1 shows random number settings that are functions of the number of inputs to a layer N_{in}, and the number of outputs of that layer N_{out}. These initializations are known as Xavier initialization, or (for the ReLU activation) He initialization [161].

Activation	Uniform: $[-r, r]$	Normal: σ
Logistic	$r = \sqrt{6}/r_0$	$\sigma = \sqrt{2}/\sigma_0$
tanh	$r = 4\sqrt{6}/r_0$	$\sigma = 4\sqrt{2}/\sigma_0$
ReLU	$r = \sqrt{22}/r_0$	$\sigma = 2/\sigma_0$
	$r_0^2 \equiv N_{\text{in}} + N_{\text{out}}$	$\sigma_0^2 \equiv N_{\text{in}} + N_{\text{out}}$

Table 12.1: Random number generator settings for network initialization.

12.10 Regularization

Overfitting is a major issue in neural networks when the neural network is trained to achieve excellent performance (such as low classification error) on the training data, but in doing so it becomes so specialized that it is unable to do well on any other data set [162]. An overfit neural network loses the ability to generalize, i.e., it is unable to apply what it has learned from the training data to other data it has never been exposed to.

As a practical matter, in any regression or classification problem, it is important to partition the data into two portions: a training portion and a test and validation portion. Typically, about 80–90% of the available data is used for training, with the remaining 10–20% for testing and validation. The important criterion is not so much how well the neural network performs on the training data, but how it does on the test data. The real validation in neural network training is then performed on the test data, not on the training data. In order to reliably measure the network's ability to generalize, we should never mix training and test data. Methods to reduce overfitting are known as regularization.

When training a neural network we often see a continual decrease in training loss accompanied by a drop in test data loss for some time before the testing loss begins to plateau or increase, as illustrated in figure 12.15. The gap between training and validation performance is sometimes referred to as the *generalization gap* [163]; this is clear evidence of overfitting when the neural network has learned to represent the training data so well that it has lost the ability to generalize to holdout data.

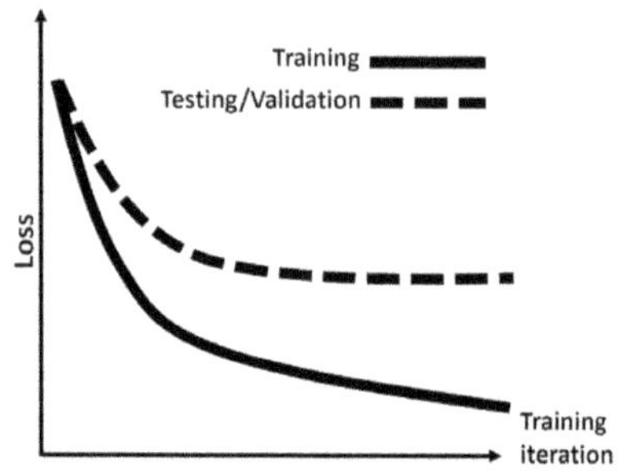

Figure 12.15: An overfit neural network.

In general controlling a neural network's ability to fit training data reduces to managing its representational capacity. Recall from theorem 51 that with enough neurons (and appropriate activation functions), standard multi-layer perceptrons are universal function approximators. Thus, sufficiently reducing the number of neurons in a neural network will reduce its representation capacity, inhibiting the model's ability to (over)fit training data. We can also control the capacity of our models by limiting the depth of the network, bounding the norm of the weight matrices with L_1 or L_2 regularization, injecting noise into the input/hidden layers of the network, and performing more general input data manipulations called *data augmentation* [164] where we perturb/transform the input data in such a way as to prevent the neural network from placing emphasis on spurious features of the data; e.g., for a cat vs dog image classifier we might perform color transformations to overcome biases in the color of the animals in the training set. These and other methods explicitly control what neural network models can learn from the data during parameter optimization (SGD and its variants) and are thus effective regularizers for practical applications.

Early stopping is another effective method to avoid overfitting [165]: when a new best loss value on the validation set is found, the current best neural network weights are saved, and training stops if the validation loss has not improved within a specified number of training iterations. The number of training iterations (or epochs) before halting is known as a "stopping criterion" and prevents model weights from continuing to improve on the training set while making little progress on the validation/test set (see figure 12.15).

Another method is the surprising idea of dropout [166, 167], when at every training step, every neuron, including input layer neurons but excluding output layer neurons, will be dropped with a probability p. A neuron that is dropped is simply ignored during this training step. The probability p is the dropout rate, and is typically set to 0.25. The idea is that the neurons that are not dropped at a given iteration must adapt to the data, becoming as useful as possible on their own and not co-adapting to their neighboring neurons. We may think of neural networks with different dropout configurations as different neural networks. If N neurons may be dropped then there are 2^N different neural networks (the set of all subsets of the given N neuron). The training with dropout gives, roughly, the average result of a large number of neural networks. This is related to the idea of boosting, in which the results of many weak classifiers are combined to produce a strong classifier result.

An improved implementation of the original concept of dropout is *inverted dropout*. For a dropout rate of p, when testing the network (without dropout), a neuron is connected to, on average, $1/(1-p)$ as many inputs as it was during training, i.e., each neuron has a total input signal that is, on average, $1/(1-p)$ as large as what the network was trained on. To compensate for this, we multiply each neuron's connection weight by $1-p$ after training.

Another way of regularizing a neural network is to tie parameters together; for instance, we may require that some weights should be equal. This provides more training data per weight. A common way of tying parameters is to use convolutional neural networks (see section 12.11).

12.11 Convolutional Neural Networks (CNNs)

In the fully connected network described in section 12.3, every input to layer l is connected to every output of the previous layer $l-1$ using the weight functions. A fully connected network with many hidden layers and many neurons in each layer can contain tens of thousands of weights which, in cases of small training data sets, can lead to overfitting issues. A far more important problem with fully connected networks in pattern recognition problems is the absence of shift (translation) invariance, or insensitivity to local distortions. For instance, spoken words can have different speed, pitch, and intonation, which can cause substantial variations in time location of important features in the input data. Although a fully-connected network can, in principle, learn to produce shift-invariant outputs, the required number of training sets and the network size may be prohibitively large. Furthermore, fully connected networks do not take advantage of the high correlations between nearby data points.

We should recall that all previous chapters in this text have used the usual definition of a convolution between a sequence $[y_1, \ldots, y_N]$ and a set of weights, or kernel, $[w_1, \ldots, w_M]$, which is the same as a correlation between $[y_1, \ldots, y_N]$ and $[w_M, \ldots, w_1]$. In what follows, we will perform correlations between two sequences, but we will refer to this as convolution so as to keep the machine learning terminology; a justification for this might be that weight vectors are found through training and so any specific initial indexing is arbitrary—if we index the weights $N, \ldots, 1$, then a convolution is the same as correlation with indices $1, \ldots, N$.

An explicitly shift-invariant architecture that learns local features first, and then recognizes spatial or temporal patterns by combining those local features, is the convolutional neural network (CNN) [168, 169], which consists of a number of layers, a number of convolutional neurons in each layer (the same as the number of outputs of that layer), a number of weight filters, or kernels, (total number of filters in each neuron equals the number of neurons in the previous convolutional layer), a bias vector for each convolutional neuron, and a *flattening layer* to produce scalar outputs in the output layer. Many convolutional networks designed to work on images include a *pooling layer* to perform down sampling after a chosen convolutional layer; the down sampling is sometimes achieved through convolution with strides greater than 1, when a large input into one layer produces a smaller input to the next layer. It is, however, more common to use a down sampling method known as *max-pooling*.

The convolutional network architecture is believed to approximate the human visual cortex, which extracts local features, e.g., horizontal edges, vertical edges, and local patterns [169]. The hierarchical viewpoint suggests the importance of multi-layer networks for pattern recognition. A CNN commonly has inputs with a depth dimension; for instance, imagery in RGB channels provide input with depth 3, or IR time series measurements in 2 frequency bands provide input with depth 2. In section 12.12 we will discuss a time series problem to classify six human activities based on nine measurements that provide time series input with depth 9. For time series inputs with a depth dimension, outputs of each hidden layer (before activation) are called *feature vectors*; when inputs are two dimensional images with depth, then outputs of each hidden layer (before activation) are called *feature maps*. As explained below, outputs of a convolutional hidden layer include layer biases as additive quantities.

Figure 12.16 illustrates the concept of a CNN for time series classification with an input depth dimension of 5, i.e., the input layer consists of five (equal dimension) time series $\boldsymbol{x}_k$, $1 \leq k \leq 5$, representing 5 different measurements for classification. There are three hidden layers with 3, 2, and 4 convolutional units (neurons), respectively, a flattening layer with input vectors $\boldsymbol{y}_k$, $1 \leq k \leq 4$, with a single $Q \times 1$ vector output, that leads to a single scalar y in the output layer. Each of the hidden layers have weight filters, or kernels, denoted by $\boldsymbol{w}_{ij}^{(l)}$, where $1 \leq l \leq 3$ is the layer number, j is the convolutional unit (neuron) in that layer whose input is denoted by index i. The output of convolutional neuron j in hidden layer $1 \leq l \leq 3$ before (nonlinear) activation is the feature vector for neuron j and layer l, and is denoted by the vector $\boldsymbol{v}_j^{(l)}$. The bias for a single convolutional neuron is a single scalar value which facilitates the neuron's ability to learn patterns in a predictable and consistent manner when presented with similar input data. Therefore, the bias vector $\boldsymbol{b}_j^{(l)}$ is always taken to be $\boldsymbol{b}_j^{(l)} = b_j^{(l)} \times [1, \ldots, 1]$.

The output of the (nonlinear) activation for neuron j and layer l is denoted by $\boldsymbol{y}_j^{(l)}$. The number of convolutional neurons in each layer determines the depth dimension of the layer's feature vectors and the layer's output vectors. For instance, $\boldsymbol{v}_1^{(1)}$, $\boldsymbol{v}_2^{(1)}$, and $\boldsymbol{v}_3^{(1)}$ are the feature vectors at hidden layer 1 while $\boldsymbol{y}_1^{(1)}$, $\boldsymbol{y}_2^{(1)}$, and $\boldsymbol{y}_3^{(1)}$ are the final outputs of layer 1 (after the nonlinearity).

The output vectors $\boldsymbol{v}$ of each hidden layer, namely, the layer's feature vectors, are computed by correlating the input vectors with their corresponding weight vectors, summing the resulting vectors and adding the bias term. The final outputs $\boldsymbol{y}$ of the layer are found by passing the result through the point-by-point nonlinear activation function σ, as illustrated on the bottom portion of figure 12.16. For instance, the output time series of the convolutional unit 2 in hidden layer 2, namely $\boldsymbol{y}_2^{(2)}$, is found from

$$\boldsymbol{y}_2^{(2)} = \sigma\big(\boldsymbol{v}_2^{(2)}\big), \ \boldsymbol{v}_2^{(2)} = \boldsymbol{w}_{12}^{(2)} * \boldsymbol{y}_1^{(1)} + \boldsymbol{w}_{22}^{(2)} * \boldsymbol{y}_2^{(1)} + \boldsymbol{w}_{32}^{(2)} * \boldsymbol{y}_3^{(1)} + \boldsymbol{b}_2^{(2)}, \tag{12.32}$$

where "$*$" denotes correlation between the associated time series and $\boldsymbol{b}_2^{(2)} = b_2^{(2)} \times [1, \ldots, 1]$. In the rest of this section we will omit the bias term for notational convenience.

If we denote the filter (kernel) dimension at hidden layer l by M_l (which we assume to be odd), $l = 1, 2, 3$, and denote by N the dimension of each of the 5 input layer vectors, then the correlation

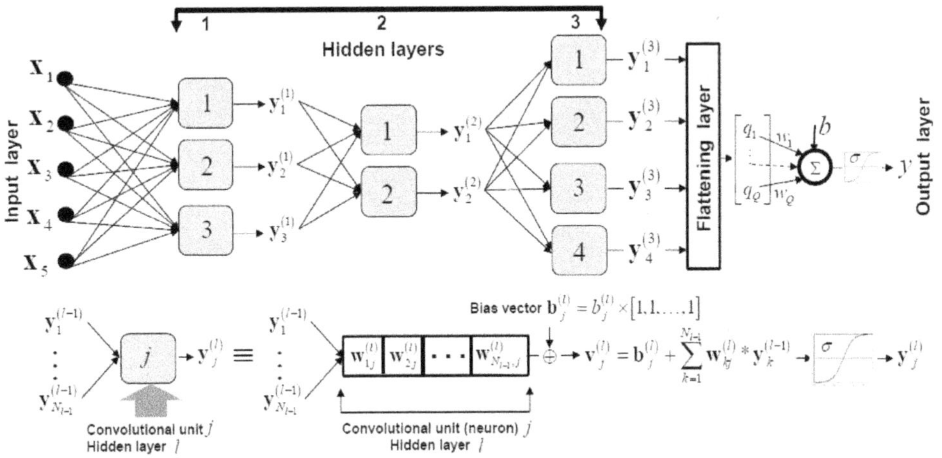

Figure 12.16: A CNN with 5 time series of equal dimension in the input layer, 3 hidden layers, and an output layer with 1 scalar output. The bottom portion describes each convolutional unit (neuron), where $\boldsymbol{w} * \boldsymbol{y}$ indicates correlation between the two sequences $\boldsymbol{w}$ and $\boldsymbol{y}$. The flattening layer in this example produces 1 scalar output for a 2-class classification network.

equation is performed according to the centered output prescription described in section 3.9, and illustrated in figure 3.10. Thus, we pad the time series $\boldsymbol{y}$ on either side by $(M_l - 1)/2$ zeros and denote the resulting $(N + M_l - 1) \times 1$ vector by $\boldsymbol{y}_0$. Next we align the filter weights $\boldsymbol{w}$ with $\boldsymbol{y}_0$ so that their first elements match and we multiply the elements 1 through M_l and sum them to obtain the first element of the correlation vector. Then we slide the filter along by one element and repeat the process to obtain a total of N correlation values; the last correlation value is found when the last element of the filter is aligned with the last element of $\boldsymbol{y}_0$. This procedure is illustrated in figure 12.17 for an input sequence with 4 elements and a filter with 3 elements; the input sequence is padded with one zero on either side and 4 correlation sequence values are calculated.

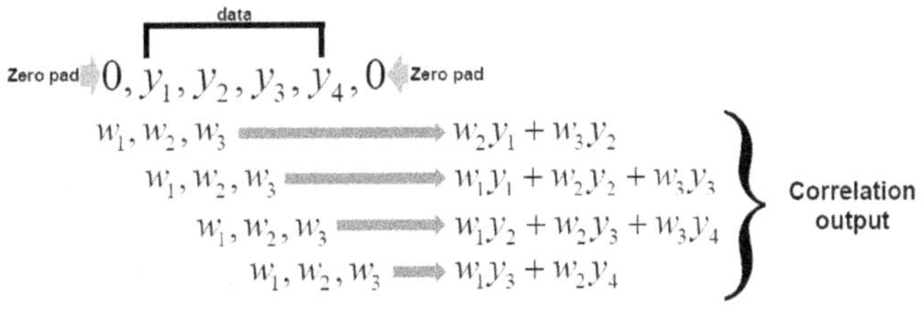

Figure 12.17: Correlation procedure for an input sequence with 4 elements and a filter with 3 elements, resulting in an output with 4 elements.

The correlations shown in figure 12.16 are, in practice, carried out as matrix multiplications with zero-padded vectors. Figure 12.18 shows the procedure to calculate the sum $\boldsymbol{w}_1 * \boldsymbol{x}_1 + \boldsymbol{w}_2 * \boldsymbol{x}_2 + \boldsymbol{w}_3 * \boldsymbol{x}_3$ for a set of three 4×1 input vectors, and three filters (kernels) with 3 elements each, with the final result having 4 elements. The 3×3 section of the input data that is being multiplied by the 3 elements of the weight vector is known as a receptive field. The weight matrix, at each correlation lag, produces a single output, and weights in different layers provide local receptive field connections.

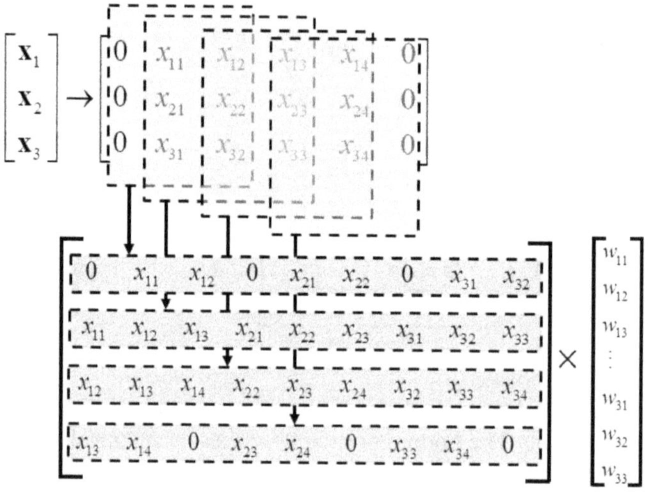

Figure 12.18: Zero padding and matrix multiplication procedure to calculate the sum $\boldsymbol{w}_1 * \boldsymbol{x}_1 + \boldsymbol{w}_2 * \boldsymbol{x}_2 + \boldsymbol{w}_3 * \boldsymbol{x}_3$.

The flattening layer, in general, produces a single vector output of some dimension (a hyper-parameter of the model), which can then be fed into a fully connected layer with **softmax** activation for classification, or if the dimension of the output vector is the same as the number of classes, then it can be fed directly into a **softmax** activation. One way to flatten is to simply stack all the vector inputs into a long vector; for instance, the quantities q_k, $1 \leq k \leq Q$ in figure 12.16 could be equivalent to the vector $[\boldsymbol{y}_1^{(3)}, \ldots, \boldsymbol{y}_4^{(3)}]$. Another method, known as global average pooling (GAP) [170] is to average each of the input vectors, and then producing a single vector of all the averages; for example, in figure 12.16, q_k, $1 \leq k \leq 4$, could be averages of each of the 4 feature vectors $\boldsymbol{y}_1^{(3)}, \ldots, \boldsymbol{y}_4^{(3)}$.

When two dimensional arrays (images) are used as inputs to a CNN, then for each convolutional neuron weight matrix (filter or kernel) the corresponding images are zero-padded around their boundaries, and correlation results for all images are summed and added to a bias array (image of the same dimensions) to produce an output image (also known as feature maps—see later in this section). Figure 12.19 shows two 5×5 images being correlated with two distinct 3×3 weight matrices (filters or kernels) and the correlation results are added to produce the 5×5 output image (we are neglecting the additional 5×5 bias image). The two images in figure 12.19 might represent input with a depth of 2 at the input layer of a neural network while the weight matrices might be those associated with one of the neurons in the first hidden layer. Alternatively, the two images might be two feature maps (outputs of a hidden layer) going into a neuron in the next hidden layer.

Figure 12.20 shows an input of depth 3 RGB images into a CNN. The first hidden layer has 5 convolutional neurons, each of which has 3 (the same number of input image depth) weight matrices (kernels) of size 3×3. There are, therefore, 5 feature maps, i.e., 5 image outputs from the hidden layer, whose size will be the same as the input image size for unit stride, or smaller if strides greater than 1 are used. Let us denote a feature map by $F_{ij}^{(d)}$, $d = 1, \ldots, 5$. Given the input image matrices R_{ij}, G_{ij}, B_{ij}, and weight

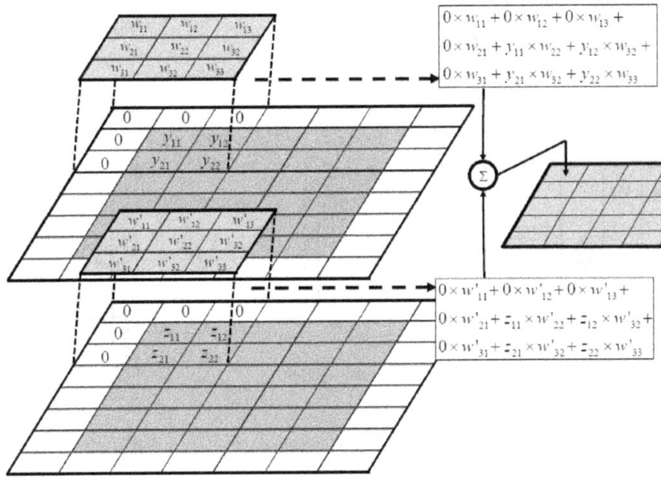

Figure 12.19: Zero padding 5 × 5 images and correlating with two 3 × 3 filter matrices.

matrices (kernels) by $w_{ij}^{(d,n)}$, $n = 1, 2, 3$, then we have

$$F_{km}^{(d)} = \sum_{i,j=0}^{2} \left(w_{ij}^{(d,1)} R_{i+k,j+m} + w_{ij}^{(d,2)} G_{i+k,j+m} + w_{ij}^{(d,3)} B_{i+k,j+m} \right), \tag{12.33}$$

where $k, m = 0, \ldots$ refer to the elements of each feature map. The outputs of a hidden layer are, of course, the feature maps plus layer biases that will form the input to a nonlinearity.

If the input images are square with dimension $N \times N$, weight matrices have odd dimension $M \times M$, and correlations are performed with stride $S = 1$ and padding with $(M-1)/2$ zeros, then $k, m = 0, \ldots, N-1$ and feature maps have dimension $N \times N$; if correlation stride $S > 1$, then feature maps have dimension $((N-1)/S+1) \times ((N-1)/S+1)$. Figure 12.20 is drawn to indicate $S > 1$ (feature map sizes are smaller than input images).

Figure 12.21 describes the sum of correlations of the three weight matrices of the first neuron with the three input images to produce feature map 1: if the 3 × 3 squares below the three weight matrices $\boldsymbol{w}^{(1,1)}$, $\boldsymbol{w}^{(1,2)}$, $\boldsymbol{w}^{(1,3)}$, are denoted by $\tilde{\boldsymbol{R}}$, $\tilde{\boldsymbol{G}}$, and $\tilde{\boldsymbol{B}}$ (portions of the appropriately zero-padded matrices), respectively, then the sum of simple dot products, namely, $\boldsymbol{w}^{(1,1)} \cdot \tilde{\boldsymbol{R}} + \boldsymbol{w}^{(1,2)} \cdot \tilde{\boldsymbol{G}} + \boldsymbol{w}^{(1,3)} \cdot \tilde{\boldsymbol{B}}$ produces the indicated element of the feature map 1 matrix.

In image classification applications a weight matrix (kernel) may be thought of as being matched to some particular feature of the input image. For example, two matrices below show sensitivity to horizontal structures and to vertical structures, respectively; i.e., using the first matrix in an input to one layer will emphasize horizontal features of that image in the input to the next layer.

$$\text{Horizontal structures} \leftrightarrow \begin{bmatrix} 0 & 0 & 0 \\ 1 & 1 & 1 \\ 0 & 0 & 0 \end{bmatrix} \quad \text{Vertical structures} \leftrightarrow \begin{bmatrix} 0 & 1 & 0 \\ 0 & 1 & 0 \\ 0 & 1 & 0 \end{bmatrix}$$

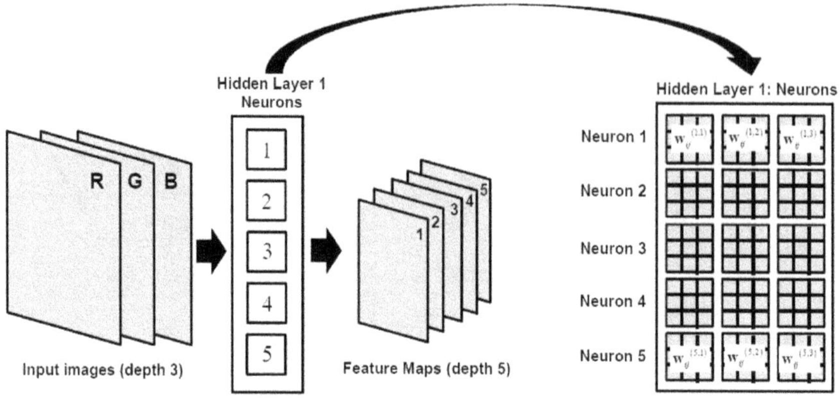

Figure 12.20: Input layer images with depth 3, first hidden layer with 5 convolutional neurons, and 5 feature maps. The box on the right shows a detailed picture of all the weight matrices for all 5 neurons.

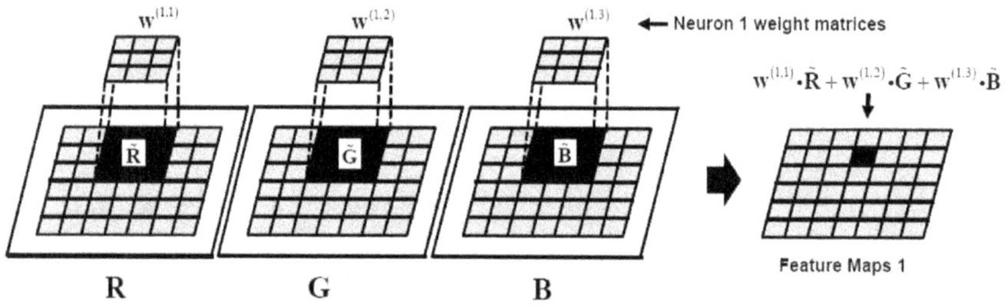

Figure 12.21: Feature map 1 generated from stride 1 sliding correlation values obtained by sum of dot products of neuron 1 weight matrices with correspondingly aligned 3×3 image sections (white sections around RGB images indicate zero padding).

12.12 Time Series Classification with a Convolutional Neural Network

Let us consider the multi-class classification problem of Human Activity Recognition (HAR) data set from the University of California at Irvine (UCI) machine learning repository [171]. Data consists of $10,299$ instances of triaxial accelerometer and gyroscope time series measurements representing six activities, namely, walking, walking upstairs, walking downstairs, sitting, standing, and laying. Each activity has 9 associated 128-point time series: body acceleration, body gyro, and total acceleration, for all three axes. Figure 12.22 shows data for one instance of three of the most dissimilar of the 6 activities, namely, walking, sitting and standing; thus, the inputs to a classification CNN are 9 time series (3 sets of triaxial measurements).

An example CNN to classify the six activities of the HAR data is depicted in figure 12.23. It consists of an input layer of nine 128×1 vectors representing the accelerometer and gyro measurements, four hidden layers, a flattening layer with 6 outputs that go through a **softmax** nonlinearity to produce the final 6 class probabilities for classifying the six activities. Hidden layer 1 has 32 convolutional neurons that each have 9 filters with 3 elements each. Hidden layers 2, 3 have 32 convolutional neurons that each have 32 filters with

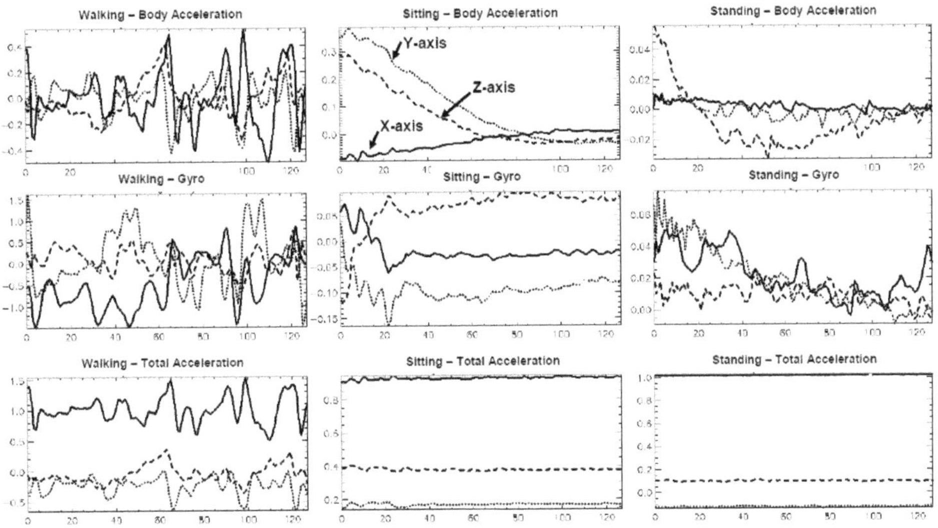

Figure 12.22: Vector time series (three spatial axes) for three of six activities of the HAR data from UC Irvine: body acceleration, gyro, and total acceleration for Walking, Sitting, and Standing.

3 elements each (recall that the number of filters in each convolutional neuron is the same as the number of input vectors into the layer). Hidden layer 4 has 6 convolutional neurons (to match the number of classes) that each have 32 filters with 3 elements each.

The GAP flattening layer has 6 scalar outputs that are passed through a **softmax** nonlinearity to produce the final output layer of 6 scalar values representing the class probabilities. Training was performed on 7,352 instances and validation was done on the remaining 2,949 instances.

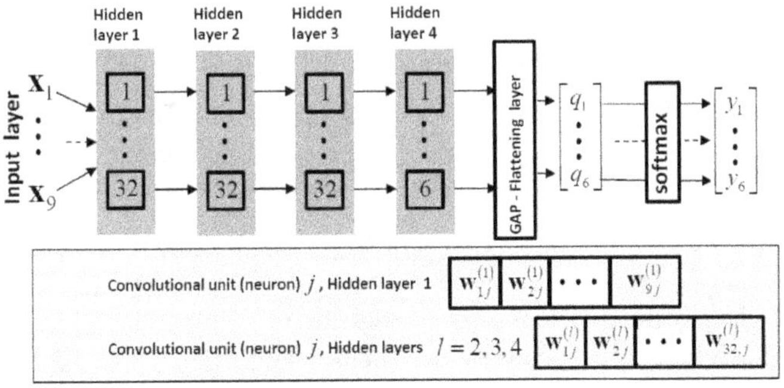

Figure 12.23: CNN to classify six activities of the HAR data set.

Figure 12.24 shows the confusion matrix for the classification of the six activities, together with a table summarizing the results. *Precision* is the probability of correct classification; it is calculated by dividing each diagonal number by the sum of the elements of the row passing through that diagonal element. *Recall* is the probability of correct prediction and is found by dividing the diagonal element by the sum of the elements of the column through that element. *F1 score* is the harmonic mean of precision and recall, and *support* is the number of instances of each activity used for validation (with each instance associated with nine 128×1 vectors). The model has $7,506$ parameters with an overall accuracy of 95.24%, which compares well with the best reported result of 96.7% achieved using deep recurrent neural networks (with an unknown number of parameters). A set of 561 hand-engineered features derived from the time series (a reduction of more than 50% from the 9×128 time series values) have been classified using a logistic regression network with an overall classification accuracy of 96.2%, while a fully connected network with 3-hidden layers and $\approx 350,000$ parameters achieved 96% overall accuracy using the same 561 features. It is possible to improve the CNN accuracy by using more complicated architectures known as *ResNets*, i.e., fully convolutional networks with residual connections, to achieve 96.9% overall accuracy, but at the cost of increasing the network parameters to $\approx 1,000,000$.

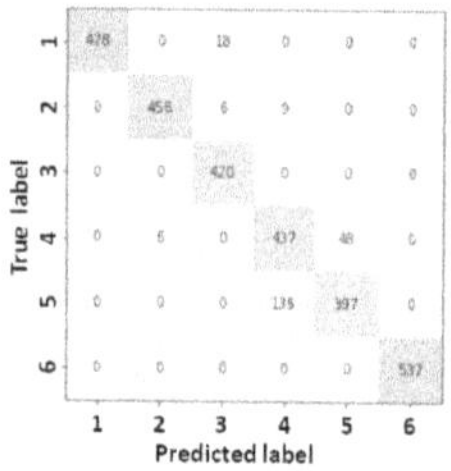

	precision	recall	F1 score	support
1. walking	1.00	0.96	0.98	496
2. walking up	0.99	0.97	0.98	471
3. walking down	0.95	1.00	0.97	420
4. sitting	0.75	0.89	0.82	491
5. standing	0.89	0.75	0.81	532
6. laying	1.00	1.00	1.00	537

Figure 12.24: Summary of results for CNN classification of HAR data.

12.13 Image Classification with a Convolutional Neural Network

The MNIST (Modified National Institute of Standards and Technology) data [172] consists of handwritten gray-scale 28×28-pixel images of digits from 0 to 9, ten samples of which are shown in figure 12.25.

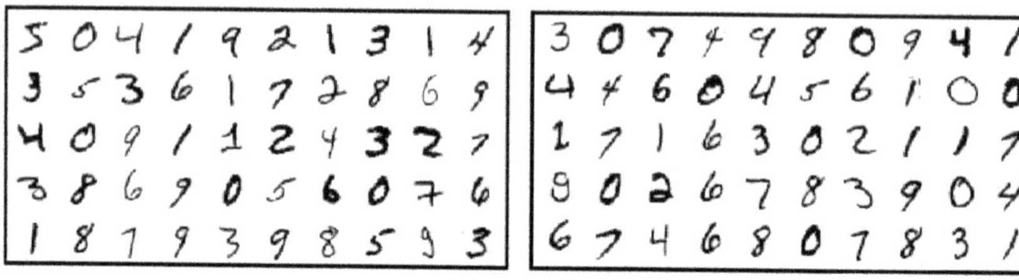

Figure 12.25: Samples of handwritten images from MNIST data.

We used a CNN to classify handwritten digits consisting of 4 hidden layers, a flattening layer with 10 outputs that go through a **softmax** nonlinearity to produce the final 10 class probabilities. Hidden layer 1 has 32 convolutional neurons that each have a single filter, while layers 2, 3 have 32 neurons each of which has 32 filters. The last hidden Layer 4 has 10 neurons with 32 filters each. Hidden layer 4 has 10 convolutional neurons that each have 32 filters; all filters have size 3×3. Figure 12.26 shows the CNN that we used. We chose a mini-batch size of 2^5, which corresponds to $1,875$ weight updates for each epoch, and we trained the network for 25 epochs. This network had a total of $19,358$ trainable parameters and achieved 97.8% accuracy on the test set. More sophisticated network architectures with considerably more parameters have achieved accuracies in excess of 99%.

Figures 12.27, 12.28, 12.29 and 12.30 show the activation outputs for convolutional layers 1 through 4.

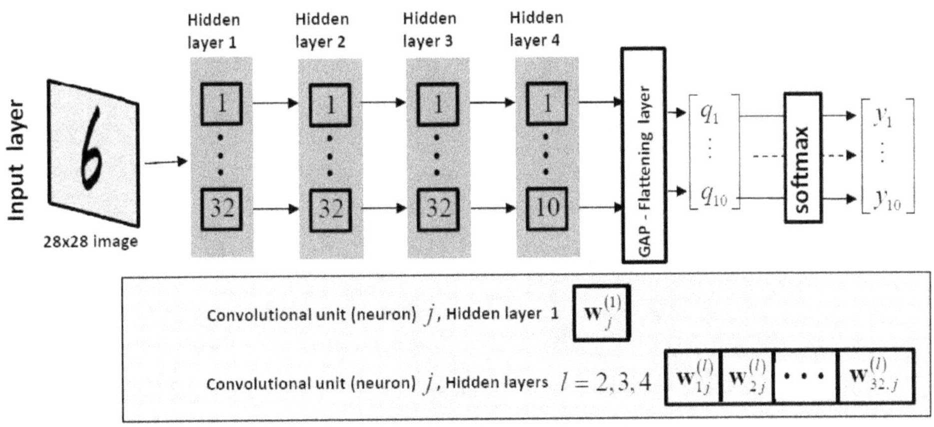

Figure 12.26: CNN to classify the MNIST handwritten images of digits 0–9.

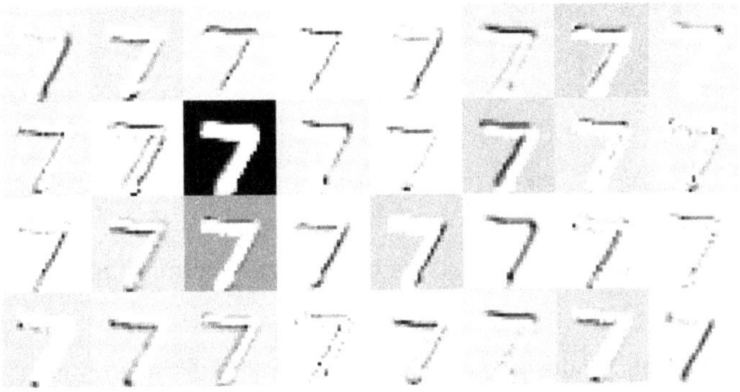

Figure 12.27: Activation outputs for class label 7 for convolutional layer 1.

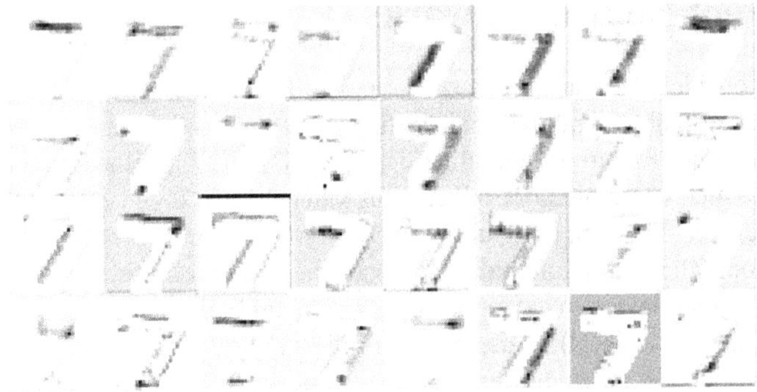

Figure 12.28: Activation outputs for class label 7 for convolutional layer 2.

Earlier layers tend to detect more primitive aspects of the image and subsequent layers tend to learn more complex features. The final convolutional layer learns a representation that maximizes the probability of class membership. For instance, in figure 12.30, the image that looks more like the correct class (in this

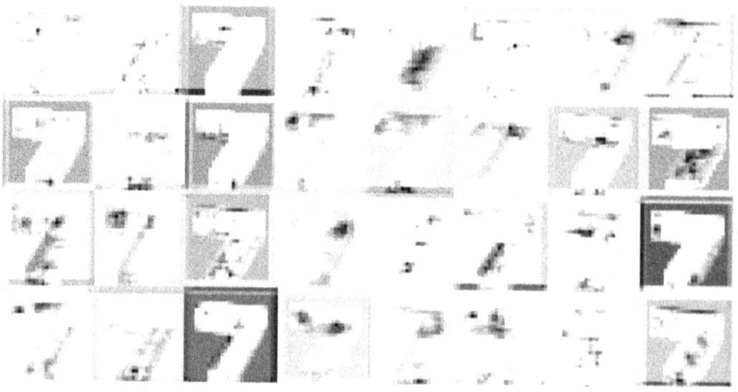

Figure 12.29: Activation outputs for class label 7 for convolutional layer 3.

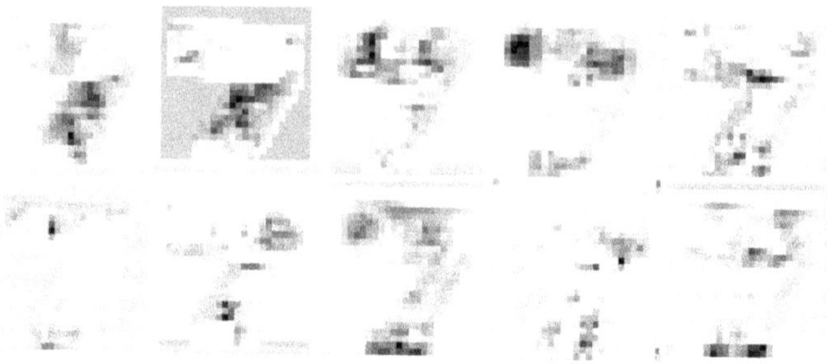

Figure 12.30: Activation outputs for class label 7 for convolutional layer 4.

case, the number 7) is, in fact, produced by neuron 8. Note that the neuron number corresponds to the actual digit plus one, since classes are numbered 0 to 9.

12.14 Recurrent Neural Networks (RNNs)

The neural networks considered so far have been static, i.e., for a given input vector, they produce a single output, or a vector; they resemble combinatorial digital logic composed of gates, and are incapable of keeping track of passage of time. Recurrent neural networks (RNNs) [173], like digital circuits with memory, have an internal state which holds memory of previous values; these together with the current input value are used to make a decision. They are particularly useful for tasks such as speech recognition or connected handwriting recognition, when modeling data sequences that depend on previous values, and they can be combined with convolutional layers to extend the effective neighborhood of pixels.

Consider a recurrent neuron (RN) as depicted on the left hand side of figure 12.31: at time step t it receives an input $\boldsymbol{x}_t$, which we assume to be a $d \times 1$ vector, in addition to the scalar valued hidden state output of the previous step h_{t-1}. The final output of the RN y_t at time step t is found from its hidden state h_t through a simple feed forward neuron with a scalar weight w_{hy} (the subscript indicates "hidden to y") and scalar bias b_y,

$$y_t = w_{hy} h_t + b_y. \tag{12.34}$$

The behavior of an RN, including only the input and the hidden state, can be understood by unrolling it through time as shown on the right hand side of figure 12.31, which also introduces the notation for a summation unit followed by an activation function. If we introduce an "x to hidden" weight vector $\boldsymbol{w}_{xh}$ of dimension d and a scalar "hidden to hidden" weight w_{hh} then the output of the recurrent neuron at time step t is

$$h_t = \sigma\left(\boldsymbol{w}_{xh}^T \boldsymbol{x}_t + w_{hh} h_{t-1} + b\right), \tag{12.35}$$

where b is the recurrent neuron bias and σ denotes the activation function. Note that the weights and bias are shared at all time steps, i.e., they do not change as a function of time step t. The hidden state h_t at one time step t is often the only quantity of interest, and so figure 12.31 shows a scalar valued hidden state on the left hand side. Clearly, the process of calculating h_t yields all previous hidden states $\ldots, h_{t-2}, h_{t-1}$ too. Putting all these values into an $N \times 1$ vector (the number of time steps) produces an $N \times 1$ vector $\boldsymbol{h}^{[t]}$; the superscript is to prevent confusion with the notation $\boldsymbol{x}_t$, which is a $d \times 1$ vector denoting the data vector at time step t, and whose elements are not necessarily a time series. For instance, $\boldsymbol{x}_t$ could denote a collection of economic indices at time step t that are used to predict the Dow Jones industrial average y_t from the hidden state h_t. Thus, in our discussion of a single RN we consider the scalar hidden value h_t and not the vector consisting of hidden states at t and all previous time steps.

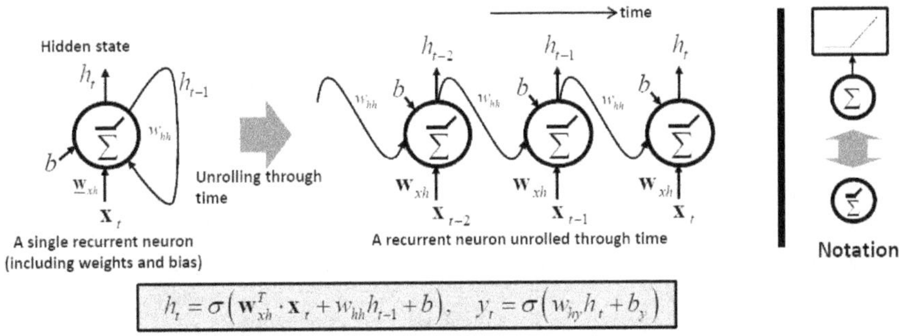

Figure 12.31: An RN and its unrolling through time.

The basic recurrent neuron concept can be generalized to a layer of J recurrent neurons as illustrated in figure 12.32. The outputs of individual recurrent neurons in each layer form the vector output of the layer, e.g., if $h_{t-1}^{(j)}$ is the scalar output of recurrent neuron j, then

$$\boldsymbol{h}_{t-1} = \left[h_{t-1}^{(1)}, \ldots, h_{t-1}^{(J)}\right]^T, \tag{12.36}$$

is the vector output of the layer at time step $t-1$, which is fed into the layer at time step t. Thus, at time step t the hidden state is

$$\boldsymbol{h}_t = \sigma\left(\boldsymbol{w}_{xh} \boldsymbol{x}_t + \boldsymbol{w}_{hh} \boldsymbol{h}_{t-1} + \boldsymbol{b}\right), \tag{12.37}$$

where $\boldsymbol{x}_t$ is a $d \times 1$ vector, $\boldsymbol{h}_t$ and $\boldsymbol{b}$ are $J \times 1$ vectors, $\boldsymbol{w}_{xh}$ is a $J \times d$ matrix, and $\boldsymbol{w}_{hh}$ is a $J \times J$ matrix. Each neuron in the layer has its own "x to hidden" weight vector and all these weight vectors form the rows of the full weight matrix $\boldsymbol{w}_{xh}$. Similarly, the "hidden to hidden" weight matrix is constructed from those associated with each neuron while $\boldsymbol{b}$ is the vector of individual biases for each RN in the layer. The final output of this layer is given by a feed forward neuron with the hidden state as input,

$$\boldsymbol{y}_t = \boldsymbol{w}_{hy} \boldsymbol{h}_t + \boldsymbol{b}_y, \tag{12.38}$$

where $\boldsymbol{w}_{hy}$ has dimension $J \times J$ and $\boldsymbol{y}_t$ and $\boldsymbol{b}_y$ have dimension $J \times 1$.

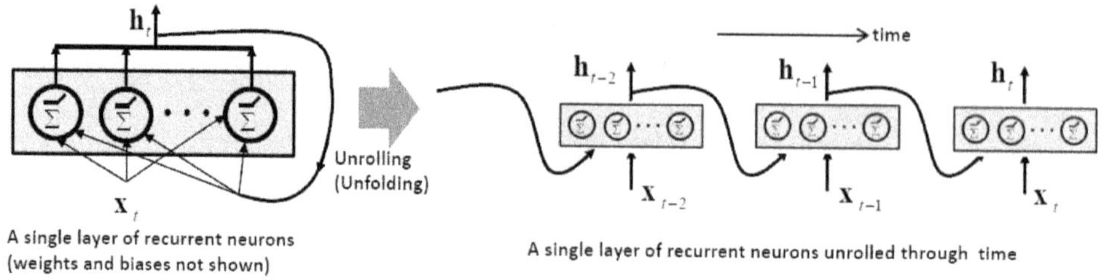

$$\mathbf{h}_t = \sigma\left(\mathbf{w}_{xh}\mathbf{x}_t + \mathbf{w}_{hh}\mathbf{h}_{t-1} + \mathbf{b}\right), \quad \mathbf{y}_t = \sigma\left(\mathbf{w}_{hy}\mathbf{h}_t + \mathbf{b}_y\right)$$

Figure 12.32: A layer of recurrent neurons and its unrolling through time.

Recurrent neural networks are used in some distinct architectures as illustrated in figure 12.33; here we show the final output of each RN without the intermediate hidden state, i.e., each box in the figure indicates an RN layer together with a final feed forward NN to produce the output $\boldsymbol{y}_t$:

- (a) shows the basic configuration in which a sequence of input vectors, in this case a sequence of 4 elements, produces a sequence of output vectors. This architecture can be used, for example, to reproduce or predict a time series.
- (b) shows a configuration with a sequence of input vectors and only a single output. This might be used, for example, in a time series prediction application, where the input is, say, a sequence of market related information and the output is the 1-day prediction of a particular stock value.
- (c) shows a sequence with a single input and a sequence of outputs. An application might use an image as input image, to produce a sequence of words composing a caption for the image.
- (d) has a sequence of inputs and a sequence of outputs, but there is a delay between input and output. This configuration is referred to as a decoder. A suggested application is language translation, in which the input and output are sequences of words. This architecture delays the producing of outputs, i.e., the translation, until a few words have been processed.

The key to training an RNN is to unroll through time and then apply backpropagation in the usual way [174]. For instance, consider the delayed sequence to sequence structure shown in figure 12.34 (once again, we show the final output and not the hidden states). At time step t training data $\boldsymbol{x}_{t-4}$, $\boldsymbol{x}_{t-3}$ (or a mini-batch at that time) are presented, resulting in outputs $\boldsymbol{y}_{t-2}$, $\boldsymbol{y}_{t-1}$, $\boldsymbol{y}_t$, based on the current weights and biases (dotted lines). Once the cost function for the current data is computed, its gradient with respect to the weights and biases is backpropagated (solid lines) through all neurons that influence the outputs (in the present example, the neuron with input $\boldsymbol{x}_{t-4}$).

RNNs are particularly susceptible to the vanishing gradient problem: states that are too far from the current state contribute nothing to the learning, yet the network must learn long-term dependencies in the data. A solution to this problem is based on the concept of a long short-term memory (LSTM) recurrent neuron [175] with a $d \times 1$ input $\boldsymbol{x}_t$ and hidden state vector $\boldsymbol{h}_t$.

The central idea to remembering inputs over a long time is that of a gated cell state (or gated memory unit) $\boldsymbol{c}_t$, which contains all the information up to time step t; it is gated so as to give the cell the ability to store information (opening the gate) or deleting it (closing the gate). An LSTM RN has additional gates that control the flow of data to update the cell state, i.e., how old memory and new memory are to be

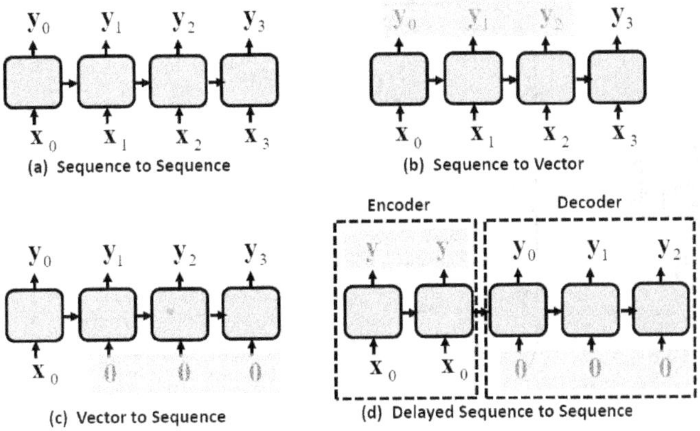

Figure 12.33: Different recurrent network architectures.

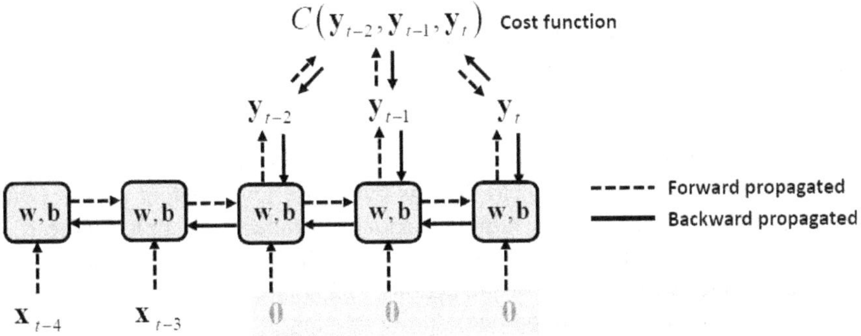

Figure 12.34: Data flow for backpropagation training of an RNN.

combined. The gates are: input, input modulation, forget, and output. The full operation of an LSTM neuron, illustrated in figure 12.35, is described by:

$$\boldsymbol{i}_t = \sigma(\boldsymbol{w}_{xi}\boldsymbol{x}_t + \boldsymbol{w}_{hi}\boldsymbol{h}_{t-1} + \boldsymbol{b}_i), \quad \boldsymbol{g}_t = \tanh(\boldsymbol{w}_{xg}\boldsymbol{x}_t + \boldsymbol{w}_{hg}\boldsymbol{h}_{t-1} + \boldsymbol{b}_g),$$
$$\boldsymbol{f}_t = \sigma(\boldsymbol{w}_{xf}\boldsymbol{x}_t + \boldsymbol{w}_{hf}\boldsymbol{h}_{t-1} + \boldsymbol{b}_f), \quad \boldsymbol{o}_t = \sigma(\boldsymbol{w}_{xo}\boldsymbol{x}_t + \boldsymbol{w}_{ho}\boldsymbol{h}_{t-1} + \boldsymbol{b}_o),$$

each defined with their own weight matrices and bias vectors. The quantities $\boldsymbol{i}_t$, $\boldsymbol{g}_t$, and $\boldsymbol{f}_t$ are computed first, and together with the previous cell state $\boldsymbol{c}_{t-1}$ are used to obtain the present cell state, which together with $\boldsymbol{o}_t$ is used to compute the LSTM hidden state output, as illustrated in figure 12.35,

$$\boldsymbol{c}_t = \boldsymbol{f}_t \otimes \boldsymbol{c}_{t-1} + \boldsymbol{i}_t \otimes \boldsymbol{g}_t, \quad \boldsymbol{h}_t = \boldsymbol{o}_t \otimes \tanh(\boldsymbol{c}_t), \tag{12.39}$$

where $\otimes$ indicates element-by-element multiplication. The final **softmax** output is

$$y_t = \textbf{softmax}(w_h^T \boldsymbol{h}_t + b_h). \tag{12.40}$$

We now show an LSTM regression analysis using the airline passenger data [107], which is an example of a non-stationary seasonal time series. Our network consists of a single LSTM layer with $J = 300$

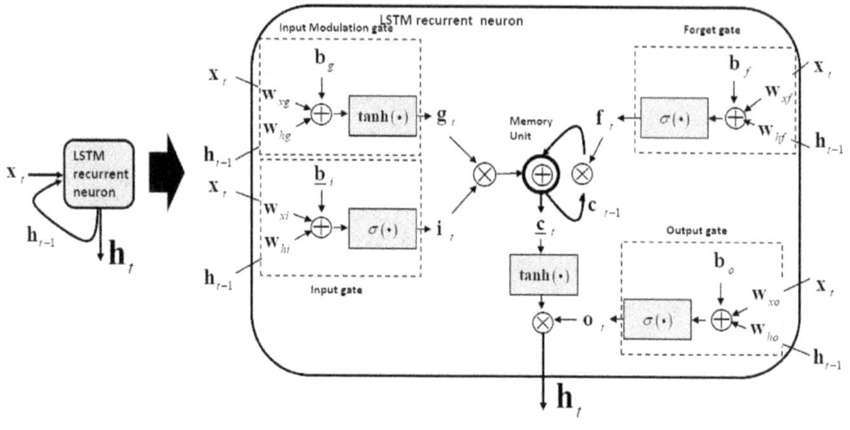

Figure 12.35: A single LSTM recurrent neuron: $\boldsymbol{i}_t$ and $\boldsymbol{g}_t$ and $\boldsymbol{f}_t$, together with the previous cell state $\boldsymbol{c}_{t-1}$, are used to calculate the present cell state $\boldsymbol{c}_t$ which together with $\boldsymbol{o}_t$ produce the LSTM output $\boldsymbol{y}_t$; element by element multiplication and addition are denoted by $\otimes$ and $\oplus$, respectively. The previous output $\boldsymbol{y}_{t-1}$ and present data $\boldsymbol{x}_t$ are input to all four gates.

memory units, and feature dimension $d = 16$, i.e., we use the previous 16 values $x_{t-16}, \ldots, x_{t-1}$ to predict the present value x_t. The output of the LSTM layer is fed into a single output neuron (with hyperbolic tangent activation) and is optimized with the minimum squared error cost function. Figure 12.36 shows the actual passenger data, the portion used for LSTM training and the training predictions, and the validation portion of the data together with the LSTM predictions. Although the network was never exposed to the validation data (to the right of the dotted line) during training, it clearly learnt the seasonal and non-stationary characteristics of the data.

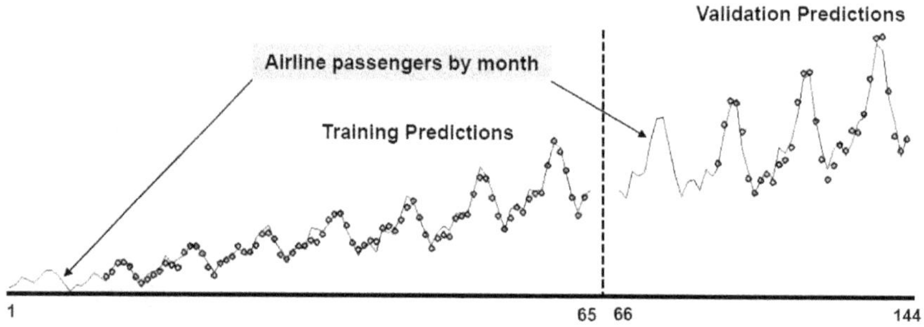

Figure 12.36: LSTM prediction of airline passenger data.

More recently, gated recurrent units (GRU) have been found to have similar performance to LSTM RNNs but with fewer parameters. GRUs have two gates:

$$\boldsymbol{i}_t = \sigma\left(\boldsymbol{w}_{xi}\boldsymbol{x}_t + \boldsymbol{w}_{hi}\boldsymbol{h}_{t-1} + \boldsymbol{b}_i\right), \quad \boldsymbol{u}_t = \sigma\left(\boldsymbol{w}_{xu}\boldsymbol{x}_t + \boldsymbol{w}_{hu}\boldsymbol{h}_{t-1} + \boldsymbol{b}_u\right), \tag{12.41}$$

and

$$\hat{\boldsymbol{h}}_t = \tanh\left(\boldsymbol{w}_{xh}\boldsymbol{x}_t + \boldsymbol{w}_{hh}\,\boldsymbol{h}_{t-1} \otimes \boldsymbol{i}_t + \boldsymbol{b}_h\right), \quad \boldsymbol{h}_t = (1 - \boldsymbol{u}_t) \otimes \boldsymbol{h}_{t-1} + \boldsymbol{u}_t \otimes \hat{\boldsymbol{h}}_t. \tag{12.42}$$

12.15 Unsupervised Learning

All neural networks described so far fit the *Supervised Learning* category, i.e., networks are trained to learn a function that maps input data to associated output labels with the goal of generalizing to new samples. The output labels act as a guide for training, i.e., we minimize the difference between actual output labels and estimated output labels. Thus, backpropagation learning is used to train neural networks with desired output labels corresponding to every training input.

Interestingly, there are neural network applications that do not need output labels corresponding to input data. Such applications of neural networks fall under the category of *Unsupervised Learning* [176]. While less common than supervised networks, unsupervised networks are used in a variety of applications. In this section we use unsupervised neural networks for data compression, reconstruction, and generative modeling. We demonstrate these capabilities with a class of neural network models known as *autoencoders* [177].

The basic configuration of an autoencoder is shown in figure 12.37. The training data is presented as both input and output data, i.e., the input data act as "targets" for the network, and the network is trained to reproduce the input data as accurately as possible. This raises the question of the usefulness of such a network if the output is essentially the same as the input. The key to the utility of the autoencoder is in the hidden layers. Typically, as suggested by figure 12.37, the hidden layers become successively smaller as the inner-most hidden layer, often called the *bottleneck layer*, is approached; the bottleneck layer corresponds to the layer with the lowest dimension in the network. Having fewer neurons to work with in each successive layer, the neural network learns to find an efficient representation of the data, i.e., it learns to perform data compression. The successive decrease in dimensionality of the hidden layers forces the network to send maximally useful information through the bottleneck layer since the network is tasked with reconstructing the input data as accurately as possible. Thus, the bottleneck layer represents the input data in an optimally compressed form and in much smaller dimension than the original input dimension.

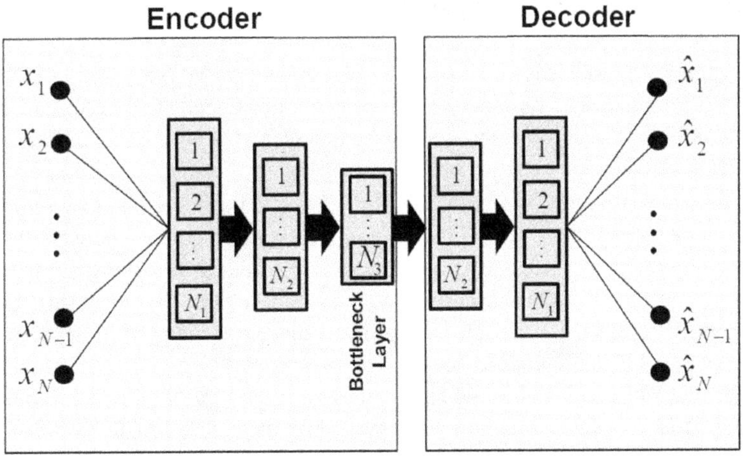

Figure 12.37: Basic autoencoder configuration with input $N \times 1$ data vector. The encoder portion has three hidden layers with $N_1 > N_2 > N_3$ neurons, with the last hidden layer known as the bottleneck layer. The decoder portion has two hidden layers with N_2 and N_1 neurons, respectively. The output vector has the same dimension N as the input vector.

For instance, consider the MNIST data set of 28×28 images of digits $0-9$, each of which is represented as a vector $\boldsymbol{x}_k$ with 784 elements. Since the images represent 10 digits, it seems unlikely that there are 784 independent dimensions in the data set. Although there are a number of swoops, loops, and strokes in all the digits, it seems reasonable to expect to represent the data with a smaller number of dimensions than 784. For example, we could train an autoencoder whose bottleneck layer has 100 neurons. If it turns out that this autoencoder is able to reproduce the input data accurately, then we say that the intrinsic data dimension is closer to 100 rather than the ambient dimension of 784. Exploring with a different number of neurons in the bottleneck layer, a sense of the intrinsic dimension of the original data can be obtained.

We may view an autoencoder as a model with two distinct pieces as illustrated in figure 12.37: the layers from the input to the bottleneck layer form the *encoder* while the layers after the bottleneck layer, all the way to reconstructed output layer, form the *decoder*. The encoder encodes the input data by embedding it into a new *latent space* whose dimension is close to the intrinsic dimension of the data. The encoder output, often referred to as *latent code*, is then passed into the decoder for reconstruction. The process is exactly the same as data compression and reconstruction but now we view the latent code $\boldsymbol{z}$ from a different perspective that will show its utility as more than a compressed representation of the original input $\boldsymbol{x}$: if we knew the distribution function $f(\boldsymbol{z})$ of the latent code $\boldsymbol{z}$, then we could generate data with the same (unknown) distribution of the input by simply feeding into the decoder realizations from the distribution $f(\boldsymbol{z})$. We shall see how this can be accomplished using a form of unsupervised learning known as a generative model. Whereas in discriminative models (supervised learning) networks learn the conditional probabilities $P(\boldsymbol{y}|\boldsymbol{x})$, where $\boldsymbol{y}$ is a set of classification labels, generative network models learn the joint probability density function $f(\boldsymbol{x})$ of the input vector $\boldsymbol{x}$. Thus, generative models allow us to create data realizations $\boldsymbol{x}$ once the joint density $f(\boldsymbol{x})$ has been learnt. The ability to create unlimited data realizations from a comparatively small sample is of enormous significance in data analysis when actual data being modeled is difficult to obtain because of processing or acquisition constraints.

12.16 Generative Adversarial Networks

A flexible class of generative models is known as Generative Adversarial Networks or GAN(s) [178], which typically consist of two competing neural networks, the generator **G** and the discriminator **D**, in a zero-sum game (a game in which the algebraic sum of all participants' gains and losses equals zero) [179] where each neural network has objectives counter to the other.

The generator network **G** has the goal of generating "fake" samples $\boldsymbol{x}_f$ that accurately mimic realizations from the distribution function $f(\boldsymbol{x})$, while the goal of the discriminator network **D** is to differentiate between "genuine" data realizations $\boldsymbol{x}$ and generated ones $\boldsymbol{x}_f$, and ultimately reject the fake realizations. Thus, **G** tries to generate fake samples from the true distribution $f(\boldsymbol{x})$ that look sufficiently genuine that **D** fails to reject them. The discriminator **D**, on the other hand, has the goal of accurately differentiating between generated samples $\boldsymbol{x}_f$ and genuine ones $\boldsymbol{x}$. The two models "learn" by minimizing their respective losses: the generator **G** minimizes its loss when it can successfully "fool" the discriminator **D** into classifying one of its generated samples as a genuine sample, while the discriminator **D** minimizes its loss when it can correctly reject the generated samples.

Figure 12.38 illustrates the GAN architecture in more detail; the generator and discriminator networks can be fully connected or convolutional. GANs have two distinct modes of training to facilitate the zero-sum game between **G** and **D**: the Generator and the Discriminator train in sequence and each network attempts to minimize its own loss. Note that the loss is always calculated at the output of **D** but the calculation could use different functions depending on the network that is being trained. Typical losses used to optimize GANs are Wasserstein and least-squares loss functions [180, 181].

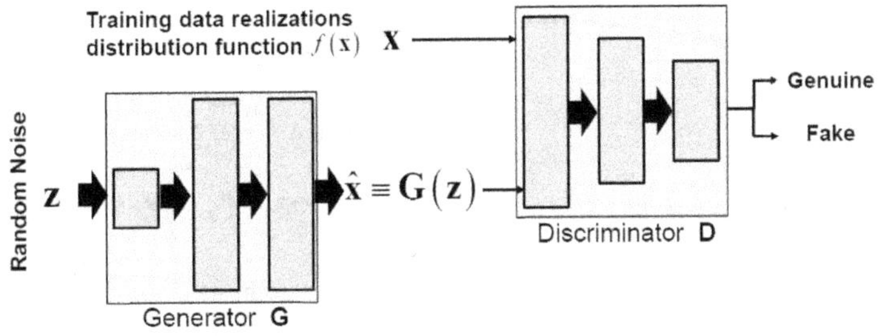

Figure 12.38: A typical Generative Adversarial Network (GAN). Each rectangle represents a hidden layer of arbitrary size except for: the output layer of **G**, which must match the dimension of $\boldsymbol{x}$, and the output of **D**, which must allow for binary classification (i.e., single sigmoid node or two **softmax** nodes).

In the first training phase, **D** attempts to minimize its error (e.g., cross entropy, MSE, etc.) by discriminating between the output of **G** and the real samples $\boldsymbol{x}$. Initially, **D** easily rejects the output of **G** since **G** has not updated its weights; it produces samples that look like random noise. In the next training phase we optimize **G** so that it can generate samples that **D** will fail to classify as fake samples. During this phase the samples generated by **G** are sent to **D** for classification and the resulting cost is used to update the weights of **G** via the backpropagation algorithm we have described in section 12.4. This gives **G** a chance to improve itself in order to fool **D** more effectively, but now **D** needs to improve itself in order to deal with the better samples now being generated by **G**. The training phases alternate allowing **G** and **D** to compete until their respective losses stabilize and **G** produces samples that are sufficiently good that **D** fails to reject them. At this point **G** can generate samples that mimic the training data and can be used purely as a generative model.

It is important to note that when optimizing **D** the label on the output of **G** is "fake" and the label on true samples is "genuine", but when optimizing **G** the label on the output of **G** is "genuine" and so is the label on $\boldsymbol{x}$ — this is necessary in order to gauge how *genuine* the fake samples look to **D**, as shown in figure 12.39.

GANs are difficult to train; they are typically very sensitive to hyper-parameter tuning [182]. While progress has been made in this regard, often GANs converge to solutions that are not very good, or can diverge altogether. Sometimes when GANs do converge, they do so in ways that do not accurately model the underlying distribution of the data. One of the most common issues encountered in GAN training is *mode collapse* [182] when the generator learns to only generate a single (or near single) mode in an underlying multi-modal distribution. Essentially, GANs can arrive at solutions that maximally fool the discriminator (i.e., images look like sufficiently realistic samples from $\boldsymbol{x}$ when evaluated by **D**) but lack the actual variation present in the underlying data. If we consider the MNIST digits, this would be the equivalent of a generator **G** learning to only generate realistic 1's and 7's, leaving all the other digits unrepresented, regardless of the input $\boldsymbol{z}$ into **G**.

Given the above issues with GAN training we wish to learn a generative model that avoids them. Autoencoders seem to be a good class of models since they explicitly model the entire data and converge using modern training techniques. Although "vanilla" autoencoders are not generative models, they can be sufficiently enhanced with ideas from the GAN paradigm to turn them into generative models. This class of unsupervised models is called adversarial autoencoders or AAE. AAEs provide the generative capabilities of GANs with the training stability and convergence properties of autoencoders, i.e., they explicitly control for mode collapse [183].

Figure 12.37 shows how an autoencoder can be viewed as the combination of two distinct parts: an

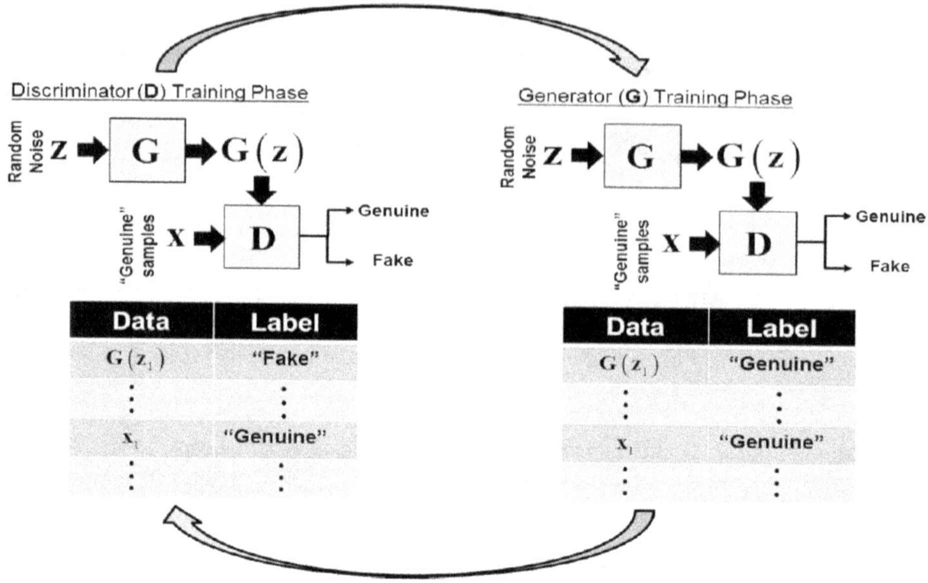

Figure 12.39: The two phases of the GAN training process.

encoder **enc** and a decoder **dec**. The encoder takes some input vector $\boldsymbol{x}$ and outputs a latent code $\mathbf{enc}(\boldsymbol{x})$ that is fed into the decoder producing the reconstructed output $\hat{\boldsymbol{x}} \equiv \mathbf{dec}(\mathbf{enc}(\boldsymbol{x}))$. As stated above, if $\mathbf{enc}(\boldsymbol{x})$ followed some probability distribution $f(\boldsymbol{x})$, then we could generate samples from this distribution and feed them to the decoder for data generation. This is precisely what adversarial autoencoders learn to do; they learn to optimally compress and reconstruct the input while simultaneously forcing $\mathbf{enc}(\boldsymbol{x})$ to follow an arbitrary prior distribution $f(\boldsymbol{x})$ (typically a joint Gaussian distribution). Once the network converges we can sample $f(\boldsymbol{x})$ and feed that data to the decoder and generate data that mimic the variation in the training set without worrying about mode collapse.

To train an AAE, the latent code representation must be regularized in such a way as to *force* it to follow a prior probability distribution. We do this by incorporating a discriminator $\mathrm{D}_{\mathrm{AAE}}$ into the autoencoder learning process. Much like regular GANs the discriminator in this setup takes two inputs. One input comes from a multi-dimensional probability distribution (which has the same dimension as the autoencoder latent code $enc(x)$) called the *real samples*, and the other input to the $\mathrm{D}_{\mathrm{AAE}}$ comes from the bottleneck layer output code $\mathbf{enc}(\boldsymbol{x})$ called the *fake samples*. To train the AAE, we alternate between two training phases much like with GANs: the first phase consists of training $\mathbf{D}_{\mathrm{AAE}}$ to discriminate between samples from a real probability distribution and the latent code of our autoencoder $\mathbf{enc}(\boldsymbol{x})$. The next phase consists of training the autoencoder to perform its regular reconstruction task while simultaneously minimizing its loss with respect to how believable the latent codes it produces are to $\mathbf{D}_{\mathrm{AAE}}$. In other words, it learns to do reconstruction while ensuring that the latent code produced by the encoder follows the specified prior distribution. Figure 12.40 shows an example of a vanilla AAE as a traditional autoencoder coupled with a discriminator that takes $f(\boldsymbol{x})$ and $\mathbf{enc}(\boldsymbol{x})$ as inputs. Forcing the AAE to learn to reconstruct the data while producing $\mathbf{enc}(\boldsymbol{x})$ samples that fool $\mathbf{D}_{\mathrm{AAE}}$ is the reason models in this class are called *adversarial*; $\mathbf{D}_{\mathrm{AAE}}$ does not want to be fooled by the output of the encoder, while the encoder tries to fool the discriminator into classifying its output $\mathbf{enc}(\boldsymbol{x})$ as coming from the real prior distribution $f(\boldsymbol{x})$.

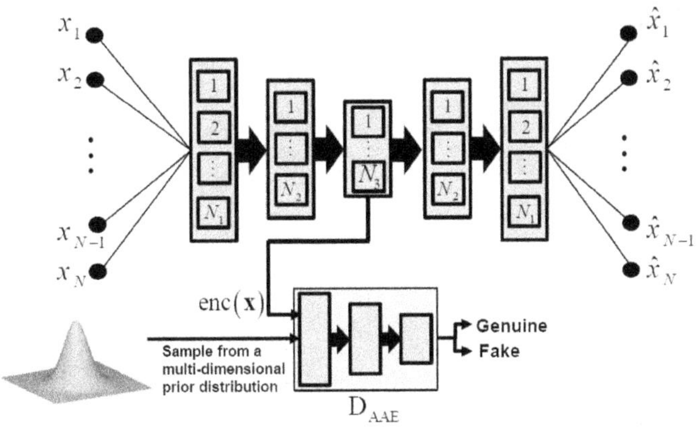

Figure 12.40: Adversarial Autoencoder (AAE): an autoencoder coupled with a discriminator, trained using the same alternating phases as GANs.

We demonstrate an AAE application using the MNIST data set. Rather than using the vanilla AAE, we use a supervised version that learns to disentangle "style" (variation) from "content" (class label) in the training images. In this setup we can hold a class label fixed and observe the variation present in class by sampling from our prior distribution and passing that as input to our decoder for image generation. We can also hold variation constant and observe how the factors of variation we have isolated affect the different classes in our data. To train this supervised variant of an AAE, the only modification we make is to add a one-hot label to the input of our decoder; otherwise the training process is the same as with the standard AAE formulation we described above. The supervised AAE allows the generator to associate the data it reconstructs with class labels. Once the model converges we can fix the one-hot label (i.e., content input) and sample our prior distribution (i.e., style input), or vice versa, and generate new data samples. Figure 12.41 illustrates a supervised AAE architecture.

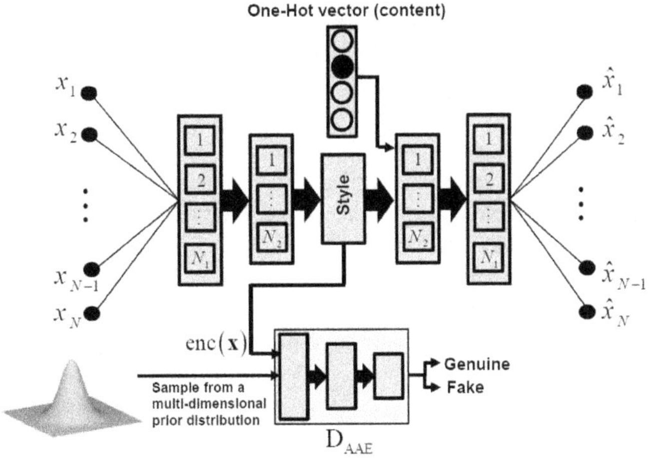

Figure 12.41: Supervised Adversarial Autoencoder (AAE).

Figure 12.42 (a replication of the results in the original AAE paper [183]) shows the output of a trained supervised AAE that clearly demonstrates the ability of the model to separate content and style in the images.

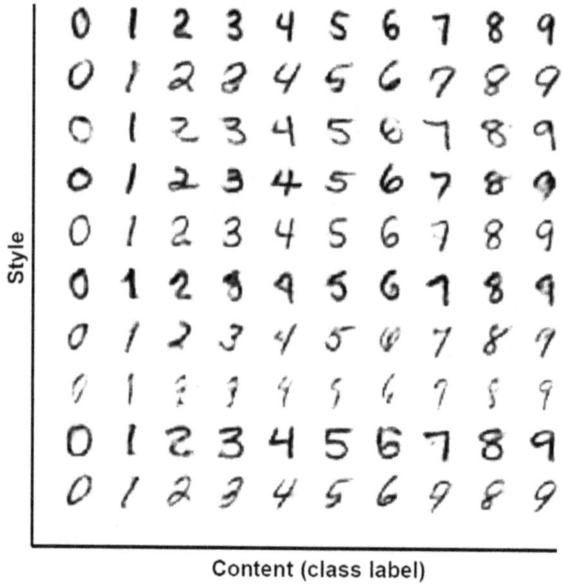

Figure 12.42: Trained supervised AAE separating content and style of handwritten digit images.

12.17 Perspective

Deep neural networks have been tremendously successful in many areas from speech and image recognition to genomics. In this chapter we have attempted to give an overview of the subject avoiding asking why deep networks work so well, and in particular, why the non-convex optimization during deep network training works as well as it does or why more layers tend to work better.

Gaining insights into the mathematical foundations of deep neural networks is an area of active research. For instance, under certain conditions, the loss function of the fully decoupled deep neural network of some depth H is intimately connected to the Hamiltonian of the H-spin spherical spin-glass model [184]; most local minima in this highly non-convex optimization problem have similar performance, and although the probability of finding a "bad" local minimum is non-zero for small networks, the probability decreases quickly with network size. In addition, it appears that deep networks are particularly efficient in approximating exceptionally simple polynomials that are sparse, symmetric and low-order [185]—polynomials that arise in describing Hamiltonians of physical systems. Finally, it has been shown [186] that for an important class of problems, namely those that compute functions of functions (compositional functions), and an important subset of these problems, namely, hierarchically local compositional functions, deep networks avoid the curse of dimensionality.

Despite the difficulties in developing a complete mathematical theory of deep neural networks, it is clear that deep networks have instigated the collection of large data sets that will hopefully provide the "experimental" insights that are essential in formulating a proper theory.

References

[1] A. V. Oppenheim and R. W. Schafer, *Digital Signal Processing*. Englewood Cliffs, NJ: Prentice-Hall, 1975.

[2] P. R. Halmos, *Finits-Dimensional Vector Spaces*. New York: Springer Verlag, 1974.

[3] G. Bachman and L. Narici, *Functional Analysis*. New York: Academic Press, 1966.

[4] J. P. Keener, *Principles of Appled Mathematics: Transformation and Approximation*. Reading, MA: Addison-Wesley, 1988.

[5] F. Riesz and B. Nagy, *Functional Analysis*. F. Ungar, 1955.

[6] G. H. Golub and C. F. V. Loan, *Matrix Computations*. Baltimore: Johns Hopkins University Press, 1996.

[7] J. W. Demmel, *Applied Numerical Linear Algebra*. Philadelphia: SIAM, 1997.

[8] N. Aronszajn, "Theory of reproducing kernels," *Trans. Am. Math. Soc.*, pp. 337–404, 1950.

[9] C. E. Shannon, "Communication in the presence of noise," *Proc. Institute of Radio Engineers*, vol. 37, no. 1, pp. 10–21, 1949.

[10] D. Slepian, "Prolate spheroidal wave functions, Fourier analysis, and uncertainty — V: The discrete case," *The Bell System Technical Journal*, vol. 57, no. 5, pp. 1371–1429, 1978.

[11] M. Reed and B. Simon, *Methods of Modern Mathematical Physics I: Functional Analysis*. Academic Press, 1980.

[12] F. Tricomi, *Integral Equations*. New York: Dover, 1985.

[13] P. M. Morse and H. Feshbach, *Methods of Theoretical Physics — I*. New York: McGraw-Hill, 1953.

[14] A. H. Najmi, *Wavelets: A Concise Guide*. Baltimore: Johns Hopkins University Press, 2012.

[15] S. Mallat, "Multiresolution approximations and wavelet orthonormal bases of $L_2(\mathbb{R})$," *Trans. Amer. Math. Soc.*, vol. 315, pp. 69–87, 1989.

[16] S. Boyd and L. Vandenberghe, *Convex Optimization*. Cambridge, U.K.: CUP, 2004.

[17] E. J. Candes, J. K. Romberg, and T. Tao, "Robust uncertainty principles: exact signal reconstruction from highly incomplete frequency information," *IEEE Trans. Inform. Theory*, vol. 52, pp. 489–509, 2006.

[18] E. J. Candes, J. K. Romberg, and T. Tao, "Stable signal recovery from incomplete and inaccurate measurements," *Comm. Pure and Applied Math.*, vol. 59, pp. 1207–1223, 2006.

[19] E. J. Candes and T. Tao, "Decoding by linear programming," *IEEE Trans. Inform. Theory*, vol. 51, pp. 4203–4215, 2005.

[20] E. J. Candes and T. Tao, "Near optimal signal recovery from random projections: universal encoding strategies," *IEEE Trans. Inform. Theory*, vol. 52, pp. 5406–5425, 2006.

[21] S. Foucart and H. Rauhut, *A Mathematical Introduction to Compressive Sensing*. New York: Springer, 2013.

[22] W. H. Press, S. A. Teukolsky, W. T. Vetterling, and B. P. Flannery, *Numerical Recipes: The Art of Scientific Computing*. Cambridge University Press, 1992.

[23] A. S. Householder, *The Theory of Matrices in Numerical Analysis*. New York: Dover, 1975.

[24] C. Eckart and G. Young, "The approximation of one matrix by another of lower rank," *Psychometrika*, vol. 1, pp. 211–218, 1936.

[25] R. M. Johnson, "On a theorem stated by Eckart and Young," *Psychometrika*, vol. 28, pp. 259–263, 1963.

[26] R. Adcock, "Note on the method of least squares," *The Analyst*, vol. 4, pp. 183–184, 1877.

[27] S. V. Huffel and J. Vandewalle, "The total least squares problem: Computational aspects and analysis," *SIAM, Philadelphia*, 1991.

[28] G. H. Golub and C. F. V. Loan, "An analysis of the total least squares problem," *SIAM J. Numer. Analysis*, vol. 17, pp. 883–983, 1970.

[29] P. H. Schönemann, "A generalized solution of the orthogonal procrustes problem," *Psychometrika*, vol. 31, pp. 1–10, 1966.

[30] W. Rudin, *Real and Complex Analysis*. New York: McGraw-Hill, 1987.

[31] K. Hoffman, *Banach Spaces of Analytic Functions*. Englewood Cliffs: Prentice-Hall, 1965.

[32] K. Kodera, R. Gendrin, and C. de Villedary, "Analysis of time-varying signals with small BT values," *IEEE Trans. Acoustics, Speech, Sig. Proc.*, vol. 26, pp. 64–76, 1978.

[33] P. Flandrin, F. Auger, and E. Chassande-Mottin, "Time-frequency reassignment: from principles to algorithms," in *Applications in Time-Frequency Signal Processing* (A. Papandreou-Suppappola, ed.), ch. 5, pp. 179–203, CRC Press, 2003.

[34] L. R. Rabiner, R. W. Schafer, and C. M. Rader, "The chirp z-transform algorithm and its application," *Bell System Technical Journal*, vol. 48, pp. 1249–1292, 1969.

[35] V. Sukhoy and A. Stoytchev, "Generalizing the inverse FFT off the unit circle," *Nature Scientific Reports*, vol. 9, no. 14443, 2019.

[36] K. F. Lee, H. W. Hon, and R. Reddy, "An overview of the SPHINX recognition system," *IEEE Trans. Sig. Proc.*, pp. 35–45, 1990.

[37] S. B. Davis and P. Mermelstein, "Comparison of parametric representations for monosyllabic word recognition in continuously spoken sentences," *IEEE Trans. Acous. Speech Sig. Proc.*, pp. 357–366, 1980.

[38] L. A. Zadeh and C. A. Desoer, *Linear System Theory*. New York: McGraw-Hill, 1969.

[39] D. Gabor, "Theory of communication," *Journal of IEE*, vol. 93, pp. 429–457, 1946.

[40] C. H. Page, "Instantaneous power spectra," *Jour. Appl. Phys.*, vol. 23, pp. 103–106, 1952.

[41] L. Cohen, *Time-Frequency Analysis*. Englewood Cliffs, N.J.: Prentice Hall, 1995.

[42] W. Yourgrau and A. van der Merwe, eds., *Perspectives in Quantum Theory*. MIT Press, 1971.

[43] A. H. Nuttall, "Wigner distribution function: Relation to short-term spectral estimation, smoothing and performance in noise," *Naval Underwater Systems Center*, vol. 8225, 1989.

[44] A. H. Najmi, "The Wigner distribution: A time-frequency analysis tool," *Johns Hopkins APL Technical Digest*, vol. 15, pp. 298–305, 1994.

[45] W. Gander, G. H. Golub, and U. von Matt, "A constrained eigenvalue problem," *Linear Algebra and Its Applications*, vol. 114, pp. 815–839, 1989.

[46] A. Tkacenko, P. P. Vaidyanathan, and T. Nguyen, "On the eigenfilter design method and its applications: A tutorial," *IEEE Trans. Circuits & Systems*, vol. 50, pp. 497–517, 2003.

[47] H. C. Cho, "On time delay estimation using an FIR filter." `http://www.ee.cityu.edu.hk/~hcso/sp_8_01.pdf`. 3 Jun 2012.

[48] P. Billingsley, *Probability and Measure*. New York: Wiley, 1995.

[49] F. D. Neeser and J. L. Massey, "Proper complex random processes with applications to information theory," *IEEE Trans. Information Theory*, vol. 39, pp. 1293–1302, 1993.

[50] W. A. Gardner, *Introduction to Random Processes (2nd edition)*. McGraw-Hill, 1990.

[51] P. L. Meyer, *Introductory Probability and Statistical Applications*. New York: Addison-Wesley, 1966.

[52] C. Maccone, *Deep Space Flight and Communications*. Chichester, UK: Springer and Praxis, 2009.

[53] D. Dini and D. P. Mandic, "Class of widely linear complex Kalman filters," *IEEE Trans. Neural Networks*, vol. 23, pp. 775–786, 2012.

[54] A. M. Mood, F. A. Graybill, and D. C. Boes, *Introduction to the Theory of Statistics*. New York: McGraw-Hill, 1974.

[55] M. Kendall and A. Stuart, *The Advanced Theory of Statistics*. New York, NY: Macmillan, 1979.

[56] P. Stoica, R. L. Moses, B. Friedlander, and T. Soderstrom, "Maximum likelihood estimation of the parameters of multiple sinusoids from noisy measurements," *IEEE Trans. Acoust., Speech, Signal Process.*, vol. 37, pp. 378–392, 1989.

[57] C. E. Shannon, "A mathematical theory of communication," *Bell System Technical Journal*, vol. 27, pp. 379–423, 1948.

[58] A. Papoulis, *Probability, Random Variables, and Stochastic Processes*. New York: McGraw-Hill, 1991.

[59] S. Kotz, T. J. Kozubowski, and K. Podgorski, *The Laplace Distribution and Generalizations: A Revisit with Applications to Comunications, Economics, Engineering and Finance*. Boston: Birkhöuser, 2001.

[60] T. M. Cover and J. A. Thomas, *Elements of Information Theory*. New York: Johns Wiley, 1991.

[61] S. Kullback, *Information Theory and Statistic*. John Wiley and Sons, 1959.

[62] S. Kullback and R. A. Leibler, "On information and sufficiency," *Anns. Math. Statist.*, vol. 22, pp. 79–86, 1951.

[63] E. L. Lehmann and G. Casella, *Thoery of Point Estimation*. New York: Springer, 1998.

[64] A. Hyvärinen, J. Karhunen, and E. Oja, *Independent Component Analysis*. New York: Johns Wiley, 2001.

[65] T. Adali, P. Schreier, and L. L. Scharf, "Complex-valued signal processing: The proper way to deal with impropriety," *IEEE Trans. Signal Processing*, pp. 5101–5125, 2011.

[66] T. Adali, M. Anderson, and G.-S. Fu, "Diversity in independent component and vector analyses," *IEEE Signal Processing Magazine*, pp. 18–33, 2014.

[67] D. C. Champeney, *A Handbook of Fourier Theorems*. Cambridge: Cambridge University Press, 1987.

[68] C. Chatfield, *Time Series: Theory and Methods*. London: Chapman and Hall, 1989.

[69] R. O. Schmidt, "Multiple emitter location and signal parameter estimation," *IEEE Trans. Antennas Propagation*, vol. 34, pp. 276–280, 1986.

[70] V. F. Pisarenko, "The retrieval of harmonics from a covariance function," *Geophys. J. Roy. Astron. Soc.*, vol. 33, pp. 347–366, 1973.

[71] R. Roy and T. Kailath, "ESPRIT – Estimation of Signal Parameters via Rotational Invariance Techniques," *IEEE Trans. Acoust. Speech, Sig. Proc.*, vol. 7, pp. 984–995, 1989.

[72] N. Wiener and P. Masani, "The prediction theory of multivariate stochastic processes I," *Acta Math.*, vol. 98, pp. 111–150, 1957.

[73] N. Wiener and P. Masani, "The prediction theory of multivariate stochastic processes II," *Acta Math.*, vol. 99, pp. 93–137, 1958.

[74] R. P. Boas, *Entire Functions*. Academic Press, 1954.

[75] R. E. A. C. Paley and N. Wiener, *Fourier Transforms in the Complex Domain*. American Mathematical Society — Colloquium Publications, 1934.

[76] N. Wiener, *Extrapolation, Interpolation, and Smoothing of Stationary Time Series*. Wiley, 1949.

[77] J. F. Coales and S. J. Kahne, "The yellow peril and after," *IEEE Control Systems Magazine*, pp. 65–69, 2014.

[78] J. L. Doob, "The brownian movement and stochastic equations," *Annals of Mathematics*, vol. 43, pp. 351–369, 1942.

[79] J. S. Bendat and A. G. Piersol, *Random Data: Analysis and Measurement Procedures.* New York: Wiley-Interscience, 1986.

[80] G. C. Carter, "Coherence and time delay estimation," *Proceedings of the IEEE*, vol. 75, pp. 236–255, 1987.

[81] P. J. Brockwell and R. A. Davis, *The Analysis of Time Series — An Introduction.* Springer, 2006.

[82] M. S. Bartlett, *An Introduction to Stochastic Processes with Special Reference to Methods and Applications.* New York: Cambridge University Press, 1953.

[83] W. B. Davenport, *Probability and Random Processes.* New York: McGraw-Hill, 1970.

[84] G. M. Jenkins and D. G. Watts, *Spectral Analysis and its Applications.* San Francisco: Holden-Day, 1968.

[85] G. R. Blackman and J. Tukey, *The Measurement of Power Spectra.* New York: Dover, 1958.

[86] P. D. Welch, "The use of fast Fourier transform for the estimation of power spectra," *IEEE Trans. Audio Electroacoust*, vol. 15, pp. 70–73, 1970.

[87] D. J. Thomson, "Spectrum estimation and harmonic analysis," *Proc. IEEE*, vol. 70, no. 9, pp. 1055–1096, 1982.

[88] D. Percival and A. Walden, *Spectral Analysis for Physical Applications.* Cambridge, U.K.: Cambridge University Press, 1993.

[89] K. S. Lii and M. Rosenblatt, "Prolate spheroidal spectral estimates," *Stat. Prob. Lett.*, vol. 78, pp. 1339–1348, 2008.

[90] H. Cramér, *Mathematical Methods of Statistics.* Princeton, NJ: Princeton University Press, 1946.

[91] J. Capon, "High-resolution frequency wave-number spectrum analysis," *Proc. IEEE*, vol. 57, pp. 1408–1418, 1969.

[92] R. T. Lacoss, "Data adaptive spectral analysis methods," *Geophysics*, vol. 36, pp. 661–675, 1971.

[93] J. D. Scargle, "Phase-sensitive deconvolution to model random processes with special reference to astronomical data," in *Applied Time Series Analysis* (D. F. Findley, ed.), pp. 549–563, Academic Press, 1981.

[94] E. J. Hannan, *Multiple Time Series.* New York: John Wiley, 1970.

[95] D. R. Brillinger, *Time Series: Data Analysis and Theory.* New York, N.Y.: Holt, 1975.

[96] J. S. Lawrence Marple, *Digital Spectral Analysis with Applications.* Englewood Cliffs, N.J.: Prentice-Hall, 1987.

[97] U. Grenander and G. Szegö, *Toeplitz Forms and and Their Applications.* New York: Chelsea, 1984.

[98] A. Antoniou and W. S. Lu, *Practical Optimization: Algorithms and Engineering Applications.* New York: McGraw-Hill, 2007.

[99] E. K. P. Chong and S. H. Zak, *An Introduction to Optimization.* New York: John Wiley, 2001.

[100] J. Makhoul, "Linear prediction: A tutorial review," *Proc. IEEE*, vol. 63, pp. 561–580, 1975.

[101] J. P. Burg, "The relationship between maximum entropy and maximum likelihood spectra," *Texas Instruments*, 1971.

[102] J. V. Pendrel and D. E. Smulie, "The relationship between maximum entropy and maximum likelihood spectra," *Geophysics*, vol. 44, pp. 1738–1739, 1979.

[103] A. H. Nuttall, "Spectral analysis of a univariate process with bad points, via maximum entropy and linear predictive techniques," *Naval Underwater Systems Center*, vol. 5303, 1976.

[104] J. S. Lawrence Marple, "A new autoregressive spectrum analysis algorithm," *IEEE Trans. Acoust. Speech Signal Process.*, vol. 28, pp. 441–454, 1980.

[105] S. M. Kay, "Recursive maximum likelihood estimation of autoregressive processes," *IEEE Trans. Acoust. Speech Signal Process.*, vol. 31, pp. 56–65, 1983.

[106] P. Whittle, "The analysis of multiple stationary time series," *J. R. Statist. Soc. B*, vol. 15, pp. 125–139, 1953.

[107] G. E. P. Box, G. M. Jenkins, and G. C. Reinsel, *Time Series Analysis: Forecasting and Control*. Hoboken, N.J.: Wiley, 2008.

[108] G. E. P. Box, "Science and statistics," *J. ASA*, vol. 71, pp. 791–799, 1976.

[109] C. Hitchcock and E. Sober, "Prediction versus accommodation and the risk of overfitting," *British Journal for the Philosophy of Science*, vol. 55, pp. 1–34, 2004.

[110] H. Akaike, "Information theory and an extension of the maximum likelihood principle," in *Proceedings of the 2nd International Symposium on Information Theory* (Petrov and Czaki, eds.), pp. 267–281, 1973.

[111] C. M. Hurvich and C.-L. Tsai, "Regression and time series model selection in small samples," *Biometrika*, vol. 76, no. 2, pp. 297–307, 1989.

[112] H. Linhart and W. Zucchini, *Model Selection*. Wiley, 1986.

[113] R. E. Kass and A. E. Raftery, "Bayes factors," *J. ASA*, vol. 90, pp. 773–795, 1995.

[114] G. Claeskens and N. L. Hjort, *Model Selection and Model Averaging*. Cambridge University Press, 2008.

[115] J. J. Rissanen, "Modeling by the shortest data description," *Automatica*, vol. 14, pp. 465–471, 1978.

[116] Grunwald, Myung, and Pitt, eds., *Advances in Minimum Description Length: Theory and Applications*. MIT Press, 2004.

[117] P. D. Grunwald, *The Minimum Description Length Principle*. MIT Press, 2007.

[118] J. J. Rissanen, "Strong optimality of the normalized ML models as universal codes and information in data," *IEEE Trans. Information Theory*, vol. 47, pp. 1712–1717, 2001.

[119] S. Amari and H. Nagaoka, *Methods of Information Geometry*. Oxford University Press, 2000.

[120] J. J. Rissanen, "Fisher information and stochastic complexity," *IEEE Trans. Information Theory*, vol. 42, pp. 40–47, 1996.

[121] P. Vaidyanathan, "On the minimum phase property of prediction error polynomials," *IEEE Sig. Proc. Lett.*, pp. 126–127, 1997.

[122] D. H. Lehmer, "A machine method for solving polynomial equations," *Journal of ACM*, vol. 8, pp. 151–162, 1961.

[123] J. D. Markel and A. H. Gray, *Linear Prediction of Speech*. New York: Springer-Verlag, 1976.

[124] P. E. Papamichalis, *Practical Approaches to Speech Coding*. Englewood Cliffs, NJ: Prentice Hall, 1987.

[125] B. Widrow and S. D. Stearns, *Adaptive Signal Processing*. Englewood Cliffs, N.J.: Prentice-Hall, 1985.

[126] B. Widrow, J. McCool, and M. Ball, "The complex LMS algorithm," *Proc. IEEE*, vol. 63, pp. 719–720, 1975.

[127] M. J. Shensa, "Non-Wiener solutions of adaptive noise canceller with a noisy reference," *IEEE Trans. Acoust., Speech, Signal Process*, vol. 28, pp. 468–473, 1980.

[128] E. Walach and B. Widrow, "The LMF adaptive algorithm and its family," *IEEE Trans. Information Theory*, vol. 30, pp. 275–283, 1984.

[129] F. M. Hsu, "Square root Kalman filtering for high-speed data received over fading dispersive HF channels," *IEEE Trans. Information Theory*, vol. 28, pp. 753–763, 1982.

[130] R. Schreiber, "Implementation of adaptive array algorithms," *IEEE Trans. Acoustics, Speech and Signal Processing*, vol. 34, pp. 1038–1045, 1986.

[131] H. Cox, "Resolving power and sensitivity to mismatch of optimum array processing," *J. Acoustical Soc. Am.*, vol. 54, pp. 771–785, 1973.

[132] H. Messer, "The potential performance gain in using spectral information in passive detection/localization of wideband sources," *IEEE Trans. Sig. Proc.*, vol. 43, pp. 2964–2974, 1995.

[133] E. Tuncer and B. Friedlander, *Classical and Modern Direction of Arrival Estimation*. Burlington, MA: Elsevier, 2009.

[134] F. C. Robey, D. Fuhrmann, E. Kelly, and R. Nitzberg, "A CFAR adaptive matched filter detector," *IEEE Trans. Aerospace and Electronic Sysytems*, vol. 28, pp. 208–216, 1992.

[135] R. T. Compton, *Adaptive Antennas*. Englewood Cliffs, NJ: Prentice Hall, 1988.

[136] K. M. Buckley, "Spatial/spectral filtering with linearly constrained minimum variance beamformer," *IEEE Trans. Acoustics, Speech, Sig. Proc.*, vol. 35, pp. 249–266, 1987.

[137] O. L. Frost, "An algorithm for linearly constrained adaptive array processing," *Proc. IEEE*, vol. 60, pp. 926–935, 1972.

[138] L. J. Griffiths and C. W. Jim, "An alternative approach to linearly constrained adaptive beamforming," *IEEE Trans. Antennas and Propagation*, vol. 30, pp. 27–34, 1982.

[139] M. Zatman, "How narrow is narrowband?," *IEE Proc.-Radar, Sonar Navig.*, vol. 145, pp. 85–91, 1998.

[140] I. S. Reed, J. Mallett, and L. Brennan, "Rapid convergence rate in adaptive arrays," *IEEE Trans. Aerospace and Electronic Sysytems*, vol. 10, pp. 853–863, 1974.

[141] J. S. Goldstein, I. S. Reed, and L. L. Scharf, "A multistage representation of the Wiener filter based on orthogonal projections," *IEEE Trans. Information Theory*, vol. 44, pp. 2943–2959, 1998.

[142] D. M. Boroson, "Sample size considerations in adaptive arrays," *IEEE Trans. Aerospace and Electronic Systems*, vol. 16, pp. 446–451, 1980.

[143] E. J. Kelly, "Performance of an adaptive detection algorithm: Rejection of unwanted signals," *IEEE Trans. Aerospace and Electronic Systems*, vol. 25, pp. 122–133, 1989.

[144] B. D. Carlson, "Covariance matrix estimation errors and diagonal loading in adaptive arrays," *IEEE Trans. Aerospace and Electronic Systems*, vol. 24, pp. 397–401, 1988.

[145] H. L. V. Trees, *Optimum Array Processing*. India: Wiley Interscience, 2013.

[146] D. A. Abraham and N. L. Owsley, "Beamforming with dominant mode rejection," in *Proceedings IEEE OCEANS Conference*, pp. 470–475, 1990.

[147] H. Cox and R. Pitre, "Robust DMR and multi-rate adaptive beamforming," in *ORINCON Report*, 1998.

[148] K. E. Wage and J. R. Buck, "Snapshot performance of the Dominant Mode Rejection beamformer," *IEEE J. Oceanic Eng.*, vol. 39, pp. 212–225, 2014.

[149] S. Liu and G. Trenkler, "Hadamard, Khatri-Rao, Kronecker, and other matrix products," *Int. J. Info. & Syst. Science*, vol. 4, pp. 160–177, 2008.

[150] G. Cybenko, "Approximation by superpositions of a sigmoidal function," *Math. Control Signals Systems*, vol. 2, pp. 303–314, 1989.

[151] K. Hornik, M. Stichcombe, and H. White, "Multilayer feedforward networks are universal approximators," *Neural Networks*, vol. 2, pp. 359–366, 1989.

[152] F. Rosenblatt, "The perceptron: A probabilistic model for information storage and organization in the brain," *Psychological Review*, vol. 65, pp. 386–408, 1958.

[153] D. Pedamonti, "Comparison of non-linear activation functions for deep neural networks on MNIST classification task." `https://arxiv.org/pdf/1804.02763.pdf`. arXiv:1804.02763v1 - 8 Apr 2018.

[154] S. Ruder, "An overview of gradient descent optimization algorithms." `https://arxiv.org/pdf/1609.04747.pdf`. arXiv:1609.04747v2 - 15 Jun 2017.

[155] D. E. Rumelhart, G. E. Hinton, and R. J. Williams, "Learning representations by back-propagating errors," *Nature*, vol. 323, pp. 533–536, 1986.

[156] P. D. Wasserman and T. Schwartz, "Neural Networks, Part 2: What they are and why everybody is so interested in them now." `https://ieeexplore.ieee.org/stamp/stamp.jsp?tp=&arnumber=2091`. IEEE Expert, Spring 1988.

[157] M. Nielsen, "Why deep neural networks are hard to train." `http://neuralnetworksanddeeplearning.com/chap5.html`.

[158] P. Chaudhari and S. Soatto, "Stochastic gradient descent performs variational inference, converges to limit cycles for deep networks." `https://arxiv.org/pdf/1710.11029.pdf`. arXiv:1710.11029v2 - 16 Jun 2018.

[159] S. Ioffe and C. Szegedy, "Batch normalization: Accelerating deep network training by reducing internal covariate shift." `https://arxiv.org/pdf/1502.03167.pdf`. arXiv:1502.03167v3 - 2 Mar 2015.

[160] Y. LeCun, L. Bottou, G. B. Orr, and K.-R. Müller, "Efficient backprop." `http://yann.lecun.com/exdb/publis/pdf/lecun-98b.pdf`.

[161] S. K. Kumar, "On weight initialization in deep neural networks." `https://arxiv.org/pdf/1704.08863.pdf`. arXiv:1704.08863v2 - 2 May 2017.

[162] P. Domingos, "A few useful things to know about machine learning." `http://www.astro.caltech.edu/~george/ay122/cacm12.pdf`.

[163] Y. Jiang, D. Krishnan, H. Mobahi, and S. Bengio, "Predicting the generalization gap in deep networks with margin distributions," in *International Conference on Learning Representations*, 2019.

[164] C. Shorten and T. Khoshgoftaar, "A survey on image data augmentation for deep learning." `https://doi.org/10.1186/s40537-019-0197-0`. 2019.

[165] R. Caruana, S. Lawrence, and L. Giles, "Over-fitting in neural nets: Backpropagation, conjugate gradient, and early stopping," in *Proceedings of the 13th International Conference on Neural Information Processing Systems*, pp. 381–387, MIT Press, 2000.

[166] G. E. Hinton, N. Srivastava, A. Krizhevsky, I. Sutskever, and R. Salakhutdinov, "Improving neural networks by preventing co-adaptation of feature detectors." `https://arxiv.org/pdf/1207.0580.pdf`. arXiv:1207.0580v1 - 3 Jul 2012.

[167] N. Srivastava, G. Hinton, A. Krizhevsky, I. Sutskever, and R. Salakhutdinov, "Dropout: A simple way to prevent neural networks from over-fitting," *Journal of Machine Learning Research*, vol. 15, pp. 1929–1958, 2014.

[168] Y. LeCun, L. Bottou, Y. Bengio, and P. Haffner, "Gradient-based learning applied to document recognition," *Proc. IEEE*, vol. 86, pp. 2278–2324, 1998.

[169] K. Fukushima, "Neocognitron: A self-organizing neural network model for a mechanism of pattern recognition unaffected by shift in position," *Biol. Cybernetics*, vol. 36, pp. 193–202, 1980.

[170] M. Lin, Q. Chen, and S. Yan, "Network in network." `https://arxiv.org/pdf/1312.4400.pdf`. arXiv:1312.4400v3 - 4 Mar 2014.

[171] D. Dua and C. Graff, "UCI machine learning repository." `http://archive.ics.uci.edu/ml`, 2017.

[172] Y. LeCun, C. Cortes, and C. J. C. Burges, "The MNIST database of handwritten digits." `http://yann.lecun.com/exdb/mnist/`.

[173] Z. C. Lipton, J. Berkowitz, and C. Elkan, "A critical review of recurrent neural networks for sequence learning." `https://arxiv.org/pdf/1506.00019.pdf`. arXiv:1506.00019v4 - 17 Oct 2015.

[174] P. Werbos, "Backpropagation through time: What it does and how to do it," *Proc. IEEE*, vol. 78, pp. 1550–1560, 1990.

[175] S. Hochreiter and J. Schmidhuber, "Long short-term memory," *Neural Computation*, vol. 9, pp. 1735–1780, 1997.

[176] Z. Ghahramani, "Unsupervised learning," *ML Summer Schools — 2003 Advanced Lectures on Machine Learning*, vol. LNAI 3176, 2004.

[177] P. Baldi, "Autoencoders, unsupervised learning, and deep architectures," *JMLR: Workshop and Conference Proceedings*, vol. 27, pp. 37–50, 2012.

[178] I. J. Goodfellow, J. Pouget-Abadie, M. Mirza, B. Xu, D. Warde-Farley, S. Ozair, A. Courville, and Y. Bengio, "Generative adversarial nets." `https://arxiv.org/pdf/1406.2661.pdf`. arXiv:1406.2661v1 - 10 Jun 2014.

[179] J. von Neumann and O. Morgenstern, *Theory of Games and Economic Behavior*. Princeton, N.J.: Princeton University Press, 1944.

[180] M. Arjovsky, S. Chintala, and L. Bottou, "Wasserstein GAN." `https://arxiv.org/pdf/1701.07875.pdf`. arXiv:1701.07875v3 - 6 Dec 2017.

[181] X. Mao, Q. Li, H. Xie, R. Y. K. Lau, Z. Wang, and S. P. Smolley, "Least squares generative adversarial networks." `https://arxiv.org/pdf/1611.04076.pdf`. arXiv:1611.04076v3 - 5 Apr 2017.

[182] N. Kodali, J. Abernethy, J. Hays, and Z. Kira, "On convergence and stability of GANs." `https://arxiv.org/pdf/1705.07215.pdf`. arXiv:1705.07215v5 - 5 Apr 2017.

[183] A. Makhzani, J. Shlens, N. Jaitly, I. Goodfellow, and B. Frey, "Adversarial autoencoders." `https://arxiv.org/pdf/1511.05644.pdf`. arXiv:1511.05644v2 - 25 May 2016.

[184] A. Choromanska, M. Henaff, M. Mathieu, and G. B. A. Y. LeCun, "The loss surfaces of multilayer networks." `https://arxiv.org/pdf/1412.0233.pdf`. arXiv:1412.0233v3 - 21 Jan 2015.

[185] H. W. Lin, M. Tegmark, and D. Rolnick, "Why deep and cheap learning works so well." `https://arxiv.org/pdf/1608.08225.pdf`. arXiv:1608.08225v4 - 3 Aug 2017.

[186] T. Poggio, H. Mhaskar, L. Rosasco, B. Miranda, and Q. Liao, "Why and when can deep but not shallow networks avoid the curse of dimensionality: A review." `https://arxiv.org/pdf/1611.00740v2.pdf`. arXiv:1611.00740v2 - 22 Nov 2016.

Index

www.ingramcontent.com/pod-product-compliance
Lightning Source LLC
Chambersburg PA
CBHW061924190326
41458CB00009B/2645
* 9 7 8 1 2 6 0 4 5 8 9 3 0 *